AF411447

Artificial Intelligence-Based Learning Approaches for Remote Sensing

Artificial Intelligence-Based Learning Approaches for Remote Sensing

Editor

Gwanggil Jeon

MDPI • Basel • Beijing • Wuhan • Barcelona • Belgrade • Manchester • Tokyo • Cluj • Tianjin

Editor
Gwanggil Jeon
Department of Embedded
Systems Engineering, Incheon
National University,
Incheon, Korea

Editorial Office
MDPI
St. Alban-Anlage 66
4052 Basel, Switzerland

This is a reprint of articles from the Special Issue published online in the open access journal *Remote Sensing* (ISSN 2072-4292) (available at: https://www.mdpi.com/journal/remotesensing/special_issues/AI_Image).

For citation purposes, cite each article independently as indicated on the article page online and as indicated below:

LastName, A.A.; LastName, B.B.; LastName, C.C. Article Title. *Journal Name* **Year**, *Volume Number*, Page Range.

ISBN 978-3-0365-6083-0 (Hbk)
ISBN 978-3-0365-6084-7 (PDF)

Contents

Preface to "Artificial Intelligence-Based Learning Approaches for Remote Sensing"

Remote sensing is a method for understanding the ground and for facilitating human–ground communications. New developments in remote sensing have led to high-resolution monitoring of the ground on a global scale, giving a huge amount of ground observation data. Thus, AI-based deep learning approaches and its applied signal processing are required for remote sensing. These approaches can be universal or specific AI tools, including well-known neural networks, regression methods, decision trees, etc. In this book, I aimed to describe recent developments and trends regarding topics such as advanced AI-based deep learning techniques and remote sensing data processing. Sixteen papers were finally published in this book.

In this book, I focus my attention on Artificial Intelligence-Based Learning Approaches for Remote Sensing. The issue is comprised of original research articles as well as some reviews, on topics such as artificial intelligence, machine learning and deep learning. The manuscripts mainly contain remote sensing research, varying from theoretical work to in application studies. Some reviews summarize the current stage of ongoing application attempts. This book contains topics, i.e., for concrete bridge defects identification, localization based on classification, multi-source real landform data, multi-label graphs, UAV visual localization, multispectral and panchromatic image fusion, prediction of weather radar data, rotated SAR ship detection and classification, AIS data based on AI, vector-based spatial data, sparse-model-driven network, machine learning and fuzzy logic, boundary-aware network for SAR ship instance segmentation, GAN for spectral domain translation of remote sensing images, lightweight deep learning detector, moving target shadow detection network, detecting pine wilt disease using RGB-based UAV images.

This book collected papers that emphasis new Artificial Intelligence-Based Learning Approaches for Remote Sensing. Furthermore, this book expects to encourage more research in the field of AI-based approaches for remote sensing.

Gwanggil Jeon
Editor

Editorial

Artificial Intelligence-Based Learning Approaches for Remote Sensing

Gwanggil Jeon [1,2]

[1] Department of Embedded Systems Engineering, Incheon National University, 119 Academy-ro, Yeonsu-gu, Incheon 22012, Korea; gjeon@inu.ac.kr
[2] Energy Excellence & Smart City Lab., Incheon National University, 119 Academy-ro, Yeonsu-gu, Incheon 22012, Korea

Citation: Jeon, G. Artificial Intelligence-Based Learning Approaches for Remote Sensing. *Remote Sens.* **2022**, *14*, 5203. https://doi.org/10.3390/rs14205203

Received: 9 October 2022
Accepted: 11 October 2022
Published: 18 October 2022

Publisher's Note: MDPI stays neutral with regard to jurisdictional claims in published maps and institutional affiliations.

1. Introduction

Remote sensing (RS) is a method for understanding the ground and for facilitating human–ground communications. New developments in RS have led to HR (high-resolution) monitoring of the ground on a global scale, giving a huge amount of ground observation data. Thus, AI-based deep learning approaches and its applied signal processing are required for RS. These approaches can be universal or specific AI tools, including well-known neural networks, regression methods, decision trees, etc. In this Special Issue, we aimed to describe recent developments and trends regarding topics such as advanced AI-based deep learning techniques and RS data processing. Sixteen papers were finally published in this Special Issue.

2. Overview of Contributions

The contribution by Zoubir et al. [1] yielded a dataset with 6900 images that feature three common deficiencies of concrete bridges. The deficiencies include cracks, efflorescence, and spalling. To compensate for trials with incomplete training data, transfer learning techniques were introduced and tested to categorize the three deficiencies. In addition, two gradient-based backpropagation interpretation methods were adopted to create a pixel-level heatmap and to localize deficiencies in the test data. Objective and subjective performances were compared to give reliable information on deficiency localization.

The contribution by Yang et al. [2] introduced the forward-propagation concept and then added forecast features such as weather, temperature, terrain, and land-cover-type distributions. Based on this information, the authors could assess the CCIs of over-the-horizon communications on the intercity connection. According to the paper, based on 1300 sets of created terrains and landforms, two deep learning models were adopted to predict the PL of over-the-horizon communications among venues in a land-based ducting environment. The authors used LSTM prediction to verify their PL prediction using deep learning.

The contribution by Duan et al. [3] presented and certified a novel suggestion with un-labeled, upsampled, generated data, which could be worthless for unwarranted non-graph data. The authors proved that the feature circulation can enable deep learning of imbalanced graphs. In addition, the authors experimentally controlled collaborative data synthesis via the generation of virtual samples in the central region of a minority. Their data upsampling framework was assessed by numerous real-world learning network datasets. According to this paper, their work provided varied and reliable benchmark models with a big advantage.

In the contribution by Wen et al. [4], the authors creatively presented a spatial endurance criterion to choose 1-O features with wealthier local facts for the computation of 2-O data to guarantee the efficiency of M-O features. To minimize the consequence of inevitable positive/negative sample inequity in target finding, weight-adaptive factors were considered to adapt to the disadvantages of the cross entropy cost. In addition, the MIoU was built to perform an anchor box regression from numerous viewpoints. In this work, the authors presented an improved Wallis shadow automatic compensation approach to pre-process aerial data, presenting a basis for the subsequent data-matching processes. The authors also provided a consumer-grade unmanned aerial vehicle, obtaining a platform to collect aerial data for investigational validation. The simulation results suggested that their framework obtained outstanding results for each numerical and subjective metric.

In the contribution by Nie et al. [5], the authors presented a frequency/spatial interaction network (SSIN) for pansharpening. The main difference from conventional work is that the authors considered the features of pansharpening as being abstract and multispectral and then interrelated them repeatedly to progressively integrate frequency/spatial information. To improve frequency/spatial information fusion, the authors presented a frequency/spatial attention module to provide more efficient frequency/spatial information transfer on the network. The simulation results were conducted on the QuickBird, WorldView-4, and WorldView-2 datasets and proved that their approach overcomes conventional methods subjectively and objectively.

In the contribution by Albu et al. [6], the authors presented a CNN for weather forecasting using radar product prediction. The authors proposed the NeXtNow model, an improved version of the ResNeXt architecture. The proposed NeXtNow has an encoder/decoder convolutional architecture and plots radar results from previous moments and from the future. The authors authenticated their method utilizing radar data that were composed from the Romanian NMA and the Norwegian MET. In addition, they experimentally determined that the presence of numerous past radar results leads to more precise forecasts in the future. According to their paper, the proposed NeXtNow presented enhanced results.

In the contribution by Shao et al. [7], the authors proposed a rotated balanced feature-aligned network (RBFA-Net). In this work, the authors proposed three networks: BAFPN, AFAN, and RDN. The first network, BAFPN, is an enhanced version of FPN that improves multi-level features; therefore, it can reduce the adverse influence of feature variances. In the second network (AFAN), the authors used an arrangement convolution layer to adaptively line up the convolution features to rotated anchor boxes to alleviate the misalignment issue. In the third network (RDN), the authors presented a TDM to regulate the feature maps to handle any struggles with regression and classification tasks. The simulation results used eight SOTA rotated detection benchmark networks; among them, the proposed method presented the best performance in terms of mean average precision metric.

In the contribution by Han et al. [8], the authors presented an approach to categorizing atmospheric duct (AD) parameters by means of AIS signals in conjunction with AI. The proposed approach comprises an AD categorizing model. The categorizing model adjudicates the type of AD, and the inversion model reverses the AD parameters, giving the type of AD. Their research results suggest that the accuracy of the AD-categorizing model based on DNN outperforms others by 97% and that the AD parameter reverse model has higher inversion precision than that of the conventional approach, therefore showing the efficiency and precision of this new approach.

With growing access to data, advances in examinations carried out in the initial step of the asset procedure are imaginable. Building extraction from raster information is a significant step, particularly for urban planning and ecological research. The key issue with the semantic segmentation tool is the partial accessibility of masks. Therefore, labelling data are not perfect for the training step. To alleviate this issue, the contribution by Glinka et al. [9] proposes a solution to the automation of data classification from cadastral data from an exposed spatial dataset utilizing CNN and classifies and obtains buildings from HR from these data. The simulation results prove that the semantic ML segmentation on this spatial dataset presents satisfactory quality regarding the results.

In general, most of the current deep learning-based approaches are merely data-driven and neglect the filtering approach; therefore, they normally require adopting big data to generate training/validation/test datasets. However, a challenge is enhancing the correctness and conducting speed. To alleviate this issue, the contribution by Wang et al. [10] presented an SMD-Net for effective and high-precision InSAR filtering by opening the sparse regularization approach to handling the filtering model with a network. Different from conventional DL-based filtering approaches, the SMD-Net generates a physical filtering process in the network and covers less parameters and layers. It is thus expected to ensure accuracy of the filtering without sacrificing speed. The simulation results proved that the presented approach enhanced numerous conventional filtering methods subjectively and objectively.

The contribution by Das et al. [11] examined the built-up expansion procedure and its probability in an area of India by utilizing multi-temporal Landsat satellite data and a combination of the ML approach and a fuzzy method. To generate the built-up expansion probability model, several indices were used, such as dominance, diversity, and connectivity for every year. This information was gathered and then combined with the fuzzy method. The simulations were conducted using data from 2001 to 2021; the built-up areas were enlarged by 21.67%, while water and vegetation bodies were reduced by 4.63% and 9.28%. This study can serve as a guide to decision-makers presenting management plans for systematic urban growth without harming the environment.

SAR is a cutting-edge microwave sensor that has been broadly adopted in remote sensing surveillance, and its process is not affected by weather or light conditions. Ship instance segmentation using SAR is able to yield not only the box-level ship location but also the pixel-level ship contour. This approach plays an significant role in remote sensing surveillance in the ocean. However, most conventional approaches have limited box-positioning abilities, therefore deterring precision enhancement in instance segmentation. To alleviate this issue, the contribution by Ke et al. [12] presented a GCBANet for enhanced SAR ship instance segmentation. This approach has two novel blocks to guarantee excellent performance: GCIM-Block and BABP-Block. The authors conducted extensive simulations to prove each block's efficiency.

Remote sensing data of the Earth are influenced by numerous factors. Based on common/known hypotheses and a generation adversarial network, the contribution by Wang et al. [13] presented the SDTGAN approach to connect spectral data and to directly create target spectral remote sensing data. Additionally, more feature map information is presented to compensate for the lack of information in the spectral data and to enhance the geographic precision. By considering features such as supervised training with a balanced dataset, cycle consistency cost, and perceptual cost, the distinctiveness of the result is ensured. The simulation results prove that the presented SDTGAN approach outperforms conventional approaches.

SAR can generate microwave remote sensing data without weather restraints. Deep learning-based SAR ship detection approaches are hard to achieve on satellite data because deep learning typically has complex models and huge computations. To alleviate this issue, the contribution by Xu et al. [14] used the YOLOv5 method and presented a lightweight on-board SAR ship detector. In this work, the authors researched three topics: how to minimize the volume of the model, how to decrease the number of floating point operations, and how to understand on-board ship detection without losing precision. To test the proposed method, an evaluation was tested on an embedded platform NVIDIA Jetson TX2. The simulation results proved that the proposed methods outperformed conventional approaches subjectively and objectively.

Current conventional SAR moving target shadow detectors have low precision due to their incomplete feature-extraction capacities between complex scenes. To alleviate this issue, the contribution by Bao et al. [15] presented a new DLN called ShadowDeNet. This network was invented for better shadow detection of moving ground targets on signal SAR data. The authors proposed five approaches to ensure its performance: (1) HESE for better shadow saliency, (2) TSAM for focusing on ROI, (3) SDAL for deep learning on a moving target, (4) SGAAL for creating adjusted anchors, and (5) OHEM for choosing distinctive hard negative data. The authors tested the simulation on public SNL signal SAR data. The simulation results were tested with the SOTA benchmarks, and the proposed ShadowDeNet outperformed conventional approaches subjectively and objectively.

Pine wilt is an overwhelming disease that kills a vast number of affected pine trees within a short time. In the contribution by You et al. [16], the authors challenged this issue by detecting pine wilt disease. The main issue in detecting this disease is that data with low resolutions are used; therefore, there is high vagueness due to poor image resolution. In this work, the authors proposed two steps: (1) gathering the disease and hard negative data utilizing a CNN and (2) adopting an object-detection method to localize the disease. The authors adopted numerous image augmentation approaches to boost performance and to avoid overfitting. The simulation results were tested with the SOTA benchmarks, and the proposed method outperformed conventional approaches subjectively and objectively.

3. Conclusions

This Special Issue collected papers that emphasis new Artificial Intelligence-Based Learning Approaches for Remote Sensing. Furthermore, this Special Issue expects to encourage more research in the field of AI-based approaches for RS.

Funding: This research received no external funding.

Conflicts of Interest: The author declares no conflict of interest.

References

1. Zoubir, H.; Rguig, M.; El Aroussi, M.; Chehri, A.; Saadane, R.; Jeon, G. Concrete Bridge Defects Identification and Localization Based on Classification Deep Convolutional Neural Networks and Transfer Learning. *Remote Sens.* **2022**, *14*, 4882. [CrossRef]
2. Yang, K.; Guo, X.; Wu, Z.; Wu, J.; Wu, T.; Zhao, K.; Qu, T.; Linghu, L. Using Multi-Source Real Landform Data to Predict and Analyze Intercity Remote Interference of 5G Communication with Ducting and Troposcatter Effects. *Remote Sens.* **2022**, *14*, 4515. [CrossRef]
3. Duan, Y.; Liu, X.; Jatowt, A.; Yu, H.-t.; Lynden, S.; Kim, K.-S.; Matono, A. SORAG: Synthetic Data Over-Sampling Strategy on Multi-Label Graphs. *Remote Sens.* **2022**, *14*, 4479. [CrossRef]
4. Wen, K.; Chu, J.; Chen, J.; Chen, Y.; Cai, J. M-O SiamRPN with Weight Adaptive Joint MIoU for UAV Visual Localization. *Remote Sens.* **2022**, *14*, 4467. [CrossRef]
5. Nie, Z.; Chen, L.; Jeon, S.; Yang, X. Spectral-Spatial Interaction Network for Multispectral Image and Panchromatic Image Fusion. *Remote Sens.* **2022**, *14*, 4100. [CrossRef]
6. Albu, A.-I.; Czibula, G.; Mihai, A.; Czibula, I.G.; Burcea, S.; Mezghani, A. NeXtNow: A Convolutional Deep Learning Model for the Prediction of Weather Radar Data for Nowcasting Purposes. *Remote Sens.* **2022**, *14*, 3890. [CrossRef]
7. Shao, Z.; Zhang, X.; Zhang, T.; Xu, X.; Zeng, T. RBFA-Net: A Rotated Balanced Feature-Aligned Network for Rotated SAR Ship Detection and Classification. *Remote Sens.* **2022**, *14*, 3345. [CrossRef]
8. Han, J.; Wu, J.; Zhang, L.; Wang, H.; Zhu, Q.; Zhang, C.; Zhao, H.; Zhang, S. A Classifying-Inversion Method of Offshore Atmospheric Duct Parameters Using AIS Data Based on Artificial Intelligence. *Remote Sens.* **2022**, *14*, 3197. [CrossRef]
9. Glinka, S.; Owerko, T.; Tomaszkiewicz, K. Using Open Vector-Based Spatial Data to Create Semantic Datasets for Building Segmentation for Raster Data. *Remote Sens.* **2022**, *14*, 2745. [CrossRef]
10. Wang, N.; Zhang, X.; Zhang, T.; Pu, L.; Zhan, X.; Xu, X.; Hu, Y.; Shi, J.; Wei, S. A Sparse-Model-Driven Network for Efficient and High-Accuracy InSAR Phase Filtering. *Remote Sens.* **2022**, *14*, 2614. [CrossRef]
11. Das, T.; Shahfahad; Naikoo, M.W.; Talukdar, S.; Parvez, A.; Rahman, A.; Pal, S.; Asgher, M.S.; Islam, A.R.M.T.; Mosavi, A. Analysing Process and Probability of Built-Up Expansion Using Machine Learning and Fuzzy Logic in English Bazar, West Bengal. *Remote Sens.* **2022**, *14*, 2349. [CrossRef]
12. Ke, X.; Zhang, X.; Zhang, T. GCBANet: A Global Context Boundary-Aware Network for SAR Ship Instance Segmentation. *Remote Sens.* **2022**, *14*, 2165. [CrossRef]
13. Wang, B.; Zhu, L.; Guo, X.; Wang, X.; Wu, J. SDTGAN: Generation Adversarial Network for Spectral Domain Translation of Remote Sensing Images of the Earth Background Based on Shared Latent Domain. *Remote Sens.* **2022**, *14*, 1359. [CrossRef]
14. Xu, X.; Zhang, X.; Zhang, T. Lite-YOLOv5: A Lightweight Deep Learning Detector for On-Board Ship Detection in Large-Scene Sentinel-1 SAR Images. *Remote Sens.* **2022**, *14*, 1018. [CrossRef]
15. Bao, J.; Zhang, X.; Zhang, T.; Xu, X. ShadowDeNet: A Moving Target Shadow Detection Network for Video SAR. *Remote Sens.* **2022**, *14*, 320. [CrossRef]
16. You, J.; Zhang, R.; Lee, J. A Deep Learning-Based Generalized System for Detecting Pine Wilt Disease Using RGB-Based UAV Images. *Remote Sens.* **2022**, *14*, 150. [CrossRef]

 *remote sensing*

Article

A Deep Learning-Based Generalized System for Detecting Pine Wilt Disease Using RGB-Based UAV Images

Jie You [1], Ruirui Zhang [2] and Joonwhoan Lee [1,*]

[1] Department of Computer Engineering, Jeonbuk National University, Jeonju-si 54896, Korea; youjie80@gmail.com
[2] School of Computer Science and Engineering, Cangzhou Normal University, Cangzhou 061001, China; czwxz@caztc.edu.cn
* Correspondence: chlee@jbnu.ac.kr

Abstract: Pine wilt is a devastating disease that typically kills affected pine trees within a few months. In this paper, we confront the problem of detecting pine wilt disease. In the image samples that have been used for pine wilt disease detection, there is high ambiguity due to poor image resolution and the presence of "disease-like" objects. We therefore created a new dataset using large-sized orthophotographs collected from 32 cities, 167 regions, and 6121 pine wilt disease hotspots in South Korea. In our system, pine wilt disease was detected in two stages: n the first stage, the disease and hard negative samples were collected using a convolutional neural network. Because the diseased areas varied in size and color, and as the disease manifests differently from the early stage to the late stage, hard negative samples were further categorized into six different classes to simplify the complexity of the dataset. Then, in the second stage, we used an object detection model to localize the disease and "disease-like" hard negative samples. We used several image augmentation methods to boost system performance and avoid overfitting. The test process was divided into two phases: a patch-based test and a real-world test. During the patch-based test, we used the test-time augmentation method to obtain the average prediction of our system across multiple augmented samples of data, and the prediction results showed a mean average precision of 89.44% in five-fold cross validation, thus representing an increase of around 5% over the alternative system. In the real-world test, we collected 10 orthophotographs in various resolutions and areas, and our system successfully detected 711 out of 730 potential disease spots.

Keywords: pine wilt disease dataset; GIS application visualization; test-time augmentation; object detection; hard negative mining

Citation: You, J.; Zhang, R.; Lee, J. A Deep Learning-Based Generalized System for Detecting Pine Wilt Disease Using RGB-Based UAV Images. *Remote Sens.* **2022**, *14*, 150. https://doi.org/10.3390/rs14010150

Academic Editor: Karem Chokmani

Received: 19 December 2021
Accepted: 23 December 2021
Published: 30 December 2021

1. Introduction

Pinewood nematode (*Bursaphelenchus xylophilus*) is a microscopic worm-like creature that causes pine wilt disease (PWD), which poses a serious threat to pine forests, as infected trees die within a few months [?]. Pinewood nematodes can quickly pass from sick to healthy trees via biologic vectors or human activities. The disease is responsible for substantial environmental and economic losses in the pine forests of Europe, the Americas, and Asia [?].

Remote sensing technology is very powerful and widely used to monitor criminal activity or geographical changes, forecast weather, scan using airborne lasers, and plan urban developments. Unmanned Aerial Vehicles (UAVs) are capable of capturing high-quality aerial photographs or videos through various high-precision sensors and automated GPS (the global positioning system) navigation. Researchers [? ? ? ?] have recently used UAVs for tree species classification. In early 2005, ref. [?] successfully utilized remote-piloted vehicles to collect viable spores of Gibberella Zeae (anamorph Fusarium graminearum) and evaluate the impact of their transport. Some studies [? ? ?] have focused on plant disease identification based on spectral and texture features captured by aerial images.

Despite these achievements, it is still difficult to accurately detect PWD in high accuracy using UAV images for the following reasons: (1) PWD data collection is time-consuming and costly. Data must be collected from August through September. As PWD-infected trees typically start to die and appear red in late August, it is best to collect images after August, once these symptoms have appeared. However, after October, broad-leaved trees (such as maple trees) change color and show a similar appearance to PWD-infected trees. (2) It is difficult to obtain high-quality orthophotographs, because they require the careful selection of proper settings in terms of image resolution, shooting perspectives, exposure time, and weather. Typically, captured patch images can be overlapped, and orthophotograph can be created based on these overlapped patch images. However, such orthophotographs often suffer from poor image registration due to the elevation differences in forest areas. (3) PWD symptoms vary between stages. In the early stage, PWD-infected tree has a similar appearance to a healthy tree. In the late stage, infected trees show visual symptoms with features that appear close to those of yellow land, bare branches, or maple trees. Color-based algorithms typically show poor detection performance for this issue. (4) Annotation of these data are a challenge and time-consuming task; mis-annotation often occurs due to the poor image resolution and the background similarity.

A common method of locating PWD is based on the handcrafted features of texture and color; specifically, identifying their corresponding relationship to find infected trees [? ? ?]. Their results have been based on a limited number of samples (Table ??), and it is more desirable to analyze PWD using a large number of data samples. Deep learning technology has a powerful ability to process complicated GIS (geographic information system) data. While the encoder of a deep convolutional network automatically extracts inherent feature information from a given input data X, the decoder tries to approximate the desired outputs Y as closely as possible to solve complex classification and regression problems. Previous studies have used deep learning methods [?] to detect interesting objects. In this paper, we propose a deep learning-based PWD detection system which is verified to be effective and can be generalized for various object detection models using RGB-based images.

To summarize, our major contributions include:

- We collected and annotated a large dataset for PWD detection using orthophotographs taken from different areas in South Korea. The dataset has 6121 PWD hotspots in total. The obtained PWD-infected trees have arbitrary sizes and resolutions, and they show various symptoms during the different stages of infection.
- In our work, large number of easy negative samples from healthy area cause unbalanced positive and negative samples, which prevents model from effective learning. Besides, some "disease-like" objects (hard negative samples)—such as maple trees are hard to be correctly recognized. We overcome those difficulties using hard negative example mining [? ?]. The hard negative samples are selected in trained network, and merged with genuine PWD infected object to retrain the network. For simplification, we accumulate all the "disease-like" objects and categorize them into six negative categories ("white branch (wb)", "white green (wg)", "yellow land", "maple", "oak", and something "yellow") to be learned along with the positive PWD objects. This simple method improves the discriminatory power of the networks and easily exclude negative objects from the final detection result.
- Drone captured data has varying image resolutions as well as differing levels of illumination and sharpness. To achieve successful detection, our network was learned with augmented training data considering the diverse imaging conditions.
- The test-time augmentation method was used to make robust predictions. To our knowledge, this method has never been applied to PWD detection problems.
- We deployed the proposed system for real-world scenarios and tested it on several orthophotographs. We found that the proposed system successfully detected 711

out of 730 PWD-infected trees. The predicted model is made freely available as an open-source module for further field investigation.

Table 1. Information about the data sets in various approaches. Symbol "-" denote the missing data in reference paper.

Data Source	GSD (cm/pixel)	The Number of Bands	Collected Cities	PWD Infected Trees
WV/2 + WV/3 [?]	45	8	2	311 + 344
Jian [?]	8.97	3	1	340
MS + HS [?]	5	5	5	44 + 40
Anbi [?]	-	3	2	59
Qingkou [?]	4	4	3	42
Huangshan + Wuhan + Yantai [?]	-	5	3	1706
Taishan [?]	-	3	3	3670
Ours+Real environment test	3.2~10.2	3	32	6121 + 730

2. Related Work

2.1. Pine Wilt Disease Classification

The damage to pine forests caused by PWD is a serious social issue, and several attempts have been made to track the disease using orthophotographs. This method of PWD detection is challenging because it is only possible to collect the data within a limited time. An infected pine tree shows its disease symptoms through changes to the colors of its pine needles; the pine needles gradually change in appearance from solid green to yellow and then brown, and the dead tree color finally turns to ash grey.

Reference [?] analyzed the spatial distribution pattern of damaged trees while introducing the Classification and Regression Trees (CART) model. Another study [?] effectively extracted ecological information to predict risk rates of infected trees using a self-organizing map (SOM) and random forest models.

In later studies, spectral sensors were used to capture hyper-spectral images for PWD analysis [? ? ?]. Spectral sensors can provide different surface reflections of PWD that can help evaluate the regions of interest better than the naked eye. For example, near infrared (NIR) is a subset of the infrared band which covers the wavelength range from 780 to 1400 nm. The fusion of the NIR and the red band can reflect the changes in photosynthesis [?]. However, multi-spectral cameras are costly and more unstable than general RGB cameras, as their image quality is affected by various environmental conditions.

As PWD becomes a more serious problem worldwide, machine learning technology is increasingly being used for its automatic detection. The most common method for the early diagnosis of the disease depends on UAV images. In this method [?], a high-quality orthophotograph is generally cropped into small pieces, and a conventional supervised classifier is applied to recognize the disease location in a limited region. In another study, ref. [?] introduced a method using simple classifiers like multi-layer perceptron (MLP) and Support Vector Machine (SVM) to distinguish the regions of healthy or PWD-infected trees.

2.2. Deep Learning-Based PWD Detection

Several deep neural architectures have been proposed in recent years, and these have achieved significant breakthroughs in diverse problem domains. Deep learning-based object detection is a challenging multi-task problem which involves assigning a class label to an object of interest and then learning to locate the object's position. There have been two main streams of research related to the object detection model. The first is the region of interest (ROI) driven method, where the DNN (deep neural network) filters out the irrelevant background and then leaves the rest for refined classification. The typical example is Faster R-CNN [?], which consists of an encoder (CNN layer), Region Proposal Network (RPN), ROI pooling, and a classifier. The encoder network extracts various feature maps by convolution mapping of the input sample into latent space. The RPN first generates thousands of candidate bounding boxes from the feature map, then a simple classifier filters out the negative bounding boxes (i.e., background) while retaining the more probable positive bounding boxes (i.e., foreground). ROI pooling [?] collects feature maps in the positive bounding box and pools them into the same size. Finally, the bounding box regression block captures the precise location, and the classifier blocks divide the ROIs into specific categories. These two-stage object detection methods are relatively slow for real-time problems. That is why another types of single-stage object detection methods have been proposed. The well-known single-stage architectures include the YOLO series [? ? ? ?] and the SSD series [? ?]. In the single-stage method, there is no RPN, and the DNN has the ability to simultaneously localize and classify the target object. Therefore, using a single-stage detector network significantly reduces network complexity and speeds up the inference process. However, it still fails to provide more accurate position and classification information than the two-stage ROI driven method.

Some researchers have used a deep learning-based object detection for PWD. Reference [?] presented a two-stage object detection method that uses UAV remote sensing images to locate PWD-infected trees. Reference [?] compared the performance of single and two-stage object detection methods for PWD. In addition, Reference [?] presented a dataset with multi-band images and built a spatial-context-attention network (SCANet) with an expanded receptive field to better utilize context information. Further, Reference [?] proposed a faster disease filtering method that employs a lightweight one-stage detection model that discards a large number of irrelevant images before classifying the rest.

The previous studies devoted to PWD detection have suffered from a lack of data to train the DNN. By contrast, the DNN in our PWD detection method was trained on a large and previously-unused dataset, and our method is therefore more generalized and robust. A large number of "disease-like" objects push our object detector to build clear boundaries to distinguish them. Compared to the existing methods, our proposed method is more precise and rigorous in a real-world scenario.

3. Methodology

3.1. System Overview

The main challenges associated with PWD detection are a lack of data for training the DNN in various stages of infection, and the existence of background objects that are similar to the object of interest. We collected a large dataset consisting of observations from various districts in South Korea. The suppression of ambiguous background objects is particularly important in this domain [?]. The hard negative samples for PWD were further divided into six categories according to their appearance and texture information. The problem-solving strategy was found to be beneficial for guiding an object detector, particularly in instances where the object of interest was highly correlated with other background objects.

As shown in Figure ??, we first augmented the training samples with suitable augmentation methods and then trained our first DNN, which was designed to distinguish the background and PWD objects. The detector was robust at detecting most background regions, including healthy trees, land, buildings, lakes, etc., but PWD-infected trees confused with "disease-like" areas (FP) such as yellow land or maple trees. In this phase, it is easy to collect a lot of "disease-like" areas, which is referred to as the mining of hard negative samples [?]. We further categorized those ambiguous FP samples into six distinct categories (Figure ??). The disease and "disease-like" samples were passed through the DNN to perform a fine-level object detection. The UAV images in our system included disease regions that varied in size from 12×8 to 360×300 pixels. Accordingly, we applied a feature pyramid network (FPN) to capture arbitrary scale features based on both bottom-up and top-down connections. ResNet was selected as a backbone network, and the features in each residual block were processed for pyramidal representation. The bottom-up pathway produced the feature map hierarchy and the top-down pathway fused higher resolution features through upsampling the spatially coarse ones. This combined bottom-up and top-down process helps generate semantically stronger features. For the feature maps in each hierarchical stage, we appended a 3×3 convolution layer to reduce the aliasing effect of upsampling. Then, the set of merged features of each FPN stage was finally used for predictions. In the inference stage, we cropped the 800×800 image patch from the large sized orthophotograph (with "*.tif" extension). We assembled the results of augmented inference images and fused them by weighted boxes fusion algorithm to enhance the location and classification accuracies. A more detailed description of each module is provided later.

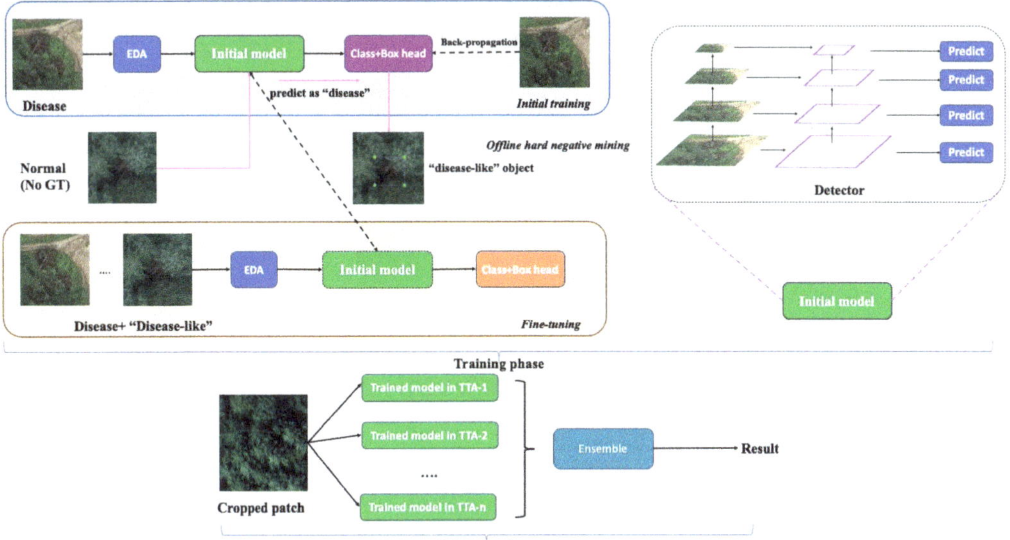

Figure 1. Pipeline overview.

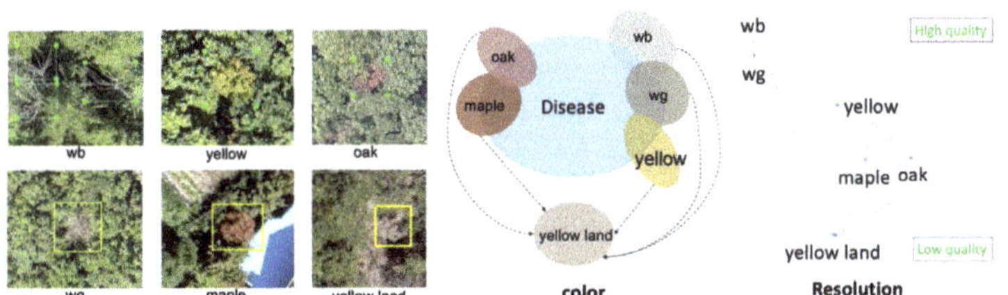

Figure 2. Left figure: the object of hard negative samples. **Middle figure**: the appearance relationship among "disease-like" objects and ground truth disease. **Right figure**: the reasons that the network predicted the different types of false negative samples. In high-quality image, branches and leaves are clear and easily identifiable. However, in low-resolution image, the real PWD-infected trees appear similar to the maple tree and the yellow land.

3.2. Efficient Data Augmentation (EDA)

Data augmentation is a well-known strategy for significantly increasing the amount of data available for training models without having to collect new data. It acts as a regularizer to reduce bias and generalize the system capability. Real-world UAV images are highly sensitive and collecting them is time-consuming. The image quality of UAV images is dependent on several environmental factors, including reflected light, contrast effects, and camera shake. Meanwhile, detection accuracy is affected by natural weather phenomena such as clouds or thick haze. We used several advanced augmentation methods to improve the performance for locating the PWD in different imaging situations.

Geometric transformation: We first cropped 800×800 patches from large "*.tif" images (more than 6×10^8 pixels) and applied random horizontal and vertical flip, rotation ($0\sim90$ degree), and resizing ($0.9\sim2.0$ times zooming) to augment the images and strengthen the model's ability to handle the various resolutions/shooting angles.

Color space augmentation: The outward appearance of PWD varies based on the stage of the disease; diseased trees are grayish-green in color in the early stage, with the needles turning brown, and eventually ash grey. In the middle stage of the disease, the color of the diseased leaf (brown) resembles the color of a maple leaf. We carried out random gamma, brightness, and contrast adjustments as well as PCA color augmentation [?] to generate synthetic images from real ones. The trained model with the augmented data including these synthetic images was less sensitive to color and focused more on texture discriminative features.

Noise injection: Poor-quality photography devices mounted on UAV often have unavoidable shot noise from unwanted electrical fluctuation that occurs when taking the pictures. We simulated this condition by adding random Gaussian noise (mean of $0\sim0.3$, std set to 1) to real images. Moreover, the device can be affected by radio interference, and a damaged image may randomly lose information, where this missing data appears as irregular black blocks. To address this problem, we augmented the training data by adopting the robust regularization technique [?] (cutout) to randomly remove various regions of input.

Other augmentations: Clouds may block sunlight and create a dark region (shadow) in an image. We added a mask to the original image to change the brightness of the local area of the image so as to simulate shadows. Furthermore, haze is more commonly seen in mountainous areas is another special natural phenomenon that reduces visibility on PWD-infected trees. Changing the opacity of image can generate synthetic haze. We augmented our images to maintain the performance of the network under poor weather conditions.

3.3. The Hard Negative Mining Algorithm

Hard negative mining (HNM) is a bootstrapping method that has been widely used in the classification field and which improves network performance by focusing on hard training samples. We improved the algorithm and applied it to the PWD detection problem. The modified method proceeds according to the following steps: (a) The object detection network is trained on the supervised training dataset. (b) The trained neural network is used to predict the unseen samples (no PWD samples) that are not included in the training samples. (c) The network predicts the objects of interest including "disease-like" objects which can be relabeled into several categories. (d) The "disease-like" objects are merged with genuine PWD-infected objects, and the neural network is retrained on the new dataset. The workflow of selecting "disease-like" objects is shown Figure **??**. Initially, the large-sized orthophotograph ("*.tif") is divided into small pieces, where the image without GT samples is kept and fed into the trained detector. Then, the model automatically filters out the easily recognizable objects (confidence score < 0.7), and the expert manually relabels the remaining hard objects according to the texture information.

The network with only three categories (background, disease, and hard negative samples) did not perform well due to the ambiguous classification decision boundaries. We manually annotated the hard negative samples into different categories. Figure **??** shows the division of "disease-like" objects and their relationships with the disease in terms of resolution and color. White branch (wb) denotes a dead tree that has a radial umbrella shape. The white-green (wg), yellow, and maple trees have a homogeneous color similar to PWD in its early and middle stages. The oak category indicates the presence of oak tree disease symptoms that resemble PWD-infected trees in UAV images. We also categorize yellow land into a separate category because this can appear similar to PWD, especially in low-resolution images.

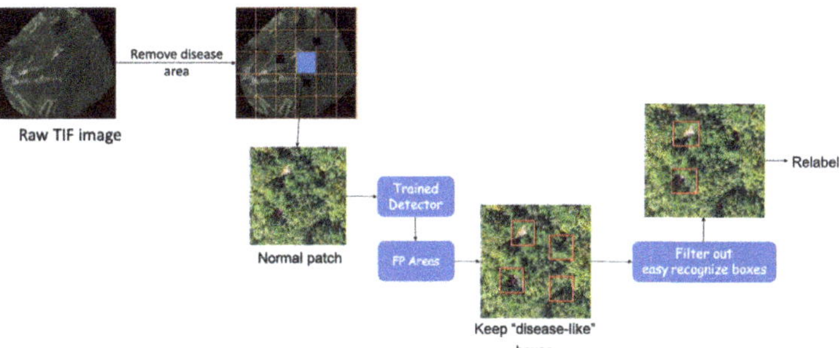

Figure 3. Extraction of "disease-like" objects.

3.4. Test Time Augmentation

Data augmentation is a common technique for increasing the size of a training dataset and reducing the chances of overfitting. Meanwhile, test time augmentation (TTA) is a data augmentation method applied during test time to improve the prediction capability of a neural network. As shown in the pipeline in Figure **??**, we created multiple augmented copies from a sample image and then made predictions for the original and synthetic samples. The prediction results show the bounding box coordinates with corresponding confidence scores from different augmented samples as well as their original test images. TTA operation is an ensemble method in which multiple augmented samples of a test image are evaluated based on a trained model. The final decision is made by weighted boxes fusion algorithm. While pursuing a balance between computation complexity and performance, we pick three augmentation methods (horizontal flip, vertical flip, and 90-degree rotation) to evaluate the improvement achieved by the TTA method.

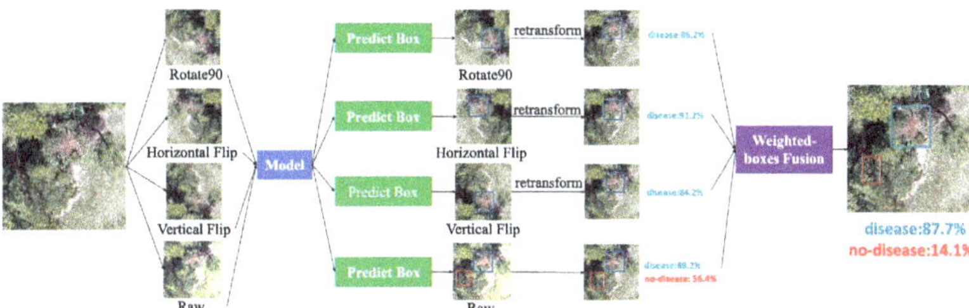

Figure 4. Test time augmentation applied in trained model. The bounding box prediction results are fused through weighted boxes fusion to highlight the important region and suppress the false detection.

3.5. Weighted Boxes Fusion Algorithm

Weighted boxes fusion (WBF) [**?**] is a key step while efficiently merging the predict position and the confidence score in TTA. Some common bounding box fusion algorithms such as NMS and soft-NMS [**?**] also work well for selecting bounding boxes by removing low-threshold overlapping boxes. However, they fail to consider the importance of different predicted bounding boxes. The WBF method will not discard any bounding boxes; instead, it uses the classification confidence scores of each predicted box to produce a combined predicted rectangle with high quality. The detailed WBF algorithm is described in Table **??**, and the notations used in the table are summarized as follows:

- **IoU**: Intersection over union.
- BBox: Bounding box.
- **N**: Augmentation methods (i.e., horizontal flip, vertical flip).
- **B**: Empty list to store predicted boxes from N augmentation methods.
- **C**: Confidence score of predicted boxes.

- **THR**: Mini bounding box overlap threshold.
- **L**: Empty list to save cluster predicted boxes in one position "**pos**", "**pos**" in one position, and **IoU** of two predicted BBox > **THR** (largely overlap).
- **F**: Empty list to store fused L boxes in different "**pos**".
- **T**: Total number of predicted bounding boxes.

Table 2. Weighted boxes fusion for merging BBox in test time augmentation method.

Step 1: Initialize the list **L**, **B**, **F**, inference **N** methods and store predicted BBox into **B**.
Step 2: Iterate BBox in **B**, find matching BBox in F, if **IoU** (F_i, B_i) > **THR**.
– **Step 2.1:** If no matching BBox, dequeue the BBox in **B** and add it into **L**, **F**.
– **Step 2.2:** If the match BBox is found, dequeue the BBox in **B** and add it into **L**. This BBox represents the matching box in "**pos**" position in **F**.
Step 3: In each "**pos** " in F, recalculate the all the BBox confidence score and coordinate by formula Equations (1)–(5), weighted sum of the coordinates of the boxes, in here, weight is the confidence score for each BBox. The weight of high confidence sore large than small one.

$$C_i = \frac{\sum_{i=1}^{T} C_i}{T} \tag{1}$$

$$x_{min} = \frac{\sum_{i=1}^{T} C_i * x_{min}}{\sum_{i=1}^{T} C_i} \tag{2}$$

$$x_{max} = \frac{\sum_{i=1}^{T} C_i * x_{max}}{\sum_{i=1}^{T} C_i} \tag{3}$$

$$y_{min} = \frac{\sum_{i=1}^{T} C_i * y_{min}}{\sum_{i=1}^{T} C_i} \tag{4}$$

$$y_{max} = \frac{\sum_{i=1}^{T} C_i * y_{max}}{\sum_{i=1}^{T} C_i} \tag{5}$$

Step 4: Readjust the confidence score as $C = C * \frac{T}{N}$, If there are few BBox in "**pos** " in the same position, the detected region is less like GT categories.

---

4. Experiment

4.1. Pine Wilt Disease-Infected Image Acquisition

Data acquisition is a primary task in training a robust DNN for classification or regression tasks. For this step, we collected diverse PWD data samples representing various stages of infection, as shown in Figure **??**. Pine wilt can kill a pine tree between 40 days to a few months after infection, and an infected tree shows different symptoms according to its stage of infection. In the early stage, the needles will remain green, but the accumulation of terpenes in xylem tissue results in cavitation, which interrupts water flux in the pine trees. In the second stage, the tree can no longer move water upward, thus causing it to wilt and the needles to turn yellow. The pinewood nematodes then grow in number, and all needles turn yellow-brown or reddish-brown. The disease progresses uniformly branch by branch. After the whole tree is fully dead, the tree needles still remain in place without falling, and show white bare color.

In this work, we obtained UAV images using drone technology. Between August and September, we took orthophotographs with UAVs in the disease-prone region. We tried to capture high-resolution images to best observe the changes in color caused by the disease. We captured high-resolution images from a low altitude using a CMOS camera (Sony Rx R12). The terrain level changes frequently within a hilly region, which affects the resolution quality of an orthophotograph. The images were sequentially taken while overlapping the area. We used drone mapping software (Pixel4Dmapper) to recorrect the GIS coordinates and ensure that the orthophotograph GSD was between 3.2 cm/pixel and 10.2 cm/pixel, where all overlapped patches composed a large orthophotograph ("*.tif"). Furthermore, each large orthophotograph had an ESRI (Environmental Systems Research Institute) format output with geographic coordinates that specify trees that have been deemed to be potentially infected with PWD by experts. During the training, we converted those geographic coordinates to a bounding box annotation. Then, we conducted a large orthophotograph crop to small 800×800 patches and sent the data to train the network with GT. After the data cleaning, we obtain a total of 4836 images with 6121 PWD damaged tree points as well as 265,694 normal patches (river, roof, field, etc. no PWD images) used to extract the "disease-like" objects. We used five-fold cross-validation to evaluate our proposed system, where each fold included balanced samples of various resolutions. In the real-world scenarios test, we captured another 10 real-world orthophotographs and compared the results with the expert-labeled ground truth points (730 PWD-infected trees).

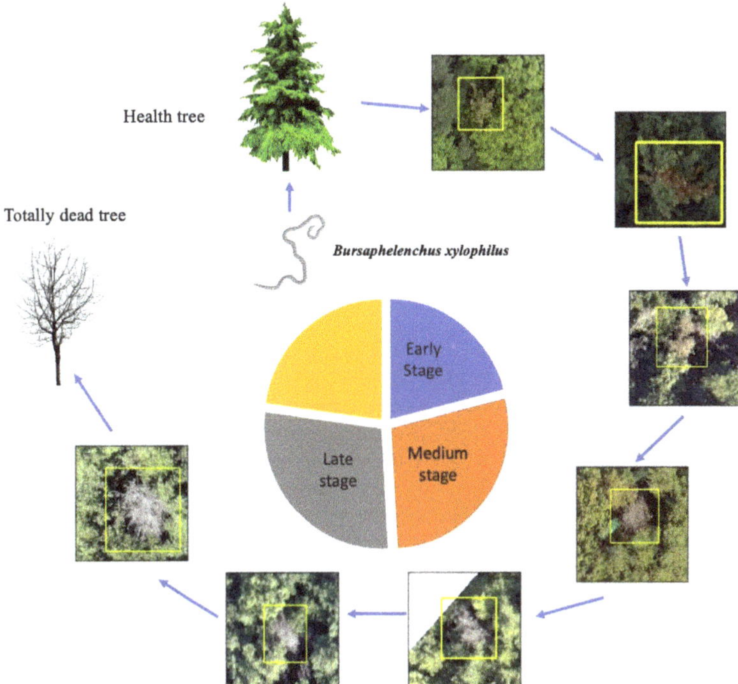

Figure 5. Symptoms of each stage of PWD infection.

4.2. Training Strategy

Our code was written in Python and the network models were implemented with PyTorch. A workstation with multiple TITAN XP GPUs and parallel processing was used to speed up the training. The COCO pre-trained model [?] was used for transfer learning, and the network was warmed up with a 1.0×10^{-6} learning rate for the first epoch to reduce the primacy effect. Then, we set a 0.001 learning rate and gradually reduced it by 50% every 50 epochs. In the experiment, the batch size was set to 8, the optimizer was SGD with 0.9 momentum for 200 training epochs. The K-means algorithm [?] was used to ensure the anchor scales and aspect ratios for fully covering the arbitrary disease shape. We use 5-fold cross-validation to split the training/validation dataset. In the first and second training stages, the strategies were the same except for the difference between the final classification headers. In the first stage, we trained the detection network with two neural nodes (categories) to distinguish between background and PWD-infected trees in the image. In the second stage, we fine-tuned the whole architecture for 200 epochs to distinguish actual PWD damaged trees as well as the other six categories of "disease-like" objects (e.g., wg, maple, wb, etc.).

4.3. Accuracy Estimation

Detector evaluation: We reported the average test accuracy of 5-fold cross-validation. The standard object detection metrics: mean average precision (*mAP*) and Recall were used to evaluate the performance of the model, and these are, respectively, expressed as Equations (**??**) and (**??**).

$$mAP = \frac{1}{|Q|} \sum_{q=1}^{Q} AP(q) \tag{6}$$

$$Recall = \frac{tp}{tp + fn} \tag{7}$$

where Q is the number of queries in the set and $AP(q)$ is the mean of the average precision scores for the given query. Our goal is to identify potential PWD-infected trees as much as possible, and we use Recall as an improvement metric to evaluate how much potential disease area has been located. Recall is assessed as the number of correctly detect diseases (TP) divided by the number of total diseases.

The setup for real-word environment evaluation: Operating in the real-world is even more challenging, as the cropped patches contain not only the PWD-infected trees but also various background content (including "disease-like" objects). Real-world PWD objects are small, irregular, and distributed across a wide area, and the GT is point-wise annotation (x, y geographic coordinates). Converting the point-wise annotation into bounding box annotation is challenging. Our goal was to correctly identify as many disease objects as possible along with their precise positions. In the field investigation to locate the disease-infected tree, an offset error of less than 8 m was deemed acceptable, so we use the x, y geographic coordinates of disease to generate 8×8 m GT bounding box annotation for each hotspot. We assume that the disease was found correctly when the overlapped area of GT and predicted bounding box (IoU: Intersection of Union) is larger than 0.3; otherwise, the system prediction is considered to be a false detection.

Specially, we use the overlap strategy to scan the orthophotograph and crop the test patches with 25% overlap. The IoU threshold is set to 0.4 for NMS, and 0.6 for merging the predicted bounding boxes in TTA process. Due to the insufficient context information, FP typically appears on the edge of the test patch. Therefore, we remove the bounding boxes that are less than 5 pixels from the boundary to reduce false alarms; for more information, please refer to Section **??**.

5. Results

We compared our proposed system to various network architectures. Table **??** compares the performance of our proposed method to those of alternative structures and backbone networks. The mAP accuracy refers to the average mean of five-fold cross validation in the same hyperparameter setting. We found that FPN with a ResNet101 backbone outperformed other structures, as it obtained the best accuracy of 89.44%; this is attributed to the fact that the fusion of bottom-up and top-down features helps capture disease in an arbitrary shape. The EDA scheme improved system performance as much as 2%. EDA overcomes the issue of data shortages and reduces the environmental effects of imaging. However, other findings were more surprising: the HNM achieved 88.35% accuracy in the FPN + Res101 architecture, and it improved by more than 3% compared to no-HNM approach. This phenomenon also occurs in other structures, which implies that the process in our HNM of splitting "disease-like" objects is important in learning the discriminative features. Another strategy we adopted to improve the inference power was the TTA method. TTA helped eliminate incorrect predictions and boost system performance as much as 1%. The improvement was noticeable in all network structures, as presented in Table **??**. The RetinaNet architecture performed worse than the FPN structure. We suspect that this was caused by the hyperparameters selected in focal loss [?]. Focal loss has two hyperparameters, $\alpha, \gamma, FL(p_t) = -\alpha(1 - p_t)^\gamma log(p_t)$, where α controls the balance of positive and negative samples and γ adjusts the weight of learning easy and difficult samples; increasing γ makes the model pay more attention to difficult samples. Tuning the two hyperparameter is a challenge task. The goal of this experiment was to evaluate the merit of our proposed strategies, and the results show that they consistently increase the performance of conventional architecture. Further, researchers can apply our proposed strategies in their designed network architectures and obtain better performance. For saving time, For saving time, we used FPN + Res101 structure in the following experiments.

Table 3. The mAP accuracy while using different backbone structures.

Backbone	Structure	Baseline	Baseline + EDA	Baseline + EDA + HNM	Baseline + EDA + HNM + TTA
Res50	Faster R-CNN	0.8248	0.8365	0.8609	0.8689
Res50	FPN	0.8286	0.8419	0.8688	0.8734
Res101	FPN	0.8312	0.8521	0.8835	0.8944
Res50	SSD	0.8169	0.8316	0.8522	0.8619
Res50	RetinaNet	0.8255	0.8445	0.8658	0.8712

5.1. Result of Real-World Environment Evaluation

5.1.1. Software Integration

The real-world inference pipeline of our disease detection system is demonstrated in Figure **??**. The UAV with an RGB camera first captured an overlapped image in the survey area. We then processed these images and integrated them into a large orthophotograph. The geographic coordinates are possibly varied due to changes in camera distance, so they had to be carefully recorrected for each object to be kept in its proper position. The resulting 8-bit "*.tif" image was further processed by the GDAL (https://gdal.org, accessed on 20 December 2021) library and cropped into 800×800 patches with overlap. The overlap is essential because the boundary region of the patch has few context information which leads to wrong detection results. We then located the presence of disease in each patch and indicated their location with a bounding box through the inference process.

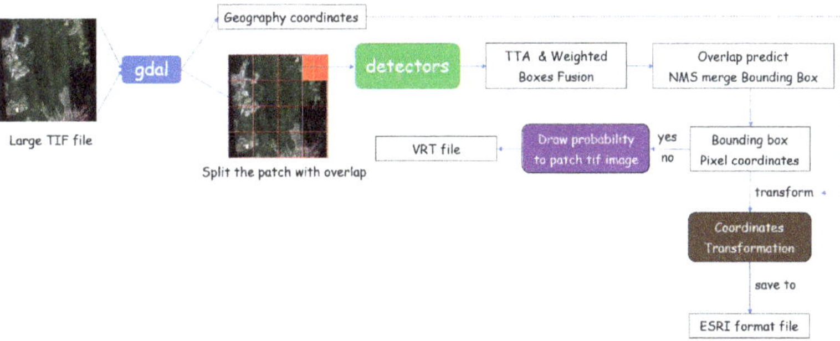

Figure 6. Pipeline of the automatic PWD diagnosis system.

For each input image patch, we generated augmented patches and returned the combined prediction result from TTA. The models provided the predicted bounding box, which had four points to construct a rectangle: the x and y pixel coordinates for the top left corner as well as the corresponding coordinates for the bottom right corner. Next, we used WBF to select the proper bounding box as well as reduce the number of irrelevant detected boxes in multiple results. After obtaining the precise bounding box location for the disease areas, we transformed the pixel coordinates to geographic coordinates based on the coordinate reference system, and stored the results to an output file. Our program produces the output in ESRI format, which includes four types of files—*.dbf, *.prj, *.shp,

*.shx and can be imported into a GIS application for visualization. This application helps experts find the proper GPS coordinates of a potential disease outbreak, and the latitude and longitude information are useful for further field investigation.

5.1.2. The Software Integration Hyperparameter Selection

The selection of proper hyperparameters is critical in the pursuit of better detection software. For real environment evaluations, the important hyperparameters are the stride size, overlap ratio, IoU threshold for NMS, number of augmented methods during TTA, IoU threshold for WBF, and bounding box distance (RBD). Table **??** lists the hyperparameter settings we used to find potential diseases in Goomisi Goaeup (Table **??**). The stride is the number of pixels that shift over for cropped patches in the next inference time; for example, 3/4 means that, in the next patch, 3/4 pixels (800 × 3/4 = 600) were moved, thus leaving 200 pixels overlapping in the horizontal and vertical directions. The small stride contributes to the increased overlap areas and potentially increases the inference time. RBD refers to the length of the predicted bounding box all the way up to the edge of the cropped patch. The bounding box was removed if the distance between the bounding boxes and the edge of the patch was less than RBD. In our experiment, we obtained a performance improvement of around 7% after removing the bounding boxes near the edge. The reason for this is a lack of context information in the marginal area, which caused the detector to frequently mislabel "disease-like" objects (wg, wb, yellow land, etc.) as disease. Our overlap strategy ensures that there is an overlapped area between the current patch and the next crop patch, so that the next crop patch contains rich context information by removing the bounding boxes located at the edge of the cropping area. Another hyperparameter was the threshold value for NMS. The two-stage object detector needed to generate a large number of candidate bounding boxes to locate the ROI regions, and the NMS was responsible for selecting the best one by filtering out the low confidence scored boxes which have higher IoU values than threshold. The next one was the proper selection of augmented methods in TTA; we used three geometric augmentation methods (horizontal flip, vertical flip, and 90-degree rotation) to evaluate the trade-off between performance and computation complexity. We also tested different IoU thresholds in WBF. In general, we achieved an improvement of 9% by employing proper parameters.

Table 4. Hyperparameter selection in software integration.

Stride (Image)	RBD	NMS (IoU Thre)	TTA	WBF (IoU_Thre)	mAP
1	-	-	-	-	0.7494
	Not Remove	0.50	-	-	0.7650
3/4	0	0.50	-	-	0.8021
	5	0.50	-	-	0.8315
	10	0.50	-	-	0.8306
	5	0.40	-	-	0.8320
3/4	5	0.45	-	-	0.8315
	5	0.60	-	-	0.8297
1/2	5	0.40	-	-	0.8275
1/4	5	0.40	-	-	0.8184
	5	0.40	1	0.70	0.8337
	5	0.40	1	0.60	0.8373
3/4	5	0.40	1	0.50	0.8037
	5	0.40	2	0.60	0.8352
	5	0.40	3	0.60	0.8336

Table 5. Real environment evaluation results.

Name	Resolution (pixel)	GSD (cm/pixel)	GT	Predict	FP	Recall
① Namyangjusi Sudongmyeon	31,495 × 30,527	6.79	34	33	21	0.97
② Namyangjusi Joanmyeon	22,317 × 32,501	6.19	50	47	70	0.94
③ Kwangjusi Docheokmyeon	33,775 × 34,618	6.09	7	5	43	0.71
④ Kwangjusi Chowoleup	35,310 × 25,731	6.12	9	9	33	1.00
⑤ Yangpyeonggun Yangdongmyeon	54,178 × 67,325	3.84	40	36	175	0.90
⑥ Goomisi Geoyidong	41,940 × 44,949	5.22	50	49	80	0.98
⑦ Goomisi Goaeup	50,689 × 37,427	5.21	110	108	90	0.98
⑧ Goomisi Sangdongmyeon	60,438 × 78,747	4.91	305	302	210	0.99
⑨ Goomisi myeon	23,630 × 27,099	7.32	48	47	62	0.98
⑩ Milyangsi Yongpyeongdong	47,612 × 51,334	5.40	77	75	150	0.97
			730	711	934	0.97

5.1.3. Evaluation in Real-World Dataset

We captured another 10 orthophotographs (Appendix **??**) from different cities in South Korea to test the reliability of the proposed system. Each orthophotograph has more than $30,000 \times 20,000$ pixels ($360,000$ m^2 in GSD 6.00 cm/pixel) area (Table **??**). We followed the same preprocessing method as that described in the "software integration" section, and the PWD detector captured 711 out of 730 PWD-infected trees in various resolutions (Table **??**). To show the robustness of our network, we draw a portion of an orthophotograph in Figure **??**. The potential infected pine trees (GT) are labeled by red dots. The blue bounding box denotes the inference result obtained by the trained detector. The green panel shows the sample of TPs, which has various symptoms of PWD-infected trees across the early to late stage. The red panel shows the false-detected PWD-infected trees. Distinguishing those "disease-like" objects in RGB channels remains a challenging task due to the ambiguity in both shape and color. Further investigations using either multi-spectral images or field investigations are needed.

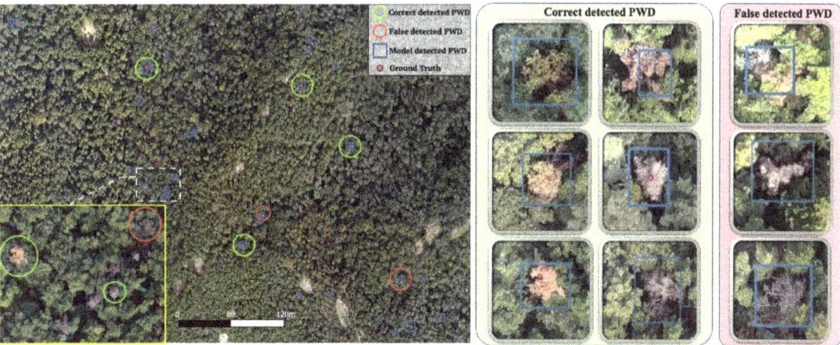

Figure 7. Prediction result using the Goomisi Goaeup dataset described in Table **??**. The bottom left corner shows a magnification of the content in the white dash box. We selected some predicted bounding boxes and marked them with circles. The instances of enlarged disease are shown in the right figure; the green panel shows correctly detected PWD while the red panel represents the false alarms that show an appearance similar to that of a diseased tree.

5.2. The Effect of Hard Negative Mining

HNM was proposed to alleviate the high variance and irreducible error in cases of limited training samples. When the model only sees a few types of disease symptom from a limited number of samples, a large number of background regions makes the detector liable to over-study, whereby it tends to map the true target into no disease. HNM provide many ambiguous "disease-like" objects which share a similar pattern with the disease in the middle and late stages. Including these ambiguous objects helps the model build clear boundaries by learning more discriminative features. In addition, the generated hard negative samples contain a lot of "out of interest regions" (background) such as highways, broadleaf forests, farmland, etc.; this diversity in the training data generalizes the system's ability to correctly classify real-world problems.

Figure **??** shows some examples of the stark contrast that occurred when we applied finetuning with six "disease-like" categories. HNM successfully suppresses the confidence score of ambiguous objects while locating real disease well. The "wb" category (first column) signifies no PWD-infected dead trees; "wb" and PWD-infected dead trees differ only slightly in the branches. Including "wb" reduces the number of FP bounding boxes and possibly well guides the network to update the network for real PWD-infected dead tree by lowering the confidence score of FPs. Moreover, low-resolution images leads to confusion in identifying the yellow land and small PWD objects. The new category "yellow land" alleviates the effect of ground (confidence score 0.960 –> 0.003 in column (2)). The same phenomenon happens with the "maple" category. Due to the loss of water in the late disease stage, PWD-infected trees tend to show a red-brown color, thus appearing similar to maple trees. Without HNM, the network predicted maple trees (column 6, first row) as PWD-infected disease (high confidence score), but including the additional maple samples promoted the learning ability to filter errors (confidence score 0.893 –> 0.630).

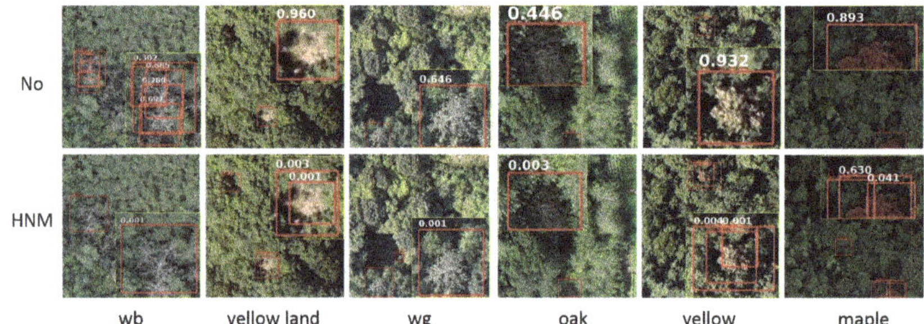

Figure 8. Advantage of using hard negative mining. The detector bounding boxes are the "disease-like" object (FP). For convenience, we zoomed out from the region of interest to highlight the difference when hard negative mining is applied.

6. GIS Application Visualization

We developed a system with which to process a large orthophotograph ("*.tif") to predict potential PWD-infected trees, as shown in Figure **??**. The system automatically detects PWD location and saves the location information in standard ESRI format files. The developed system was integrated with QGIS (https://qgis.org, accessed on 20 December 2021) to visualize the input "*.tif" images and potential disease locations (Figure **??**). Potential disease regions with a high confidence score are represented by yellow bounding boxes while "disease-like" objects are illustrated by blue bounding boxes. The output file preserves class label, predicted confidence score, as well as left-top and right-bottom position for the target bounding box. As shown in Figure **??**b, the column score represents the confidence score within [0,1], and it indicates how much a tree looks like a PWD-infected tree. The expert can reduce the threshold to find more potential disease spots. The columns lr_i and rb_i include the coordinate values of a bounding box in the image coordinate system. The columns lt and rb represent the GPS coordinates in a coordinate reference system [?] (EPSG: 5186 Korean 2000/Central Belt 2010). Using the lt and rb values, expert can locate PWD-infected trees in field investigation.

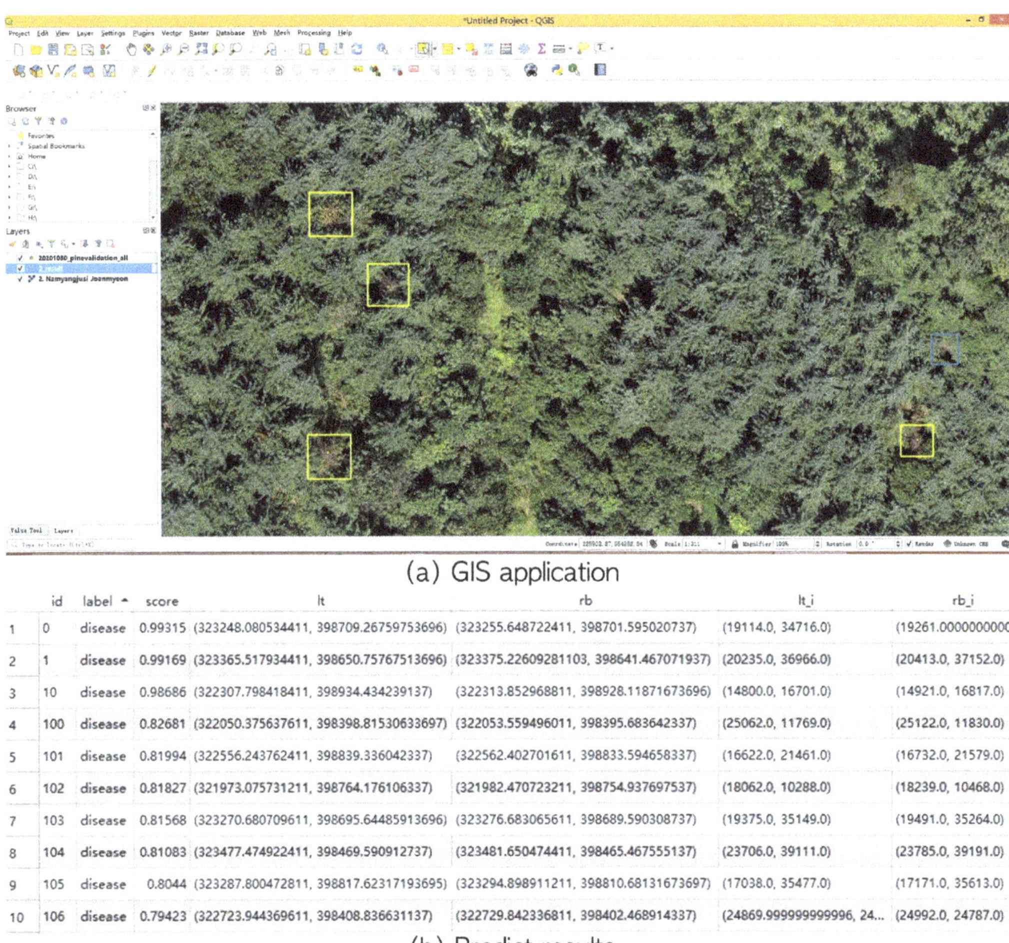

(a) GIS application

	id	label ▲	score	lt	rb	lt_i	rb_i
1	0	disease	0.99315	(323248.080534411, 398709.26759753696)	(323255.648722411, 398701.595020737)	(19114.0, 34716.0)	(19261.0000000000.
2	1	disease	0.99169	(323365.517934411, 398650.75767513696)	(323375.22609281103, 398641.467071937)	(20235.0, 36966.0)	(20413.0, 37152.0)
3	10	disease	0.98686	(322307.798418411, 398934.434239137)	(322313.852968811, 398928.11871673696)	(14800.0, 16701.0)	(14921.0, 16817.0)
4	100	disease	0.82681	(322050.375637611, 398398.81530633697)	(322053.559496011, 398395.683642337)	(25062.0, 11769.0)	(25122.0, 11830.0)
5	101	disease	0.81994	(322556.243762411, 398839.336042337)	(322562.402701611, 398833.594658337)	(16622.0, 21461.0)	(16732.0, 21579.0)
6	102	disease	0.81827	(321973.075731211, 398764.176106337)	(321982.470723211, 398754.937697537)	(18062.0, 10288.0)	(18239.0, 10468.0)
7	103	disease	0.81568	(323270.680709611, 398695.64485913696)	(323276.683065611, 398689.590308737)	(19375.0, 35149.0)	(19491.0, 35264.0)
8	104	disease	0.81083	(323477.474922411, 398469.590912737)	(323481.650474411, 398465.467555137)	(23706.0, 39111.0)	(23785.0, 39191.0)
9	105	disease	0.8044	(323287.800472811, 398817.62317193695)	(323294.898911211, 398810.68131673697)	(17038.0, 35477.0)	(17171.0, 35613.0)
10	106	disease	0.79423	(322723.944369611, 398408.836631137)	(322729.842336811, 398402.468914337)	(24869.999999999996, 24...	(24992.0, 24787.0)

(b) Predict results

Figure 9. Visualization of the inference result in the QGIS program.

7. Discussion and Future Work

In this paper, we proposed a system for improving the performance of the object detection model that detects PWD-infected trees using RGB-based UAV images. To learn a robust network, we created a large dataset which contains a total of 6121 disease spots from various infected stages and areas. The comparison results show that our proposed system has great consistency across different backbone structures. HNM can select "disease-like" objects from six categories. Trained and fine-tuned networks successfully built better decision boundaries with which to distinguish true PWD objects from the six "disease-like" ones. EDA and TTA achieved significant gains by alleviating the data bias problem. In addition, 711 out of 730 PWD-infected trees were identified in 10 large size orthophotographs, indicating that this method shows great potential in locating PWD in various pine forest resources. Finally, the integrated software can automatically locate potential PWD-infected locations and save to ESRI format. It is also convenient to visualize the results in the GIS application for field investigation.

However, there is still work to be done. For example, the best way to utilize context information during the training remains unclear. PWD only infects the pine family, so tree species classification will help reduce inference time and make it easier to precisely locate infected regions. The other problem to be overcome is the method of filtering low-quality images. In UAV images, it is difficult to ensure a consistent resolution, implying that poor-resolution images with ambiguous features decrease the performance. Further, RGB-based PWD detection method still has limitations, as it confuses PWD-infected trees with "disease like" objects in the early and later stages. Reference [?] has demonstrated that PWD infected trees exhibit a reduction in normalized difference vegetation index (NDVI), and [?] showed the effectiveness of conifer broadleaf classification with multi-spectral image. These studies provide insight into the usage of multispectral information to aid in recognition. However, multi-spectral images typically have a lower resolution than RGB images, so the best way to collect and efficiently use multi-spectral images remains a question of interest. We believe our proposed method can be used as preprocessing stage to filter the irrelevant region as well as to find fuzzy PWD hotspot. It is only necessary to reanalyze suspected images with the multi-spectral image, which greatly reduces the time for data collection.

Author Contributions: J.Y. designed the study, conducted the experiments and prepared the original manuscript. R.Z. prepared the dataset. J.L. supervised the research and edited the manuscript. All authors have read and agreed to the published version of the manuscript.

Funding: This work was supported in Brain Korea 21 PLUS Project.

Institutional Review Board Statement: Not applicable.

Informed Consent Statement: Not applicable.

Data Availability Statement: Data sharing is not applicable to this article.

Acknowledgments: The authors express special thanks to Korea Forestry Promotion Institute which has provided the data for experiments in the paper. I would like to give my sincere gratitude to Yagya Raj Pandeya and Bhuwan bhattarai, gave me great help for grammer correction.

Conflicts of Interest: The authors declare that there is no conflict of interest.

Appendix A

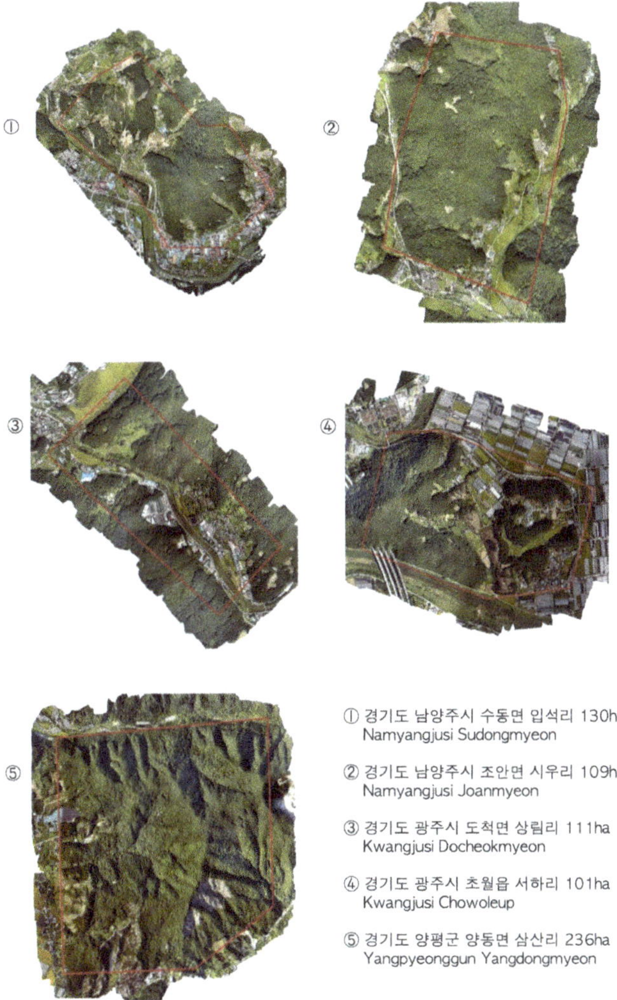

① 경기도 남양주시 수동면 입석리 130ha
Namyangjusi Sudongmyeon

② 경기도 남양주시 조안면 시우리 109ha
Namyangjusi Joanmyeon

③ 경기도 광주시 도척면 상림리 111ha
Kwangjusi Docheokmyeon

④ 경기도 광주시 초월읍 서하리 101ha
Kwangjusi Chowoleup

⑤ 경기도 양평군 양동면 삼산리 236ha
Yangpyeonggun Yangdongmyeon

References

- Khan, M.A.; Ahmed, L.; Mandal, P.K.; Smith, R.; Haque, M. Modelling the dynamics of Pine Wilt Disease with asymptomatic carriers and optimal control. *Sci. Rep.* **2020**, *10*, 1–15. [CrossRef] [PubMed]
- Hirata, A.; Nakamura, K.; Nakao, K.; Kominami, Y.; Tanaka, N.; Ohashi, H.; Takano, K.T.; Takeuchi, W.; Matsui, T. Potential distribution of pine wilt disease under future climate change scenarios. *PLoS ONE* **2017**, *12*, e0182837. [CrossRef] [PubMed]
- Nevalainen, O.; Honkavaara, E.; Tuominen, S.; Viljanen, N.; Hakala, T.; Yu, X.; Hyyppä, J.; Saari, H.; Pölönen, I.; Imai, N.N. Individual tree detection and classification with UAV-based photogrammetric point clouds and hyperspectral imaging. *Remote Sens.* **2017**, *9*, 185. [CrossRef]
- Nezami, S.; Khoramshahi, E.; Nevalainen, O.; Pölönen, I.; Honkavaara, E. Tree species classification of drone hyperspectral and rgb imagery with deep learning convolutional neural networks. *Remote Sens.* **2020**, *12*, 1070. [CrossRef]
- Sothe, C.; Dalponte, M.; de Almeida, C.M.; Schimalski, M.B.; Lima, C.L.; Liesenberg, V.; Miyoshi, G.T.; Tommaselli, A.M.G. Tree species classification in a highly diverse subtropical forest integrating UAV-based photogrammetric point cloud and hyperspectral data. *Remote Sens.* **2019**, *11*, 1338. [CrossRef]
- Egli, S.; Höpke, M. CNN-Based Tree Species Classification Using High Resolution RGB Image Data from Automated UAV Observations. *Remote Sens.* **2020**, *12*, 3892. [CrossRef]

. Maldonado-Ramirez, S.L.; Schmale, D.G., III; Shields, E.J.; Bergstrom, G.C. The relative abundance of viable spores of Gibberella zeae in the planetary boundary layer suggests the role of long-distance transport in regional epidemics of Fusarium head blight. *Agric. For. Meteorol.* **2005**, *132*, 20–27. [CrossRef]

. Nguyen, H.T.; Caceres, M.L.L.; Moritake, K.; Kentsch, S.; Shu, H.; Diez, Y. Individual Sick Fir Tree (Abies mariesii) Identification in Insect Infested Forests by Means of UAV Images and Deep Learning. *Remote Sens.* **2021**, *13*, 260. [CrossRef]

. Arantes, B.H.T.; Moraes, V.H.; Geraldine, A.M.; Alves, T.M.; Albert, A.M.; da Silva, G.J.; Castoldi, G. Spectral detection of nematodes in soybean at flowering growth stage using unmanned aerial vehicles. *Ciência Rural* **2021**, *51*. [CrossRef]

. Xiao, Y.; Dong, Y.; Huang, W.; Liu, L.; Ma, H. Wheat Fusarium Head Blight Detection Using UAV-Based Spectral and Texture Features in Optimal Window Size. *Remote Sens.* **2021**, *13*, 2437. [CrossRef]

. Takenaka, Y.; Katoh, M.; Denga, S.; Cheunga, K. Detecting forests damaged by pine wilt disease at the individual tree level using airborne laser data and worldview-2/3 images over two seasons. In Proceedings of the International Archives of the Photogrammetry, Remote Sensing and Spatial Information Sciences, Volume XLII-3/W3, Jyväskylä, Finland, 25–27 October 2017.

. Yu, R.; Luo, Y.; Zhou, Q.; Zhang, X.; Wu, D.; Ren, L. A machine learning algorithm to detect pine wilt disease using UAV-based hyperspectral imagery and LiDAR data at the tree level. *Int. J. Appl. Earth Obs. Geoinf.* **2021**, *101*, 102363. [CrossRef]

. Deng, X.; Tong, Z.; Lan, Y.; Huang, Z. Detection and Location of Dead Trees with Pine Wilt Disease Based on Deep Learning and UAV Remote Sensing. *AgriEngineering* **2020**, *2*, 294–307. [CrossRef]

. Zhao, Z.Q.; Zheng, P.; tao Xu, S.; Wu, X. Object detection with deep learning: A review. *IEEE Trans. Neural Netw. Learn. Syst.* **2019**, *30*, 3212–3232. [CrossRef] [PubMed]

. Sung, K.K.; Poggio, T. Example-based learning for view-based human face detection. *IEEE Trans. Pattern Anal. Mach. Intell.* **1998**, *20*, 39–51. [CrossRef]

. Felzenszwalb, P.F.; Girshick, R.B.; McAllester, D.; Ramanan, D. Object detection with discriminatively trained part-based models. *IEEE Trans. Pattern Anal. Mach. Intell.* **2009**, *32*, 1627–1645. [CrossRef] [PubMed]

. Iordache, M.D.; Mantas, V.; Baltazar, E.; Pauly, K.; Lewyckyj, N. A Machine Learning Approach to Detecting Pine Wilt Disease Using Airborne Spectral Imagery. *Remote Sens.* **2020**, *12*, 2280. [CrossRef]

. Syifa, M.; Park, S.J.; Lee, C.W. Detection of the Pine Wilt Disease Tree Candidates for Drone Remote Sensing Using Artificial Intelligence Techniques. *Engineering* **2020**, *6*, 919–926. [CrossRef]

. Wu, B.; Liang, A.; Zhang, H.; Zhu, T.; Zou, Z.; Yang, D.; Tang, W.; Li, J.; Su, J. Application of conventional UAV-based high-throughput object detection to the early diagnosis of pine wilt disease by deep learning. *For. Ecol. Manag.* **2021**, *486*, 118986. [CrossRef]

. Qin, J.; Wang, B.; Wu, Y.; Lu, Q.; Zhu, H. Identifying Pine Wood Nematode Disease Using UAV Images and Deep Learning Algorithms. *Remote Sens.* **2021**, *13*, 162. [CrossRef]

. Li, F.; Liu, Z.; Shen, W.; Wang, Y.; Wang, Y.; Ge, C.; Sun, F.; Lan, P. A Remote Sensing and Airborne Edge-Computing Based Detection System for Pine Wilt Disease. *IEEE Access* **2021**, *9*, 66346–66360. [CrossRef]

. Son, M.H.; Lee, W.K.; Lee, S.H.; Cho, H.K.; Lee, J.H. Natural spread pattern of damaged area by pine wilt disease using geostatistical analysis. *J. Korean Soc. For. Sci.* **2006**, *95*, 240–249.

. Park, Y.S.; Chung, Y.J.; Moon, Y.S. Hazard ratings of pine forests to a pine wilt disease at two spatial scales (individual trees and stands) using self-organizing map and random forest. *Ecol. Inform.* **2013**, *13*, 40–46. [CrossRef]

. Lee, J.B.; Kim, E.S.; Lee, S.H. An analysis of spectral pattern for detecting pine wilt disease using ground-based hyperspectral camera. *Korean J. Remote Sens.* **2014**, *30*, 665–675. [CrossRef]

. Kim, S.R.; Lee, W.K.; Lim, C.H.; Kim, M.; Kafatos, M.C.; Lee, S.H.; Lee, S.S. Hyperspectral analysis of pine wilt disease to determine an optimal detection index. *Forests* **2018**, *9*, 115. [CrossRef]

. Lee, S.; Park, S.J.; Baek, G.; Kim, H.; Lee, C.W. Detection of damaged pine tree by the pine wilt disease using UAV Image. *Korean J. Remote Sens.* **2019**, *35*, 359–373.

. Ren, S.; He, K.; Girshick, R.; Sun, J. Faster R-CNN: Towards real-time object detection with region proposal networks. *IEEE Trans. Pattern Anal. Mach. Intell.* **2016**, *39*, 1137–1149. [CrossRef] [PubMed]

. Girshick, R. Fast r-cnn. In Proceedings of the IEEE International Conference on Computer Vision, Santiago, Chile, 7–13 December 2015; pp. 1440–1448.

. Redmon, J.; Divvala, S.; Girshick, R.; Farhadi, A. You only look once: Unified, real-time object detection. In Proceedings of the IEEE Conference on Computer Vision and Pattern Recognition (CVPR), Las Vegas, NV, USA, 30 June 2016.

. Redmon, J.; Farhadi, A. YOLO9000: Better, faster, stronger. In Proceedings of the IEEE Conference on Computer Vision and Pattern Recognition, Honolulu, HI, USA, 26 July 2017.

. Redmon, J.; Farhadi, A. Yolov3: An incremental improvement. *arXiv* **2018**, arXiv:1804.02767.

. Bochkovskiy, A.; Wang, C.Y.; Liao, H.Y.M. Yolov4: Optimal speed and accuracy of object detection. *arXiv* **2020**, arXiv:2004.10934.

. Liu, W.; Anguelov, D.; Erhan, D.; Szegedy, C.; Reed, S.; Fu, C.Y.; Berg, A.C. Ssd: Single shot multibox detector. In *European Conference on Computer Vision*; Springer: Cham, Switzerland, 2016.

. Liu, S.; Huang, D. Receptive field block net for accurate and fast object detection. In Proceedings of the European Conference on Computer Vision (ECCV), Munich, Germany, 8–14 September 2018.

. Shrivastava, A.; Gupta, A.; Girshick, R. Training region-based object detectors with online hard example mining. In Proceedings of the IEEE Conference on Computer Vision and Pattern Recognition, Las Vegas, NV, USA, 30 June 2016.

Krizhevsky, A.; Sutskever, I.; Hinton, G.E. ImageNet Classification with Deep Convolutional Neural Networks. *Commun. ACM* **2017**, *60*, 84–90. [CrossRef]

DeVries, T.; Taylor, G.W. Improved regularization of convolutional neural networks with cutout. *arXiv* **2017**, arXiv:1708.04552.

Solovyev, R.; Wang, W.; Gabruseva, T. Weighted boxes fusion: Ensembling boxes from different object detection models. *Image Vis. Comput.* **2021**, *107*, 104117. [CrossRef]

Bodla, N.; Singh, B.; Chellappa, R.; Davis, L.S. Soft-NMS–improving object detection with one line of code. In Proceedings of the IEEE International Conference on Computer Vision, Venice, Italy, 22–29 October 2017.

Lin, T.Y.; Maire, M.; Belongie, S.; Hays, J.; Perona, P.; Ramanan, D.; Dollár, P.; Zitnick, C.L. Microsoft coco: Common objects in context. In *European Conference on Computer Vision*; Springer: Cham, Switzerland, 2014.

Lin, T.Y.; Goyal, P.; Girshick, R.; He, K.; Dollár, P. Focal Loss for Dense Object Detection. In Proceedings of the IEEE International Conference on Computer Vision, Venice, Italy, 22–29 October 2017.

Janssen, V. Understanding coordinate reference systems, datums and transformations. *Int. J. Geoinform.* **2009**, *5*, 41–53.

Abdollahnejad, A.; Panagiotidis, D. Tree Species Classification and Health Status Assessment for a Mixed Broadleaf-Conifer Forest with UAS Multispectral Imaging. *Remote Sens.* **2020**, *12*, 3722. [CrossRef]

remote sensing

Article

ShadowDeNet: A Moving Target Shadow Detection Network for Video SAR

Jinyu Bao, Xiaoling Zhang *, Tianwen Zhang and Xiaowo Xu

School of Information and Communication Engineering, University of Electronic Science and Technology of China, Chengdu 611731, China; 201811011909@std.uestc.edu.cn (J.B.); twzhang@std.uestc.edu.cn (T.Z.); xuxiaowo@std.uestc.edu.cn (X.X.)
* Correspondence: xlzhang@uestc.edu.cn

Abstract: Most existing SAR moving target shadow detectors not only tend to generate missed detections because of their limited feature extraction capacity among complex scenes, but also tend to bring about numerous perishing false alarms due to their poor foreground–background discrimination capacity. Therefore, to solve these problems, this paper proposes a novel deep learning network called "ShadowDeNet" for better shadow detection of moving ground targets on video synthetic aperture radar (SAR) images. It utilizes five major tools to guarantee its superior detection performance, i.e., (1) histogram equalization shadow enhancement (HESE) for enhancing shadow saliency to facilitate feature extraction, (2) transformer self-attention mechanism (TSAM) for focusing on regions of interests to suppress clutter interferences, (3) shape deformation adaptive learning (SDAL) for learning moving target deformed shadows to conquer motion speed variations, (4) semantic-guided anchor-adaptive learning (SGAAL) for generating optimized anchors to match shadow location and shape, and (5) online hard-example mining (OHEM) for selecting typical difficult negative samples to improve background discrimination capacity. We conduct extensive ablation studies to confirm the effectiveness of the above each contribution. We perform experiments on the public Sandia National Laboratories (SNL) video SAR data. Experimental results reveal the state-of-the-art performance of ShadowDeNet, with a 66.01% best $f1$ accuracy, in contrast to the other five competitive methods. Specifically, ShadowDeNet is superior to the experimental baseline Faster R-CNN by a 9.00% $f1$ accuracy, and superior to the existing first-best model by a 4.96% $f1$ accuracy. Furthermore, ShadowDeNet merely sacrifices a slight detection speed in an acceptable range.

Keywords: video synthetic aperture radar (SAR); moving target; shadow detection; deep learning; false alarms; missed detections

Citation: Bao, J.; Zhang, X.; Zhang, T.; Xu, X. ShadowDeNet: A Moving Target Shadow Detection Network for Video SAR. *Remote Sens.* **2022**, *14*, 320. https://doi.org/10.3390/rs14020320

Academic Editor: Gwanggil Jeon

Received: 2 December 2021
Accepted: 24 December 2021
Published: 11 January 2022

Publisher's Note: MDPI stays neutral with regard to jurisdictional claims in published maps and institutional affiliations.

1. Introduction

Synthetic aperture radar (SAR) is an advanced Earth observation remote sensing tool. Its active radar-based remote sensing ensures its all-day and all-weather working advantage compared with optical sensors [1–3]. Thus, so far, it has been widely applied in civil fields, such as marine exploration, forestry census, topographic mapping, land resources survey, and traffic control, as well as military fields, such as battlefield reconnaissance, war situation monitoring, radar guidance, and strike effect evaluation [4–6]. Video SAR provides continuous multi-SAR images of the target imaging area to dynamically monitor the target scene in real time. It can continuously record the changes of the target area and exhibit the information from the time dimension through the form of visual active images, conducive to the intuitive interpretation of human eyes [7]. Thus, it is receiving extensive attention from increasing scholars [8–10].

Moving target tracking is one of the most significant applications using video SAR. It can provide the important information such as the geographical location, moving direction [11], moving route, and speed of high-value targets in real time [12]. Obviously, it contributes to ground traffic management and accurate attack of military targets. Thus, it has become a research hotspot in recent years [7,13]. So far, some scholars [11–30] have proposed various methods for video SAR moving target tracking that offered competitive results.

Notably, it is interesting that the commonality of the above video SAR moving target tracking methods is to indicate the real moving target with the help of the target's shadow. This is because in video SAR, the Doppler modulation of the moving target echo is rather sensitive to target motion due to the extremely high working frequency, so a slight motion will lead to the large target location offset and target defocus in SAR images, as shown in Figure 1. However, the above phenomena do not happen on the shadow of the moving target [7], thus the shadow reflects the real position and motion state information of the moving target. More formation mechanisms of the moving target shadows in video SAR can be found in [17].

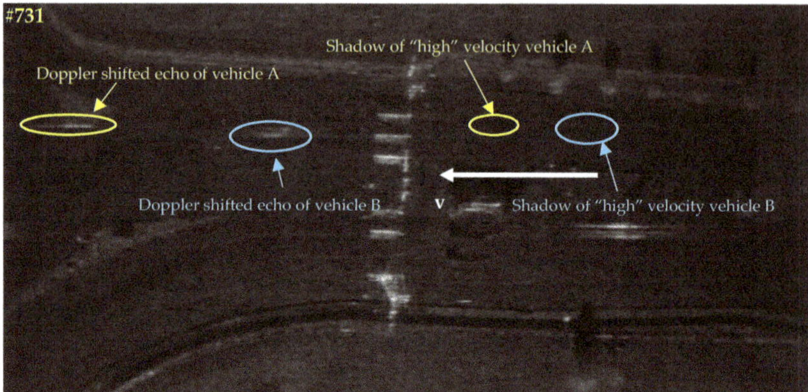

Figure 1. Relative positions between the targets and corresponding shadows. This video SAR image is the 731st frame in the SNL data.

Especially, moving target shadows are very informative from the following two aspects [7]. On the one hand, the contrast between the moving target shadow and its background area, and the gradient information of the shadow intensity along the moving direction, are both closely related to the target speed. On the other hand, because the synthetic aperture time of a single frame image is short, the dynamic shadow also reflects the instantaneous position of the moving target in the scene [7]. Thus, using shadows to complete the video SAR moving target detection and tracking task has become a new research pathway. Furthermore, combined with the Doppler processing technology, the shadow detection can also greatly expand the detectable velocity range of moving targets and improve the robustness of trackers further.

To summarize, moving target shadow detection in video SAR is extremely important and valuable. It is a fundamental and significant prerequisite of the moving target tracking. Only after the shadow is detected successfully can a subsequent series of tasks be carried out smoothly, such as trajectory filtering/reconstruction [14], data association (i.e., target ID allocation), transformation discrimination between old target disappearance and new one appearance (i.e., target ID switching), velocity estimation [31], SAR image refocusing [11,12], and so on. More descriptions about the relationship between detection and tracking can be found in [32,33]. Thus, this paper will research this valuable work emphatically, that is, video SAR moving target shadow detection. So far, various methods or algorithms [16,17] have been proposed for video SAR moving target shadow detection. These methods can be summarized as two types—(1) traditional feature extraction methods and (2) modern deep learning methods.

The traditional feature extraction methods are based on hand-designed features using expert experience. Wang et al. [10] used a constant false alarm rate (CFAR) detector to detect the moving target shadow, but CFAR is very sensitive to ground clutters, resulting in poor migration ability. Zhong et al. [14] designed a cell-average CFAR based on the mean filtering for the shadow detection, but their method relied heavily on the manual model parameter adjustment. Worse still, their detector was provided with weak capacity to suppress false alarms, which brought huge burdens to the follow-up tracker. Zhao et al. [15] proposed a visual saliency-based detection mechanism based on the image contrast to enhance target shadow to improve discrimination performance by using an adaptive threshold. However, the shadow of the moving target is very dim [16], and easy to submerge by surrounding clutters, leading to its less-salient features. Tian et al. [16] proposed a region-partitioning-based algorithm to search for shadows, but this algorithm suffered from too-complex mathematical theories, with poor flexibility, extensibility, and adaptability. Liu et al. [17] proposed a local feature analysis method based on the OTSU's method [18] to detect moving target shadows, but their method needed to model background clutters which is challenging for various backgrounds. Zhang et al. [19] proposed a Tsallis-entropy-based [34] segmentation threshold algorithm to classify background pixels and shadow pixels but obtaining the optimal threshold in the complex mathematical equations is rather time-consuming. Shang et al. [20] leveraged the idea of change detection to detect moving target shadows in THz video-SAR images based on their own private terahertz radar system, but change detection (i.e., background subtraction) worked only on strictly static backgrounds and was sensitive to clutters. He et al. [21] proposed an improved difference-based moving target shadow detection method where the morphological filtering was used to suppressed false alarms. However, their approach required a series of well-designed and complicated preprocessing techniques, reducing the application scope of the method. They improved their previous method [21] in the report of [22] using the speeded-up robust features (SURF) algorithm [35], but computational costs were greatly increased. In short, the above traditional methods are all heavy in computation, weak in generalization, and troublesome in manual feature extraction. Moreover, they are both time-consuming and labor-consuming.

Modern deep learning methods mainly draw support from multilayer neural networks to automatically extract features based on given training samples. In the computer vision (CV) community, many deep learning-based methods using convolutional neural networks (CNNs) have boosted object detection performance greatly, e.g., Faster R-CNN [36], feature pyramid network (FPN) [37], you only look once (YOLO) [38], RetinaNet [39], and CenterNet [40]. Some scholars [41,42] have applied them to detection and classification. Nowadays, many scholars in the video SAR community also have applied them for moving target shadow detection. Ding et al. [24] applied Faster R-CNN to detect shadows, but the raw Faster R-CNN is designed for generic objection detection in optical natural images, so their direct use without critical thinking might be controversial if not considering the targeted video SAR task. Wen et al. [25] adopted dual Faster R-CNN detectors to simultaneously detect shadows in the image spatial domain and range-Doppler (RD) spectrum domain. However, the shadow features in image spatial domain were not comprehensively mined by them, which led to missed detections and false alarms once the raw video SAR echo was not available. Huang et al. [26] proposed an improved Faster R-CNN to boost the per-frame features by incorporating the spatiotemporal information extracted from multiple adjacent frames using 3D CNNs, which improved shadow detection performance. However, for the online shadow detection, it is impossible to draw support from the future image sequences to establish a spatiotemporal information space so as to enhance the past image sequences. Moreover, this method must require an accurate registration, a fixed scene, and a constant number of sequence images [17]. These strict requirements are bound to limit algorithm applications in the velocity-independent continuous tracking radar mode of video SAR, which often has a constantly changing scene [17]. Therefore, to achieve more flexible moving detection and tracking, one should better detect shadows using single-frame images [17]. Yan et al. [27] adopted FPN to detect shadows using their self-developed video MiniSAR system. They used the k-means to cluster video SAR targets, and then regarded the results as the basis for setting anchor box scales, so as to speed up network convergence and improve accuracy. However, their preset anchor box cannot resist shadow deformation once the motion speed is changed. Therefore, their model is powerless for noncooperative enemy moving targets. Zhang et al. [28] also used FPN to detect shadows and added a dense local regression module to boost shadow location performance. However, their experimental dataset only contains some simple scenes, which is not enough to confirm the universality of the proposed method. Hu et al. [29] adopted YOLOv3 equipped with FPN to provide initial shadow detection results for the follow-up tracker on the basis of the joint detector embedding model (JDE) [33]. However, YOLOv3 may be not robust enough for more complex scenes. Additionally, in SAR surveillance videos, moving target shadows usually occupy relatively few pixels resulting in their small shape appearance, which is rather challenging to capture with YOLOv3 due to its poor small detection performance [40,41]. Wang et al. [30] adopted CenterNet [40] to detect shadows inspired by FairMOT [43] and CenterTrack [44], but this kind of anchor-free detector still lacks the capacity to deal with complex scenes and cases [45], bringing about many missed detections and false alarms. It should be noted that Lu et al. [46] proposed a RetinaTrack for online single-stage joint detection and tracking where RetinaNet [39] was used to detect targets. Future scholars can also use RetinaNet to detect moving target shadows, because it solves the problem of extreme imbalance between foregrounds and backgrounds by introducing a focal loss. This imbalance is universal for SAR images. Thus, we will apply this focal loss for moving target shadow detection, for the first time, in this paper. To sum up, although the above existing deep learning-based moving target shadow detectors have achieved competitive detection results, their provided detection performance is still limited. For one thing, they tend to generate missed detections due to their limited feature-extraction capacity among complex scenes. For another thing, they also tend to bring about numerous perishing false alarms due to their poor foreground–background discrimination capacity.

Therefore, to handle the above problems, this paper proposes a novel deep learning network named ShadowDeNet for better moving target shadow detection in video SAR images. There are five core contributions to ensure the excellent performance of Shadow-DeNet. These are (1) a histogram equalization shadow enhancement (HESE) preprocessing technique is used for enhancing shadow saliency (i.e., contrast ratio) to facilitate the follow-up feature extraction, (2) a transformer self-attention mechanism (TSAM) is proposed for paying more attention to regions of interests to suppress clutter interferences, (3) a shape deformation adaptive learning (SDAL) network is designed based on deformable convolutions [47] for learning moving target deformed shadows to conquer motion speed variations, (4) a semantic-guided anchor-adaptive learning (SGAAL) network is designed for achieving optimized anchors to adaptively match shadow location and shape, and (5) an online hard-example mining (OHEM) training strategy [48] is adopted for selecting typical difficult negative samples to boost background discrimination capacity. We conduct extensive ablation studies to confirm the effectiveness of the above each contribution. Finally, the experimental results on the open Sandia National Laboratories (SNL) video SAR data [49] reveal the state-of-the-art moving target shadow performance of ShadowDeNet, compared with the other five competitive methods. Specifically, ShadowDeNet is better than the experimental baseline Faster R-CNN by a 9.00% $f1$ accuracy, and it is also superior to the existing first-best model by a 4.96% $f1$ accuracy. Furthermore, ShadowDeNet merely sacrifices a slight detection speed in an acceptable range.

2. Methodology

ShadowDeNet is based on the mainstream two-stage framework Faster R-CNN [36]. A two-stage detector usually has better detection accuracy than a one-stage one [40], so we select the former as our experimental baseline in the paper. Figure 2 shows the shadow detection framework of ShadowDeNet. Faster R-CNN consists of a backbone network, a region proposal network (RPN), and a Fast R-CNN [36]. HESE is a preprocessing tool. TSAM and SDAL are used to improve the feature extraction ability of the backbone network. SGAAL is used to improve the proposal generation ability of RPN. OHEM is used to improve the detection ability of Fast R-CNN.

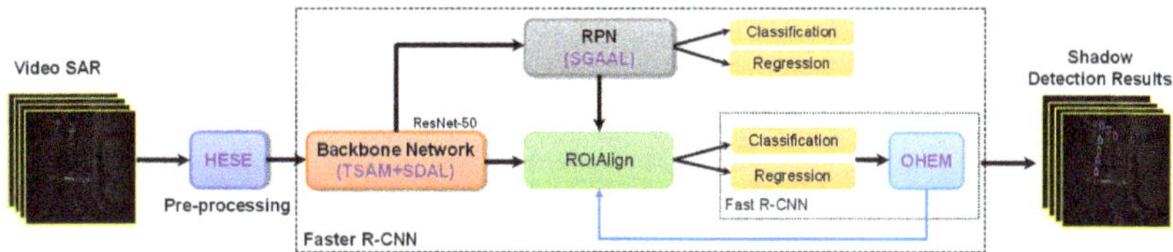

Figure 2. Shadow detection framework of ShadowDeNet. HESE denotes the histogram equalization shadow enhancement. TSAM denotes the transformer self-attention mechanism. SDAL denotes the shape deformation adaptive learning. SGAAL denotes the semantic-guided anchor-adaptive learning. OHEM denotes the online hard-example mining. In ShadowDeNet, without losing generality, we select the commonly-used ResNet-50 [50] as the backbone network.

From Figure 2, we first preprocess the input video SAR images using the proposed HESE technique to enhance shadow's saliency or contrast ratio. The detailed descriptions are introduced in Section 2.1. Then, a backbone network is used to extract shadow features. In ShadowDeNet, without losing generality, we select the commonly-used ResNet-50 [50] as the backbone network. One can leverage more advanced backbone network which may achieve better performance, but this is beyond the scope of this article. In the backbone network, the proposed TSAM and SDAL are embedded, which can both enable better feature extraction. The former is used to pay more attention to regions of interests to suppress clutter interferences based on the attention mechanism [51], which is introduced in detail in Section 2.2. The latter is used to adapt to moving target deformed shadows to overcome motion speed variations based on deformable convolutions [47], which is introduced in detail in Section 2.3.

Immediately, the feature maps are inputted into an RPN to generate regions of interests (ROIs) or proposals. In RPN, a classification network outputs a $2k$-dimension vector to represent a proposal category, i.e., a positive or negative sample. Here, k denotes the number of anchor boxes, which is set to nine in line with the raw Faster R-CNN. The determination of positive and negative samples is based on the intersection over union (IOU), also called Jaccard distance [52,53], with the corresponding ground truth (GT). Similar to the original Faster R-CNN, IOU > 0.70 means positive samples, while IOU < 0.30 means negative samples. Samples with 0.30 < IOU < 0.70 are discarded. Moreover, a regression network outputs a $4k$-dimension vector to represent a proposal location and shape, i.e., (x, y, w, h), where (x, y) denotes the proposal central coordinate, w denotes the width, and h denotes the height. Regression is performed to locate shadows in essence, whose inputs are the feature maps extracted by the backbone network, and outputs are the possible locations of shadows. In RPN, the proposed SGAAL is inserted to improve the quality of proposals. It can generate optimized anchors to adaptively match shadow location and shape inspired by the works of [23,45], which are introduced in detail in Section 2.4.

Afterwards, one ROIAlign layer [54] is used to map the proposals to the feature maps in the backbone network for the subsequent refined classification and regression. Note that the raw Faster R-CNN used one ROIPooling layer to reach such aim, but we replace it with ROIAlign because ROIAlign can address the problem of misalignments caused by twice-quantization [54] so as to avoid a feature loss. Finally, the refined classification and regression are completed by Fast R-CNN [55] to output the final shadow detection results. Moreover, in training, in Fast R-CNN, OHEM is applied to select typical difficult negative samples to boost background discrimination capacity inspired by the works of [48,53], which are introduced in detail in Section 2.5.

Next, we introduce HESE, TSAM, SDAL, SGAAL, and OHEM in detail in the following subsections.

2.1. Histogram Equalization Shadow Enhancement (HESE)

Moving target shadows in video SAR images are rather dim [16] and are always easy to be submerged by surrounding clutters, leading to their less-salient features from the human vision perspective. Figure 3 shows a video SAR image and the corresponding shadow ground truths. In Figure 3a, it is difficult for human vision to find the shadow of a moving target quickly and clearly if one does not refer to the ground truths in Figure 3b. The contrast between the shadow and the surrounding is very low, resulting in their unclear appearances. Moreover, the patrol gate made of metal materials also poses serious negative effects to shadow detection. This experimental data is introduced in detail in Section 3.1. Therefore, to perform some image preprocessing means is necessary, otherwise the learning benefits of features would be reduced.

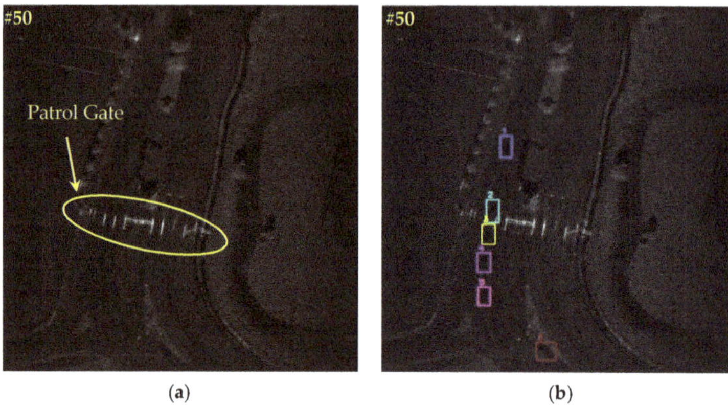

(a) (b)

Figure 3. A video SAR image. (**a**) The raw video SAR image; (**b**) the shadow corresponding ground truths. Here, different vehicles are marked in boxes with different colors and numbers for an intuitive visual observation. This video SAR image is the 50th frame in the SNL data.

Many previous scholars proposed various techniques for image preprocessing, e.g., denoising [8], pixel density clustering [24], morphological filtering [17], visual saliency-based enhancement [15], etc. However, they all rely heavily on expert experience with a series of cumbersome steps, reducing model flexibility. Therefore, we come up with the simple but effective histogram equalization to preprocess video SAR images. For brevity, we denote this process as the histogram equalization shadow enhancement (HESE).

For a video SAR image I, if n_i denotes the number of occurrences of the gray value i $0 \leq i < 256$, then the occurrence probability of pixels with the gray value i is

$$p_I(i) = \frac{n_i}{n} \tag{1}$$

where n denotes the number of all pixels in the image, and $p_I(i)$ denotes, actually, the histogram of the image with pixel value i, normalized to [0, 1]. The HESE is described by

$$HESE(i) = \sum_{i=0}^{k} p_I(i) \ k = 0, 1, 2, \cdots, 255 \tag{2}$$

Figure 4 shows the image histogram of the image in Figure 3a before HESE and after HESE. From Figure 4, after HESE, the whole gray value distribution (marked in red) is similar to the uniform distribution, so this image has a large gray dynamic range and high contrast, and the details of the image are richer. In essence, HESE is used to stretch the image nonlinearly, and redistribute the image pixel values so that the number of pixel values in a certain gray range is roughly equal. In this way, the contrast of the peak part in the middle of the original histogram is enhanced, while the contrast of the valley bottom part on both sides is reduced. Finally, the contrast of the entire image increases.

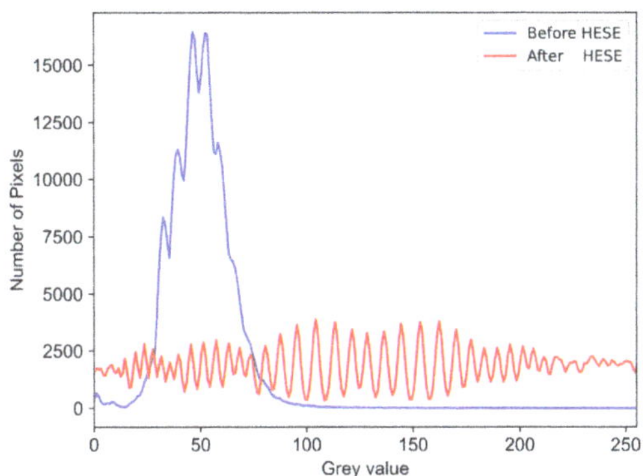

Figure 4. Image pixel histogram before HESE and after HESE.

Figure 5 shows the moving target shadow enhancement results. From Figure 5, one can clearly find that after HESE, the shadow in the zoom region becomes clearer. In Figure 5a, the raw shadow is hardly captured by human eye vision, but in Figure 5b, anyone can find the shadow quickly and easily. We also evaluate the shadow quality in the zoom region by using the classic 4-neighborhood method [56]. The evaluation results are shown in Table 1. From Table 1, the shadow contrast with HESE is far larger than that without HESE (29,215.43 >> 20,979.31). The shadow contrast enhancement reaches up to ~40%, i.e., (29,215.43–20,979.31)/20,979.31. Moreover, the running time is just 13.06 ms, i.e., 7.66 images per second. This seems to be acceptable. Compared to many previous preprocessing means [8,15,17,24], HESE is rather fast with a rather simple theory and workflow. Readers can find more shadow enhancement results of other frames (#3, #13, #23, #33) in Figure 6.

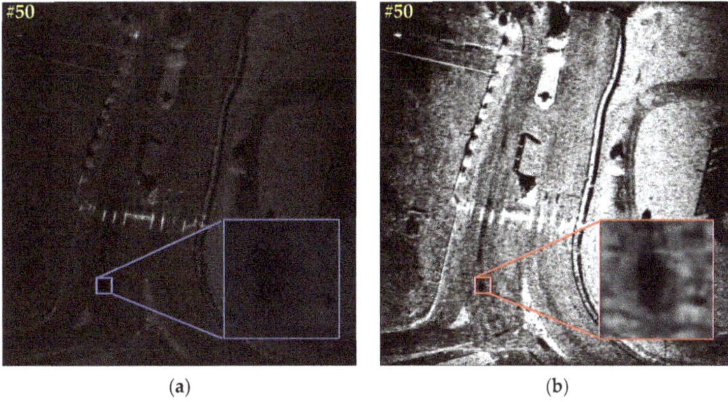

(a)   (b)

Figure 5. Moving target shadow before HESE and after HESE. (a) Before HESE; (b) after HESE. The raw video SAR image is in Figure 3a.

Table 1. Shadow quality evaluation results with and without HESE. The running time is obtained on the Intel(R) Core(TM) i9-10900KF CPU.

HESE	Shadow Contrast	Running Time (ms)
✗	20,979.31	-
√	29,215.43 (↑8236.12)	13.06

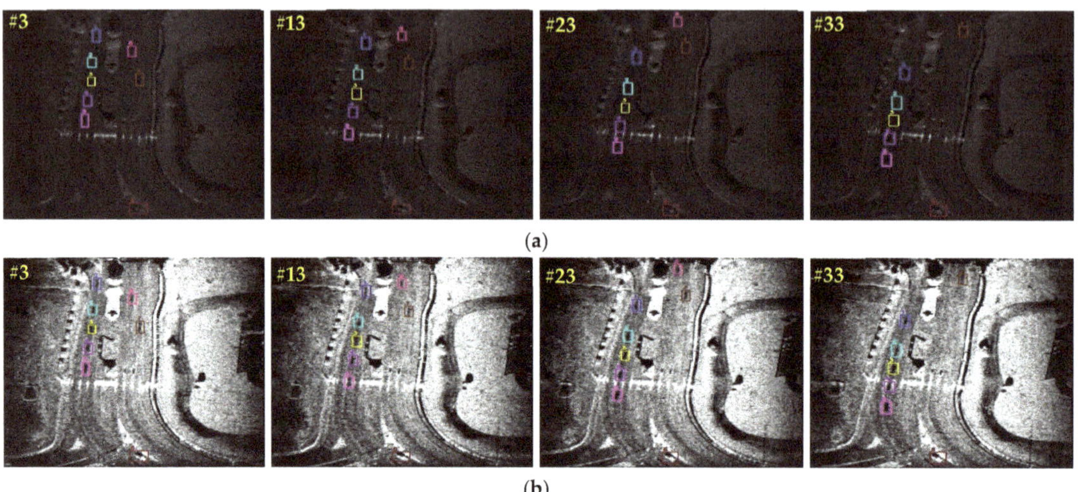

(a)

(b)

Figure 6. More results of the histogram equalization shadow enhancement (HESE). (**a**) Before HESE; (**b**) after HESE. Different vehicles are marked in boxes with different colors and numbers for an intuitive visual observation. #*N* denotes the *N*-th frame. The white arrows indicate the moving direction.

2.2. Transformer Self-Attention Mechanism (TSAM)

Attention mechanisms are widely used in the CV community that can adaptively learn feature weights to focus on important information and suppress the useless. So far, scholars from the SAR community have applied it to various applications, e.g., SAR automatic target recognition (ATR) [57,58], SAR target detection [59,60] and classification [61,62], and so on. Recently, transformer detectors [63,64] have received increasing concerns in the CV community. The remarkable characteristic of transformer models is the internal self-attention mechanism which is able to effectively capture some important long-range dependencies among the entire location space [65]. In video SAR images, there are many clutter interferences from Figure 3a, so we adopt such self-attention mechanism to suppress them so as to focus on more valuable regions of interests. We call this process transformer self-attention mechanism (TSAM). Specifically, we insert TSAM to the backbone network which enables efficient information flow to extract more representative features. As mentioned before, we selected ResNet-50 as our backbone network in Figure 2, thus we insert TSAM to the residual block to promote better residual learning, as is shown in Figure 7.

In Figure 7, the first 1×1 convolution (conv) is used to reduce the input channel dimension, and the second 1×1 conv is used to increase the output channel dimension for the follow-up adding operation. The 3×3 conv is used to extract shadow features. TSAM is used behind the 3×3 conv, meaning that the extracted shadow features are prescreened by TSAM. In this way, the important features are retained while the useless interferences are suppressed. The above practice is similar to that in the convolutional block attention module (CBAM) [66] and squeeze-and-excitation (SE) [67], which can be described as

$$F' = F + f_{1\times1}\{TSAM(f_{3\times3}(f_{1\times1}(F)))\} \tag{3}$$

where F denotes the input of a residual block, F' denotes the output, $f_{1\times1}(\cdot)$ denotes the 1×1 conv operation, $f_{3\times3}(\cdot)$ denotes the 3×3 conv operation, and $TSAM(\cdot)$ denotes the $TSAM$ operation.

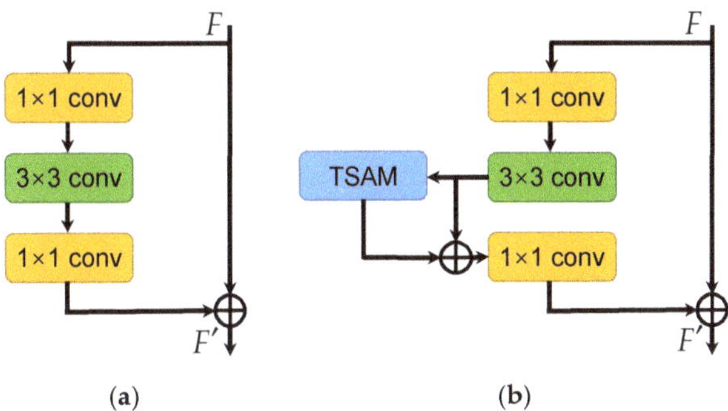

Figure 7. Residual block in the backbone network. (**a**) The raw residual block in ResNet-50; (**b**) the improved residual block with TSAM.

Figure 8 shows the detailed implementation process of $TSAM$. In Figure 8, H denotes the height of the input feature map X, H denotes the width, and C denotes the channel number. According to Wang et al. [68], the general transformer self-attention can be summarized as

$$\mathbf{y}_i = \frac{1}{C(\mathbf{x})}\sum_{\forall j} f(\mathbf{x}_i, \mathbf{x}_j) g(\mathbf{x}_j) \tag{4}$$

where i is the index of the required output location (i.e., the response of the i-th location is to be calculated), and j is the index that enumerates all possible locations, i.e., $\forall j$. $\mathbf{x}$ denotes the input feature map, and y denotes the output feature map with the same dimension as $\mathbf{x}$. The paired function f computes the relationship between i and all j. The unary function g calculates the representation of the input feature map at the j-th location. Finally, the response is normalized by a factor $C(\mathbf{x})$.

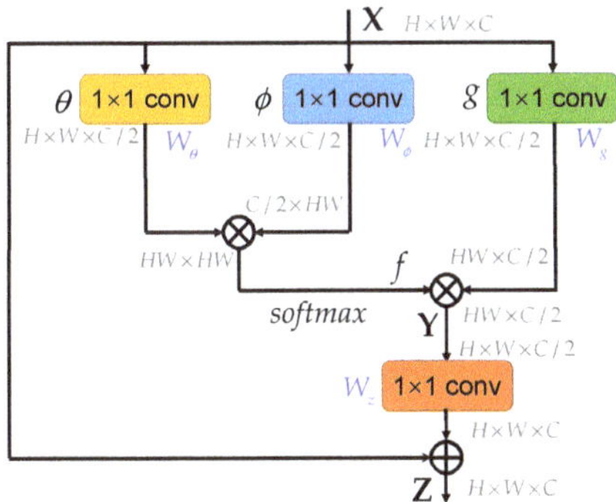

Figure 8. Detailed implementation process of TSAM.

The paired function f can be achieved by an embedded Gaussian function so as to compute similarity between i and all j in an embedding space, i.e.,

$$f(\mathbf{x}_i, \mathbf{x}_j) = e^{\theta(\mathbf{x}_i)^T \phi(\mathbf{x}_j)} \tag{5}$$

where θ denotes the embedding of $\mathbf{x}_i$ and ϕ denotes the embedding of $\mathbf{x}_j$. From Figure 8, they are implemented by using two 1×1 convs W_θ and W_ϕ. That is, $\theta(\mathbf{x}_i) = W_\theta \mathbf{x}_i$ and $\phi(\mathbf{x}_j) = W_\phi \mathbf{x}_j$. Here, to reduce the computation cost, their kernel numbers are set to $C/2$ if the input channel number is C. The normalization factor is set as $C(\mathbf{x}) = \Sigma_{\forall j} f(\mathbf{x}_i, \mathbf{x}_j)$. Thus, for a given i, $f(\mathbf{x}_i, \mathbf{x}_j)/C(\mathbf{x})$ will become the *softmax* computation along the dimension j where *softmax* is defined by $e^{\mathbf{x}_i} / \sum_j e^{\mathbf{x}_j}$ [69]. Here, the *softmax* computation is responsible for generating the weight (i.e., importance level) of each location. Similarly, the representation of the input feature map at the j-th location is also calculated in an embedding space by using another one 1×1 conv W_g. With the matrix multiplication, the output of the self-attention $\mathbf{Y}$ is obtained. Finally, in order to complete the residual operation (i.e., adding), one 1×1 conv W_z is used to increase the channel number from $C/2$ to C, i.e.,

$$\mathbf{Z} = W_z \mathbf{Y} + \mathbf{X} \tag{6}$$

In essence, TSAM is able to calculate the interaction between any two positions and also directly captures the remote dependence without being limited to adjacent points. It is equivalent to constructing a convolution kernel as large as the size of the feature map, so that more background context information can be maintained. In this way, the network can focus on important regions of interests to suppress clutter interferences or other negative effects of useless backgrounds.

2.3. Shape Deformation Adaptive Learning (SDAL)

The contrast between the shadow generated by the moving target and its background area, the gradient information of the shadow intensity along the moving direction, and the shape of the shadow are all closely related to the moving speed of the target [7]. On the premise that the shadow can be formed, the smaller the moving speed of the target, the greater the shadow extension [70], and the clearer the shadow contour of the moving target. However, the larger the moving speed of the target, the lower the shadow extension, and the more blurred the shadow contour of the moving target. In other words, when the motion speed of the moving target changes, the shadow shape will change. When the motion speed changes continuously in multiframe video SAR images, the shadow of the same target will become deformed. This challenges the robustness of the detector. Readers can refer to [11] for more details about the shadow size relationship with the speed.

Figure 9 shows the moving target shadow deformation with the change of moving speed. In Figure 9, due to the stopping signal of the traffic light, the vehicle-A's speed is becoming smaller and smaller. One can find that the same vehicle-A exhibits shadows with different shapes (the zoom region) in the different frames in the video SAR. Thus, a good shadow detector should resist such shadow deformation. However, the feature extraction process of classical convolution neural networks mainly depends on convolution kernels, but the geometric structure of traditional convolution kernel is fixed. Only fixed local feature information is extracted each time when a convolution operation is performed. Thus, the classical convolution cannot solve the shape deformation problem. Fortunately, the recent deformation convolution proposed by Dai et al. [47] can overcome this problem because its convolution kernel can produce free deformation to adapt to the geometric deformation of the target. The deformable convolution changes the sampling position of the standard convolution kernel by adding additional offsets at the sampling points. The compensation obtained can be learned through training without additional supervision. We call the above shape deformation-adaptive learning (SDAL).

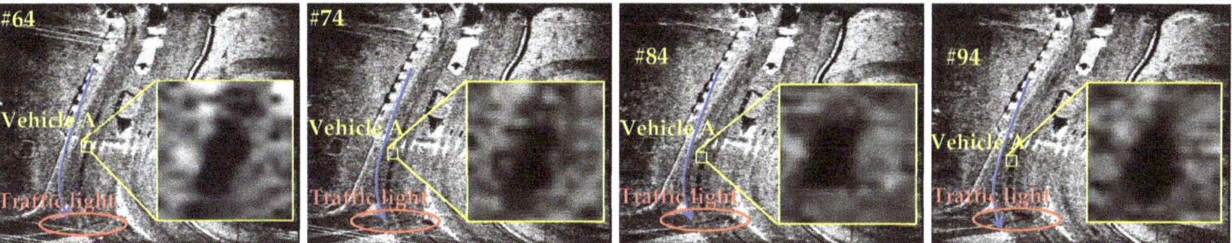

Figure 9. Moving target shadow deformation with the change of moving speed. From left to right (#64 → #74 → #84 → #94), the speed becomes smaller and smaller. The blue arrow indicates the moving direction.

Figure 10 is the sketch map of different convolutions. From Figure 10, the deformation convolution can resist shadow deformation effectively by the learned location offsets. Figure 11 shows the implementation of SDAL.

(a)   (b)

Figure 10. Different convolutions. (**a**) Classical convolution; (**b**) deformation convolution.

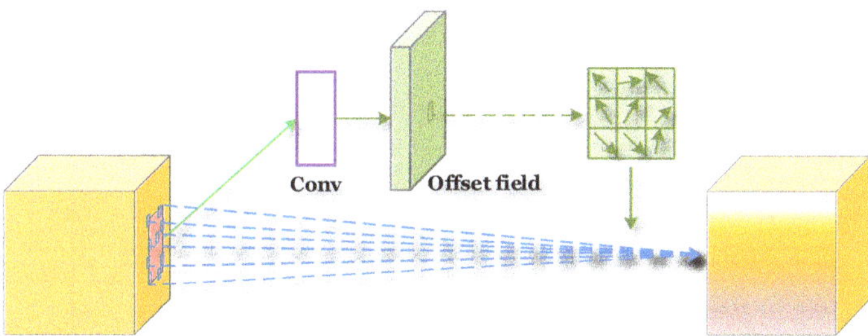

Figure 11. Detailed implementation process of SDAL.

The standard convolution kernel is augmented with offsets $\Delta\mathbf{p}_n$ which are adaptively learned in training to model various shape features, i.e.,

$$\mathbf{y}(\mathbf{p}_0) = \sum_{\mathbf{p}_n \in \Re} \mathbf{w}(\mathbf{p}_n) \cdot \mathbf{x}(\mathbf{p}_0 + \mathbf{p}_n + \Delta\mathbf{p}_n) \tag{7}$$

where $\mathbf{p}_0$ denotes each location, $\Re$ denotes the convolution region, $\mathbf{w}$ denotes the weight parameters, $\mathbf{x}$ denotes the input, $\mathbf{y}$ denotes the output, and $\Delta\mathbf{p}_n$ denotes the learned offsets in the n-th location. $\Delta\mathbf{p}_n$ is typically fractional, so the bilinear interpolation is used to ensure the smooth implementation of convolution, i.e.,

$$G(\mathbf{q}, \mathbf{p}) = g(q_x, p_x) \cdot g(q_y, p_y) \tag{8}$$

where $g(a,b) = \max(0, 1 - |a - b|)$. We add another convolution layer (marked in purple in Figure 11) to learn the offsets $\Delta\mathbf{p}_n$, and then, the standard convolution combining

$\Delta \mathbf{p}_n$ is performed on the input feature maps. Moreover, inspired by [47], the traditional convolutions of the high-level layers, i.e., conv3_x, conv4_x, conv5_x in ResNet-50, are replaced with deformation ones to extract more robust shadow features. This is because the slightest change in the receptive field size among the high-level layers is able to pose a remarkable difference to the following networks, thus obtaining a better geometric modeling capacity in transformation of shape-changeable moving target shadows [71].

2.4. Semantic-Guided Anchor-Adaptive Learning (SGAAL)

Anchors are the basis in modern object detection, which are usually a set of artificially designed boxes, used as the benchmark for classification and bounding box regression. However, previous video SAR moving target shadow detectors mostly adopt preset fixed-shape or fixed-size or fixed-scale anchors. In other words, they have changeless scales and aspect ratios, potentially declining the feature learning benefits of moving target shadows. Moreover, the raw anchors are arranged in the feature map densely and uniformly and are not in line with the video SAR image characteristic, as in Figure 12a. This is because moving target shadows in video SAR images are distributed sparsely and unevenly; if the dense and uniform anchors are used, there will be many false alarms generated. Therefore, inspired by Wang et al. [45], we design a novel semantic-guided anchor-adaptive learning (SGAAL) tool to generate high-quality location-adaptive and shape-adaptive anchors in the RPN, as in Figure 12b. Here, we adopt the high-level deep semantic features to guide anchor generation which can ensure higher anchor quality [45].

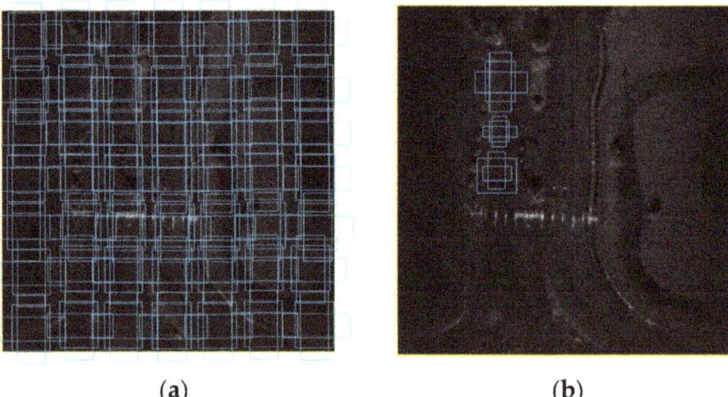

(**a**)   (**b**)

Figure 12. Sketch map of different anchor distributions. (**a**) The raw distribution; (**b**) the improved distribution with SGAAL. Anchors are marked in blue boxes.

The aim of SGAAL is to adaptively obtain the anchor location and the corresponding shape, that is, the parameters (x, y, w, h) of anchors in the image I, where (x, y) denotes the spatial coordinate of the anchor center, w denotes the width of the anchor box, and h denotes the height of the anchor box. Therefore, SGAAL can be described by

$$p(x, y, w, h | I) = p(x, y | I) \cdot p(w, h | x, y, I) \tag{9}$$

where $p(x, y | I)$ denotes the prediction process of the anchor location for a given image I, and $p(w, h | x, y, I)$ denotes the prediction process of the anchor shape for a given image I and the corresponding known location. In other words, SGAAL will first adaptively predict the location (x, y) of anchors, and then adaptively predict the shape (w, h) of anchors. Figure 13 shows the detailed implementation process of SGAAL. In Figure 13, the input semantic feature is denoted by Q. Its height is denoted by H, its width is denoted by W, and its channel number is denoted by C.

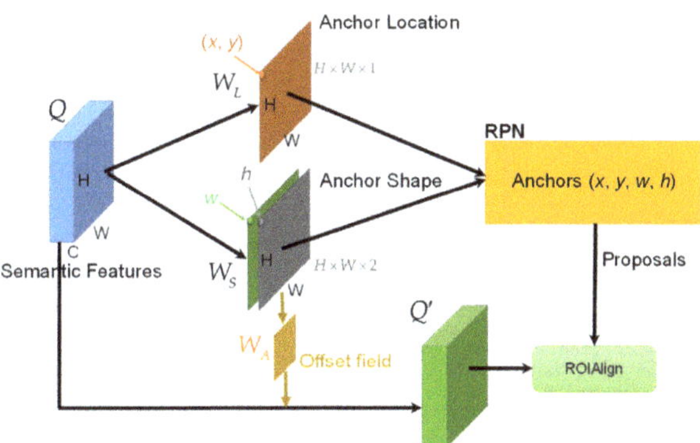

Figure 13. Detailed implementation process of SGAAL.

From Figure 13, we use a 1×1 conv W_L to predict the anchor location whose channel number is set to 1, which will encode the whole $H \times W$ location space. This 1×1 conv layer is followed by a *sigmod* activation function that is defined by $1/(1 + e^{-x})$ to represent the occurrence probability of the shadow location. Here, the location threshold is denoted by ε_L. That is, when the value of the location (x, y) is bigger than ε_L, then this location is assigned by a positive "1" label (i.e., the network should generate anchors at this location); otherwise, it is assigned by a negative "0" label (i.e., the network should not generate anchors at this location). As locations with shadows occupy a small portion of the whole feature map, we adopt the focal loss (FL) of RetinaNet [39] to train this anchor location prediction network so as to avoid falling into a large number of negative samples, i.e.,

$$loss_{FL}(p_t) = -\alpha_t(1 - p_t)^\gamma \log(p_t)$$
$$p_t = \begin{cases} p & if\ y = 1 \\ 1 - p & otherwise \end{cases} \tag{10}$$

where y denotes the ground-truth class. $y = 1$ means the positive, otherwise it is the negative. p denotes the predicted probability ranging from 0 to 1. γ denotes the focusing parameter, set to 2 empirically, and α_t denotes the weighting factor, set to 0.25 empirically.

We use a 1×1 conv W_S to predict the anchor shape whose channel number is set to 2 because we need to obtain the anchor width w and height h. The anchor shape prediction is across the whole $H \times W$ location space. However, the shape predictions whose corresponding location predictions are lower than the threshold ε_L are filtered. This threshold ε_L will be determined experimentally in Section 5.4. The bounded IOU loss [72] is used to train this anchor shape prediction network because it is more sensitive to box spatial locations, i.e.,

$$loss_{BIOU} = -\log(1 - \frac{G \cap P}{G \cup P}) \tag{11}$$

where G denotes the ground-truth box and P denotes the prediction box.

As a result, the anchor location and shape are obtained combined with W_L and W_S. Note that Wang et al. [45] pointed out that the feature for a large anchor should encode the content over a large region, while those for small anchors should have smaller scopes accordingly, thus, following their practice, we also devise an anchor-guided feature adaptation component, which will transform the feature at each individual location i based on the underlying anchor shape, i.e.,

$$q_i' = A(q_i, w_i, h_i) \tag{12}$$

where q_i denotes the i-th location element of the raw feature map Q, and q_i' denotes the i-th location element of the transformed feature map Q', and w_i and h_i denote the width and height of anchors at the i-th location. Moreover, A is a 3×3 deformable convolutional layer W_A which is used to predict the offset field from the output of the anchor shape prediction branch, and then apply the learned offset to the original feature map to obtain the final feature map Q'.

Finally, based on the adaptively learned anchors, the high-quality proposals are generated, and then they are mapped to the transformed feature map Q' by ROIAlign to extract their corresponding feature regions for the subsequent classification and regression in Fast R-CNN, as in Figure 2. In this way, the obtained optimized anchors will be able to adaptively match shadow location and shape so as to enable better false-alarm suppression ability and ensure more attentive shadow feature learning.

2.5. Online Hard-Example Mining (OHEM)

SGAAL can remove many negative samples by the location judgment, but it still does not solve the imbalance problem between positive samples and negative samples. For a location in the feature map, the number of the generated negative samples is still far more than that of positive ones, because background pixels usually occupy a larger proportion. Among a large number of negative samples, it is necessary to select more typical difficult negative samples and abandon easy ones to enhance background discrimination capacity. Online hard-example mining (OHEM) is an advanced difficult-identified negative sample mining method during training. It was proposed by Shrivastava et al. [48] in 2016 and mainly selects some difficult negative samples as training samples in the training process of the target detection model, so as to improve the model parameters and make it converge to a better effect. Difficult samples refer to the samples which are difficult to distinguish, with a large training loss value. For a simple sample that is easy to correctly classify, it is difficult for the model to learning more effective information from it. Thus, hard samples are more valuable for model optimization and are more worth mining and utilization.

Figure 14 shows the detailed implementation process of OHEM. In Figure 14, the classification loss of Fast R-CNN is denoted $loss_{cls}$, and the regression loss is denoted $loss_{reg}$. We sum the $loss_{cls}$ and $loss_{reg}$ of the negative samples. Their sum loss $loss_{sum}$ is then ranked, and the top k samples are selected into the hard-negative sample pool, where k is set to 256, inspired by [48]. Moreover, the positive sample number is set to 256 to avoid falling into the local optimization of a certain positive or negative category. When the sample number in the pool reaches a batch size, they are mapped into the feature map by ROIAlign again to be trained repeatedly and emphatically. In this way, Fast R-CNN is able to learn more representative background features to further suppress false alarms. More details can be found in [48].

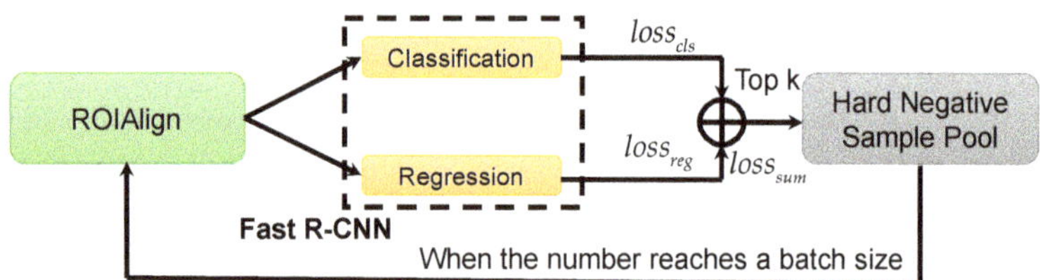

Figure 14. Detailed implementation process of OHEM.

3. Experiments

Our experiments are all performed on a personal computer (PC) which is equipped with an NVIDIA GeForce RTX 3090 GPU, an Intel(R) Core(TM) i9-10900KF CPU, and 32 G memory based on the Pytorch [73] and MMDetection [74] software development environment using the Python language. Furthermore, CUDA-11.1 and CuDNN-8.0.5 are both used for the training and inference GPU acceleration.

3.1. Data

We conduct experiments on the public data released by Sandia National Laboratories [49]. This video SAR data is the only publicly available, thus it is selected. It is the video SAR footage of a gate at the Kirtland Air Force Base Eubank Gate. Table 2 shows the radar system parameters of this data. Figure 15 shows the experimental working environment of the optical image and the corresponding SAR image.

Table 2. Radar system parameters of the public Sandia National Laboratories (SNL) video SAR data.

Parameter	Value
Mode	Spotlight
Center Frequency	16.7 GHz (Ku band)
Wavelength	1.8 cm
Incidence Angle	65°
Platform Height	2 km
Platform Speed	245 km/h
Cross Range Resolution	0.1 m
Total Imaging Time	232 s
Total Rotation Angle	200°

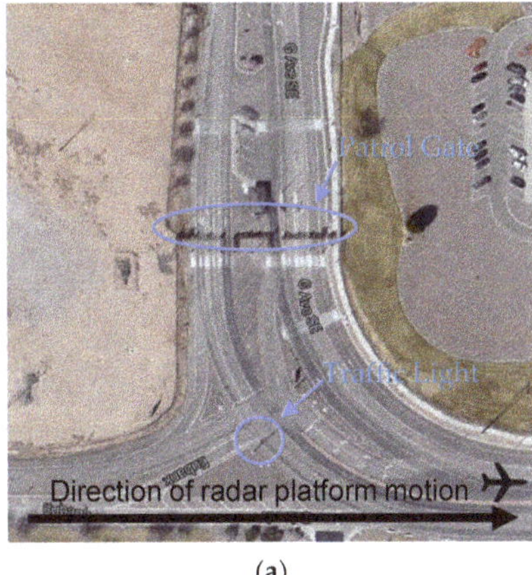

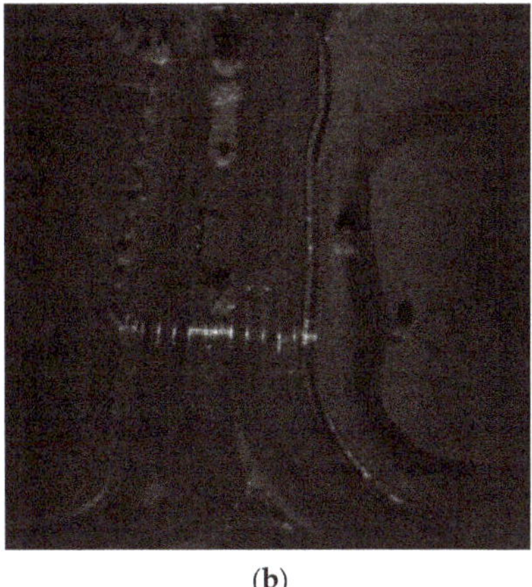

(a)   (b)

Figure 15. Experimental working environment of the SNL video SAR data at the Kirtland Airforce Base Eubank Gate. (**a**) The optical image; (**b**) the corresponding SAR image.

There are 900 frames of continuous SAR images in the SNL video data. The image size is 660 pixel × 720 pixel. We divide the raw video into nine sub-videos, i.e., one sub-video contains 100 continuous SAR image sequences. The moving target shadow ground truths are labeled by professional experts. The first six sub-videos, i.e., 600 continuous SAR images, are selected as the training dataset. The remaining three sub-videos, i.e., 300 continuous SAR images, are selected as the test dataset. This data is available https://www.sandia.gov/app/uploads/sites/124/2021/08/eubankgateandtrafficvideosar.mp4 (accessed on 20 November 2021) where related scholars can download it for free for scientific research.

3.2. Experimental Details

We train ShadowDeNet based on the stochastic gradient descent (SGD) optimizer [75] by 12 epochs. The learning rate is set to 0.008, the momentum is set to 0.9, and the weight decay is set to 0.0001. The training warmup is the linear mode with 500 iterations. The learning rate is reduced by 10 times at the 8th and the 11th epoch. The training batch size is set to four because of the limited GPU memory. The input image size of ShadowDeNet is 660 pixel $\times$ 720 pixel. To avoid overfitting, the ImageNet pretrained weights [76] of ResNet-50 [50] are loaded for transfer learning. Other network parameters are initialized using the Kaiming method [77]. Other hyperparameters not mentioned are same as Faster R-CNN. During inference, the nonmaximum suppression (NMS) [78] is used to suppress duplicate detection boxes with an IOU threshold of 0.50.

3.3. Evaluation Indices

The PASCAL VOC evaluation indices [79] are adopted whose detection IOU threshold is 0.50.

The recall (r) is defined by

$$r = \frac{\#tp}{\#tp + \#fn} \times 100\% \tag{13}$$

where tp denotes the true positives (i.e., correct detections), fn denotes the false negatives (i.e., missed detections), and # denotes the number. In essence, r is equal to the detection rate P_d.

The precision (p) is defined by

$$p = \frac{\#tp}{\#tp + \#fp} \times 100\% \tag{14}$$

where fp denotes the false positives (i.e., false alarms). In essence, p is equal to $1 - P_f$ (the false alarm rate).

The average precision (ap) is defined by

$$ap = \int_0^1 p(r) \cdot dr \tag{15}$$

where $p(r)$ denotes the precision-recall curve.

The $f1$-score is defined by

$$f1 = 2 \times \frac{r \times p}{r + p} \tag{16}$$

In this paper, we mainly use $f1$ as the core accuracy index because it can make a trade-off between the detection rate and the false alarm rate regardless of if in the traditional machine learning community or in the modern deep learning community.

We use the frames per second (FPS) to measure the detection speed, defined by

$$FPS = \frac{1}{T} \tag{17}$$

where T denotes the time consumed to complete an image detection.

4. Results

4.1. Quantitative Results

Table 3 shows the quantitative results of ShadowDeNet on the SNL video SAR data. In Table 3, we show the quantitative results by the mean of progressively adding the proposed improvements to the experimental baseline Faster R-CNN [36]. More ablation studies about the impact of each improvement on the whole ShadowDeNet model are introduced in Section 5 by the mean of each installation and removal.

Table 3. Quantitative results of ShadowDeNet on the SNL video SAR data. #*gt*: ground truths. #*tp*: the higher means the better (i.e., correct detections). #*fp*: the lower means the better (i.e., false alarms). #*fn*: the lower means the better (i.e., missed detections). r (%): recall, the higher means the better. p (%): precision, the higher means the better. ap (%): average precision, the higher means the better. $f1$ (%): $f1$-score, the higher means the better. #: the number.

HESE [1]	TSAM [2]	SDAL [3]	SGAAL [4]	OHEM [5]	#*gt*	#*tp*	#*fp*	#*fn*	r (%)	p (%)	ap (%)	$f1$ (%)
-	-	-	-	-	1581	884	636	697	55.91	58.16	46.33	57.01
√					1581	898	540	683	56.80	62.45	49.83	59.49
√	√				1581	903	516	678	57.12	63.64	49.84	60.20
√	√	√			1581	908	392	673	57.43	69.85	50.91	63.03
√	√	√	√		1581	904	308	677	57.18	74.59	50.33	64.73
√	√	√	√	√	1581	902	250	679	57.05	78.30	**51.87**	**66.01**
-	-	-	-	-	-	+18	−386	−18	+1.14	+20.14	**+5.54**	**+9.00**

[1] HESE denotes the histogram equalization shadow enhancement. [2] TSAM denotes the transformer self-attention mechanism. [3] SDAL denotes the shape deformation adaptive learning. [4] SGAAL denotes the semantic-guided anchor-adaptive learning. [5] OHEM denotes the online hard example mining.

From Table 3, one can draw the following conclusions:

1. The detection accuracy presents a rising trend by progressively adding the proposed improvements to the experimental baseline Faster R-CNN. This confirms the effectiveness of each technique. In terms of the $f1$ accuracy, it has increased from the initial 57.01% to 66.01%, up to a 9.00% huge improvement. In terms of the ap accuracy, it has increased from the initial 46.33% to 51.87%, up to a 5.54% observable improvement. Other accuracy indexes are also increased, more or less. This fully reveals the extraordinary moving target shadow detection performance.

2. HESE improves the ap accuracy by ~3.5% and the $f1$ accuracy by ~2.5%, showing its effectiveness. It can improve the detection rate (r is increased from 55.91% to 56.80%), meanwhile suppressing false alarms (p is increased from 58.16% to 62.45%), because it can enhance shadow saliency for robust shadow feature extraction and strong background discrimination. As presented before in Section 2.1, the enhanced shadow becomes more prominent and clearer and becomes easier and more intuitive to capture for human eye vision. This resulting benefit is also helpful for the network feature adaptive learning.

3. TSAM improves the ap accuracy marginally, but the $f1$ accuracy is still improved by ~0.7%. It is still useful for shadow detection because it can detect more real shadows, i.e., the number of correct detections is increased from 898 to 903 (that is, the previous five missed detections are detected successfully again by using TSAM), meanwhile it still can suppress many false alarms, i.e., the number of false alarms is reduced from 540 to 516 (that is, the previous 24 false alarms are suppressed thoroughly by using TSAM). From the comparison results, TSAM seems to be able to enable better false alarm suppress ability, which is in fact in line with its internal working mechanism introduced in Section 2.2. That is, it can receive more contextual information to focus more on regions of interests and suppress clutter interferences. As a result, the exhibited results indicate the lower false alarm rate when TSAM is used.

4. SDAL improves the ap accuracy by ~1.1% and the $f1$ accuracy by ~3.0%. The latter's improvement is larger than the former's. This is because the latter is more sensitive to the false alarm rate, which is exactly in line with the fact that SDAL is able to suppress false alarms greatly, i.e., the p is increased from 63.64% to 69.85%. SDAL can also detect another five real shadows, because it can find the shadows with intense deformations again when the motion speed is changed, as introduced in Section 2.3. Moreover, we hold the view that when SDAL is used, the network generalization ability can be enhanced, because the network can adapt to "moving" target shadow deformations so as to reduce the possibility of falling into the local optimization

among a large number of the "static" dark shadow-like background negative samples. Consequently, the false alarm rate is decreased greatly.

5. SGAAL offers an almost similar *ap* accuracy, but it still improves the *f*1 accuracy by ~1.75%. This *f*1 accuracy improvement is from the increase of *p*, that is, from 69.85% to 74.59%. This phenomenon exactly accords with the theoretical analysis in Section 2.4. SGAAL can filter many dense anchors using its anchor location prediction network. As a result, the anchors become sparser in line with the sparse distribution of moving shadows in the video SAR images. Moreover, it should be noted that SGAAL seems not to improve the detection rate from the above results but declines it a little, but this does not mean that its anchor shape prediction network is useless, because another experiment in Section 5.4 shows that it still plays an important role in the whole fully-equipped ShadowDeNet (that is, when the other four improvements are used, only SGAAL is removed).

6. OHEM further decreases the false alarm rate, i.e., the *p* is improved from 74.59% to 78.30%. As a result, the *ap* accuracy is improved from the previous 50.33% to the final 51.87%, and the *f*1 accuracy is improved from the previous 64.73% to the final 66.01%. This is because OHEM can mine difficult negative samples and train them repeatedly and emphatically, as introduced in Section 2.5. Finally, the foreground–background discrimination capacity of the model can be boosted.

7. Each improvement offers different accuracy gains. The accuracy growth rate is also different when one of them is inserted. The overall change trend of accuracy is constantly upward. However, the internal synergistic effects between different improvements are difficult to figure out thoroughly. This will need extensive experiments to study further. It is impossible for this paper to exhaustively complete all combination of experiments in the current state. They can be arranged in the future. Despite all this, the scientific value of this paper is not completely affected.

Table 4 shows the performance comparison with the other five state-of-the-art detectors. From Table 4, this paper mainly selects the Faster R-CNN, FPN, YOLOv3, RetinaNet, and CenterNet to conduct performance comparison. They are all trained using the video SAR data of this paper again. Their implementations are basically the same as their original reports. Their backbone networks also load the ImageNet pretrained weights so as to ensure the comparison fairness. Moreover, we indeed know that there might be more other advanced generic deep learning object detection models in the CV community; however, they have not been applied for video SAR moving shadow detection so far by scholars in the SAR community. Therefore, we do not compare ShadowDeNet with them because this paper is limited in the SAR community. On the contrary, Faster R-CNN, FPN, YOLOv3, RetinaNet, and CenterNet have been applied for video SAR moving shadow detection by Ding et al. [24], Wen et al. [25], Huang et al. [26], Yan et al. [27], Zhang et al. [28], Hu et al. [29], Wang et al. [30], and so on, so they are selected. Moreover, various signs of the existing state indicate that deep learning methods have far exceeded most traditional methods, so we leave out the comparison with traditional methods that are usually based on excessive manual features.

Table 4. Performance comparison with the other five state-of-the-art detectors. These detectors have been applied for video SAR moving target shadow detection in the SAR community. The best model is marked in bold. The second-best model is marked by underline. The IOU threshold is 0.50.

Method	#gt	#tp	#fp	#fn	r (%)	p (%)	ap (%)	f1 (%)	FPS
Faster R-CNN [25]	1581	884	636	697	55.91	58.16	46.33	57.01	15.79
FPN [37]	1581	916	504	665	57.94	64.51	51.55	<u>61.05</u>	14.26
YOLOv3 [38]	1581	723	216	858	45.73	77.00	40.08	57.38	21.43
RetinaNet [46]	1581	789	575	792	49.91	57.84	38.19	53.58	16.67
CenterNet [43,44]	1581	519	981	1062	32.83	34.60	18.09	33.69	27.27
ShadowDeNet (Ours)	1581	902	250	679	57.05	78.30	51.87	**66.01**	11.11

From Table 4, one can draw the following conclusions:

1. ShadowDeNet achieves the best accuracy for the video SAR moving shadow detection regardless of the *ap* index or the *f*1 index. Although the *ap* index of ShadowDeNet is slightly better than FPN, its *f*1 index is far superior to FPN, i.e., 66.01% >> 61.05%, about a ~5% accuracy improvement. Notably, FPN generates much more false alarms than ShadowDeNet, i.e., 504 >> 250. As a result, the *p* index of FPN is far lower than that of ShadowDeNet, i.e., 64.51% << 78.30%, about such a huge 14% gap. Therefore, the above reveals the state-of-the-art moving target shadow performance of ShadowDeNet.

2. ShadowDeNet offers a huge accuracy gain based on the experimental baseline Faster R-CNN. The total accuracy gain is up to ~1.14% in terms of the *r* index, ~20.14% in terms of the *p* index, ~5.54% in terms of the *ap* index, and ~9.00% in terms of the *f*1 index. These all benefit from the five improvement methods used as introduced before.

3. ShadowDeNet merely sacrifices a slight detection speed compared with the experimental baseline Faster R-CNN, i.e., from 15.79 FPS to 11.11 FPS. Therefore, ShadowDeNet is cost-effective. It can sacrifice a relatively weak speed in exchange for a huge increase in accuracy, demonstrating its advanced nature.

4. CenterNet offers the fastest detection speed, i.e., 27.27 FPS, but its detection accuracy is too poor to satisfy actual application requirements, i.e., its 33.69% *f*1 << ShadowDeNet's 66.01% *f*1. Moreover, the poor detection performance of CenterNet might be from its anchor-free mechanism based on the keypoint detection, because in video SAR images, the energy of the moving target shadow is rather weak, resulting in the shadow's rather few keypoints.

5. The two-stage detectors (Faster R-CNN and FPN) achieve the better accuracy performance than the one-stage ones (YOLOv3, RetinaNet, and CenterNet). However, the detection speed of the two-stage detectors is universally slower than that of the one-stage ones. This phenomenon is in line with the common sense in the CV community, because RPNs in two-stage detectors usually consume more time while enabling better box detection accuracy.

Figure 16 shows the accuracy changing curves (*r*, *p*, *ap*, and *f*1) of different methods when the IOU threshold is set increasingly. From Figure 16, when the IOU threshold becomes bigger and bigger, the accuracy becomes poorer and poorer. This is in line with common sense, because a larger IOU threshold will challenge the box regression performance seriously. Furthermore, except the recall–IOU curve in Figure 16a, most accuracy–IOU curves of ShadowDeNet are on the upper of all other curves, especially for the *f*1–IOU curve in Figure 16d, which shows the better detection performance of ShadowDeNet. The results in Table 4 are at an IOU threshold of 0.50, the same as the PASCAL VOC criterion [79]. Finally, from Figure 16, one can find that it is rather necessary to further improve the box location performance. In the future, scholars should better design stronger coordinate positioning networks to ensure better performance at a large IOU threshold.

Figure 17 shows the precision–recall (*p*–*r*) curves of different methods. In Figure 17, the curve of ShadowDeNet is on the top of all other curves, almost all across the whole horizontal axis. This also shows the better detection performance of ShadowDeNet.

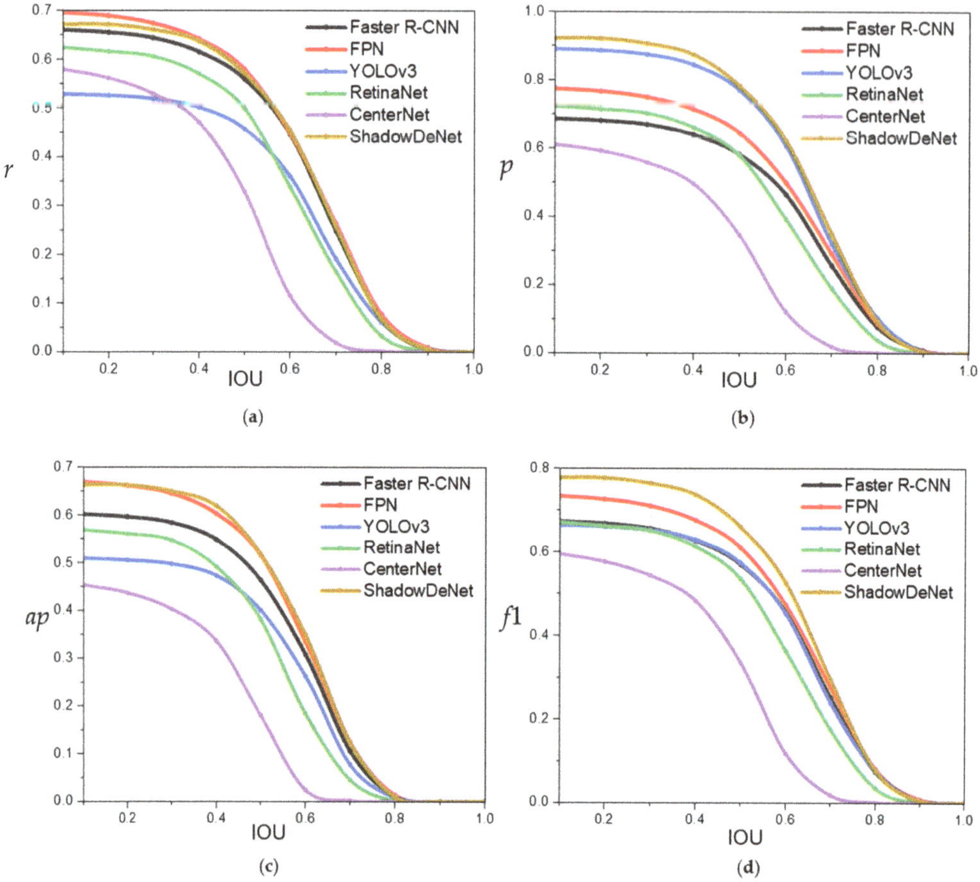

Figure 16. The accuracy changing curves with different IOU thresholds of different methods. (**a**) The curve between recall (*r*) and IOU; (**b**) the curve between precision (*p*) and IOU; (**c**) the curve between average precision (*ap*) and IOU; (**d**) the curve between $f1$ and IOU.

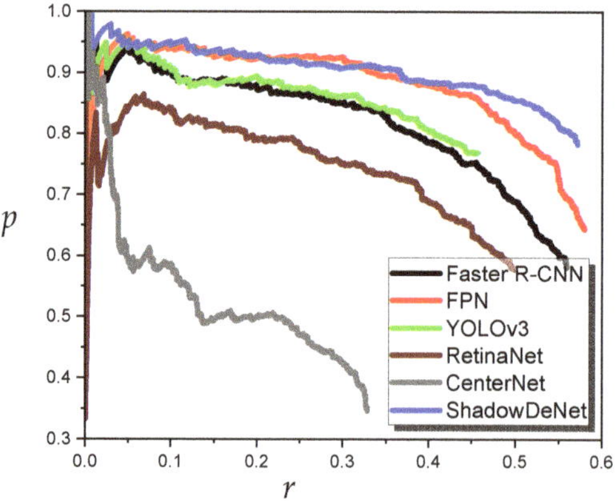

Figure 17. Precision–recall (*p–r*) curves of different methods.

4.2. Qualitative Results

Figure 18 shows the qualitative moving target shadow detection results of different methods. In Figure 18, the display confidence threshold of Faster R-CNN, FPN, YOLOv3, RetinaNet, and ShadowDeNet is 0.50. However, CenterNet does not offer the similar classification confidence scores. In Figure 18f, the numbers above boxes denote CenterNet's Gaussian heatmap probabilities of the top five keypoints. Here, we exhibit the shadow detection results of the 601st, 772nd, 806th, and 900th frames.

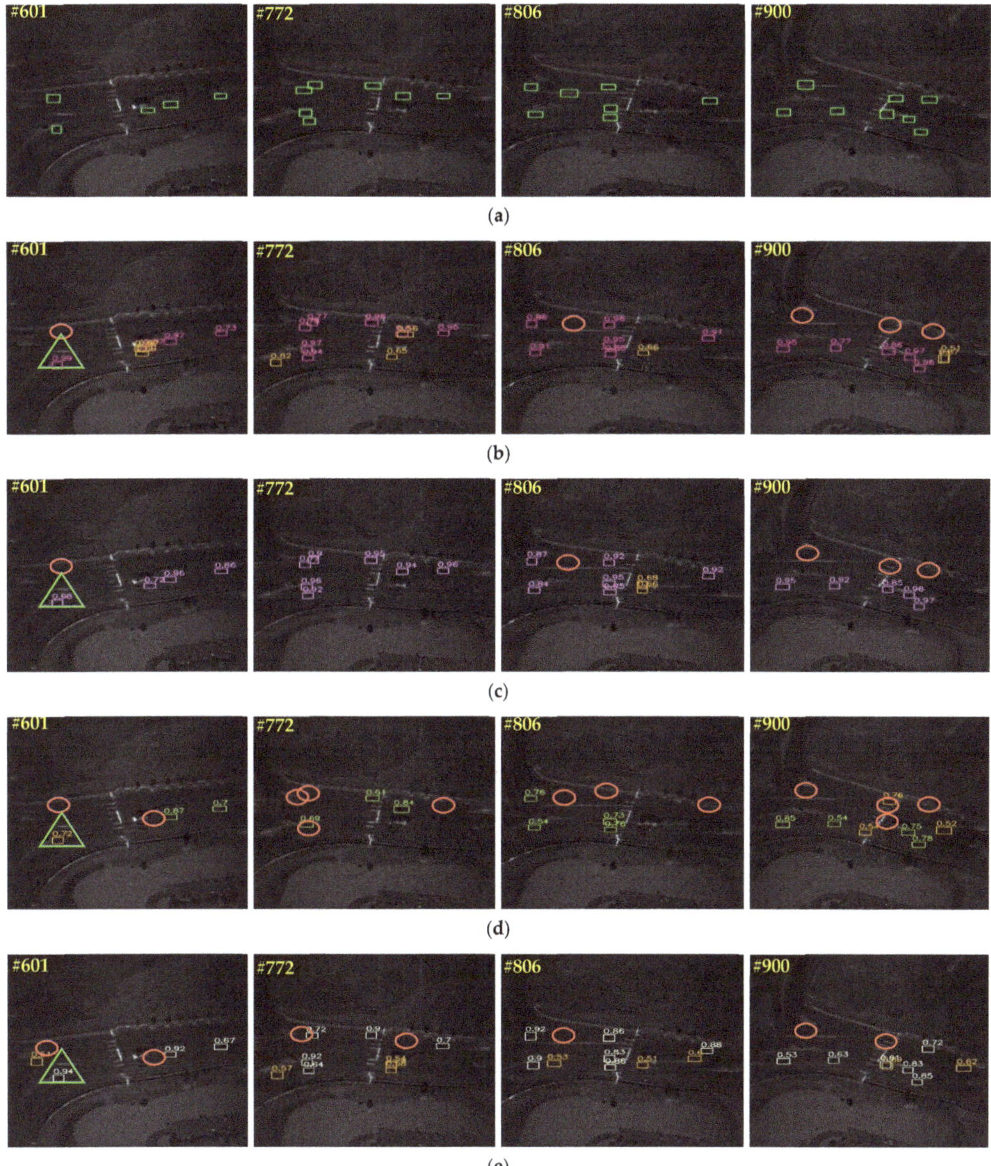

Figure 18. *Cont.*

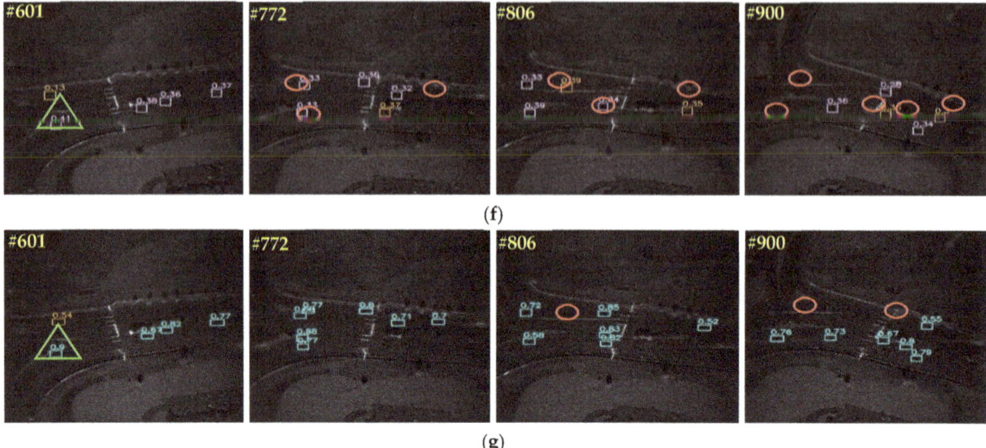

Figure 18. Qualitative video SAR moving target shadow detection results of different methods. (**a**) Ground truth; (**b**) Faster R-CNN; (**c**) FPN; (**d**) YOLOv3; (**e**) RetinaNet; (**f**) CenterNet; (**g**) ShadowDeNet. The false alarms are marked by orange boxes. The missed detections are marked by red ellipses. Apart from CenterNet, the numbers above boxes are the confidences. The numbers above boxes in (**f**) denote CenterNet's Gaussian heatmap probabilities of the top five keypoints. The IOU threshold is 0.50, the same as the PASCAL VOC criterion [79].

From Figure 18, one can draw the following conclusions:

1. ShadowDeNet can avoid many missed detections (marked by red ellipses). If taking the 900th frame as an example, most other methods always miss more shadows than ShadowDeNet, e.g., three missed detections of Faster R-CNN, 3, three ones of FPN, four ones of YOLOv3, and so on, which are all more than that of ShadowDeNet (only two missed detections).
2. ShadowDeNet can suppress many false alarms (marked by orange boxes). If taking the 806th frame as an example, most other methods generate more false alarms than ShadowDeNet, e.g., one false alarm of Faster R-CNN, two ones of FPN, three ones of RetinaNet, and so on, which are all more than that of ShadowDeNet (no false alarms).
3. ShadowDeNet can offer comparable confidences of shadows. For example, the vehicle marked by a green triangle has a 0.99 confidence for Faster R-CNN, a 0.98 confidence for FPN, a 0.72 confidence for YOLOv3, a 0.94 confidence for RetinaNet, a 0.41 confidence for CenterNet, and a 0.90 confidence for ShadowDeNet. As CenterNet does not offer classification confidences, we omit the confidence comparisons of different confidence thresholds. Otherwise, it is unfair for CenterNet. For example, if we regard the Gaussian heatmap probabilities of CenterNet as the classification confidences, and set the confidence threshold to 0.50, CenterNet will miss all vehicles. This is obviously unreasonable. Moreover, ShadowDeNet's confidences are slightly lower than Faster R-CNN and FPN. This disadvantage will be solved in the future.
4. ShadowDeNet offers the most advanced video SAR moving target shadow detection performance compared to all the five state-of-the-art methods.

5. Ablation Study

In this section, we carry out some ablation studies of each characteristic in ShadowDeNet. The impact of each characteristic on the whole ShadowDeNet model performance will be introduced by the mean of each installation and removal.

5.1. Ablation Study on HESE

Table 5 shows the quantitative detection results of ShadowDeNet with and without HESE. From Table 5, HESE can improve the overall performance of ShadowDeNet by ~4.0% $f1$ accuracy, which shows its effectiveness. This is because HESE can enhance the shadow saliency and its contrast ratio, as shown in Figure 5 and Table 1, so as to enable the better follow-up feature extraction of the backbone network. When the shadow is enhanced, the discrimination between foreground and background is bound to become more relaxing. Of course, one may adopt more advanced techniques to further highlight the shadow for better performance, but HESE might be the most direct and easiest tool without complex theory and cumbersome steps.

Table 5. Quantitative results of ShadowDeNet with and without HESE.

HESE	#gt	#tp	#fp	#fn	r (%)	p (%)	ap (%)	$f1$ (%)
✗	1581	868	337	713	54.90	72.03	49.69	62.31
✓	1581	902	250	679	57.05	78.30	**51.87**	**66.01**

Moreover, we also select the adaptive histogram equalization shadow enhancement (AHESE) to discuss the shadow detection performance impacts. There are two hyperparameters in AHESE, i.e., the limited contrast ratio, and the size of blocks. We set the limited contrast ratio to 2.0, and set the size of blocks to 8, which are both default values in OpenCV [80]. Figure 19 shows the comparison results of HESE and AHESE. From Figure 19, AHESE does not make backgrounds brighter than HESE. It seems that HESE performs a little bit better than AHESE from human eye visual observation, because the shadows enhanced by HESE are clearer than those enhanced by AHESE. Moreover, from the yellow ellipse regions in Figure 19, the shadows enhanced by AHESE are more easily submerged by the similar black backgrounds. This is one reason why we use HESE.

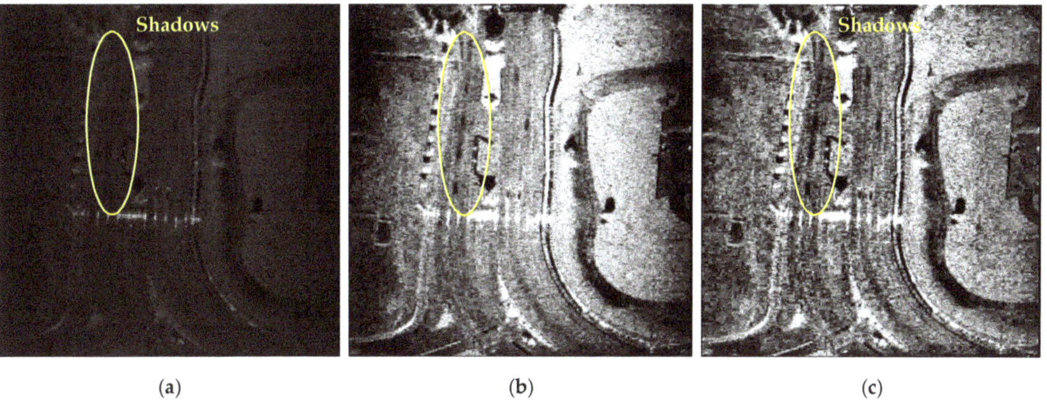

Figure 19. Different histogram equalization shadow enhancements. (**a**) Shadows in the raw video SAR image; (**b**) shadows enhanced by HESE; (**c**) shadows enhanced by AHESE.

We make a quantitative assessment of HESE and AHESE in Table 6. From Table 6, AHESE does not offer better $f1$ accuracy than HESE (64.87% < 66.01%), but it offers better ap accuracy than HESE (53.33% > 51.87%). Therefore, HESE is comparable to AHESE on our experimental data. Considering that AHESE not only consumes more time than HESE but also adds another two hyperparameters requiring trouble-manual adjustments [81], we select the most classic HESE as the preprocessing tool.

Table 6. Quantitative results of ShadowDeNet with HESE or AHESE.

Type	#gt	#tp	#fp	#fn	r (%)	p (%)	ap (%)	f1 (%)
AHESE	1581	937	371	644	59.27	71.64	53.33	64.87
HESE	1581	902	**250**	679	57.05	78.30	51.87	**66.01**

5.2. Ablation Study on TSAM

Table 7 shows the quantitative detection results of ShadowDeNet with and without TSAM. From Table 7, TSAM can improve the overall performance of ShadowDeNet by ~3.2% $f1$ accuracy, which shows its effectiveness. Combined with TSAM, the network can adaptively learn differential spatial information weight to pay more attention to shadow regions rather than background ones, which will enable to detect more shadows and suppress false alarms. As a result, the detection rate is increased by ~1.0% and the false alarm rate is decrease by ~8.0% from Table 7. Furthermore, TSAM can, in fact, calculate the interaction relationship between any two positions and also directly captures the remote dependence without being limited to adjacent points. In this way, more background context information can be maintained. Note that related scholars can adopt other more advanced attention mechanisms [82] to boost detection performance further. Yet, this paper does not study it in depth any more at present. We can continue this task in the future.

Table 7. Quantitative results of ShadowDeNet with and without TSAM.

TSAM	#gt	#tp	#fp	#fn	r (%)	p (%)	ap (%)	f1 (%)
✗	1581	891	366	690	56.36	70.88	49.83	62.79
√	1581	902	**250**	679	57.05	**78.30**	51.87	**66.01**

5.3. Ablation Study on SDAL

Table 8 shows the quantitative detection results of ShadowDeNet with and without SDAL. From Table 8, SDAL can improve the overall performance of ShadowDeNet by ~3.8% $f1$ accuracy, which shows its effectiveness. This is because SDAL can model the geometric deformations of the moving target shadow to adapt to motion speed variations. Finally, ShadowDeNet can discriminate the static backgrounds and the moving target shadows more effectively. Of course, SDAL is not free, and it usually needs more time to learn the kernel offsets of Equation (7) in training. However, once the kernel offsets have been learned, the speed sacrifice in the inference or test process is insignificant.

Table 8. Quantitative results of ShadowDeNet with and without SDAL.

SDAL	#gt	#tp	#fp	#fn	r (%)	p (%)	ap (%)	f1 (%)
✗	1581	934	488	647	59.08	65.68	50.33	62.20
√	1581	902	**250**	679	57.05	78.30	51.87	**66.01**

5.4. Ablation Study on SGAAL

Table 9 shows the quantitative detection results of ShadowDeNet with and without SGAAL. From Table 9, SDAL can improve the overall performance of ShadowDeNet by ~2.8% $f1$ accuracy and ~3% ap accuracy, which shows its effectiveness. For one thing, its internal anchor location network can predict shadow-like regions adaptively to suppress false alarms. For another thing, its internal anchor shape network can predict better shapes to match moving target shadows adaptively to improve the detection rate. As a result, the false alarm rate is dropped by ~2.3%; meanwhile, the detection rate is increased by ~3.0%.

Table 9. Quantitative results of ShadowDeNet with and without SGAAL.

SGAAL	#gt	#tp	#fp	#fn	r (%)	p (%)	ap (%)	f1 (%)
✗	1581	856	270	725	54.14	76.02	48.68	63.24
✓	1581	902	250	679	57.05	78.30	**51.87**	**66.01**

We perform another experiment to study the impact of the location filter threshold ε_L. The quantitative results are shown in Table 10. In Table 10, the value range of ε_L is suggested by Wang et al. [45]. From Table 10, when ε_L is set to 0.01, the accuracy reaches the best. Thus, the final ε_L is set to 0.01 in ShadowDeNet. Our experimental results are in line with [45] where ε_L is also set to 0.01. According to their findings, a small location filter threshold has already removed many false positives. However, a too-large location filter threshold is bound to remove many true positives, resulting in a lower detection rate. We find that such phenomenon is also in line with video SAR images.

Table 10. Quantitative results of ShadowDeNet with different location filter thresholds.

ε_L	#gt	#tp	#fp	#fn	r (%)	p (%)	ap (%)	f1 (%)
0.00	1581	873	275	708	55.22	76.05	48.82	63.98
0.01	1581	902	250	679	57.05	78.30	51.87	**66.01**
0.02	1581	922	381	659	58.32	70.76	**52.02**	63.94
0.03	1581	893	386	688	56.48	69.82	49.72	62.45
0.04	1581	869	343	712	54.97	71.70	49.02	62.23
0.05	1581	860	326	721	54.40	72.51	47.24	62.16

5.5. Ablation Study on OHEM

Table 11 shows the quantitative detection results of ShadowDeNet with and without OHEM. From Table 11, SDAL can improve the overall performance of ShadowDeNet by ~1.3% $f1$ accuracy and ~1.5% ap accuracy, which shows its effectiveness. Although the detection rate is not very sensitive to OHEM, the false alarm rate is decreased greatly by ~4.0% when OHEM is used. This is because OHEM can mine hard negative samples and train them repeatedly and emphatically. Finally, the foreground–background discrimination ability of ShadowDeNet can be boosted further.

Table 11. Quantitative results of ShadowDeNet with and without OHEM.

OHEM	#gt	#tp	#fp	#fn	r (%)	p (%)	ap (%)	f1 (%)
✗	1581	904	308	677	57.18	74.59	50.33	64.73
✓	1581	902	250	679	57.05	78.30	**51.87**	**66.01**

6. Discussion

We also discuss the shadow detection performance of ShadowDeNet on another video SAR dataset to reveal its universal effectiveness and excellent migration ability. This data is provided by Zhou et al. [12] and China Aerospace Science and Industry Corporation (CASIC) 23 research institute. This data was obtained from the spotlight SAR that was equipped on an aircraft and flew along a circular trajectory at a height of 3000 m. The carrier frequency, resolution, and velocity of aircraft are Ka-band, 0.15 m, and 80 m/s. There are 369 images with 1000×1000 pixel size. Among them, 246 images are selected as the training set, and the remaining 123 images are selected as the test set.

Figure 20 shows the qualitative results of ShadowDeNet on the CASIC 23 research institute video SAR data, and the quantitative results are shown in Table 12. From Figure 20, many moving target shadows can be detected by ShadowDeNet successfully, showing the excellent migration ability of ShadowDeNet. From Table 12, ShadowDeNet offers comparable shadow detection performance on the CASIC 23 research institute video SAR data with the SNL video SAR data, i.e., 66.01% $f1$ vs. 66.28% $f1$ and 52.39% ap vs. 51.87% ap. Therefore, ShadowDeNet is effective on video SAR data, showing its universal effectiveness.

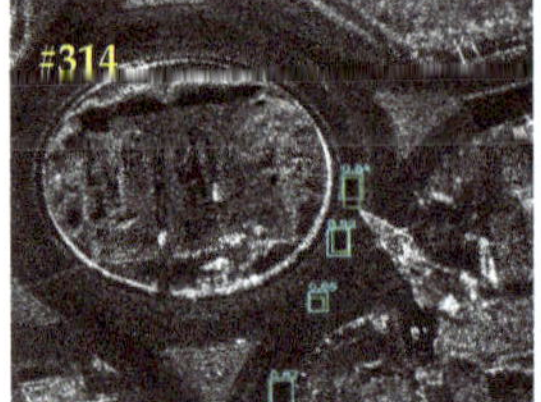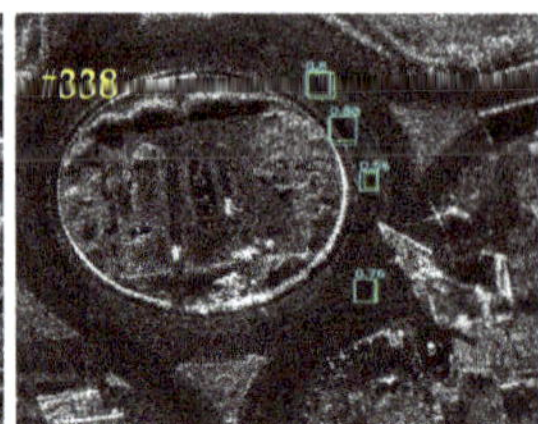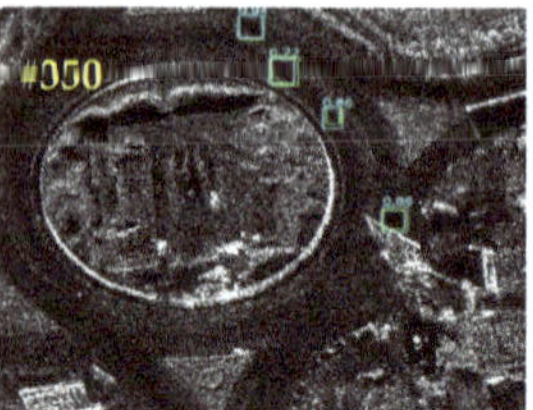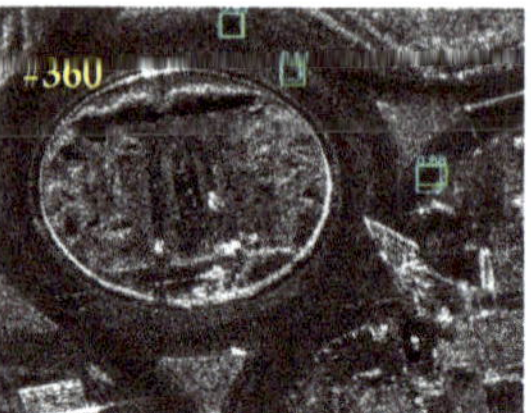

Figure 20. Qualitative video SAR moving target shadow detection results of ShadowDeNet on the CASIC 23 research institute data. The ground truths are marked by green boxes. The numbers above boxes are the confidences. The IOU threshold is 0.50, the same as the PASCAL VOC criterion [79].

Table 12. Quantitative results of ShadowDeNet on the CASIC 23 research institute video SAR data.

#gt	#tp	#fp	#fn	r (%)	p (%)	ap (%)	$f1$ (%)
521	286	56	235	54.89	83.63	52.39	**66.28**

7. Conclusions

This paper proposes a novel deep learning network ShadowDeNet for the moving target shadow detection from video SAR images. Five characteristics are used to guarantee ShadowDeNet's superior detection performance, i.e., (1) HESE, which is used to enhance shadow saliency to facilitate feature extraction, (2) TSAM, which is used to focus on regions of interests to suppress clutter interferences, (3) SDAL, which is used to learn moving target deformed shadows adaptively to conquer motion speed variations, (4) SGAAL, which is used to generate optimized anchors to match shadow location and shape, and (5) OHEM, which is used to select typical difficult negative samples to improve background discrimination capacity. Finally, the quantitative and qualitative results reveal the state-of-the-art detection performance of ShadowDeNet with a 66.01% best $f1$ accuracy. Specifically, it is superior to the experimental baseline Faster R-CNN by a 9.00% $f1$ accuracy. It is also superior to the existing best model by a 4.96% $f1$ accuracy. Moreover, the detection speed sacrifice is very slight. Last but not least, we also conduct extensive ablation studies on the public SNL video SAR data to confirm the effectiveness of each characteristic. We also use ShadowDeNet to detect shadows on another one video SAR data, and the results reveal its universal effectiveness and excellent migration ability. In short, ShadowDeNet can provide high-quality predetection results for subsequent trackers, of great value.

Our future work is as follows:

1. We will improve the detection rate of ShadowDeNet further in the future.
2. We will conduct the subsequent target association and the trajectory reconstruction tasks.
3. We will postprocess the current obtained detection results to remove missed detections and suppress false alarms further by using the multiframe relationship.
4. Traditional features can be considered to be injected into CNN-based models to improve performance further.

Author Contributions: Conceptualization, T.Z.; methodology, T.Z.; software, J.B.; validation, X.X.; formal analysis, J.B.; investigation, J.B.; resources, T.Z.; data curation, J.B.; writing—original draft preparation, T.Z. and J.B.; writing—review and editing, T.Z., J.B. and X.Z.; visualization, X.Z.; supervision, X.Z.; project administration, T.Z.; funding acquisition, X.Z. All authors have read and agreed to the published version of the manuscript.

Funding: This work was supported in part by the National Key R&D Program of China under Grant 2017YFB0502700 and in part by the National Natural Science Foundation of China under Grants 61571099, 61501098, and 61671113.

Institutional Review Board Statement: Not applicable.

Informed Consent Statement: Not applicable.

Data Availability Statement: No new data were created or analyzed in this study. Data sharing is not applicable to this article. The public video SAR data provided by Sandia National Laboratories (SNL) is available from https://www.sandia.gov/app/uploads/sites/124/2021/08/eubankgateandtrafficvideosar.mp4 (accessed on 20 November 2021) to download for scientific research. China Aerospace Science and Industry Corporation (CASIC) 23 research institute video SAR data is not public because of copyright restrictions.

Acknowledgments: The authors would like to thank the editors and anonymous reviewers for their valuable comments that greatly improve our manuscript. The authors would like to thank Sandia National Laboratories (SNL) and China Aerospace Science and Industry Corporation (CASIC) 23 research institute for providing the video SAR data.

Conflicts of Interest: The authors declare no conflict of interest.

References

1. Moreira, A.; Prats-Iraola, P.; Younis, M.; Krieger, G.; Hajnsek, I.; Papathanassiou, K.P. A Tutorial on Synthetic Aperture Radar. *IEEE Geosci. Remote Sens. Mag.* **2013**, *1*, 6–43. [CrossRef]
2. Zhang, T.; Zhang, X.; Shi, J.; Wei, S. Depthwise Separable Convolution Neural Network for High-Speed SAR Ship Detection. *Remote Sens.* **2019**, *11*, 2483. [CrossRef]
3. Zhang, T.; Zhang, X. High-Speed Ship Detection in SAR Images Based on a Grid Convolutional Neural Network. *Remote Sens.* **2019**, *11*, 1206. [CrossRef]
4. Zhang, T.; Zhang, X. Shipdenet-20: An Only 20 Convolution Layers and <1-Mb Lightweight SAR Ship Detector. *IEEE Geosci. Remote Sens. Lett.* **2021**, *18*, 1234–1238. [CrossRef]
5. Zhang, T.; Zhang, X.; Shi, J.; Wei, S.; Wang, J.; Li, J.; Su, H.; Zhou, Y. Balance Scene Learning Mechanism for Offshore and Inshore Ship Detection in SAR Images. *IEEE Geosci. Remote Sens. Lett.* **2020**, *19*, 4004905. [CrossRef]
6. Zhang, T.; Zhang, X. Squeeze-and-Excitation Laplacian Pyramid Network with Dual-Polarization Feature Fusion for Ship Classification in SAR Images. *IEEE Geosci. Remote Sens. Lett.* **2021**, *19*, 4019905. [CrossRef]
7. Ding, J. Focusing Algorithms and Moving Target Detection Based on Video SAR. *J. Radars* **2020**, *9*, 321–334.
8. Huang, X.; Xu, Z.; Ding, J. Video SAR Image Despeckling by Unsupervised Learning. *IEEE Trans. Geosci. Remote Sens.* **2020**, *59*, 1–10. [CrossRef]
9. Huang, X.; Ding, J.; Guo, Q. Unsupervised Image Registration for Video SAR. *IEEE J. Sel. Top. Appl. Earth Obs. Remote Sens.* **2021**, *14*, 1075–1083. [CrossRef]
10. Wang, H.; Chen, Z.; Zheng, S. Preliminary Research of Low-RCS Moving Target Detection Based on Ka-Band Video SAR. *IEEE Geosci. Remote Sens. Lett.* **2017**, *14*, 811–815. [CrossRef]
11. Yang, X.; Shi, J.; Zhou, Y.; Wang, C.; Hu, Y.; Zhang, X.; Wei, S. Ground Moving Target Tracking and Refocusing Using Shadow in Video-SAR. *Remote Sens.* **2020**, *12*, 3083. [CrossRef]
12. Zhou, Y.; Shi, J.; Wang, C.; Hu, Y.; Zhou, Z.; Yang, X.; Wei, S. SAR Ground Moving Target Refocusing by Combining Mre^3 Network and Tvβ-Lstm. *IEEE Trans. Geosci. Remote Sens.* **2020**, *60*, 1–14. [CrossRef]
13. Damini, A.; Balaji, B.; Parry, C.; Mantle, V. A videoSAR mode for the X-band wideband experimental airborne radar. In *Algorithms for Synthetic Aperture Radar Imagery*, 17th ed.; International Society for Optics and Photonics: Bellingham, WA, USA, 2010; p. 76990. [CrossRef]
14. Zhong, C.; Ding, J.; Zhang, Y. Joint Tracking of Moving Target in Single-Channel Video SAR. *IEEE Trans. Geosci. Remote Sens.* **2021**. [CrossRef]
15. Zhao, B.; Han, Y.; Wang, H.; Tang, L.; Liu, X.; Wang, T. Robust Shadow Tracking for Video SAR. *IEEE Geosci. Remote Sens. Lett.* **2020**, *18*, 821–825. [CrossRef]
16. Tian, X.; Liu, J.; Mallick, M.; Huang, K. Simultaneous Detection and Tracking of Moving-Target Shadows in ViSAR Imagery. *IEEE Trans. Geosci. Remote Sens.* **2021**, *59*, 1182–1199. [CrossRef]
17. Liu, Z.; An, D.; Huang, X. Moving Target Shadow Detection and Global Background Reconstruction for VideoSAR Based on Single-Frame Imagery. *IEEE Access.* **2019**, *7*, 42418–42425. [CrossRef]
18. Otsu, N. A Threshold Selection Method from Gray-Level Histograms. *IEEE Trans. Syst. Man. Cybern. Syst.* **1979**, *9*, 62–66. [CrossRef]

19. Zhang, Y.; Mao, X.; Yan, H.; Zhu, D.; Hu, X. A Novel Approach to Moving Targets Shadow Detection in VideoSAR Imagery Sequence. In Proceedings of the IEEE International Geoscience and Remote Sensing Symposium (IGARSS), Quebec City, QC, Canada, 13–18 July 2014; pp. 606–609.
20. Shang, S.; Wu, F.; Liu, Z.; Yang, Y.; Li, D.; Jin, L. Moving Target Shadow Detection and Tracking Based on Thz Video-SAR. In Proceedings of the International Conference on Infrared, Millimeter and Terahertz Waves (IRMMW-THz), Houston, TX, USA, 2–7 October 2011; pp. 1–2.
21. He, Z.; Chen, X.; Yu, C.; Li, Z.; Yu, A.; Dong, Z. A Robust Moving Target Shadow Detection and Tracking Method for VideoSAR. *J. Electron. Inf. Technol.* **2021**, *7*, 1–9.
22. He, Z.; Chen, X.; Yi, T.; He, F.; Dong, Z.; Zhang, Y. Moving Target Shadow Analysis and Detection for ViSAR Imagery. *Remote Sens.* **2021**, *13*, 3012. [CrossRef]
23. Bao, J.; Zhang, X.; Zhang, T.; Shi, J.; Wei, S. A Novel Guided Anchor Siamese Network for Arbitrary Target-of-Interest Tracking in Video-SAR. *Remote Sens.* **2021**, *13*, 4504. [CrossRef]
24. Ding, J.; Wen, L.; Zhong, C.; Loffeld, O. Video SAR Moving Target Indication Using Deep Neural Network. *IEEE Trans. Geosci. Remote Sens.* **2020**, *58*, 7194–7204. [CrossRef]
25. Wen, L.; Ding, J.; Loffeld, O. Video SAR Moving Target Detection Using Dual Faster R-CNN. *IEEE J. Sel. Top. Appl. Earth Obs. Remote Sens.* **2021**, *14*, 2984–2994. [CrossRef]
26. Huang, X.; Liang, D.; Ding, J. Moving Target Detection in Video SAR Based on Improved Faster R-CNN. In Proceedings of the European Conference on Synthetic Aperture Radar (EUSAR), Online Event, 29 March–1 April 2021; pp. 1–5.
27. Yan, H.; Huang, J.; Li, R.; Wang, X.; Zhang, J.; Zhu, D. Research on Video SAR Moving Target Detection Algorithm Based on Improved Faster Region-based CNN. *J. Electron. Inf. Technol.* **2021**, *43*, 615–622.
28. Zhang, H.; Liu, Z. Moving Target Shadow Detection Based on Deep Learning in Video SAR. In Proceedings of the IEEE International Geoscience and Remote Sensing Symposium (IGARSS), Quebec City, QC, Canada, 13–18 July 2021; pp. 4155–4158.
29. Hu, Y. Research on Shadow-Based SAR Multi-Target Tracking Method. Master's Thesis, University of Electronic Science and Technology of China, Chengdu, China, 2021.
30. Wang, W.; Hu, Y.; Zou, Z.; Zhou, Y.; Wang, C. Video SAR Ground Moving Target Indication Based on Multi-Target Tracking Neural Network. In Proceedings of the IEEE International Geoscience and Remote Sensing Symposium (IGARSS), Quebec City, QC, Canada, 13–18 July 2014; pp. 4584–4587.
31. Shang, S.; Wu, F.; Zhou, Y.; Liu, Z. Moving Target Velocity Estimation of Video SAR Based on Shadow Detection. In Proceedings of the Cross Strait Radio Science & Wireless Technology Conference (CSRSWTC), Fuzhou, China, 13–16 December 2020; pp. 1–3.
32. Yu, F.; Li, W.; Li, Q.; Liu, Y.; Shi, X.; Yan, J. POI: Multiple Object Tracking with High Performance Detection and Appearance Feature. In Proceedings of the European Conference on Computer Vision Workshops, Amsterdam, The Netherlands, 8–10 and 15–16 October 2016; pp. 36–42.
33. Wang, Z.; Zheng, L.; Liu, Y.; Li, Y.; Wang, S. Towards Real-Time Multi-Object Tracking. In Proceedings of the European Conference on Computer Vision, Glasgow, UK, 23–28 August 2020; pp. 107–122.
34. Anastasiadis, A. Special Issue: Tsallis Entropy. *Entropy* **2012**, *14*, 174–176. [CrossRef]
35. Bay, H.; Ess, A.; Tuytelaars, T.; Gool, L.V. Speeded-Up Robust Features (SURF). *Comput. Vis. Image Underst.* **2008**, *110*, 346–359. [CrossRef]
36. Ren, S.; He, K.; Girshick, R.; Sun, J. Faster R-CNN: Towards Real-Time Object Detection with Region Proposal Networks. *IEEE Trans. Pattern Anal. Mach. Intell.* **2017**, *39*, 1137–1149. [CrossRef]
37. Lin, T.-Y.; Dollar, P.; Girshick, R.; He, K.; Hariharan, B.; Belongie, S. Feature Pyramid Networks for Object Detection. In Proceedings of the IEEE Conference on Computer Vision and Pattern Recognition (CVPR), Honolulu, HI, USA, 21–26 July 2017; pp. 936–944.
38. Redmon, J.; Farhadi, A. YOLOv3: An Incremental Improvement. *arXiv* **2018**, arXiv:1804.02767.
39. Lin, T.-Y.; Goyal, P.; Girshick, R.; He, K.; Dollar, P. Focal Loss for Dense Object Detection. *IEEE Trans. Pattern Anal. Mach. Intell.* **2020**, *42*, 318–327. [CrossRef]
40. Duan, K.; Bai, S.; Xie, L.; Qi, H.; Huang, Q.; Tian, Q. CenterNet: Keypoint Triplets for Object Detection. In Proceedings of the IEEE International Conference on Computer Vision (ICCV), Seoul, Korea, 27–28 October 2019; pp. 6568–6577.
41. Zhang, T.; Zhang, X.; Liu, C.; Shi, J.; Wei, S.; Ahmad, I.; Zhan, X.; Zhou, Y.; Pan, D.; Li, J.; et al. Balance Learning for Ship Detection from Synthetic Aperture Radar Remote Sensing Imagery. *ISPRS J. Photogramm. Remote Sens.* **2021**, *182*, 190–207. [CrossRef]
42. Zhang, T.; Zhang, X. Injection of Traditional Hand-Crafted Features into Modern CNN-Based Models for SAR Ship Classification: What, Why, Where, and How. *Remote Sens.* **2021**, *13*, 2091. [CrossRef]
43. Zhang, Y.; Wang, C.; Wang, X.; Zeng, W.; Liu, W. FairMOT: On the Fairness of Detection and Re-Identification in Multiple Object Tracking. *Int. J. Comput. Vis.* **2021**, *129*, 3069–3087. [CrossRef]
44. Zhou, X.; Koltun, V.; Krähenbühl, P. Tracking Objects as Points. In Proceedings of the European Conference on Computer Vision (ECCV), Graz, Austria, 7–13 May 2006; pp. 474–490.
45. Wang, J.; Chen, K.; Yang, S.; Loy, C.C.; Lin, D. Region Proposal by Guided Anchoring. In Proceedings of the IEEE Conference on Computer Vision and Pattern Recognition (CVPR), Long Beach, CA, USA, 15–20 June 2019; pp. 2960–2969.
46. Lu, Z.; Rathod, V.; Votel, R.; Huang, J. RetinaTrack: Online Single Stage Joint Detection and Tracking. In Proceedings of the IEEE Conference on Computer Vision and Pattern Recognition (CVPR), Seattle, WA, USA, 14–19 June 2020; pp. 14656–14666.

47. Dai, J.; Qi, H.; Xiong, Y.; Li, Y.; Zhang, G. Deformable Convolutional Networks. In Proceedings of the IEEE International Conference on Computer Vision (ICCV), Venice, Italy, 22–29 October 2017; pp. 764–773.
48. Shrivastava, A.; Gupta, A.; Girshick, R. Training Region-Based Object Detectors with Online Hard Example Mining. In Proceedings of the IEEE Conference on Computer Vision and Pattern Recognition (CVPR), Las Vegas, NV, USA, 27–30 June 2016; pp. 761–769.
49. National Technology and Engineering Solutions of Sandia. Pathfinder Radar ISR & SAR Systems. Eubank Gate and Traffic VideoSAR. 2021. Available online: http://www.sandia.gov/radar/video (accessed on 30 November 2021).
50. He, K.; Zhang, X.; Ren, S.; Sun, J. Deep Residual Learning for Image Recognition. In Proceedings of the IEEE Conference on Computer Vision and Pattern Recognition (CVPR), Las Vegas, NV, USA, 27–30 June 2016; pp. 770–778.
51. Zhu, X.; Cheng, D.; Zhang, Z.; Lin, S.; Dai, J. An Empirical Study of Spatial Attention Mechanisms in Deep Networks. In Proceedings of the IEEE International Conference on Computer Vision (ICCV), Seoul, Korea, 27–28 October 2019; pp. 6687–6696.
52. Kosub, S. A Note on the Triangle Inequality for the Jaccard Distance. *Pattern Recognit Lett.* **2019**, *120*, 36–38. [CrossRef]
53. Li, J.; Qu, C.; Shao, J. Ship Detection in SAR Images Based on an Improved Faster R-CNN. In Proceedings of the SAR in Big Data Era: Models, Methods and Applications (BIGSARDATA), Beijing, China, 13–14 November 2017; pp. 1–6.
54. He, K.; Gkioxari, G.; Dollar, P.; Girshick, R. Mask R-CNN. In Proceedings of the IEEE International Conference on Computer Vision (ICCV), Venice, Italy, 22–29 October 2017; pp. 2980–2988.
55. Girshick, R. Fast R-CNN. In Proceedings of the IEEE International Conference on Computer Vision (ICCV), Santiago, Chile, 7–13 December 2015; pp. 1440–1448.
56. Ibrahim, H.; Kong, N.S.P. Brightness Preserving Dynamic Histogram Equalization for Image Contrast Enhancement. *IEEE Trans. Consum. Electron.* **2007**, *53*, 1752–1758. [CrossRef]
57. Zhang, M.; An, J.; Yu, D.H.; Yang, L.D.; Wu, L.; Lu, X.Q. Convolutional Neural Network with Attention Mechanism for SAR Automatic Target Recognition. *IEEE Geosci. Remote Sens. Lett.* **2020**, *19*, 4004205. [CrossRef]
58. Li, R.; Wang, X.; Wang, J.; Song, Y.; Lei, L. SAR Target Recognition Based on Efficient Fully Convolutional Attention Block CNN. *IEEE Geosci. Remote Sens. Lett.* **2020**, 1–5. [CrossRef]
59. Zhang, T.; Zhang, X.; Shi, J.; Wei, S. Hyperli-Net: A Hyper-Light Deep Learning Network for High-Accurate and High-Speed Ship Detection from Synthetic Aperture Radar Imagery. *ISPRS J. Photogramm. Remote Sens.* **2020**, *167*, 123–153. [CrossRef]
60. Gao, F.; He, Y.; Wang, J.; Hussain, A.; Zhou, H. Anchor-Free Convolutional Network with Dense Attention Feature Aggregation for Ship Detection in SAR Images. *Remote Sens.* **2020**, *12*, 2619. [CrossRef]
61. Zhang, T.; Zhang, X. A Polarization Fusion Network with Geometric Feature Embedding for SAR Ship Classification. *Pattern Recognit.* **2021**, *123*, 108365. [CrossRef]
62. Zhang, T.; Zhang, X.; Ke, X.; Liu, C.; Xu, X.; Zhan, X.; Wei, S. HOG-ShipCLSNet: A Novel Deep Learning Network with Hog Feature Fusion for SAR Ship Classification. *IEEE Trans. Geosci. Remote Sens.* **2021**, 1–22. [CrossRef]
63. Carion, N.; Massa, F.; Synnaeve, G.; Usunier, N.; Kirillov, A.; Zagoruyko, S. End-to-End Object Detection with Transformers. In Proceedings of the European Conference on Computer Vision (ECCV), Heraklion, Greece, 5–11 September 2010; pp. 213–229.
64. Zhu, X.; Su, W.; Lu, L.; Li, B.; Wang, X.; Dai, J. Deformable Detr: Deformable Transformers for End-to-End Object Detection. In Proceedings of the International Conference on Learning Representations (ICLR), Addis Ababa, Ethiopia, 26–30 April 2020; p. 12.
65. Vaidwan, H.; Seth, N.; Parihar, A.S.; Singh, K. A Study on Transformer-Based Object Detection. In Proceedings of the 2021 International Conference on Intelligent Technologies (CONIT), Hubli, India, 25–27 June 2021; pp. 1–6.
66. Woo, S.; Park, J.; Lee, J.-Y.; Kweon, I.S. Cbam: Convolutional Block Attention Module. In Proceedings of the European Conference on Computer Vision (ECCV), Heraklion, Greece, 5–11 September 2018; pp. 3–19.
67. Hu, J.; Shen, L.; Albanie, S.; Sun, G.; Wu, E. Squeeze-and-Excitation Networks. *IEEE Trans. Pattern Anal. Mach. Intell.* **2020**, *42*, 2011–2023. [CrossRef]
68. Wang, X.; Girshick, R.; Gupta, A.; He, K. Non-Local Neural Networks. In Proceedings of the IEEE Conference on Computer Vision and Pattern Recognition (CVPR), Salt Lake City, UT, USA, 18–22 June 2018; pp. 7794–7803.
69. Bishop, M.C. *Pattern Recognition and Machine Learning (Information Science and Statistics)*; Springer: Berlin, Germany, 2006.
70. Ann Marie, R.; Douglas, L.B.; Armin, W.D. Stationary and Moving Target Shadow Characteristics in Synthetic Aperture Radar. In Proceedings of the Radar Sensor Technology XVIII, Baltimore, MD, USA, 29 May 2014; pp. 1–15.
71. Ke, X.; Zhang, X.; Zhang, T.; Shi, J.; Wei, S. SAR Ship Detection Based on an Improved Faster R-CNN Using Deformable Convolution. In Proceedings of the IEEE International Geoscience and Remote Sensing Symposium (IGARSS), Seoul, Kore, 25–29 July 2021; pp. 3565–3568.
72. Tychsen-Smith, L.; Petersson, L. Improving Object Localization with Fitness Nms and Bounded Iou Loss. In Proceedings of the IEEE Conference on Computer Vision and Pattern Recognition (CVPR), Salt Lake City, UT, USA, 18–22 June 2018; pp. 6877–6885.
73. Ketkar, N. Introduction to Pytorch. *Deep Learning with Python: A Hands-On Introduction*; Apress: Berkeley, CA, USA, 2017; pp. 195–208. Available online: https://link.springer.com/chapter/10.1007/978-1-4842-2766-4_12 (accessed on 1 December 2021).
74. Chen, K.; Wang, J.; Pang, J.; Cao, Y.; Xiong, Y.; Li, X.; Lin, D. MMDetection: Open MMLAB Detection Toolbox and Benchmark. *arXiv* **2019**, arXiv:1906.07155.
75. Goyal, P.; Dollár, P.; Girshick, R.; Noordhuis, P.; Wesolowski, L.; Kyrola, A.; He, K. Accurate, Large Minibatch SGD: Training ImageNet in 1 Hour. *arXiv* **2017**, arXiv:1706.02677.
76. He, K.; Girshick, R.; Doll´ar, P. Rethinking ImageNet Pre-Training. In Proceedings of the IEEE International Conference on Computer Vision (ICCV), Seoul, Korea, 27–28 October 2019; pp. 4917–4926.

77. He, K.; Zhang, X.; Ren, S.; Sun, J. Delving Deep into Rectifiers: Surpassing Human-Level Performance on Imagenet Classification. In Proceedings of the IEEE International Conference on Computer Vision (ICCV), Santiago, Chile, 7–13 December 2015; pp. 1026–1034.
78. Hosang, J.; Benenson, R.; Schiele, B. Learning Non-Maximum Suppression. In Proceedings of the IEEE Conference on Computer Vision and Pattern Recognition (CVPR), Honolulu, HI, USA, 21–26 July 2017; pp. 6469–6477.
79. Everingham, M.; Eslami, S.M.A.; Gool, L.V.; Williams, C.K.I.; Winn, J.; Zisserman, A. The Pascal Visual Object Classes Challenge: A Retrospective. *Int. J. Comput. Vis.* **2015**, *111*, 98–136. [CrossRef]
80. OpenCV. Available online: https://opencv.org/ (accessed on 30 November 2021).
81. Stark, J.A. Adaptive Image Contrast Enhancement Using Generalizations of Histogram Equalization. *IEEE Trans. Image Process.* **2000**, *9*, 889–896. [CrossRef]
82. Niu, Z.; Zhong, G.; Yu, H. A Review on the Attention Mechanism of Deep Learning. *Neurocomputing* **2021**, *452*, 48–62. [CrossRef]

Article

Lite-YOLOv5: A Lightweight Deep Learning Detector for On-Board Ship Detection in Large-Scene Sentinel-1 SAR Images

Xiaowo Xu, Xiaoling Zhang * and Tianwen Zhang

School of Information and Communication Engineering, University of Electronic Science and Technology of China, Chengdu 611731, China; xuxiaowo@std.uestc.edu.cn (X.X.); twzhang@std.uestc.edu.cn (T.Z.)
* Correspondence: xlzhang@uestc.edu.cn

Abstract: Synthetic aperture radar (SAR) satellites can provide microwave remote sensing images without weather and light constraints, so they are widely applied in the maritime monitoring field. Current SAR ship detection methods based on deep learning (DL) are difficult to deploy on satellites, because these methods usually have complex models and huge calculations. To solve this problem, based on the You Only Look Once version 5 (YOLOv5) algorithm, we propose a lightweight on-board SAR ship detector called Lite-YOLOv5, which (1) reduces the model volume; (2) decreases the floating-point operations (FLOPs); and (3) realizes the on-board ship detection without sacrificing accuracy. First, in order to obtain a lightweight network, we design a lightweight cross stage partial (L-CSP) module to reduce the amount of calculation and we apply network pruning for a more compact detector. Then, in order to ensure the excellent detection performance, we integrate a histogram-based pure backgrounds classification (HPBC) module, a shape distance clustering (SDC) module, a channel and spatial attention (CSA) module, and a hybrid spatial pyramid pooling (H-SPP) module to improve detection performance. To evaluate the on-board SAR ship detection ability of Lite-YOLOv5, we also transplant it to the embedded platform NVIDIA Jetson TX2. Experimental results on the Large-Scale SAR Ship Detection Dataset-v1.0 (LS-SSDD-v1.0) show that Lite-YOLOv5 can realize lightweight architecture with a 2.38 M model volume (14.18% of model size of YOLOv5), on-board ship detection with a low computation cost (26.59% of FLOPs of YOLOv5), and superior detection accuracy (1.51% F1 improvement compared with YOLOv5).

Keywords: synthetic aperture radar (SAR); on-board; ship detection; YOLOv5; lightweight detector

Citation: Xu, X.; Zhang, X.; Zhang, T. Lite-YOLOv5: A Lightweight Deep Learning Detector for On-Board Ship Detection in Large-Scene Sentinel-1 SAR Images. *Remote Sens.* **2022**, *14*, 1018. https://doi.org/10.3390/rs14041018

Academic Editor: Gwanggil Jeon

Received: 30 December 2021
Accepted: 7 February 2022
Published: 20 February 2022

Publisher's Note: MDPI stays neutral with regard to jurisdictional claims in published maps and institutional affiliations.

1. Introduction

In recent years, an increasing number of high-quality microwave remote sensing images have been provided by synthetic aperture radar (SAR) satellites. Due to the all-day and all-weather ability of SAR, SAR remote sensing images have been widely applied in the field of ship detection. Currently, increasing number of scholars are paying attention to ship detection in SAR images due to its potential application in environmental monitoring, shipwreck rescue, oil leakage detection, marine shipping control [1–4], etc. Thus, it is of great significance to obtain the real-time and accurate ship detection results.

Recently, there have been great breakthroughs of deep learning (DL) in several fields, including computer vision (CV), natural language processing (NLP), communications, and networking [5,6]. An increasing amount of attention has been focused on SAR-based processing based on convolutional neural networks (CNNs) [7,8], especially for ship detection in SAR images. For example, Kang et al. [9] utilized contextual region-based CNN based on multilayer fusion in the field of SAR ship detection. Jiao et al. [10] proposed a densely connected end-to-end neural network to solve the problem of multi-scale and multi-scene SAR ship detection. Cui et al. [11] used a dense attention pyramid network (DAPN) for multi-scene SAR ship detection, where a pyramid structure and convolutional block attention module were adopted. Liu et al. [12] used multi-scale proposal generation for SAR ship detection, with a framework that mainly contained hierarchical grouping and proposal scoring. Wang et al. [13] proposed a rotatable bounding box ship detection method fused with an attention module and angle regression. An et al. [14] proposed an improved rotatable bounding box SAR ship detection framework, where a feature pyramid network (FPN), a modified encoding scheme, and a focal loss (FL) combined with hard negative mining (HNM) technique were adopted. Chen et al. [15] proposed a ship detection network combined with an attention module which can accurately locate ships in complex scenes. Dai et al. [16] proposed a novel CNN for multi-scale SAR ship detection, which is composed of a fusion feature extraction network (FFEN), a region proposal network (RPN), and a refined detection network (RDN). Wei et al. [17] offered a precise and robust ship detector based on a high-resolution ship detection network (HR-SDNet). The above methods have achieved fairish performance in the SAR ship detection field. However, they all have complex models and huge calculations, which can be a significant obstacle to deploy on satellites with limited memory and computation resources for on-board detection. The issue of obtaining high model detection performance with a low model volume remains to be tackled.

Many researches are dedicated to proposing lightweight SAR ship detectors. Chang et al. [18] designed a brand-new SAR ship detector with fewer parameters based on YOLOv2. They have achieved a competitive detection speed, but lack a theoretical explanation. Zhang et al. [19] established a depth-wise separable convolution neural network (DS-CNN) by integrating a multi-scale detection mechanism, concatenation mechanism, and anchor box mechanism to achieve high-speed SAR ship detection. However, their model still contains a partial heavy traditional convolution layer, decreasing detection speed. Mao et al. [20] proposed an effective and low-cost SAR ship detector, where a simplified U-Net and an anchor-free detection frame are integrated. However, while lightening the network architecture, it also sacrifices the detection accuracy. In addition, Zhang et al. [21] offered a lightweight feature optimization network (LFO-Net) based on SSD, but their model tends to ignore some offshore ships during the detection stage. Wang et al. [22] explored the application of RetinaNet in ship detection from multi-resolution Gaofen-3 imagery, but the detection accuracy of ships near harbors is still unsatisfactory. Later, to maximize the greatest advantage of deep learning, Wang et al. [23] also constructed a large volume of labeled SAR ship detection datasets named SAR-Ship-Dataset that consists of 43,819 ship chips of 256 pixels in both range and azimuth, collected from 102 Chinese Gaofen-3 and 108 Sentinel-1 SAR images. In this work, they also proposed a modified SSD-300 and a modified SSD-512 to reduce detection time as a research baseline. However, their dataset does not accord with the characteristics of large scenes of SAR images [24]. Moreover, their modified SSD models also lack sufficient theoretical supports in their reports. The above methods have made a reasonable contribution to lightening models in the SAR ship detection field. Unfortunately, there are few studies successfully designing a detector for on-board SAR ship detection. Table 1 shows the details of the above related works.

Table 1. The details of the related works.

Category	Related Works	Main Distinctive Characteristics
DL-based SAR ship detectors	[9–17]	Fairish detection performance ✔ Competitive detection speed ✘ Designed for on-board platform ✘
DL-based lightweight SAR ship detectors	[18–24]	Fairish detection performance ✔ Competitive detection speed ✔ designed for on-board platform ✘

Most of above methods are all designed for use high-power GPUs on ground stations to detect ships. However, the traditional mode of collecting data via satellite and processing data in ground stations could be time-consuming, and the longer the time from when the satellite generates the SAR images to when SAR ship information is extracted on the ground, the less useful that SAR images will be [25]. In order to fully shorten the time delay of ship information extraction, it is necessary to migrate the ship detection algorithm from the ground to the on-board computing platform (i.e., NVIDIA Jetson TX2) [25]. In addition, under the limited memory (i.e., memory size of 8 G) and computation resources (i.e., memory bandwidth of 59.7 GB/S) of the satellite processing platform, it is a challenge for the on-board ship detection of a lightweight SAR satellite to realize the accurate and fast detection performance.

Therefore, in this paper, we propose an end-to-end and elegant on-board SAR ship detector called Lite-YOLOv5. First, to obtain a lightweight network, inspired by Han et al. [26], a lightweight cross stage partial (L-CSP) module is inserted into the backbone network of the You Only Look Once version 5 (YOLOv5) algorithm [27] for reducing the amount of calculation; motivated by the network slimming algorithm proposed by Liu et al. [28], we apply network pruning for a more compact model. Then, in order to compensate the detection accuracy, we (1) propose a histogram-based pure backgrounds classification (HPBC) module to effectively exclude pure background samples and suppress false alarms; (2) propose a shape distance clustering (SDC) model to generate superior priori anchors; (3) apply a channel and spatial attention (CSA) model to enhance the SAR ships semantic feature extraction ability, inspired by Woo et al. [29]; and (4) propose a hybrid spatial pyramid pooling (H-SPP) model to increase the context information of the receptive field, inspired by He et al. [30]. Finally, to evaluate the on-board SAR ship detection ability of Lite-YOLOv5, the detector is transplanted to NVIDIA Jetson TX2 and implements the on-board ship detection without sacrificing accuracy.

Our main contributions are as follows:

1. In order to obtain a lightweight network, we (1) design a lightweight cross stage partial (L-CSP) module for reducing the amount of calculation and (2) apply network pruning for a more compact detector.
2. In order to ensure the detection performance, we (1) propose a histogram-based pure backgrounds classification (HPBC) module for excluding pure background samples to effectively suppress false alarms; (2) propose a shape distance clustering (SDC) model for generating superior priori anchors to match ship shape better; (3) apply a channel and spatial attention (CSA) model for paying more attention to regions of interest to enhance ships feature extraction capacity; and (4) propose a hybrid spatial pyramid pooling (H-SPP) model for increasing the context information of the receptive field to attach importance to key small ships.
3. We conduct extensive ablation studies to confirm the effectiveness of each above contribution. The experimental results on the Large-Scale SAR Ship Detection Dataset-v1.0 (LS-SSDD-v1.0) reveal the state-of-the-art on-board SAR ship detection performance of Lite-YOLOv5 compared with eight other competitive methods. In addition, we also transplant Lite-YOLOv5 on the embedded platform NVIDIA Jetson TX2 to evaluate its on-board SAR ship detection ability.

The remaining materials are arranged as follows. Section 2 introduces the methodology. Section 3 describes the experiments. Section 4 shows the quantitative and qualitative results, respectively. Section 5 describes the abundant ablation studies that were conducted. Section 6 discusses the whole scheme. Finally, Section 7 summarizes the entire article. In addition, Table A1 in Appendix A offers all the abbreviations and corresponding full names involved for the convenience of reading.

Notation: Boldfaced uppercase letters are used for matrices, $\mathbf{X}$. The operation $g(\cdot)$ denotes the L1 regularization value of the argument. The operations $GAP(\mathbf{X})$ and $GMP(\mathbf{X})$ denote the global average-pooling and global max-pooling values of a matrix $\mathbf{X}$. The operation $Conv_{1\times1}(\mathbf{X})$ denotes the new matrix obtained by the 1×1 convolution operation on $\mathbf{X}$. The operations $MaxPool\,(\mathbf{X})$ and $AvgPool\,(\mathbf{X})$ denote the average-pooling and max-pooling values of a matrix $\mathbf{X}$.

2. Methodology

This section details the main idea of Lite-YOLOv5. Section 2.1 introduces the network architecture of YOLOv5. Section 2.2 introduces the whole network architecture of Lite-YOLOv5. Sections 2.3 and 2.4, respectively, introduce the lightweight network design part and detection accuracy compensation part.

2.1. Network Structure of YOLOv5

The key to the on-board ship detection is to find a suitable lightweight detector which can balance detection accuracy and model complexity under the constraints of satellite processing platforms with limited memory and computation resources. YOLOv5 is a state-of-the-art object detection algorithm with fast inference speed and exact accuracy, which scores 72% AP@0.5 on the COCO val2017 dataset [31]. The main network structure of YOLOv5 is divided into four types, separately named as YOLOv5s, YOLOv5m, YOLOv5l, and YOLOv5x. The YOLOv5s model in particular has the advantages of a small size and fast speed, which is fit for embedded devices [32]. Thus, YOLOv5s is adopted in our implementation.

The architecture of YOLOv5 is manly composed of five parts: input, backbone, neck, prediction, and output. Figure 1 shows the architecture of YOLOv5. The CBL module is the basic module in YOLOv5, which is composed of a convolution (Conv) layer, batch normalization (BN) [33] layer, and a Leaky_ReLu (L-ReLu) activation [34] layer.

As shown in Figure 1, in the backbone part, by performing a Focus module, spatial information of the input image can be transferred into the channel dimension without losing details. The raw backbone part of YOLOv5 is CSPDarknet53, and it is used for feature extraction through a cross stage partial (CSP) module [35], which consists of a bottleneck structure and three convolutions. There are two kinds of CSP modules. The CSP in the backbone, respectively, consists of one residual unit (i.e., CSP1_1) or three residual units (i.e., CSP1_3), while the CSP in the neck (i.e., CSP2_1) replaces the residual units with the CBL modules. Moreover, the spatial pyramid pooling (SPP) [30] module offers different receptive fields to enrich the expression ability of features.

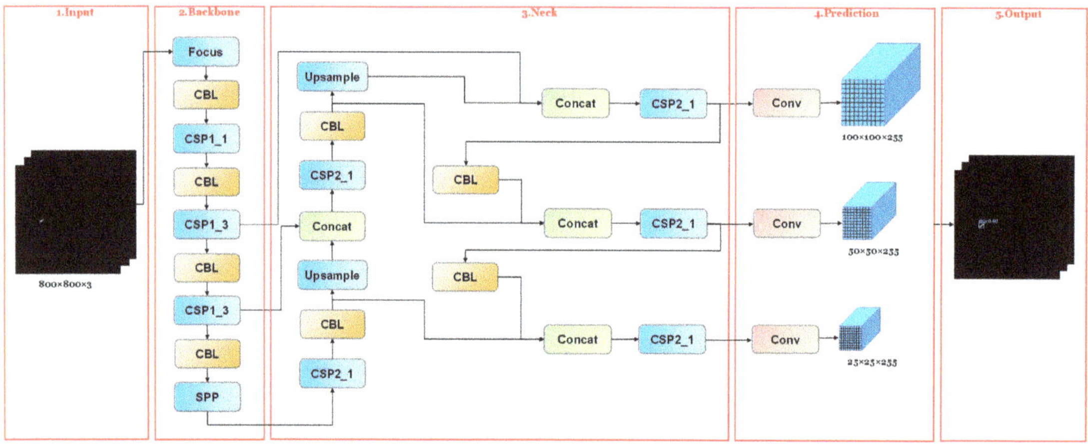

Figure 1. The architecture of YOLOv5. The network consists of five main parts: input, backbone, neck, prediction, and output.

In the neck part, the structure of a feature pyramid network (FPN) [36] and path aggregation network (PAN) [37] are adopted. The FPN module is top-down, and the high-level feature maps are fused with the low-level feature maps through up-sampling to achieve enhanced semantic features. Meanwhile, the PAN module is bottom-up; on the basis of FPN, PAN transmits the positioning information from the shallow layers to the deep layers to obtain the enhanced spatial features. The above modules jointly strengthen both the semantic information and spatial information of ships. The prediction part is composed of the three prediction layers with different scales. The small-scale detection head is suitable for detecting large ships, and the large-scale detection head is suitable for detecting small ships. The final result is obtained by means of non-maximum suppression processing during the post processing of ship detection.

2.2. Network Structure of Lite-YOLOv5

YOLOv5 is designed for optical images detection, which is not fully applicable to the field of on-board SAR ship detection. For on-board ship detection in large-scene Sentinel-1 SAR images, Lite-YOLOv5 is proposed. As shown in Figure 2, based on YOLOv5, Lite-YOLOv5 specifically injects (1) an LCB module to lighten the network without sacrificing much accuracy; (2) network pruning to obtain a much more compact network; (3) a HPBC module to effectively exclude pure background samples and suppress false alarms; (4) a SDC module to generate superior priori anchors; (5) a CSA module to enhance the SAR ships semantic feature extraction ability; and (6) a H-SPP model to increase the context information of the receptive field. Note that we use the raw SAR gray images as the input features. Following the paradigm of optical image object detection, we simply copy the single-channel SAR image to generate the three-channel SAR image. Therefore, in Figure 2, our SAR image input size is $800 \times 800 \times 3$.

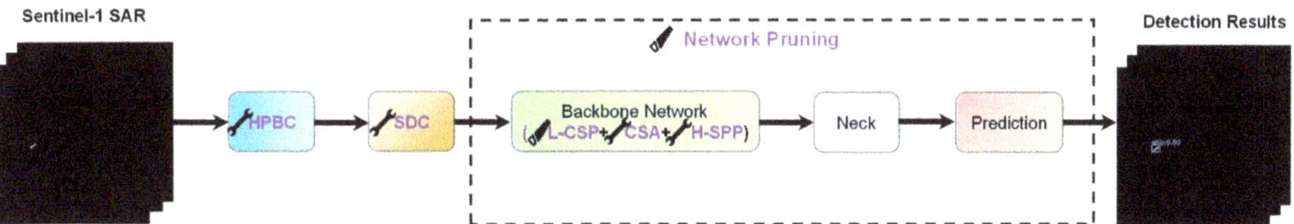

Figure 2. The ship detection framework of Lite-YOLOv5. ✔ means the lightweight network design part. ✔ means the detection accuracy compensation part. For obtaining a lightweight network, L-CSP and network pruning are inserted. For better feature extraction, HPBC, SDC, CSA, and H-SPP are embedded. L-CSP denotes the lightweight cross stage partial. Network pruning denotes the network pruning procedure. HPBC denotes the histogram-based pure backgrounds classification. SDC denotes the shape distance clustering. CSA denotes the channel and spatial attention. H-SPP denotes the hybrid spatial pyramid pooling.

For the lightweight network design, we first inject the L-CSP module into the backbone network. It is used to lighten the network with a slight accuracy sacrifice, which will be introduced in detail in Section 2.3.1. Then, for a more compact network structure, network pruning is adopted among the backbone, neck, and prediction parts. The details will be introduced in Section 2.3.2.

For better feature extraction, we first preprocess the input Sentinel-1 SAR images by the proposed HPBC module to effectively exclude pure background samples. The details will be introduced in Section 2.4.1. It should be noted that the prior anchors of the existing algorithms are generated based on optical images object distribution, which is not applicable to SAR images with characteristics of small ship size and large ship aspect ratio. We use the SDC module to generate superior priori anchors, which will be described in detail in Section 2.4.2. In the backbone, the proposed CSA module and H-SPP module are embedded. The former is used to enhance the SAR ships semantic feature extraction ability, which will be introduced in detail in Section 2.4.3. The latter is used to increase the context information of the receptive field, which will be introduced in detail in Section 2.4.4.

Next, we will introduce the L-CSP module, network pruning, HPBC module, SDC module, CSA module, and H-SPP module in detail in the following subsections.

2.3. Lightweight Network Design

2.3.1. L-CSP Module

The lightweight cross stage partial (L-CSP) module mainly consists of a bottleneck structure and three convolutions. To obtain a lightweight backbone, L-CSP module adopts LDL units (i.e., a lightweight convolution layer (L-Conv), a BN layer, and an L-Relu activation layer) instead of the original CBL units (i.e., a Conv layer, a BN layer, and an L-ReLu activation layer).

Figure 3 shows the detailed structures of an L-CSP module. When the feature map F_{in} inputs, its transmission path is divided into two parallel branches as shown in Figure 3, where the channel of F_{in} is reduced by half to generate two new feature maps. Then, two feature maps of parallel branches are concatenated as a whole feature map (i.e., output feature map F_{out}).

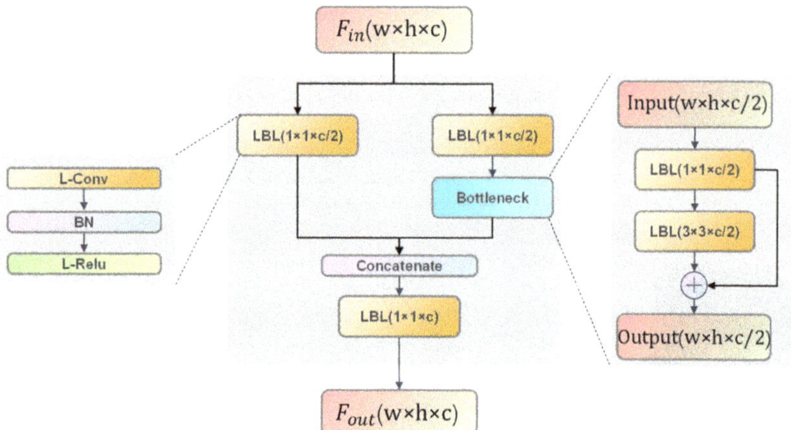

Figure 3. The detailed structure of an L-CSP module.

Figure 4 shows the detailed structures of Conv and L-Conv, respectively. Figure 4a shows the architecture of the raw Conv. In practice, the input $X \in R^{W \times H \times C}$, where W, H, and C represent the height, width, and channel of the input feature map, respectively. The convolution calculation formula is defined by

$$Y = X * f + b \tag{1}$$

where $*$ is the convolution operation; $f \in R^{C \times K \times K \times H}$ are the convolution filters, where $K \times K \times N$ denotes the kernel size and channel of the convolution filters; and b denotes the bias of the convolution filters.

Figure 4b shows the architecture of the L-Conv. As we can see, only part of the input channels is utilized to generate intermediate feature maps, and the output feature maps are calculated from the obtained intermediate feature maps with low-cost linear operations (i.e., 3×3 convolution kernels). In this way, SAR ship features extraction becomes more efficient due to fewer floating-point operations.

Specifically, given the input $X \in R^{W \times H \times C}$, M intermediate feature maps $Y' \in R^{W \times H \times C}$ are obtained by using a convolution operation, i.e.,

$$Y' = X * f' \tag{2}$$

where $f' \in R^{C \times K \times K \times H}$ are the convolution filters and $M < N$ means there are fewer convolution filters than the original convolution block. To further obtain the N output feature maps, a series of low-cost linear operations are utilized to obtain the rest of feature maps. The calculation formula is defined by

$$y_{ij} = \Phi_{i,j}\left(y_i'\right), \forall i = 1, \ldots, m; j = 1, \ldots, s \tag{3}$$

where y_i' is the i-th intermediate feature maps of Y' and $\Phi_{i,j}$ is the j-th ($j < s$) linear operation to calculate the j-th final output feature maps $y_{i,j}$ from the i-th intermediate feature maps of Y'. The last $\Phi_{i,s}$ is an identity mapping operation as shown in Figure 4b. Since there are m intermediate feature maps and s linear operations, we can obtain $n = m \times s$ final output feature maps, where the number of output feature maps is equal to the original convolution operations.

Since the computational complexity of linear operations is much less than that of ordinary convolution operations, L-Conv is a more lightweight convolution layer than Conv. Finally, by injecting the L-CSP module including L-Conv into the backbone, we can obtain a lightweight network architecture of low computational cost.

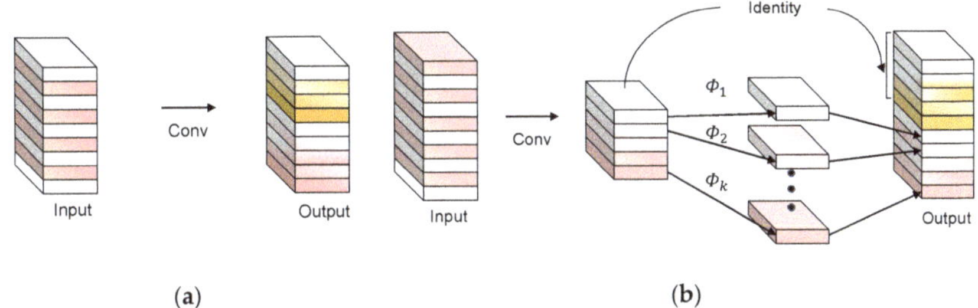

Figure 4. The detailed structures of Conv and L-Conv: (**a**) description of Conv; (**b**) description of L-Conv.

2.3.2. Network Pruning

The key of on-board ship detection is to find a lightweight on-board SAR ship detector that balances detection accuracy and model complexity under the constraints of satellites with limited memory and computation resources. Thus, it is of great significance to find a model compression method without much accuracy sacrifice. Network pruning is a mainstream model compression method. By pruning unimportant neurons, filters, or channels, it can effectively compress the parameters and computation of the model. Therefore, to further obtain a more compact detector, we follow the scheme illustrated in Figure 5 to prune the Conv and BN layers of the network.

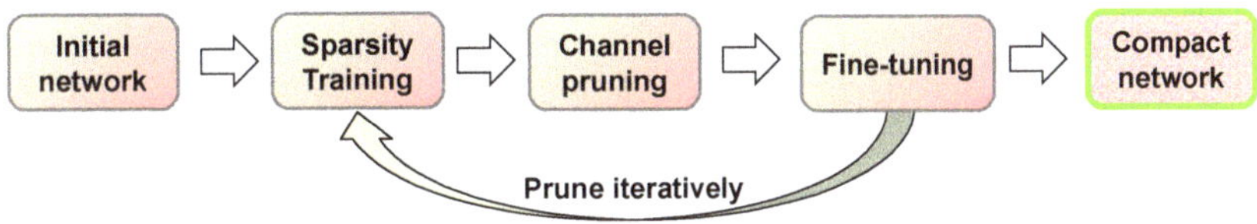

Figure 5. Flow-chart of iterative network pruning procedure.

From Figure 5, firstly, through sparse regularization training of the model, some parameters of the initial network tend to zero or equal to zero during training, and the neural network model with sparse weights is obtained. Then, the model is pruned to remove sparse channels. Next, the model is fine-tuned to restore the accuracy of the model. Finally, by iterating the above network pruning procedures, we can obtain the ultimate compact network.

The scaling factors, the sparsity training, the channel pruning, and fine-tuning will be illustrated below.

(1) Scale factors in BN layers: The Conv layer and BN layer are widely used in CNNs. In the Conv layer, reducing the number of filters can effectively reduce the amount of network parameters and computation, and accelerate the reasoning speed of the model. In the BN layer, the characteristic scaling coefficient of each BN layer corresponds to each channel, representing the activation degree of its corresponding channel. The operation formula of BN layer is formulated by

$$Z_{out} = \alpha \frac{Z_{in} - \mu_c}{\sqrt{\sigma_c + \varepsilon}} + \beta \tag{4}$$

where Z_{in} denotes input, Z_{out} denotes output, μ_c and σ_c denote the mean and variance of input activation values, respectively, and α and β denote the scaling factor and offset factor corresponding to the activation channel, respectively.

In practice, the BN layers in our network are all followed after the Conv layers. Therefore, to prune a BN layer, it is necessary to prune the convolution kernel corresponding to the upper layer and subtract the channels of the convolution kernel corresponding to the next layer.

(2) Training with the L1 sparsity constraint: The sparse effect of L1 regularization on CNNs has been proved and widely used [38,39]. In this paper, a penalty factor is added to the loss function to constrain the weight of the Conv layer and the scaling coefficient of the BN layer, and the model will be sparse. The larger the regularization coefficient λ is, the greater the constraint is. Specifically, the sparsity constraint loss function is defined by

$$L = L_{raw} + \lambda \sum_{Y \in \Gamma} g(Y) \tag{5}$$

where $Loss_{raw}$ denotes the loss function of the raw detector, $g(\gamma) = |\gamma|$ denotes L1 regularization, and λ denotes the regularization coefficient, adjusted according to the dataset.

During the sparsity training procedure of network pruning, we visualize the scale factors of BN layers. In general, the smaller the scale factor is, the sparser the channel parameters of the network are. We visualize the scale factors with three typical regularization factors (i.e., $\lambda = 0$, $\lambda = 10^{-4}$, $\lambda = 10^{-3}$), respectively, following [39]. Obviously, $\lambda = 0$ means there is no sparsity training (i.e., normal training). From Figure 6, when $\lambda = 0$, the scale factors distribution of the BN layer obeys an approximately normal distribution. When $\lambda = 10^{-4}$, some scale factors are pressed towards 0, but not enough to guarantee the sparsity of scale factors. When $\lambda = 10^{-3}$, most of the scale factors are pressed towards 0, which is enough to guarantee channel-wise pruning is being followed. Therefore, in our implementation, Lite-YOLOv5 is sparsity trained with $\lambda = 10^{-3}$ to guarantee the channel-wise sparsity.

(3) Channel pruning and fine-tuning: After sparse regularization training, the model with more sparse weights is obtained and many weights are close to zero. Then, we prune the channels of the near-zero scaling factor by deleting all incoming and outcoming connections and corresponding weights. We prune the channels across all BN layers according to the prune ratio P_r, which is defined as a certain percentile of all the scaling factor values. This ratio P_r will be determined experimentally in Section 5.2.2. Then, we adopt fine-tuning for the pruned model to restore the accuracy. Note that fine-tuning is a simple training the same as normal training. However, in this way, we can obtain a lightweight detector without sacrificing too much accuracy. In addition, an iterative network pruning procedure can lead to a more compact network [28] as can be seen in Figure 5.

2.4. Detection Accuracy Compensation

2.4.1. HPBC Module

In an on-board SAR ship detection mission, quantities of pure backgrounds images will bring additional detection burden to the detector (pure background images mean that there are no ships in images). [24]. Based on the common sense that the ocean area is much larger than the land area, most of the pure background images are pure background ocean images. On the one hand, false alarms may occur even encountering pure background images, as seen in Figure 7; on the other hand, pure background images may only increase the detection time of the detector without any benefit.

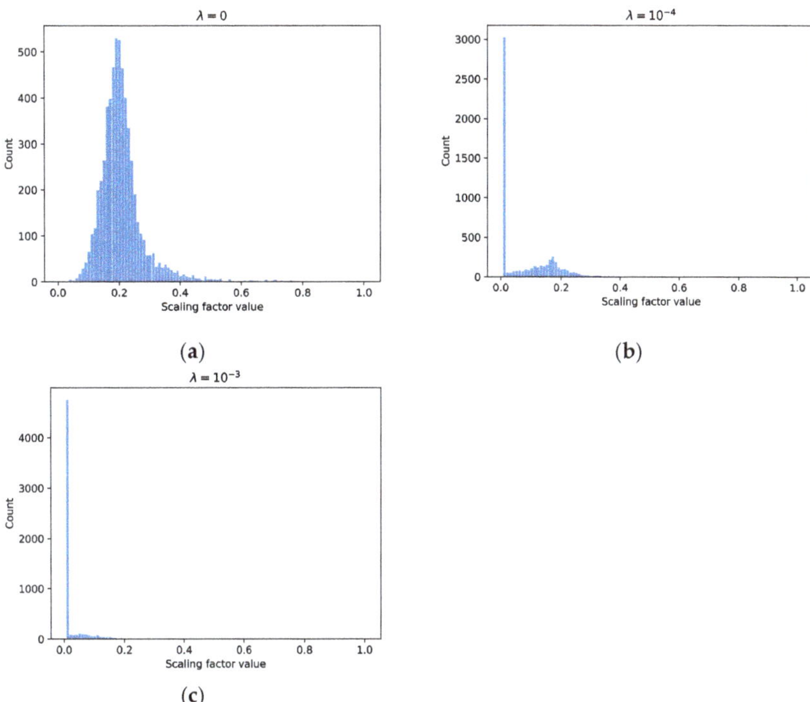

Figure 6. Distribution of scaling factors in a trained Lite-YOLOv5 under various degrees of sparsity regularization: (**a**) regularization factors equal to 0; (**b**) regularization factors equal to 10^{-4}; (**c**) regularization factors equal to 10^{-3}. The larger the λ, the sparser the scaling factors.

From Figure 7, there are some false alarm examples of the DL detector. A DL detector without prior knowledge can be fooled when encountering ghost shadows, radio frequency interference, etc. In fact, many statistical models have been developed to describe SAR image data in the constant false alarm rate (CFAR)-based algorithms [40]. The analysis of a large number of measured data shows that Gamma distribution can be well applied to sea clutter modeling [41–43].

Thus, it is necessary to integrate the traditional mature methods with rich expert experience into the preprocessing of a DL detector, otherwise on-board SAR ship detection will be time-consuming and labor intensive.

For the first time, we bring traditional sea clutter modeling method into the preprocessing of a DL detector. Inspired by the sea clutter modeling method, we propose a simple but effective histogram-based pre-classification to process the SAR images. For brevity, it is noted as the histogram-based pure backgrounds classification (HPBC) module.

For a SAR image I, its histogram is to count the frequency of all pixels in I according to the size of the gray value, which reflects the statistical characteristics of I. The histogram can be described by

$$H(i) = \sum_{i=0}^{k} \frac{n_i}{n}, k = 0, \ldots, 255 \tag{6}$$

where n_i denotes the occurrence numbers of pixels with the gray value i and n denotes the total numbers of pixels.

Note that the sea clutter sample is the ocean images. Moreover, sea clutter samples can be simply divided into pure background images and ships involved images. Figure 8a shows a typical pure background ocean image of sea clutter. Figure 8b shows its histogram and corresponding Gamma distribution curve. Figure 8c shows a typical ship image of sea clutter. Figure 8d shows its histogram and corresponding Gamma distribution curve. From Figure 8, one can conclude that sea clutter meets Gamma distribution.

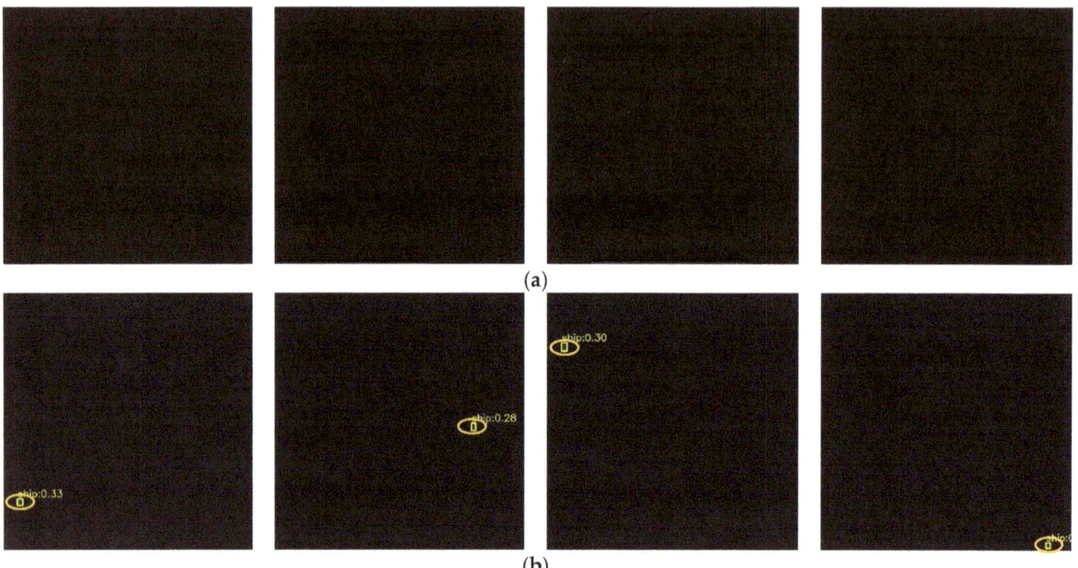

Figure 7. Some false alarms in the ocean: (**a**) the original images; (**b**) some false alarm detections. The false alarms are marked by orange ellipses.

On the one hand, a typical pure background ocean image means that the maximum of abscissa of its corresponding Gamma distribution curve is much less than 255 (i.e., A pure background sample means there is no strong scattering point, where the maximum pixels value of its histogram is much less than 255). On the other hand, a typical ship image means that the maximum of the abscissa of its corresponding Gamma distribution curve can be up to 255 (i.e., A ship target usually means a strong scattering point, where the maximum pixels value of its histogram can be up to 255).

The flow of the HPBC module is as follows.

Step 1: We simply divide the original large-scale images into 800×800 sub-images without overlap, which is kept the same as in Zhang et al. [24].

Step 2: We calculate the sub-images' histograms one by one. Once the histogram of an image is the Gamma distribution and the maximal abscissa of its corresponding Gamma distribution curve is less than the threshold ε_a, we simply judge it as a pure back ground sample. This threshold ε_a will be determined experimentally in Section 5.3.

Step 3: A pure background ocean image will not be input to the detector. As a consequence, the HPBC module can suppress the number of false alarms (i.e., there may be some false alarms in pure background ocean images as seen in Figure 7). In addition, it is helpful to reduce the detection time of the detector. The above conclusions will be confirmed in Section 5.3.

Note that when the threshold ε_a is set higher, more images will be excluded. However, since we focus on not excluding the ship images by mistake, the threshold being equal to

ε_a is fine (i.e., HPBC is only a rough preprocessing, and we would rather recognize fewer pure backgrounds than recognize the positive sample images as pure backgrounds).

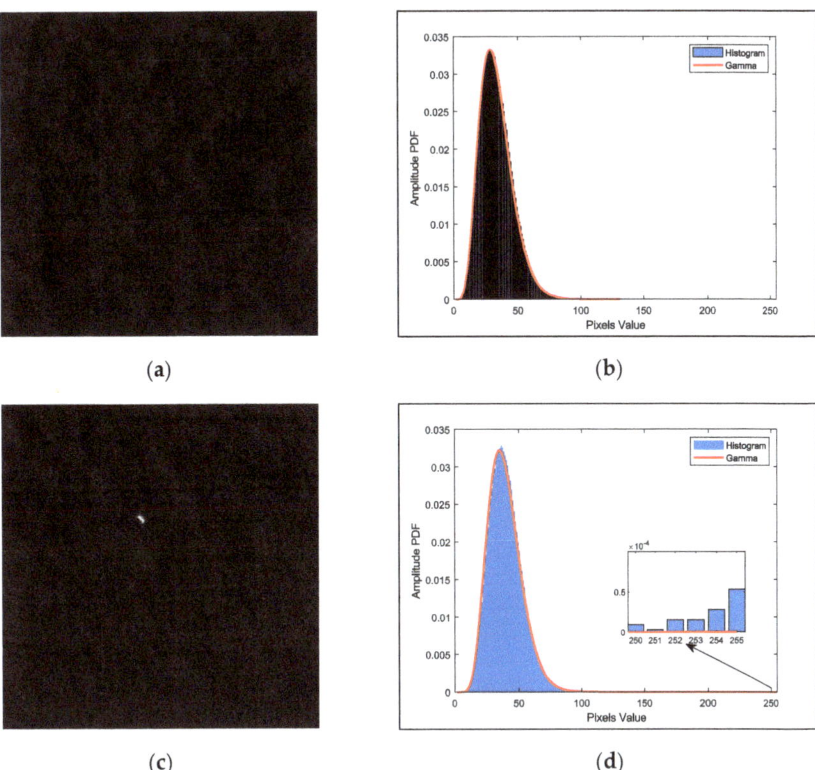

Figure 8. (**a**) A typical pure background ocean image of sea clutter; (**b**) its histogram and corresponding Gamma distribution curve; (**c**) A typical ship image of sea clutter; (**d**) its histogram and corresponding Gamma distribution curve. Amplitude PDF means probability density function of sea clutter amplitude. The range of pixels value is [0, 255].

2.4.2. SDC Module

The main idea of SDC is using the SAR ship shape distance to generate the superior prior anchors to match ship shape better. SDC is composed of the following steps.

First, k ground-truth anchors are randomly chosen as the k initialization prior anchors. Subsequently, for each ground-truth anchor, the cluster labels of the sample are calculated, which can be described by

$$label_i = \underset{1 \leq j \leq k}{\arg\min} \left(\left(L_i - L_j\right)^2 + \left(W_i - W_j\right)^2 + \left(AR_i - AR_j\right)^2 \right)^{\frac{1}{2}} \tag{7}$$

where L_i denotes the length of the i-th ground-truth anchor, L_j denotes the length of the j-th prior anchor, W_i denotes the width of the i-th ground-truth anchor, W_j denotes the width of the j-th prior anchor, AR_i denotes the aspect ratio of the i-th ground-truth anchor, AR_j denotes the aspect ratio of the j-th prior anchor, and $label_i$ denotes the label of the i-th ground-truth anchor.

Then, the length, width, and aspect ratio of the j-th prior anchor are updated by the following formulas:

$$L_j = \frac{1}{n_j}\Sigma_{i=1}^{n_j} L_i, L_i \in C_j \tag{8}$$

$$W_j = \frac{1}{n_j}\Sigma_{i=1}^{n_j} W_i, W_i \in C_j \tag{9}$$

$$AR_j = \frac{1}{n_j}\Sigma_{i=1}^{n_j} AR_i, AR_i \in C_j \tag{10}$$

where n_j denotes the number of ground-truth anchors belonging to the j-th prior anchors and C_j denotes the j-th cluster space.

Finally, iterate Formulas (7)–(10) until the following Formula (11) reaches the local optimal solution:

$$E = \Sigma_{j=1}^{k}\Sigma_{i=1}^{n_j}\left(\left(L_i - L_j\right)^2 + \left(W_i - W_j\right)^2 + \left(AR_i - AR_j\right)^2\right)^{\frac{1}{2}} \tag{11}$$

The K-means clustering algorithm is widely used in clustering problems owing to its simplicity and efficiency [44]. Therefore, in our paper, the prior anchors obtained by K-means and the SDC module are shown in Table 2. Figure 9a shows the cluster analysis of LS-SSDD-v1.0 by K-means. Figure 9b shows the cluster analysis of LS-SSDD-v1.0 by the SDC module.

Table 2. The prior anchors obtained by K-means and SDC module.

Method	Receptive Field	Prior Boxes (Width, Height)
K-means	Big	$(5, 5), (7, 8), (11, 12)$
	Medium	$(15, 15), (17, 22), (26, 19)$
	Small	$(23, 30), (35, 31), (49, 49)$
SDC Module	Big	$(5, 5), (7, 8), (9, 11)$
	Medium	$(13, 14), (14, 19), (22, 19)$
	Small	$(21, 31), (32, 22), (36, 38)$

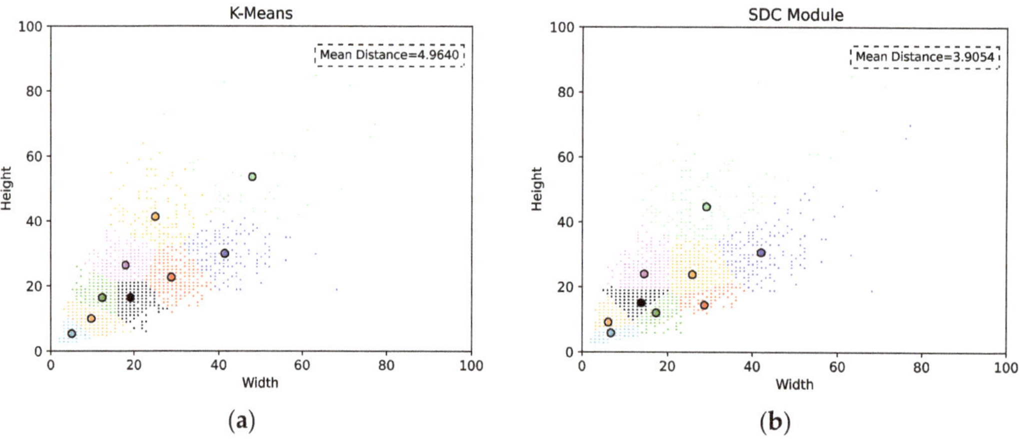

Figure 9. (**a**) Cluster analysis of LS-SSDD-v1.0 by K-means; (**b**) cluster analysis of LS-SSDD-v1.0 by SDC module. We set 9 prior anchors just as [27] did. Different colors mean different clusters. Nine cluster centroids are marked by a large circle with different colors. Mean distance means the average Euclidean distance between each ground-truth anchor and its cluster center; smaller is better.

From Figure 9, one can conclude that the proposed SDC module possesses a superior clustering performance (i.e., ~3.90 mean distance < ~4.96 mean distance compared with K-means). Subjectively, in Figure 9, the prior anchors clustered by the SDC module are smaller and roughly symmetrically distributed by the central axis, which conforms to the distribution law of LS-SSDD-v1.0 (i.e., there are numerous of small ships and the aspect ratio of ships is symmetrically distributed.). The above fully confirms the effectiveness of the SDC module.

2.4.3. CSA Module

The essence of the attention mechanism is to extract more valuable information for task objectives in the target area and to suppress or ignore some irrelevant details. Channel attention focuses on the "what" problem, (i.e., it focuses on what plays an important role in the whole image). However, in general, a ship in the image is sparsely distributed and its pixel proportion is quite small. Thus, only part of the pixel space in a ship detection task is valuable. Spatial attention focuses on the "where" problem, (i.e., where the ship is in the whole image). Spatial attention is the supplement of channel attention; each spatial feature is selectively aggregated through the weighting of spatial features. Therefore, different from SENet [45], which only focuses on channel attention, we bring channel and spatial attention simultaneously, named as the CSA Module.

To achieve this, we sequentially apply channel and spatial attention. From Figure 10, given the input feature map $F \in R^{W \times H \times C}$, channel attention can generate channel weight $W_C \in R^{C \times 1 \times 1}$ and spatial attention can generate spatial weight $W_S \in R^{H \times W \times 1}$. The different depths of color in Figure 10 represent different values of weights.

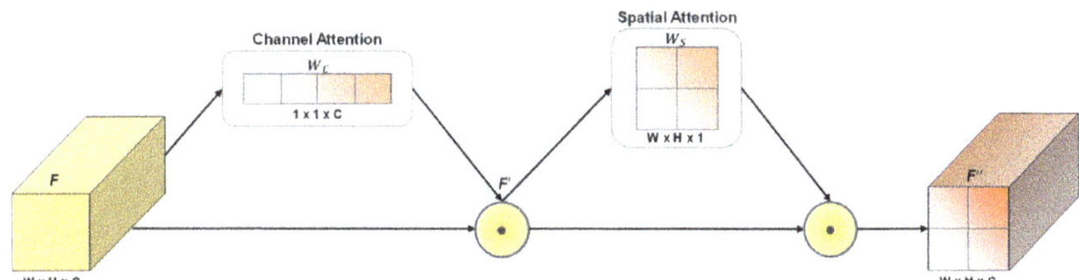

Figure 10. The detailed structure of CSA module, which is composed of channel attention and spatial attention.

Specifically, for channel attention, given the input feature map $F \in R^{W \times H \times C}$, first, global average pooling (GAP) [46] is carried out to generate the average spatial response and global max pooling (GMP) [46] to generate the maximum spatial response; then, they are transmitted, respectively, to the multi-layer perceptron (MLP) to encode the channel information, which is helpful to infer finer channel attention; next, element-wise addition is conducted between the two feature maps; then, the synthetic channel information is activated through the sigmoid function to obtain the channel weight feature map, i.e., a channel weight matrix W_C; finally, element-wise multiplication is conducted between the original feature map $F \in R^{W \times H \times C}$ and obtained channel weight matrix W_C.

In short, the above can be defined by

$$F' = F \odot W_c \tag{12}$$

where F' denotes the weighted feature map, F denotes the input feature map, $\odot$ denotes element-wise multiplication, and W_C denotes the obtained channel weight matrix, i.e.,

$$W_c = \sigma\{f_{c-encode}(GAP(F)) \oplus f_{c-encode}(GMP(F))\} \tag{13}$$

where GAP denotes the global average-pooling operation, GMP denotes the global max-pooling operation, $\oplus$ denotes element-wise summation, $f_{c\text{-}encode}$ denotes the channel coder, and σ denotes the sigmod activation function.

As for spatial attention, given the input feature map $F' \in R^{C \times W \times H}$, first, global average pooling (GAP) [46] is carried out to generate the average spatial response and global max pooling (GAP) [46] is carried it to generate the maximum spatial response; then, the two generated results are concatenated as a whole feature map; next, a space encoder of our design is used to encode the space information, which is helpful to infer finer spatial attention; then, it is activated through the sigmoid function to obtain the spatial weight feature map, i.e., a spatial weight matrix W_S; finally, element-wise multiplication is conducted between the original feature map $F' \in R^{C \times W \times H}$ and the obtained spatial weight matrix W_S.

In short, the above can be defined by

$$F'' = F' \odot W_S \tag{14}$$

where F'' denotes the weighted feature map, F' denotes the input feature map, $\odot$ denotes element-wise multiplication, and W_S denotes the obtained spatial weight matrix, i.e.,

$$W_S = \sigma\{f_{s-encode}(GAP(F)(c)GMP(F))\} \tag{15}$$

where GAP denotes the global average-pooling operation, GMP denotes the global max-pooling operation, $\copyright$ denotes concatenation operation, $f_{s\text{-}encode}$ denotes the space coder, and σ denotes the sigmod activation function.

Finally, we can obtain a finer channel and space information. With the CAS module inserted into the network, each level can extract both rich spatial and rich semantic information, which is helpful to improve the detection performance of small ships.

2.4.4. H-SPP Module

In the CV community, by capturing rich context information, the network can better understand the relationship between pixels and enhance the performance of detection. For aggregating context information, a pyramid pooling module with max-pooling has been commonly adopted so far [47–49]. The previous works believe that since the object of interest may produce the largest pixel value, adopting max-pooling is enough. We argue that average-pooling gathers another important clue about global information extraction capacity. This idea is inspired by the works of [29], which is recommended for readers. Thus, we adopt a hybrid spatial pyramid pooling (H-SPP) module to further enhance the global context information extraction capacity of the network. It can be aware of both local and global contents of feature maps, and attach importance to key small ship features.

Different from previous works [47–49], the H-SPP module mainly aggregates the feature map generated by both average-pooling and max-pooling operations with different pooling sizes. Figure 11 shows the detailed structures of the H-SPP module. Then, we will further introduce the principle of the H-SPP module.

The H-SPP module aggregates the feature map generated by both the max-pooling layer and average-pooling layer of different kernel sizes (i.e., 5×5, 9×9 and 13×13), as shown in Figure 11. Specifically, given the input feature map $F_{in} \in R^{W \times H \times C}$ generated by the backbone, it is first transmitted to a CBL module to generate a feature map $F' \in R^{W \times H \times 0.5C}$ with refined channel information. Then, (1) max-pooling of different kernel sizes (i.e., 5×5, 9×9 and 13×13) is simultaneously carried out to generate three local receptive field feature maps and (2) average-pooling of different kernel sizes (i.e., 5×5, 9×9 and 13×13) is carried out to generate three global receptive field feature maps. Next, six generated results and original ones (i.e., F_1–F_7 from Figure 11) are concatenated as a synthetic feature map. Finally, the feature map level fusion of local features and global features is realized, which enriches the expression ability of the final feature map $F_{out} \in R^{W \times H \times 3.5C}$.

In short, the above can be defined by

$$F_{out} = Conv_{1\times1}(F_{in})\,(c)\,MaxPool(Conv_{1\times1}(F_{in}))\,(c)\,AvgPool(Conv_{1\times1}(F_{in})) \qquad (16)$$

where $Conv_{1\times1}$ denotes the 1×1 convolution operation, *MaxPool* denotes the max-pooling operations (with kernel sizes of 5×5, 9×9 and 13×13, respectively), *AvgPool* denotes the average-pooling operations (with kernel sizes of 5×5, 9×9, and 13×13, respectively), and © denotes the concatenation operation.

By aggregating the feature maps of abundant receptive fields, the H-SPP module obtains different degrees of context information, and enhances the network's ability to capture both local and global information. Thus, the H-SPP module can improve the accuracy of the final prediction result of the algorithm. In this paper, the H-SPP module will be used to further improve the detection performance of Lite-YOLOv5.

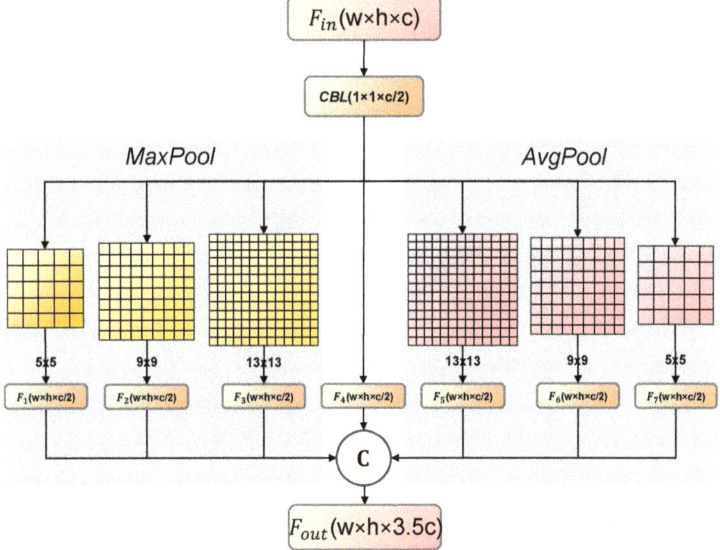

Figure 11. The detailed structure of H-SPP module. MaxPool means the max pooling layer, AvgPool means the average pooling layer, and © means the concatenation operation.

3. Experiments

Different from the traditional integrated training and testing platform, we first used a workstation with the powerful NVIDIA RTX3090 as the training experiment platform to generate the well-trained detector Lite-YOLOv5, then we used a NVIDIA Jetson TX2 as the training experiment platform to evaluate the on-board SAR ship detection ability of Lite-YOLOv5.

3.1. Experimental Platform

3.1.1. Training Experimental Platform

Considering the limited computing resources and computing power of NVIDIA Jetson TX2, we used the workstation with the GPU model of NVIDIA RTX3090, CPU model of i7-10700, and memory size of 32 G to carry out the training part of the experiment. PyTorch 1.7.0 [50] based on the Python 3.8 language was adopted as the framework of our algorithm. We also used CUDA11.1 in our experiments to call the GPU for training acceleration. Subsequently, we transplanted the trained model into the NVIDIA Jetson TX2.

3.1.2. Testing Experimental Platform

We used the NVIDIA Jetson TX2 as the development board in order to realize on-board ship detection during the testing part of experiment. The NVIDIA Jetson TX2 is an embedded vision computer system with the 256-core NVIDIA Maxwell GPU model, dual-core Denver2 CPU model, and an 8 G memory size. Meanwhile, the characteristics of low power consumption, high performance, large memory bandwidth, etc. make it very suitable for on-board satellite data processing.

3.2. Dataset

The LS-SSDD-v1.0 dataset is widely used for SAR image intelligent interpretation [51–54]. The characteristic of small ships with large-scale backgrounds in LS-SSDD is close to actual satellite images; thus, we adopted the LS-SSDD-v1.0 dataset to verify the effectiveness of Lite-YOLOv5. Table 3 shows the details of the LS-SSDD dataset.

Table 3. Details of the LS-SSDD-v1.0 dataset.

Key	Value
Sensors	Tokyo, Adriatic Sea, etc.
Polarization	VV, VH
Sensor mode	IW
Scene	land, sea
Resolution (m)	5×20
Number of images	15
Image size	$24,000 \times 16,000$
Cover width (km)	~250

From Table 3, there are 15 large-scale images (cover width ~250 km) numbered by 00.jpg to 15.jpg from different places (Tokyo, Adriatic Sea, etc.), polarizations (VV, VH), and scenes (land, sea). Considering the computing power of the GPU, we simply divided the original large-scale images into 800×800 sub-images without embellishment, keeping to the method of Zhang et al. [24]. Since there were fifteen 24,000 pixels $\times$ 16,000 pixels large-scale images, the total sub-image number was 9000. Finally, according to Zhang et al. [24], the LS-SSDD-v1.0 dataset was divided into a training set for training learning and a test set for result performance evaluation via the ratio of 2:1.

3.3. Experimental Details

We employed the stochastic gradient descent (SGD) [55] algorithm to train our network. The network input size was 800 pixels $\times$ 800 pixels and the batch size of 16 was adopted. During normal and sparsity training, we trained the network for 100 total epochs. We also set the learning rate as 0.001, the weight decay as 0.0005, and the momentum as 0.937. Other hyper-parameters not mentioned were kept the same as those in YOLOv5.

3.4. Evaluation Indices

Precision (P) is calculated by

$$P = \frac{\#TP}{\#TP + \#FP} \times 100\% \tag{17}$$

where # denotes the number, TP denotes the situation where the prediction and label are both ships, and FP denotes the situation where the prediction is a ship but the label is the background.

Recall (R) is calculated by

$$R = \frac{\#TP}{\#TP + \#FN} \times 100\% \tag{18}$$

where FN denotes the situation where the prediction is the background but the label is a ship.

The average precision (AP) is calculated by

$$AP = \int_0^1 P(R) \cdot dR \tag{19}$$

where P denotes the precision, and R denotes the recall.

$F1$ can take account of both precision and recall and is calculated by

$$F1 = 2 \times \frac{R \times P}{R + P} \tag{20}$$

Finally, t denotes the inference consuming time of a sub-image detection. As a result, the running time T of a large-scale image is equal to 600 t. Moreover, in order to evaluate the portability performance of Lite-YOLOv5 on on-board SAR ship detection, we also calculated the parameter size, FLOPs, and model volume.

4. Results

4.1. Quantitative Results

Table 4 shows the quantitative results of Lite-YOLOv5 on the LS-SSDD-v1.0 dataset. In Table 4, we can see the detection performance comparison with the raw YOLOv5. The ablation studies about the influence of each proposed module will be introduced in detail in Section 5 by the means of each installation and removal.

From Table 4, one can conclude that:

1. Compared with YOLOv5, our Lite-YOLOv5 can guarantee the model is lightweight and realize the slight improvement of detection performance at the same time.
2. On the one hand, as for accuracy indices, Lite-YOLOv5 can make a 5.97% precision improvement (i.e., from 77.04% to 83.01%), 1.12% AP improvement (i.e., from 72.03% to 73.15%), and 1.51% F1 improvement (i.e., from 72.01% to 73.52%). This fully reveals the effectiveness of the proposed HPCB, SDC, CSA, and H-SPP modules.
3. On the other hand, as for other evaluation indices, Lite-YOLOv5 can realize on-board ship detection with 37.51 s per large-scale image (73.29% of the processing time of YOLOv5), a lighter architecture with 4.44 G FLOPs (26.59% of the FLOPs of YOLOv5), and 2.38 M model volume (14.18% of the model size of YOLOv5). This fully reveals the effectiveness of the proposed L-CSP module and network pruning.

Table 4. The performance comparison with the raw YOLOv5. P: Precision, the higher the better; R: recall, the higher the better; AP: average precision, the higher the better; *F1*: F1-score, a main evaluation index, the higher the better; FLOPs: floating point operations, refer to model complexity; model volume: refers to size of model weight; T: the running time of a large-scale image, refers to detection speed (tested on the Jetson TX2).

Method	P (%)	R (%)	AP (%)	F1 (%)	T (s)	FLOPs (G)	Model Volume (M)
YOLOv5	77.04	67.60	72.03	72.01	51.18	16.70	13.70
Lite-YOLOv5(ours)	83.01	65.97	73.15	73.52	37.51	4.44	2.38

Table 5 shows the performance comparisons of Lite-YOLOv5 with eight other state-of-the-art detectors. In Table 5, we mainly select the Libra R-CNN [56], Faster R-CNN [57], EfficientDet [58], free anchor [59], FoveaBox [60], RetinaNet [61], SSD-512 [62], and YOLOv5 [27] for comparison. They were all trained on the LS-SSDD-v1.0 dataset with loading ImageNet pre-training weights. Their implementations were also kept basically the same as in the original report. In addition, it should be emphasized that there is no end-to-end on-board SAR ship detector. Thus, we selected the mainstream two-stage detector (i.e., Libra R-

CNN, Faster R-CNN) and single-stage detectors (i.e., EfficientDet, free anchor, FoveaBox, RetinaNet, SSD-512, and YOLOv5) in the CV community for comparison.

From Table 5, the following conclusions can be drawn:

1. What stands out in this table is the competitive accuracy performance with the greatly reduced model volume of Lite-YOLOv5.
2. The AP and F1 of Lite-YOLOv5 cannot reach the best performance at the same time; nevertheless, the excellent performance of the other evaluation indicators can make up for it. More prominently, with the tiny model size of ~2 M and competitive accuracy indicators, Lite-YOLOv5 can ensure a superior on-board detection performance.
3. Compared with the experimental baseline YOLOv5, Lite-YOLOv5 offers ~1.1% AP improvement (i.e., from 72.03% to 73.15%) and ~1.5% F1 improvement (i.e., from 72.01% to 73.52%). This fully reveals the effectiveness of the proposed HPBC, SDC, CSA, and H-SPP modules.
4. Compared with the experimental baseline YOLOv5, Lite-YOLOv5 offers the most lightweight network architecture with 4.44 G FLOPs (~26.6% of the FLOPs of YOLOv5), 1.04 M parameter size (~14.7% of the parameter size of YOLOv5), and ~ 2 M model volume (~14.2% of the model size of YOLOv5). This fully reveals the effectiveness of the proposed L-CSP module and network pruning.
5. Libra R-CNN offers the highest F1 (i.e., 75.93%), but its AP is rather poor to satisfy the basic detection application, i.e., its 62.90% AP << Lite-YOLOv5's 73.15%. Furthermore, its detection time, FLOPs, parameter size, and model volume are all one or two orders of magnitude than those of Lite-YOLOv5, which is a huge obstacle for on-board detection.

Table 5. The performance comparisons of Lite-YOLOv5 with eight other state-of-the-art detectors. The best model is marked in bold. Parameter Size refers to the model complexity. (tested on the RTX3090).

Method	AP (%)	F1 (%)	T (s)	FLOPs (G)	Parameter Size (M)	Model Volume (M)
Libra R-CNN [56]	62.90	**75.93**	62.28	162.18	41.62	532
Faster R-CNN [57]	63.00	69.48	124.45	134.38	33.04	320
EfficientDet [58]	61.35	64.70	131.33	107.52	39.40	302
Free anchor [59]	71.04	64.60	52.32	127.82	36.33	277
FoveaBox [60]	52.30	68.26	'52.32	126.59	36.24	277
RetinaNet [61]	54.31	70.53	52.06	127.82	36.33	277
SSD-512 [62]	40.60	57.65	23.09	87.72	24.39	186
YOLOv5 [27]	72.03	72.01	1.92	16.70	7.06	14
Lite-YOLOv5 (ours)	**73.15**	73.52	**1.41**	**4.44**	**1.04**	**2**

4.2. Qualitative Results

Figure 12 shows the visualization results on the LS-SSDD-v1.0 as an example. As we can see, Lite-YOLOv5 can carry out accurate SAR ship detection even in difficult conditions (i.e., larger scene image, multi-scale ships, and different aspect ratios of ships).

Figure 13 shows the detection results of different methods under complicated scenarios (i.e., offshore scenes of strong speckle noise and inshore scenes). Note that we only chose some lightweight models for fair comparison.

From Figure 13, one can conclude the following:

1. In the offshore scenes, Lite-YOLOv5 can offer high-quality detection results even under the environment of strong speckle noise. Most other methods always produce the missed alarms caused by speckle noise. Taking the second line of images as an example, there were four missed detections of RetinaNet and three missed detections of YOLOv5, which are both more than that of Lite-YOLOv5 (only one missed ship).
2. In the inshore scenes, Lite-YOLOv5 can offer high-quality detection results even under the environment of ship-shaped reefs and buildings near shore. Most other methods

always produce the missed alarms caused by them. Taking the fourth line of images as an example, there were two missed detections of RetinaNet and two missed detections of YOLOv5, which are both more than that of Lite-YOLOv5 (only one missed ship).

3.   Lite-YOLOv5 can offer an advanced on-board ship detection performance compared with other state-of-the-art methods.

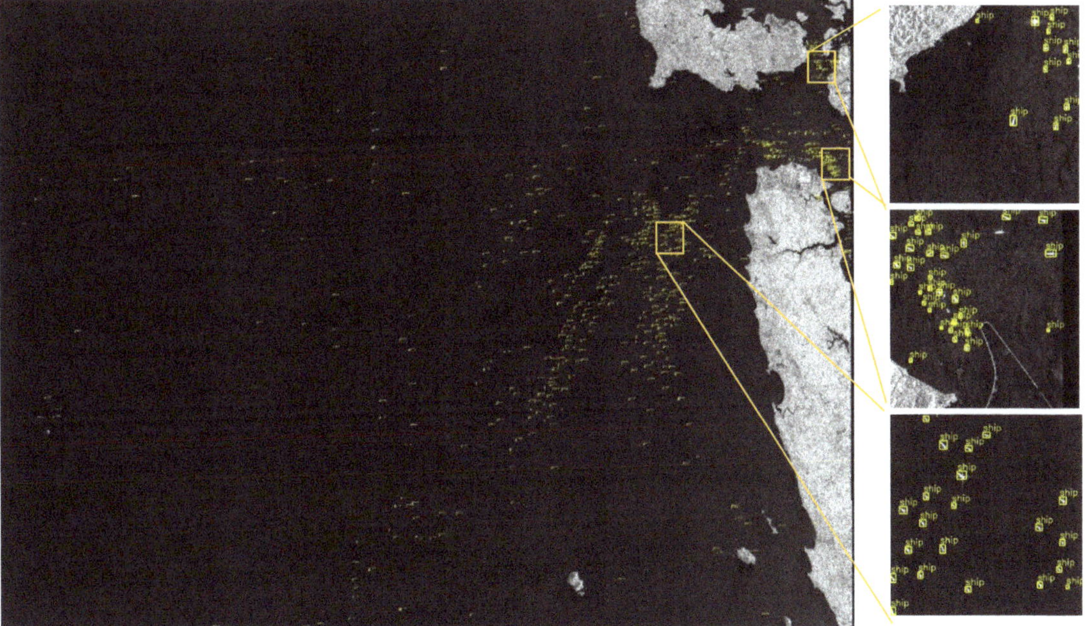

Figure 12. The qualitative SAR ship detection results of Lite-YOLOv5. A score threshold of 0.25 is used for display. Best viewed in zoom in.

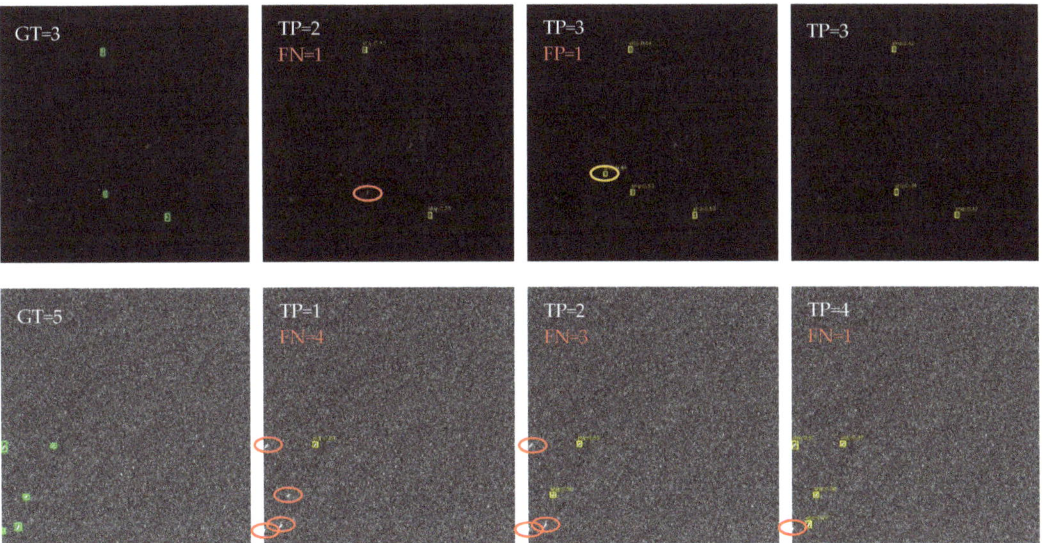

Figure 13. *Cont.*

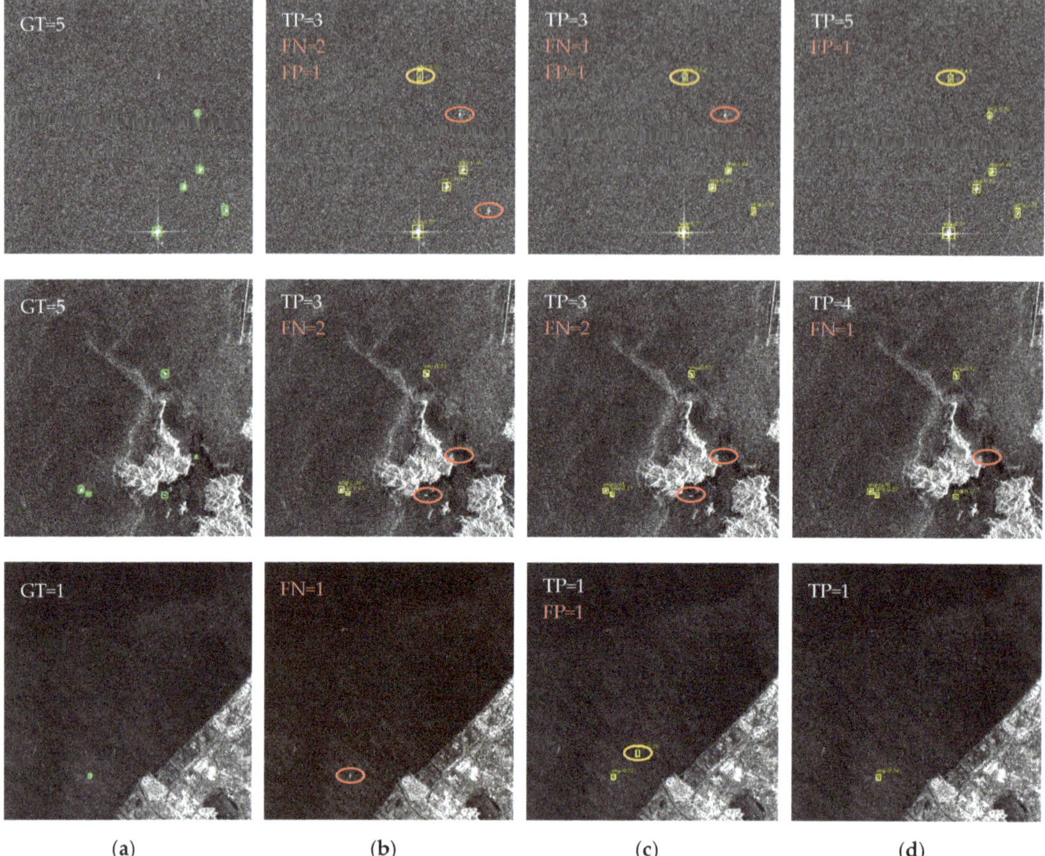

Figure 13. The qualitative SAR ship detection results of different methods: (**a**) ground truth; (**b**) RetinaNet; (**c**) YOLOv5; (**d**) Lite-YOLOv5. The ground truths are marked by green boxes. Prediction results are marked by yellow boxes with confidence scores. The false alarms are marked by orange ellipses. The missed detections are marked by red ellipses. GT means the number of ground truths. FN means the number of missed detections. FP means the number of false alarms.

5. Ablation Study

In this section, we will introduce the ablation studies to show the influence of each proposed module by the means of each installation and removal.

5.1. Ablation Study on the L-CSP Module

Table 6 shows the ablation study of Lite-YOLOv5 with and without the L-CSP module. In Table 6, "✗" means Lite-YOLOv5 without the L-CSP module (while keeping the other five modules), while "✔" means Lite-YOLOv5 with the L-CSP module (i.e., our proposed detector). From Table 6, one can find that the L-CSP module can guarantee a lighter architecture with 4.44 G FLOPs and a 2.38 M model volume (~45.6% decrease of FLOPs and ~7.0% decline of model volume when compared with Lite-YOLOv5 without the L-CSP module), which confirms that the L-CSP module can offers a model with greatly reduced computation. In addition, there is only a slight decrease to the overall detection performance. Thus, the L-CSP module can realize a model of sharply reduced computation with only a slight accuracy loss, which confirms its superior cost-effectiveness in lightweight network design.

Table 6. The ablation study of Lite-YOLOv5 with and without L-CSP module.

L-CSP	P (%)	R (%)	AP (%)	F1 (%)	FLOPs (G)	Model Volume (M)
✘	82.23	67.16	73.17	73.93	8.16	2.56
✔	83.01	65.97	73.15	73.52	**4.44**	**2.38**

5.2. Ablation Study on Network Pruning

We conducted two ablation experiments on network pruning. Experiment 1 in Section 5.2.1 shows the effectiveness of network pruning in Lite-YOLOv5. Experiment 2 in Section 5.2.2 shows the effectiveness of channel-wise pruning in network pruning.

5.2.1. Experiment 1: Effectiveness of Network Pruning

Table 7 shows the ablation study of Lite-YOLOv5 with and without network pruning. In Table 7, "✘" means Lite-YOLOv5 without network pruning (while keeping the other five modules), while "✔" means Lite-YOLOv5 with network pruning (i.e., our proposed detector). From Table 7, one can find that network pruning can realize a lighter architecture with 4.44 G FLOPs and a 2.38 M model volume (~68.6% decrease of FLOPs and ~81.5% decline of model volume when compared with Lite-YOLOv5 without network pruning). Thus, network pruning can achieve a huge compression of the model with a slight accuracy loss, which confirms its superior performance in lightweight network design. In addition, we also conducted another experiment to explore the effect of channel-wise pruning.

Table 7. The ablation study of Lite-YOLOv5 with and without network pruning.

Network Pruning	P (%)	R (%)	AP (%)	F1 (%)	FLOPs (G)	Model Volume (M)
✘	80.93	67.75	73.84	73.76	14.16	12.90
✔	83.01	65.97	73.15	73.52	**4.44**	**2.38**

5.2.2. Experiment 2: Effect of Channel-Wise Pruning

During the channel pruning procedure of network pruning, we conducted several experiments under different pruning ratios P_r. In general, the larger the pruning ratio is, the smaller the model volume of the network is while the poorer the model performance is. Thus, it is of great importance to trade off the pruning ratio and the model performance. From Figure 14, we can see the effect of choosing different pruning ratios from Lite-YOLOv5 trained on LS-SSDD-v1.0 with $\lambda = 10^{-3}$. When P_r goes beyond 0.7, the F1 of the model seriously deteriorates. Thus, in our implementation, Lite-YOLOv5 is channel-wise pruned with a P_r equal to 0.7 to trade off the model performance and model complexity.

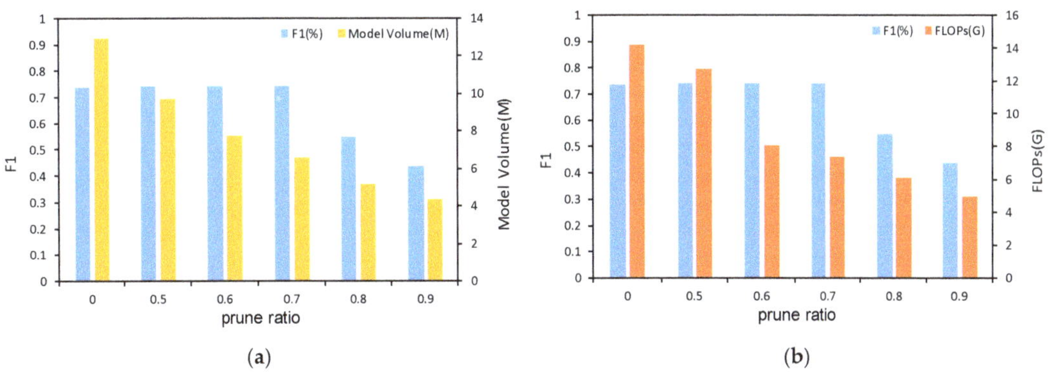

Figure 14. The effect of choosing different pruning ratios from Lite-YOLOv5 trained on LS-SSDD-v1.0 with $\lambda = 10^{-3}$: (**a**) model volume vs. F1; (**b**) FLOPs vs. F1.

5.3. Ablation Study on the HPBC Module

Table 8 shows the ablation study of Lite-YOLOv5 with and without the HPBC module. In Table 8, "✗" means Lite-YOLOv5 without the HPBC module (while keeping the other five modules), while "✔" means Lite-YOLOv5 with the HPBC module (i.e., our proposed detector). From Table 8, one can find that the HBC module can make a ~0.2% improvement with AP and F1. Note that the ~0.7% improvement in precision (i.e., the decrease of false alarms) reveals the reason of the improvement of overall detection performance (i.e., the HPBC module can effectively exclude pure background ocean images, so some false alarms in them are avoided). Furthermore, the HBC module can obtain the real-time detection performance with only a 37.51 s running time for one large-scale image (~10.4 s decrease of running time compared with Lite-YOLOv5 without the HPBC module).

Table 8. The ablation study of Lite-YOLOv5 with and without HPBC module.

HPBC	P (%)	R (%)	AP (%)	F1 (%)	T (s)
✗	82.45	65.97	72.96	73.30	47.88
✔	83.01	65.97	73.15	**73.52**	**37.51**

All of the above reveal that the HPBC module can effectively classify pure background ocean images; thus, it can (1) suppress some false alarms, and therefore the overall accuracy indices are increased and (2) decrease the detection burden of the detector, and therefore real-time detection performance is guaranteed. Significantly, one may find more powerful techniques to further classify the pure background samples, but HPBC might be one of the most direct approaches without complicated steps and obscure theories.

We performed another experiment to study the impact of the abscissa filter threshold ε_a. The experimental results are shown in Table 9. It can be concluded that when ε_a is set higher, more pure background ocean images will be excluded (i.e., fewer images remain) and detection performance will be improved. However, in the actual scene, we focus on not excluding the ship images by mistake. Thus, it is of great importance to optimize ε_a on the basis of guaranteeing the original number of ship images. In Table 9, ε_a being set to 128 is the optimal choice for the balance of the number of ship images and detection accuracy. Thus, the final ε_a is set to 128 in Lite-YOLOv5.

Table 9. The ablation study of Lite-YOLOv5 with different abscissa filter thresholds. #Images: number of test set images; #Ships: number of ships in test set images.

ε_a	#Images	#Ships	P (%)	R (%)	AP (%)	F1 (%)
0	3000	2378	82.45	65.97	72.96	73.30
96	2566	2378	82.49	65.97	73.00	73.31
112	2450	2378	82.62	65.97	73.07	73.36
128	2350	**2378**	83.01	65.97	73.15	73.52
144	2271	2372	82.02	66.94	73.28	73.72

5.4. Ablation Study on the SDC Module

Table 10 shows the ablation study of Lite-YOLOv5 with and without the SDC module. In Table 10, "✗" means Lite-YOLOv5 without the SDC module (while keeping the other five modules), while "✔" means Lite-YOLOv5 with the SDC module (i.e., our proposed detector). From Table 10, one can find that the SDC module can make an overall detection performance improvement with a ~ 0.9% F1 improvement, which confirms its effectiveness. This is because the SDC module can utilize the SAR ship shape distance (i.e., the distribution of length, width, and aspect ratio) to generate a more appropriate prior anchor. Finally, Lite-YOLOv5 can extract SAR ship information more effectively. In addition, the SDC module brings hardly any model complexity increase, which confirms its superior cost-effectiveness in detection accuracy compensation.

Table 10. The ablation study of Lite-YOLOv5 with and without SDC module.

SDC	P (%)	R (%)	AP (%)	F1 (%)	FLOPs (G)	Model Volume (M)
✘	82.37	64.89	72.53	72.59	4.33	2.33
✔	83.01	65.97	**73.15**	**73.52**	4.44	2.38

5.5. Ablation Study on the CSA Module

Table 11 shows the ablation study of Lite-YOLOv5 with and without the CSA module. In Table 11, "✘" means Lite-YOLOv5 without the CSA module (while keeping the other five modules), while "✔" means Lite-YOLOv5 with the CSA module (i.e., our proposed detector). From Table 11, one can find that the CSA module can make an overall detection performance improvement with a ~2.6% AP and ~1.9% F1 improvement, which confirms its effectiveness. This is because the CSA module can extract both rich spatial and rich semantic information. Finally, Lite-YOLOv5 can improve the ship detection performance. In addition, the CSA module only brings a slight model complexity increase, which confirms its superior cost-effectiveness in detection accuracy compensation.

Table 11. The ablation study of Lite-YOLOv5 with and without CSA module.

CSA	P (%)	R (%)	AP (%)	F1 (%)	FLOPs (G)	Model Volume (M)
✔	80.38	64.59	70.56	71.63	4.05	2.31
✔	83.01	65.97	**73.15**	**73.52**	4.44	2.38

5.6. Ablation Study on the H-SPP Module

Table 12 shows the ablation study of Lite-YOLOv5 with and without the H-SPP module. In Table 12, " ✘" means Lite-YOLOv5 without the H-SPP module (while keeping the other five modules), while "✔" means Lite-YOLOv5 with the H-SPP module (i.e., our proposed detector). From Table 12, one can find that the H-SPP module can make an overall detection performance improvement with a ~0.8% F1 improvement, which confirms its effectiveness. This is because the H-SPP module can aggregate the feature maps of abundant receptive fields and obtain different degrees of context information. Finally, Lite-YOLOv5 can effectively improve the network's capacity to capture both local and global information of SAR images. In addition, the H-SPP module only brings a slight model complexity increase, which confirms its superior cost-effectiveness in detection accuracy compensation.

Table 12. The ablation study of Lite-YOLOv5 with and without H-SPP module.

H-SPP	P (%)	R (%)	AP (%)	F1 (%)	FLOPs (G)	Model Volume (M)
✘	82.34	65.14	72.61	72.74	5.19	2.35
✔	83.01	65.97	**73.15**	**73.52**	4.44	2.38

6. Discussion

The above experiments and ablation studies verify the effectiveness of Lite-YOLOv5. We can transplant it to the embedded platform NVIDIA Jetson TX2 on the SAR satellite for on-board SAR ship detection. The combination of six optimization characters (i.e., L-CSP, network pruning, HPBC, SDC, CSA, and H-SPP) guarantee the advanced on-board ship detection performance. As for the on-board processing, firstly, we cut the large-scale SAR imagery into 800 pixels × 800 pixels image patches without embellishment. Then, we conduct ship detection using Lite-YOLOv5. Finally, the obtained detection results on patches are coordinate mapped to obtain the final large-scale SAR ship results. In this way, only the ship sub-images and corresponding coordinates will be transmitted to the ground station, which is of great significance to utilize real-time and accurate ship information, especially in emergencies.

In addition, the all of the above show that Lite-YOLOv5 possesses an advanced on-board SAR ship detection performance. In order to obtain better and faster ship detection results, the follow-up work will need to explore the reasonable hardware acceleration strategy of the platform. Aiming at giving full play to the computing power of the NVIDIA Jetson TX2 hardware, we will allocate each module to the appropriate hardware to maximize the computing efficiency and obtain more efficient detection results. In addition, there are many other feasible schemes in lightweight model design (such as knowledge distillation). Therefore, our future work will explore distillation techniques.

7. Conclusions

This paper proposes a lightweight on-board SAR ship detector called Lite-YOLOv5, which (1) reduces the model volume; (2) decreases the floating-point operations (FLOPs), and (3) guarantees the on-board ship detection without sacrificing accuracy. First, two characteristics are used to obtain a lightweight network, i.e., (1) a LCB module is inserted into the backbone network of YOLOv5 and (2) network pruning is applied to obtain a more compact model. Then, four characteristics are used to guarantee the detection accuracy, i.e., (1) an HPCB module to effectively exclude pure background samples and suppress the false alarms; (2) a SDC method to generate superior priori anchor; (3) a CSA model to enhance the SAR ships semantic feature extraction ability; an (4) an H-SPP model to increase the context information of the receptive field. To evaluate the on-board SAR ship detection ability of Lite-YOLOv5, we also transplanted it to the embedded platform NVIDIA Jetson TX2. Experimental results on the Large-Scale SAR Ship Detection Dataset-v1.0 (LS-SSDD-v1.0) show that Lite-YOLOv5 can realize a lighter architecture with a 2.38 M model volume (14.18% of the model size of YOLOv5), on-board ship detection with a low computation cost (26.59% of FLOPs of YOLOv5), and superior detection accuracy (1.51% F1 improvement compared with YOLOv5). We also conducted a large quantity of ablation experiments to verify the effectiveness of the proposed modules. Thus, Lite-YOLOv5 can provide high-performance on-board SAR ship detection, which is of great significance.

In the future, our works will be as follows:

1. We will decrease the detection time further.
2. We will lighten the detector further without sacrificing the accuracy.
3. We will explore a reasonable hardware acceleration scheme for on-board SAR ship detection.
4. We will explore other viable approaches such as distillation techniques in the following lightweight detector design.

Author Contributions: Conceptualization, X.X.; methodology, X.X.; software, X.X.; validation, T.Z.; formal analysis, T.Z.; investigation, X.X.; resources, X.X.; data curation, X.X.; writing—original draft preparation, X.X.; writing—review and editing, T.Z.; visualization, T.Z.; supervision, T.Z.; project administration, T.Z.; funding acquisition, X.Z. All authors have read and agreed to the published version of the manuscript.

Funding: This research received no external funding.

Institutional Review Board Statement: Not applicable.

Informed Consent Statement: Not applicable.

Data Availability Statement: No new data were created or analyzed in this study. Data sharing is not applicable to this article. The LS-SSDD-v1.0 dataset provided by Tianwen Zhang is available from https://github.com/TianwenZhang0825/LS-SSDD-v1.0-OPEN (accessed on 6 February 2022) to download for scientific research.

Acknowledgments: The authors would like to thank the editors and anonymous reviewers for their valuable comments that greatly improved our manuscript.

Conflicts of Interest: The authors declare no conflict of interest.

Appendix A

For the reader's convenience, in Table A1 we list all of the abbreviations and corresponding full name involved in this paper. The abbreviations are arranged in alphabetical order.

Table A1. The abbreviations and corresponding full names.

Abbreviation	Full Name
AP	average precision
BN	batch normalization
Conv	convolution
CFAR	constant false alarm rate
CNN	convolutional neural network
CSA	channel and spatial attention
CSP	cross stage partial
CV	computer vision
DAPN	dense attention pyramid network
DL	deep learning
DS-CNN	depth-wise separable convolution neural network
FFEN	fusion feature extraction network
FL	focal loss
FLOPs	floating point operations
FPN	feature pyramid network
GAP	global average pooling
GMP	global max pooling
HR-SDNet	high-resolution ship detection network
HNM	hard negative mining
HPBC	histogram-based pure backgrounds classification
H-SPP	hybrid spatial pyramid pooling
L-Conv	lightweight convolution
L-CSP	lightweight cross stage partial
LFO-Net	lightweight feature optimization network
L-Relu	Leaky_ReLu
LS-SSDD-v1.0	Large-Scale SAR Ship Detection Dataset-v1.0
MLP	multi-layer perceptron
NLP	natural language processing
PAN	path aggregation network
RDN	refined detection network
RPN	region proposal network
SAR	synthetic aperture radar
SDC	shape distance clustering
SGD	stochastic gradient descent
SPP	spatial pyramid pooling
YOLOv5	You Only Look Once version 5

References

1. Zhang, T.; Zhang, X.; Li, J.; Xu, X.; Wang, B.; Zhan, X.; Xu, Y.; Ke, X.; Zeng, T.; Su, H.; et al. SAR Ship Detection Dataset (SSDD): Official Release and Comprehensive Data Analysis. *Remote Sens.* **2021**, *13*, 3690. [CrossRef]
2. Lin, Z.; Ji, K.; Leng, X.; Kuang, G. Squeeze and Excitation Rank Faster R-CNN for Ship Detection in SAR Images. *IEEE Geosci. Remote Sens. Lett.* **2019**, *16*, 751–755. [CrossRef]
3. Xu, X.; Zhang, X.; Zhang, T. Multi-Scale SAR Ship Classification with Convolutional Neural Network. In Proceedings of the IEEE International Geoscience and Remote Sensing Symposium (IGARSS), Online Event, 11–16 July 2021; pp. 4284–4287.
4. Zhang, T.; Zhang, X.; Ke, X. Quad-FPN: A Novel Quad Feature Pyramid Network for SAR Ship Detection. *Remote Sens.* **2021**, *13*, 2771. [CrossRef]
5. O'Shea, T.; Hoydis, J. An Introduction to Deep Learning for the Physical Layer. *IEEE Trans. Cogn. Commun. Netw.* **2017**, *3*, 563–575. [CrossRef]
6. Aceto, G.; Ciuonzo, D.; Montieri, A.; Pescapé, A. Mobile Encrypted Traffic Classification Using Deep Learning: Experimental Evaluation, Lessons Learned, and Challenges. *IEEE Trans. Netw. Serv. Manag.* **2019**, *16*, 445–458. [CrossRef]

7. Liu, G.; Li, L.; Jiao, L.; Dong, Y.; Li, X. Stacked Fisher autoencoder for SAR change detection. *Pattern Recognit.* **2019**, *96*, 106971. [CrossRef]

8. Ciuonzo, D.; Carotenuto, V.; De Maio, A. On Multiple Covariance Equality Testing with Application to SAR Change Detection. *IEEE Trans. Signal Process.* **2017**, *65*, 5078–5091. [CrossRef]

9. Kang, M.; Ji, K.; Leng, X.; Lin, Z. Contextual Region-Based Convolutional Neural Network with Multilayer Fusion for SAR Ship Detection. *Remote Sens.* **2017**, *9*, 860. [CrossRef]

10. Jiao, J.; Zhang, Y.; Sun, H.; Yang, X.; Gao, X.; Hong, W.; Fu, K.; Sun, X. A Densely Connected End-to-End Neural Network for Multiscale and Multiscene SAR Ship Detection. *IEEE Access.* **2018**, *6*, 20881–20892. [CrossRef]

11. Cui, Z.; Li, Q.; Cao, Z.; Liu, N. Dense Attention Pyramid Networks for Multi-Scale Ship Detection in SAR Images. *IEEE Trans. Geosci. Remote Sens.* **2019**, *57*, 8983–8997. [CrossRef]

12. Liu, N.; Cao, Z.; Cui, Z.; Pi, Y.; Dang, S. Multi-Scale Proposal Generation for Ship Detection in SAR Images. *Remote Sens.* **2019**, *11*, 526. [CrossRef]

13. Wang, J.; Lu, C.; Jiang, W. Simultaneous Ship Detection and Orientation Estimation in SAR Images Based on Attention Module and Angle Regression. *Sensors* **2018**, *18*, 2851. [CrossRef]

14. An, Q.; Pan, Z.; Liu, L.; You, H. DRBox-v2: An Improved Detector With Rotatable Boxes for Target Detection in SAR Images. *IEEE Trans. Geosci. Remote Sens.* **2019**, *57*, 8333–8349. [CrossRef]

15. Chen, C.; Hu, C.; He, C.; Pei, H.; Pang, Z.; Zhao, T. SAR Ship Detection Under Complex Background Based on Attention Mechanism. In *Image and Graphics Technologies and Applications*; Springer: Singapore, 2019; pp. 565–578.

16. Dai, W.; Mao, Y.; Yuan, R.; Liu, Y.; Pu, X.; Li, C. A Novel Detector Based on Convolution Neural Networks for Multiscale SAR Ship Detection in Complex Background. *Sensors* **2020**, *20*, 2547. [CrossRef] [PubMed]

17. Wei, S.; Su, H.; Ming, J.; Wang, C.; Yan, M.; Kumar, D.; Shi, J.; Zhang, X. Precise and Robust Ship Detection for High-Resolution SAR Imagery Based on HR-SDNet. *Remote Sens.* **2020**, *12*, 167. [CrossRef]

18. Chang, Y.-L.; Anagaw, A.; Chang, L.; Wang, Y.C.; Hsiao, C.-Y.; Lee, W.-H. Ship Detection Based on YOLOv2 for SAR Imagery. *Remote Sens.* **2019**, *11*, 786. [CrossRef]

19. Zhang, T.; Zhang, X.; Shi, J.; Wei, S. Depthwise Separable Convolution Neural Network for High-Speed SAR Ship Detection. *Remote Sens.* **2019**, *11*, 2483. [CrossRef]

20. Mao, Y.; Yang, Y.; Ma, Z.; Li, M.; Su, H.; Zhang, J. Efficient Low-Cost Ship Detection for SAR Imagery Based on Simplified U-Net. *IEEE Access.* **2020**, *8*, 69742–69753. [CrossRef]

21. Zhang, X.; Wang, H.; Xu, C.; Lv, Y.; Fu, C.; Xiao, H.; He, Y. A Lightweight Feature Optimizing Network for Ship Detection in SAR Image. *IEEE Access.* **2019**, *7*, 141662–141678. [CrossRef]

22. Wang, Y.; Wang, C.; Zhang, H.; Dong, Y.; Wei, S. Automatic Ship Detection Based on RetinaNet Using Multi-Resolution Gaofen-3 Imagery. *Remote Sens.* **2019**, *11*, 531. [CrossRef]

23. Wang, Y.; Wang, C.; Zhang, H.; Dong, Y.; Wei, S. A SAR Dataset of Ship Detection for Deep Learning under Complex Backgrounds. *Remote Sens.* **2019**, *11*, 765. [CrossRef]

24. Zhang, T.; Zhang, X.; Ke, X.; Zhan, X.; Shi, J.; Wei, S.; Pan, D.; Li, J.; Su, H.; Zhou, Y.; et al. LS-SSDD-v1.0: A Deep Learning Dataset Dedicated to Small Ship Detection from Large-Scale Sentinel-1 SAR Images. *Remote Sens.* **2020**, *12*, 2997. [CrossRef]

25. Xu, P.; Li, Q.; Zhang, B.; Wu, F.; Zhao, K.; Du, X.; Yang, C.; Zhong, R. On-Board Real-Time Ship Detection in HISEA-1 SAR Images Based on CFAR and Lightweight Deep Learning. *Remote Sens.* **2021**, *13*, 1995. [CrossRef]

26. Han, K.; Wang, Y.H.; Tian, Q.; Guo, J.Y.; Xu, C.J.; Xu, C. GhostNet: More Features from Cheap Operations. In Proceedings of the IEEE Conference on Computer Vision and Pattern Recognition (CVPR), Seattle, WA, USA, 14–19 June 2020; pp. 1577–1586.

27. Ultralytics. YOLOv5. Available online: https://github.com/ultralytics/yolov5 (accessed on 1 November 2021).

28. Liu, Z.; Li, J.G.; Shen, Z.Q.; Huang, G.; Yan, S.M.; Zhang, C.S. Learning Efficient Convolutional Networks through Network Slimming. In Proceedings of the IEEE International Conference on Computer Vision (ICCV), Venice, Italy, 22–29 October 2017; pp. 2755–2763.

29. Woo, S.; Park, J.; Lee, J.-Y.; Kweon, I.S. CBAM: Convolutional Block Attention Module. *arXiv* **2018**, arXiv:1807.06521.

30. He, K.M.; Zhang, X.Y.; Ren, S.Q.; Sun, J. Spatial Pyramid Pooling in Deep Convolutional Networks for Visual Recognition. In Proceedings of the European Conference on Computer Vision (ECCV), Zurich, Switzerland, 6–12 September 2014; pp. 346–361.

31. Lin, T.Y.; Maire, M.; Belongie, S.; Hays, J.; Perona, P.; Ramanan, D.; Dollar, P.; Zitnick, C.L. Microsoft COCO: Common Objects in Context. In Proceedings of the 13th European Conference on Computer Vision (ECCV), Zurich, Switzerland, 6–12 September 2014; pp. 740–755.

32. Xu, R.; Lin, H.; Lu, K.; Cao, L.; Liu, Y. A Forest Fire Detection System Based on Ensemble Learning. *Forests* **2021**, *12*, 217. [CrossRef]

33. Ioffe, S.; Szegedy, C. Batch normalization: Accelerating deep network training by reducing internal covariate shift. In Proceedings of the 32nd International Conference on International Conference on Machine Learning (ICML), Lile, France, 6–11 July 2015; pp. 448–456.

34. Mastromichalakis, S. ALReLU: A different approach on Leaky ReLU activation function to improve Neural Networks Performance. *arXiv* **2020**, arXiv:2012.07564.

35. Wang, C.Y.; Liao, H.Y.M.; Wu, Y.H.; Chen, P.Y.; Hsieh, J.W.; Yeh, I.H. CSPNet: A new backbone that can enhance learning capability of CNN. In Proceedings of the IEEE/CVF Conference on Computer Vision and Pattern Recognition Workshops, Seattle, WA, USA, 14–19 June 2020; pp. 390–391.

36. Lin, T.; Dollár, P.; Girshick, R.; He, K.; Hariharan, B.; Belongie, S. Feature Pyramid Networks for Object Detection. In Proceedings of the 2017 IEEE Conference on Computer Vision and Pattern Recognition (CVPR), Honolulu, HI, USA, 21–26 July 2017; pp. 936–944.
37. Liu, S.; Qi, L.; Qin, H.; Shi, J.; Jia, J. Path Aggregation Network for Instance Segmentation. In Proceedings of the 2018 IEEE/CVF Conference on Computer Vision and Pattern Recognition, Salt Lake City, UT, USA, 18–23 June 2018; pp. 8759–8768.
38. Scardapane, S.; Comminiello, D.; Hussain, A.; Uncini, A. Group sparse regularization for deep neural networks. *Neurocomputing* **2017**, *241*, 81–89. [CrossRef]
39. Chen, S.; Zhan, R.; Wang, W.; Zhang, J. Learning Slimming SAR Ship Object Detector Through Network Pruning and Knowledge Distillation. *IEEE J. Sel. Top. Appl. Earth Obs. Remote Sens.* **2021**, *14*, 1267–1282. [CrossRef]
40. Gao, G. Statistical Modeling of SAR Images: A Survey. *Sensors* **2010**, *10*, 775–795. [CrossRef]
41. Wackerman, C.C.; Friedman, K.S.; Pichel, W.; Clemente-Colón, P.; Li, X. Automatic detection of ships in RADARSAT-1 SAR imagery. *Can. J. Remote Sens.* **2001**, *27*, 568–577. [CrossRef]
42. Ferrara, M.N.; Torre, A. Automatic moving targets detection using a rule-based system: Comparison between different study cases. In Proceedings of the IEEE International Geoscience and Remote Sensing Symposium (IGARSS), Seattle, WA, USA, 6–10 July 1998; pp. 1593–1595.
43. Gagnon, L.; Oppenheim, H.; Valin, P. R&D activities in airborne SAR image processing/analysis at Lockheed Martin Canada. *Proc. SPIE Int. Soc. Opt. Eng.* **1998**, *3491*, 998–1003.
44. Chen, P.; Li, Y.; Zhou, H.; Liu, B.; Liu, P. Detection of Small Ship Objects Using Anchor Boxes Cluster and Feature Pyramid Network Model for SAR Imagery. *J. Mar. Sci. Eng.* **2020**, *8*, 112. [CrossRef]
45. Hu, J.; Shen, L.; Sun, G. Squeeze-and-excitation networks. *arXiv* **2017**, arXiv:1709.01507.
46. Lin, M.; Chen, Q.; Yan, S. Network in Network. *arXiv* **2013**, arXiv:1312.4400.
47. Khan, A.; Sohail, A.; Zahoora, U.; Qureshi, A.S. A survey of the recent architectures of deep convolutional neural networks. *Artif. Intell. Rev.* **2020**, *53*, 5455–5516. [CrossRef]
48. Wu, X.W.; Sahoo, D.; Hoi, S.C.H. Recent advances in deep learning for object detection. *Neurocomputing* **2020**, *396*, 39–64. [CrossRef]
49. Huang, Z.C.; Wang, J.L. DC-SPP-YOLO: Dense connection and spatial pyramid pooling based YOLO for object detection. *Inf. Sci.* **2020**, *522*, 241–258. [CrossRef]
50. Ketkar, N. Introduction to Pytorch. In *Deep Learning with Python: A Hands-On Introduction*; Apress: Berkeley, CA, USA, 2017; pp. 195–208. Available online: https://link.springer.com/chapter/10.1007/978-1-4842-2766-4_12 (accessed on 1 December 2021).
51. Gao, S.; Liu, J.M.; Miao, Y.H.; He, Z.J. A High-Effective Implementation of Ship Detector for SAR Images. *IEEE Geosci. Remote Sens. Lett.* **2021**, *19*, 1–5. [CrossRef]
52. Zhang, F.; Zhou, Y.; Zhang, F.; Yin, Q.; Ma, F. Small Vessel Detection Based on Adaptive Dual-Polarimetric Sar Feature Fusion and Attention-Enhanced Feature Pyramid Network. In Proceedings of the IEEE International Geoscience and Remote Sensing Symposium (IGARSS), Online Event, 11–16 July 2021; pp. 2218–2221.
53. Zhang, L.; Liu, Y.; Guo, Q.; Yin, H.; Li, Y.; Du, P. Ship Detection in Large-scale SAR Images Based on Dense Spatial Attention and Multi-level Feature Fusion. In Proceedings of the ACM Turing Award Celebration Conference—China (ACM TURC 2021), Hefei, China, 30 July–1 August 2021; Association for Computing Machinery: Hefei, China, 2021; pp. 77–81.
54. Zhang, X.; Huo, C.; Xu, N.; Jiang, H.; Cao, Y.; Ni, L.; Pan, C. Multitask Learning for Ship Detection From Synthetic Aperture Radar Images. *IEEE J. Sel. Top. Appl. Earth Obs. Remote Sens.* **2021**, *14*, 8048–8062. [CrossRef]
55. Sergios, T. Stochastic gradient descent. *Mach. Learn.* **2015**, *5*, 161–231.
56. Pang, J.; Chen, K.; Shi, J.; Feng, H.; Ouyang, W.; Lin, D. Libra R-CNN: Towards balanced learning for object detection. *arXiv* **2019**, arXiv:1904.02701.
57. Ren, S.; He, K.; Girshick, R.; Sun, J. Faster R-CNN: Towards real-time object detection with region proposal networks. *IEEE Trans. Pattern Anal. Mach. Intell.* **2017**, *39*, 1137–1149. [CrossRef]
58. Tan, M.; Pang, R.; Le, Q.V. EfficientDet: Scalable and efficient object detection. *arXiv* **2019**, arXiv:1911.09070.
59. Zhang, X.; Wan, F.; Liu, C.; Ye, Q. FreeAnchor: Learning to match anchors for visual object detection. *arXiv* **2019**, arXiv:1909.02466.
60. Kong, T.; Sun, F.; Liu, H.; Jiang, Y.; Li, L.; Shi, J. FoveaBox: Beyond anchor-based object detector. *arXiv* **2019**, arXiv:1904.03797.
61. Lin, T.Y.; Goyal, P.; Girshick, R.; He, K.; Dollár, P. Focal loss for dense object detection. In Proceedings of the International Conference on Computer Vision (ICCV), Venice, Italy, 22–29 October 2017; pp. 2999–3007.
62. Liu, W.; Anguelov, D.; Erhan, D.; Szegedy, C.; Reed, S.; Fu, C.; Berg, A. SSD: Single shot multibox detector. In Proceedings of the 14th European Conference on Computer Vision (ECCV), Amsterdam, The Netherlands, 11–14 October 2016; pp. 21–37.

Article

SDTGAN: Generation Adversarial Network for Spectral Domain Translation of Remote Sensing Images of the Earth Background Based on Shared Latent Domain

Biao Wang [1], Lingxuan Zhu [2,*], Xing Guo [1], Xiaobing Wang [2] and Jiaji Wu [1]

[1] School of Electronic Engineering, Xidian University, Xi'an 710071, China; wangbiao@stu.xidian.edu.cn (B.W.); guox@xidian.edu.cn (X.G.); wujj@mail.xidian.edu.cn (J.W.)
[2] Science and Technology on Electromagnetic Scattering Laboratory, Shanghai 200438, China; wxb218@sina.com
*   Correspondence: yubaiscat@outlook.com

Abstract: The synthesis of spectral remote sensing images of the Earth's background is affected by various factors such as the atmosphere, illumination and terrain, which makes it difficult to simulate random disturbance and real textures. Based on the shared latent domain hypothesis and generation adversarial network, this paper proposes the SDTGAN method to mine the correlation between the spectrum and directly generate target spectral remote sensing images of the Earth's background according to the source spectral images. The introduction of shared latent domain allows multi-spectral domains connect to each other without the need to build a one-to-one model. Meanwhile, additional feature maps are introduced to fill in the lack of information in the spectrum and improve the geographic accuracy. Through supervised training with a paired dataset, cycle consistency loss, and perceptual loss, the uniqueness of the output result is guaranteed. Finally, the experiments on the Fengyun satellite observation data show that the proposed SDTGAN method performs better than the baseline models in remote sensing image spectrum translation.

Keywords: remote sensing image; spectral domain translation; generative adversarial network; paired translation

Citation: Wang, B.; Zhu, L.; Guo, X.; Wang, X.; Wu, J. SDTGAN: Generation Adversarial Network for Spectral Domain Translation of Remote Sensing Images of the Earth Background Based on Shared Latent Domain. *Remote Sens.* **2022**, *14*, 1359. https://doi.org/10.3390/rs14061359

Academic Editors: Saeid Homayouni and Claudio Piciarelli

Received: 19 January 2022
Accepted: 8 March 2022
Published: 11 March 2022

Publisher's Note: MDPI stays neutral with regard to jurisdictional claims in published maps and institutional affiliations.

1. Introduction

Remote sensing images are widely used in environmental monitoring, remote sensing analysis, and target detection and classification. However, in practical applications, it is difficult to obtain multi-spectral remote sensing data, especially high-resolution infrared remote sensing data, and spectrally poor data may be available for longer periods of time than spectrally rich data [1]. Many researchers have explored the acquisition of demanded spectral remote sensing images based on simulation methods [2–4]. The spectral characteristics are determined by the optical characteristics of the underlying surface type, atmosphere, sunlight, and terminal sensors [5]. The traditional methods based on radiation transfer models [6–8] require pre-building a large database of ground features and environmental characteristics. However, it is still difficult to model the complex and random atmosphere and clouds. When the input condition is insufficient for simulating the images of earth background, based on the correlation between the spectral domains, the known spectral images can be used to achieve target spectral image synthesis [9–11]. However, the correlation between the spectral domains is implicit and non-linear.

As deep learning technology can obtain feature correlations in complex spaces through a large amount of data to realize end-to-end image generation, generative adversarial networks (GAN) have achieved rapid development in recent years [12], from the initial supervised image translation [13–15] to the subsequent unsupervised image translation [16] and the later multi-modal image translation [17]. Domain adaptation is critical for the successful application of neural network models in new, unseen environments [18]. Many tasks that support translation from one domain to another have achieved excellent results. Spectral domain translation refers to generating an image of the target spectral do-main based on the image of the source spectral domain while ensuring that each pixel of the generated image conforms to the physical mapping relationship. In the field of spectral imaging, super-resolution [19], spectral reconstruction [20,21], and spectral fusion [22,23] have successively adopted the GAN technology. Kongxin Tang et al. [24] used generative adversarial networks to achieve RGB visualization through hyperspectral images. The method reduces the dimensionality of the spectral data from tens to hundreds to three dimensions (RGB). CHENG Wencong [25] combined satellite infrared images and numerical weather prediction (NWP) products to generate adversarial network based on conditions. Then, night satellite visible-light images were synthesized. However, this method is limited to the field of view specified by the data set, and it is difficult to express the underlying surface stably and accurately. In cross-domain research, GANs are used for image fusion of SAR images, infrared images, and visible-light images [22,23,26]. This type of method combines the source-domain data with different characteristics to synthesize a fusion image that is easy to understand.

Hyperspectral image reconstruction is an example of spectral-domain translation [20,21]. Arad et al. [27] collected hyperspectral data and built a sparse hyperspectral dictionary based on the sparse dictionary. Then, they used it as prior information to map the RGB image to the spectral image. These methods usually learn a nonlinear mapping from RGB to hyperspectral images based on a large amount of training data. Wu, J et al. [21] applied hyperspectral reconstruction based on super-resolution technology. Pengfei Liu et al. [28] proposed a generative adversarial model based on a convolution neural network for hyperspectral reconstruction from a single RGB image.

Although these methods have achieved satisfactory results in image-to-image translation, they still cannot be directly applied to the spectral domain translation of remote sensing images mainly due to the following limitations.

1. The location accuracy of the surface area: The cloud and water vapor will shield the earth's surface in the remote sensing image and affect the transmittance of the atmospheric radiation, resulting in the incompleteness of the surface boundary and misjudgment of features in the image. Based on a single source of remote sensing spectral data, it is difficult to deduce the true surface under cloud cover and atmospheric transmittance fluctuations.
2. Limitations of spectral characterization information: The physical characteristics expressed by each spectrum are different. For example, the band of 3.5~4.0 microns can filter water vapor to observe the surface, while the band of 7 microns can only show water vapor and clouds. Due to the differences in the information of different spectral images, even with spatio-temporal matching datasets, datasets, it is difficult to realize the information migration or speculation between the bands with significant differences.
3. Spectral translation accuracy: Computational vision tasks often focus on the similarity of image styles in different domains and encourage the diversity of synthesis effects. However, spectral translation tasks require the conversion of pixels under the same input conditions between different spectral images. The result is unique and conforms to physical characteristics.

To overcome the limitations mentioned above, this paper explores the spectral translation by introducing the conditional GAN, which focuses on the migration and amplification of a small amount of spectral data to multi-dimensional data.

As shown in Figure 1, the translation task into two steps: the first step is to encode the source spectral domain image and add additional feature maps to the shared latent domain through the source domain encoder. The second step is to decode the shared latent domain code to the target spectral domain through the target domain decoder.

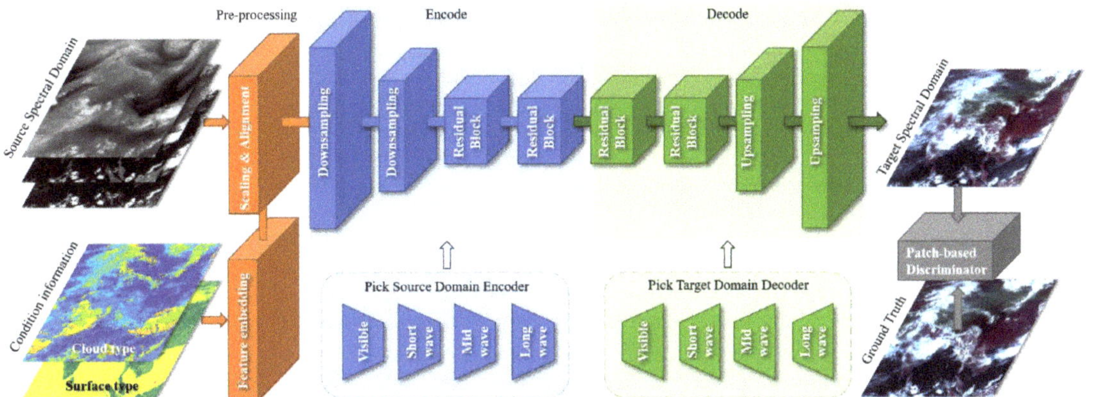

Figure 1. The framework of SDTGAN.

The main contributions of this paper are as follows:

1. The introduction of shared latent domain: Through cross domain translation and within domain self-reconstruction training, the shared latent domain fits the joint probability distribution of the multi-spectral domain, and can parse and store diversity and the characteristics of each spectral domain. It is the end of encoders and the beginning of the decoders of all spectral domains. In this way, the parameter expansion problem of many-to-many translation is avoided.

2. The introduction of multimodal feature map: By introducing discrete feature (e.g., surface type and cloud map type), and numerical feature maps (e.g., surface temperature), the location accuracy of the surface area is improved, and the limitations of spectral characterization information are overcome.

3. The training is conducted on the supervised spatio-temporal matching data sets, combined with cycle consistency loss and perceptual loss, to ensure the uniqueness of the output result and improve spectral translation accuracy.

The structure of this paper is as follows: In Section 2, the structure and loss functions of the GAN used in this study are introduced. In Section 3, the building of the datasets and the experiments to evaluate different methods are elaborated. Finally, future work and conclusions are given in Section 4.

2. Materials and Methods

We begin with an overview of the spectral domain translation method, and then, the basic assumptions and model architectures are introduced. Finally, the loss function of networks and the training process are described.

2.1. Overview of the Method

In this work, a multi-spectral domain translation generation adversarial model is proposed for remote sensing images. Following the basic framework of image-conditional GANs [12], the model has an independent encoder E, a decoder G, and a discriminator D for each spectral domain. The difference is that the model assumes the existence of a shared latent domain, which make it possible to encode each spectral domain into that space and reconstruction of information from that space.

In the training process, the shared latent domain is constructed in two ways. First, the source domain spectral image and the target domain spectral image are encoded to the feature matrix with the same size. The training with L1 loss makes the encoded feature matrix consistent across spectral domains. Second, in within domain training, the source and target domains use their encoders and decoders to achieve image reconstruction from the feature matrix. In cross domain training, the feature matrices output from the source and target domain encoders are exchanged, and then the images are reconstructed following the above steps. The purpose of this step is to enable decoders in different spectral domains to obtain the information needed for their reconstruction from the shared latent domain.

During the test, there is no need to reload all the encoders and decoders. Only the combination of encoders for the source domain spectrum and the combination of decoders for the target domain need to be loaded. The feature matrix is generated by the encoder in the source domain, and then the spectral image is generated by the decoder in the target domain.

Since all encoding and decoding is based on the shared latent domain, the set of spectral domains of the model can be continuously expanded. When a new spectral domain is added, it is only necessary to ensure that the encoder of the new spectral domain can make the image output to the shared latent domain and the decoder can recover its own image from that space.

Meanwhile, the model can add additional physical property information to improve the simulation accuracy. For remote sensing imaging, the underlying surface and clouds are the main influencing factors of optical radiation transfer. Therefore, earth surface classification data R^{GT} and cloud classification data R^{CLT} are used as feature maps to form the boundary conditions of the scene.

2.2. Shared Latent Domain Assumption

Let $x_i \in \chi_i$ be the spectral images from spectral domain χ_i, and there are N spectral domains. Let $r \in R$ be the condition information of image boundary condition R. Our goal is to estimate the conditional distribution $p\big(x_i\big|(x_j,r)\big)$ between domains i and j with a learned deterministic mapping function $p\big(x_{j\to i}\big|(x_j,r)\big)$ and the joint distribution $p(x_1,x_2,\cdots,x_N,r)$.

To translate from one spectral domain to multiple spectral domains, this study makes a fully shared latent space assumption [17,29]. It is assumed that each spectral image x_i is generated from a latent code $s \in S$ that is shared by all spectral domains and conditional information. Using the shared latent code domain as a bridge, spectral image x_i can be synthesized by decoder $G_i^*(s)$, and the joint probability distribution s can be obtained by encoder $E_i^*(x_i,r)$, so that $E_i^*(x_i,r) = \big(G_i^*(s)\big)^{-1} = s$.

2.3. Architecture

As shown in Figure 1, the encoder–decoder-discriminator pair constitutes the SDT-GAN model. Considering that the intrinsic information of an image is shared among multiple spectral domains, the output matrix dimensions of the encoder of all spectral-domain models are consistent.

In the process of image translation from source spectral domain χ_i to target spectral domain χ_j, the source-domain encoder E_i is selected from the encoder library, and the target-domain decoder G_j is selected from the decoder library. The encoder E_i maps the input matrix to the shared latent code s, and the decoder G_j reconstructs the target spectral-domain image from the latent code s. Then, the adversarial loss is calculated by the target-domain discriminator D_j. Since the latent code is shared in each spectral domain, the latent code generated by the source spectral-domain encoding can be decoded into multiple codes in the target spectral domain. Meanwhile, the input matrices need to be preprocessed, including the source spectral image matrix and the condition information matrix after feature embedding.

2.3.1. Generative Network

The generative network is based on the architecture proposed by Johnson et al. [30]. The encoder consists of a set of stride-2 downsampling and convolutional layers and several residual blocks. The decoder processes the latent code by a set of residual blocks and then restores the image size through 1/2-strided upsampling and convolutional layers.

2.3.2. Patch Based Discriminator Network

The patch discriminator with different fields of view is used [31,32]. The discriminator outputs a predicted probability value for each area (patch) of the input image. Evolving from judging whether the input is true or false, patch discriminator judges whether the input area with a size of N × N is true or false. The discriminator with a large perceptual field ensures the consistency of geographic location, and discriminator with a small perceptual field ensures the characteristics of texture details.

2.3.3. Feature Embedding

The information carried by spectral images with few bands is limited. For instance, the earth's surface is seriously obscured in the water vapor bands. In the process of spectrum translation, it is difficult for the model to accurately derive the surface structure. To address this issue, feature maps are added to the input matrix to fill the lack of information in the spectrum.

Remote sensing image features include discrete features and numerical features. The semantic labels of pixels such as land surface type and cloud cover type are discrete features; the quantitative information of pixel areas such as land surface temperature and cloud cover rate are numerical features. For discrete features, this study pre-allocates a fixed number of channels for each category, and encodes the label as a one-hot vector. For numeric features, this study pre-sets the interval of the upper and lower limits of the value, and then normalizes the value to [0, 1]. Then, the size of the feature map is adjusted to that of the spectral image. Finally, the feature matrix and the spectral matrix are combined and input to the encoder.

2.4. Loss Function

Based on the paired dataset, this study introduces the bidirectional reconstruction loss [29] to achieve the reversibility of the encoding and decoding processes and reduce the redundant function mapping space. Meanwhile, this study adopts the objective function to make all encoders output to the same latent space, and the images of various spectral domains can be reconstructed from the latent space. Figure 2 shows the training flow of the loss function for with-domain and cross-domain.

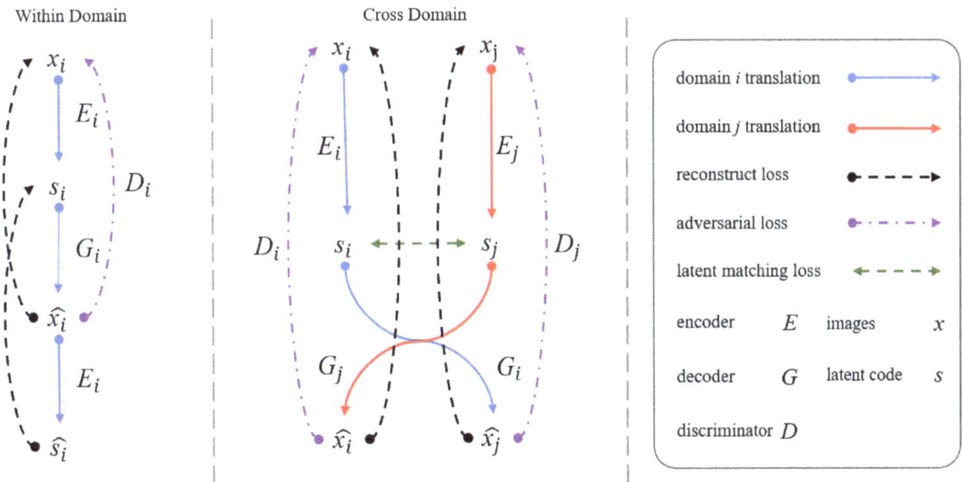

Figure 2. The within-domain and cross-domain training flowchart.

2.4.1. Reconstruction loss

Based on the reversibility of the encoder and decoder, an objective function that enables the cycle consistency of image and feature coding is constructed. The use of cross-domain reconstruction consistency constrains the spatial diversity of the encoding and decoding between multiple domains, and it stabilizes spectral-domain translation results. Previous studies have found adding reconstruction loss with L1 loss is conducive to reducing the probability of model collapse [13,29].

1. Within domain reconstruction loss

Given an image sampled from the data distribution, its spectral image and latent code after encoding and decoding can be reconstructed, and the within-domain reconstruction loss can be defined as:

$$
L_{\text{Recon}}^{x_i} = \lambda_{\text{I}} \mathbb{E}_{x_i, r \sim p(x_i, r)} \left[\| G_i(E_i(x_i, r)) - x_i \|_1 \right] + \lambda_{\text{E}} \mathbb{E}_{x_i, r \sim p(x_i, r)} \left[\| E_i(G_i(E_i(x_i, r))) - E_i(x_i, r) \|_1 \right] \tag{1}
$$

where λ_{I} and λ_{E} are the weights that control the importance of image reconstruction term and latent code reconstruction term, respectively.

2. Cross domain reconstruction loss

Given two spectra domain images sampled from the joint data distribution, their spectral images after exchanging encoding and decoding can be reconstructed, and its cross domain reconstruction loss can be defined as:

$$
L_{\text{Recon}}^{x_i, x_j} = \mathbb{E}_{x_i, x_j, r \sim p(x_i, x_j, r)} \left[\| G_i(E_j(x_j, r)) - x_i \|_1 \right] + \mathbb{E}_{x_i, x_j, r \sim p(x_i, x_j, r)} \left[\| G_j(E_i(x_i, r)) - x_j \|_1 \right] \tag{2}
$$

$$
L_{\text{Recon}} = \lambda_{\text{within}} \sum_{i=1}^{N} L_{\text{Recon}}^{x_i} + \lambda_{\text{cross}} \sum_{i=1}^{N} \sum_{j=i+1}^{N} L_{\text{Recon}}^{x_i, x_j} \tag{3}
$$

where L_{Recon} is the sum of the multi-domain reconstruction loss; λ_{within} and λ_{cross} are weights that control the importance of the within-domain reconstruction term and cross-domain reconstruction term, respectively.

2.4.2. Latent Matching loss

Given two multi-spectral images sampled from the same data distribution, the latent code should be matched after encoding. In previous work, auto-encoders and GANs use KLD loss, and adversarial loss [16,33] or implicitly constrain [29] the latent domain distribution. The present model uses the calculation of L1 Loss across domains to strongly constrain different domains to encode in the same space.

$$
L_{LM}^{x_i, x_j} = \mathbb{E}_{x_i, x_j, r \sim p(x_i, x_j, r)} \left[\| E_i(x_i, r) - E_j(x_j, r) \|_1 \right] \tag{4}
$$

$$
L_{\text{Match}} = \sum_{i=1}^{N} \sum_{j=i+1}^{N} L_{LM}^{x_i, x_j} \tag{5}
$$

where $L_{LM}^{x_i, x_j}$ is the latent matching loss that depicts the L1 loss between the latent of the i-domain and j-domain images under the joint probability distribution.

2.4.3. Adversarial loss

This study employs Patch-GANs to distinguish between the real images or the images translated from the latent space in the target domain.

$$
L_{\text{GAN}}^{x_i} = \mathbb{E}_{x_i, r \sim p(x_i, r)} \left[\log(1 - D_i(G_i(E_i(x_i, r)))) \right] \\
+ \mathbb{E}_{x_i, r \sim p(x_i, r)} \left[\log(D_i(x_i)) \right] \tag{6}
$$

$$
L_{\text{GAN}}^{x_j \to i} = \mathbb{E}_{x_i, x_j, r \sim p(x_i, x_j, r)} \left[\log(1 - D_i(G_i(E_j(x_j, r)))) \right] \tag{7}
$$

$$L_{\text{GAN}} = \lambda_{\text{cross}}\sum_{i=1}^{N}\sum_{j=i+1}^{N} L_{\text{GAN}}^{x_{j\to i}} + \lambda_{\text{within}}\sum_{i=1}^{N} L_{\text{GAN}}^{x_i} \tag{8}$$

where $L_{\text{GAN}}^{x_i}$ is the within-domain GAN loss of the images sampled from domain i; $L_{\text{GAN}}^{x_{j\to i}}$ is the cross-domain GAN loss that depicts the GAN loss of spectral image translation from domain j to domain i, and L_{GAN} is the sum of multi-domain GAN loss.

2.4.4. Total Loss

In this study, the model is optimized through joint training of encoders, decoders, and discriminators in all spectral domains. The total loss function is the weighted sum of the counter loss, reconstruction loss, and latent matching loss in each spectral domain.

When a new spectral domain is added to the trained multi-spectral domain model, the model parameters of the existing spectral domain can be fixed, and the shared feature space can be exploited to accelerate the training process and avoid the expansion of training parameters.

$$\min_{E,G}\max_{D} L_{\text{Total}}(E,G,D) = \lambda_{\text{GAN}} L_{\text{GAN}} + \lambda_{\text{Recon}} L_{\text{Recon}} + \lambda_{\text{Match}} L_{\text{Match}} \tag{9}$$

where λ_{GAN}, λ_{Recon}, and λ_{Match} are weights that control the importance of loss terms.

2.5. Traning Process

In the initial training of the model, all source and target domains are combined into a set that contains N spectral domains. Then, the SDTGAN model is trained by updating the generators and discriminators alternately, which follows the basic rule of GANs. The training of generators is the key point of the method, and the steps are illustrated in the following Algorithm 1.

Algorithm 1. Generators training process in a single iteration

---

$\textbf{for } i = 1 \text{ to N}$
    $\textbf{for } j = i + 1 \text{ to N}$
        $L_{\text{Recon}}^{x_i} \leftarrow \|G_i(E_i(x_i,r)) - x_i\|_1$
        $L_{\text{Recon}}^{x_j} \leftarrow \|G_j\left(E_j\left(x_j,r\right)\right) - x_j\|_1$
        $L_{\text{Recon}}^{x_i,x_j} \leftarrow \|G_j(E_i(x_i,r)) - x_j\|_1 \, and \, \|G_i\left(E_j\left(x_j,r\right)\right) - x_{is}\|_1$
        $L_{\text{LM}}^{x_i,x_j} \leftarrow \|E_i(x_i,r) - E_j\left(x_j,r\right)\|_1$
        $L_{\text{GAN}}^{x_i} \leftarrow D_i(G_i(E_i(x_i,r)))$
        $L_{\text{GAN}}^{x_j} \leftarrow D_j\left(G_j\left(E_j\left(x_j,r\right)\right)\right)$
        $L_{\text{GAN}}^{x_{i\to j}} \leftarrow D_j\left(G_j(E_i(x_i,r))\right)$
        $L_{\text{GAN}}^{x_{j\to i}} \leftarrow D_i\left(G_i\left(E_j\left(x_j,r\right)\right)\right)$
        L_{Total} update
        Backward gradient decent
        Optimizer update
    $\textbf{end for}$
$\textbf{end for}$

---

3. Experiment

3.1. Datasets

In the experiment, the paired data consists of spectral remote sensing images of earth background and condition information data including earth surface type data and cloud type data. For paired data of spectral images, the earth coordinates corresponding to each pixel need to be aligned since the task aims to achieve spectral translation at the pixel level of the image. The model establishes the mapping relationship between spectra by learning the intensity mapping relationship originating from the same location and time of the sampled images.

3.1.1. Remote sensing Datasets

This study takes the L1 level data obtained by the multi-channel scanning imaging radiometer of the FY-4A satellite [34,35] as the satellite spectral image data. The FY-4A satellite is a new generation of China's geostationary meteorological satellite, and it is equipped with various observation instruments including the Advanced Geosynchronous Radiation Imager (AGRI). As shown in Table 1, AGRI has 14 spectral bands from visible to infrared (0.45–13.8 μm) with a high spatial resolution (1 km for visible light channels, 2 km for near-infrared channels, and 4 km for remaining infrared channels) and temporal resolution (full-disk images at the 15-min interval).

Table 1. The description of spectral image information.

Channel ID	Description	Band (μm)	Spatial Resolution (km)	Main Application
CH01	Visible & Near-Infrared	0.45~0.49	1	Aerosol
CH02		0.55~0.75	0.5~1	Fog, Cloud
CH03		0.75~0.90	1	Vegetation
CH04	Short-Wave Infrared	1.36~1.39	2	Cirrus
CH05		1.58~1.64	2	Cloud, Snow
CH06		2.1~2.35	2~4	Cirrus, Aerosol
CH07	Mid-Wave Infrared	3.5~4.0 (High)	2	Fire
CH08		3.5~4.0 (Low)	4	Land Surface
CH09	Water Vapor	5.8~6.7	4	Water Vapor
CH10		6.9~7.3	4	Water Vapor
CH11	Long-Wave Infrared	8.0~9.0	4	Water Vapor,
CH12		10.3~11.3	4	Cloud
CH13		11.5~12.5	4	Surface Temperature
CH14		13.2~13.8	4	Surface Temperature

The daily data from January to December 2020 were used for the training process, and the daily data from June to August 2020 were used for testing. The daily data are sampled from 12:00 in the satellite's time zone, to maximize the visible area in the image.

3.1.2. Condition Information Dataset

(1)   Earth surface type

The earth surface type data is obtained from the global land cover maps (Glob-Cover [36]) developed and demonstrated by ESA. The theme legend of GlobCover is compatible with that of the UN Land Cover Classification System (LCCS).

As shown in Table 2, GlobCover is a static global gridded surface type map with a resolution of 300 m and 23 classification types. Since the surface type labels of GlobCover are encoded as one-hot vectors, they are rearranged in this study.

Table 2. The description of the earth surface type label.

Label	Type
0	Post-flooding or irrigated croplands
1	Rainfed croplands
2	Mosaic Cropland (50–70%)/Vegetation (grassland, shrubland, forest) (20–50%)
3	Mosaic Vegetation (grassland, shrubland, forest) (50–70%)/Cropland
4	Closed to open (>15%) broadleaved evergreen and/or semi-deciduous forest (>5 m)
5	Closed (>40%) broad leaved deciduous forest (>5 m)
6	Open (15–40%) broad leaved deciduous forest (>5 m)
7	Closed (>40%) needle leaved evergreen forest (>5 m)
8	Open (15–40%) needle leaved deciduous or evergreen forest (>5 m)
9	Closed to open (>15%) mixed broadleaved and needle leaved forest (>5 m)
10	Mosaic Forest/Shrubland (50–70%)/Grassland (20–50%)

Table 2. *Cont.*

Label	Type
11	Mosaic Grassland (50–70%)/Forest/Shrubland (20–50%)
12	Closed to open (>15%) shrubland (<5 m)
13	Closed to open (>15%) grassland
14	Sparse (>15%) vegetation (woody vegetation, shrubs, grassland)
15	Closed (>40%) broadleaved forest regularly flooded-Fresh water
16	Closed (>40%) broadleaved semi-deciduous and/or evergreen forest regularly flooded-Saline water
17	Closed to open (>15%) vegetation (grassland, shrubland, woody vegetation) on regularly flooded or waterlogged soil-Fresh, brackish or saline water
18	Artificial surfaces and associated areas (urban areas >50%)
19	Bare areas
20	Water bodies
21	Permanent snow and ice
22	No data

(2) Cloud type

The cloud classification data is obtained from the L2 level real-time product of the FY4A satellite. According to the microphysical structure and thermodynamic properties of the cloud, the effective absorption optical thickness ratios of the four visible light channels have different properties. As shown in Table 3, there are 10 categories of cloud type labels included in the image. The sampling moment for cloud classification is the same as that of spectral images, thus forming paired data.

Table 3. The description of the cloud type label.

Label	Type
0	Clear
1	Water Type
2	Super Cooled Type
3	Mixed Type
4	Ice Type
5	Cirrus Type
6	Overlap Type
7	Uncertain
8	Space
9	Fill Number

3.2. Implementation Details

In the experiment, the parameters of the proposed method were fixed. For the network architecture, each encoder contains three convolutional layers for downsampling and three residual blocks for feature extraction. The decoder adopts the symmetric structure of the encoder, including three layers of residual blocks and three layers of upsampling convolutional layers. The discriminators consist of stacks of convolutional layers. Besides, LeakyReLU was used for nonlinearity. The hyper-parameters were set as follows:

$$\lambda_I = 1, \ \lambda_E = 1, \ \lambda_{\text{within}} = 1, \ \lambda_{\text{cross}} = 10, \ \lambda_{\text{GAN}} = 1, \ \lambda_{\text{Recon}} = 1 \text{ and } \lambda_{\text{Match}} = 1.$$

The translation models taken for comparison include the SDTGAN model using surface type tags, the SDTGAN model using both surface type tags and cloud type tags, the pix2pixHD model, the cycleGAN model, and the UNIT model. Among these models, SDTGAN, cycleGAN, and UNIT can achieve multiple outputs for a single model, and pix2pixHD needs to exchange input and output data to train two sets of models.

For all models, the training was repeated for 200 epochs on an NVIDIA RTX3090 GPU with 24GB GPU memory. The weights were initialized with Kaiming initialization [37]. The Adam optimizer [38] was used, and the momentum was set to 0.5. The learning rate was set to 0.0001, and it linearly decayed after 100 epochs. Instance normalization [39] was used, which is more suitable for scenes with high requirements for a single pixel was used. Reflection padding was used to reduce artifacts. The size of the input and output image blocks for training was 512 × 512. Each mini-batch consisted of one image from each domain.

3.3. Visual Comparison

In this work, the first three reflection channels of AGRI (CH01, CH02, and CH03) are used for visible light spectrum combination. The combined images of the three RGB channels are more in line with the human eye observation and can effectively visualize the details of oceans, lands, and clouds. Meanwhile, the long-wave infrared band CH11 is used, and its main visual content is water vapor and cloud features. Due to the lack of features of the underlying surface of the earth in the visual results of CH11, the image translation from the infrared spectral domain to the visible spectral domain poses a challenge to the model.

Figures 3–5 illustrate two examples of the translation between the above two sets of spectral remote sensing images. Each set contains image information such as oceans, lands, and clouds. In Figure 3, the underlying surface of the earth in group (a) is mainly land, and that of the earth in group (b) is dominated by the ocean. Figure 4 shows the visual comparison of different models for translation from infrared spectrum domain to visible spectrum domain. Figure 5 shows the visual comparison of different models for translation from visible spectrum domain to infrared spectrum domain.

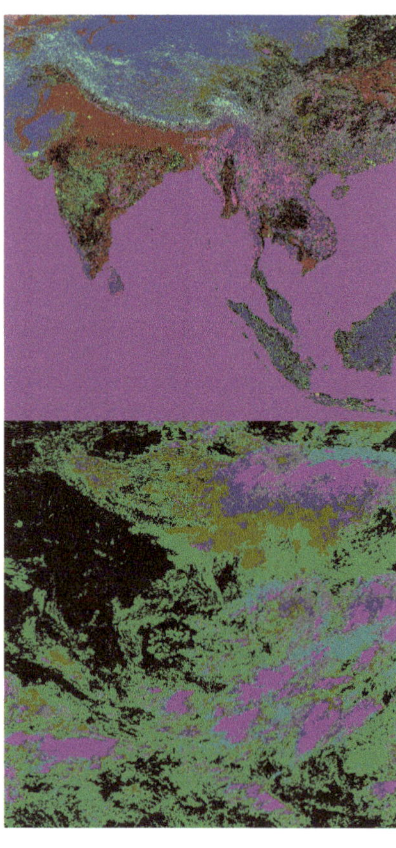

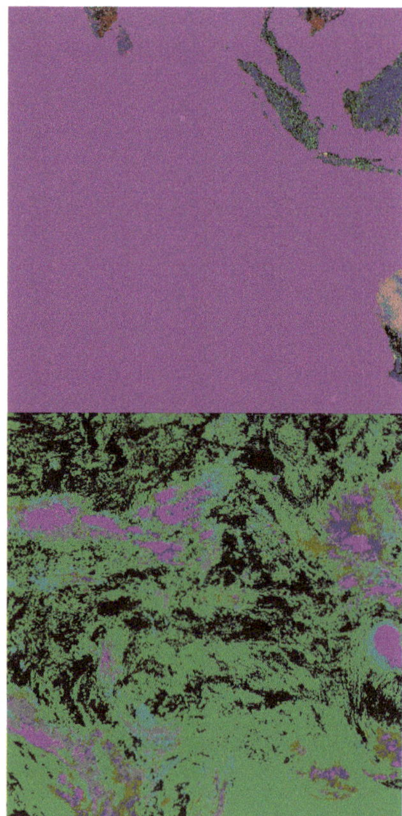

Figure 3. *Cont.*

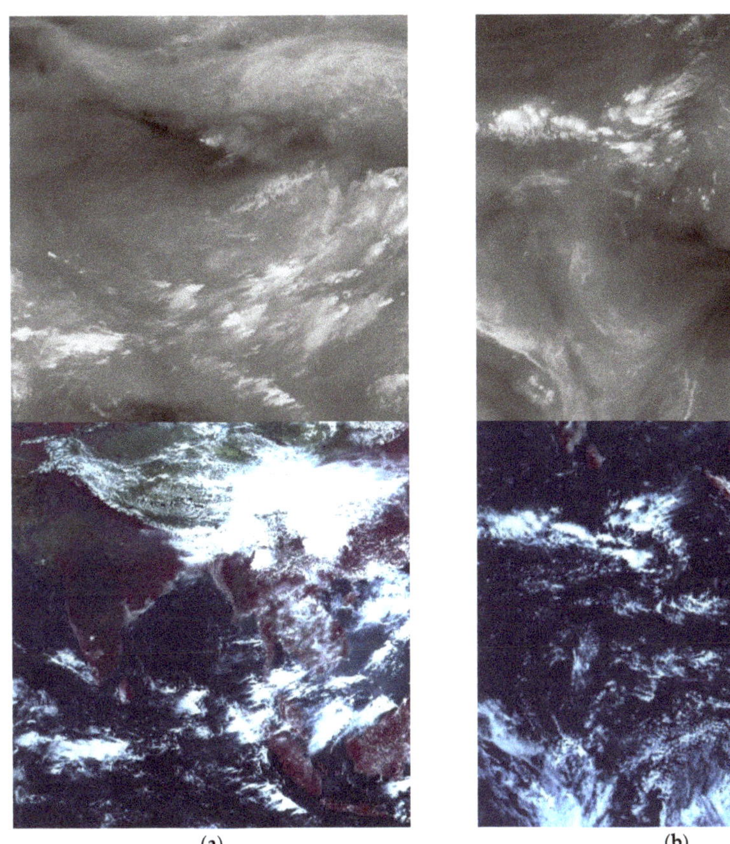

Figure 3. The input spectral-domain images and condition information of the two sets of visual comparison experiments. The first line presents label maps of the earth surface type; the second line presents cloud-type label maps; the third line presents the ground-truth of the infrared remote sensing image; and the fourth line presents the ground-truth of the visible remote sensing image. Subfigures (**a**) and subfigures (**b**) are two independent test cases.

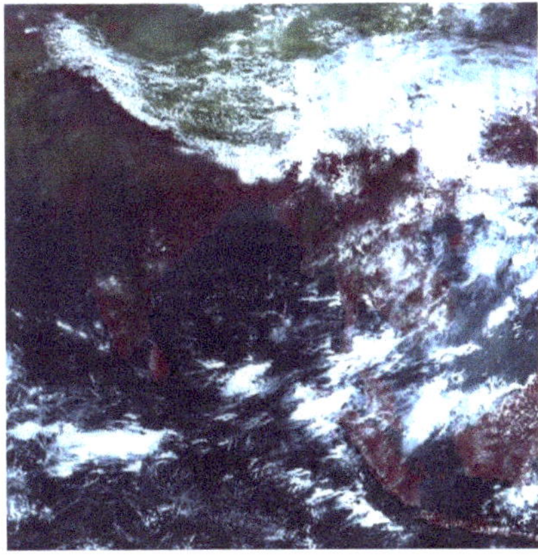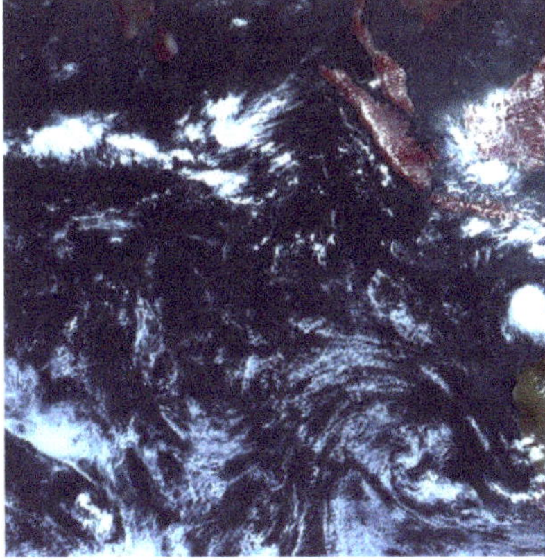

Figure 4. *Cont.*

SDTGAN
with Surface
and Cloud
Label

Pix2pix-
HD

Cycle-
GAN

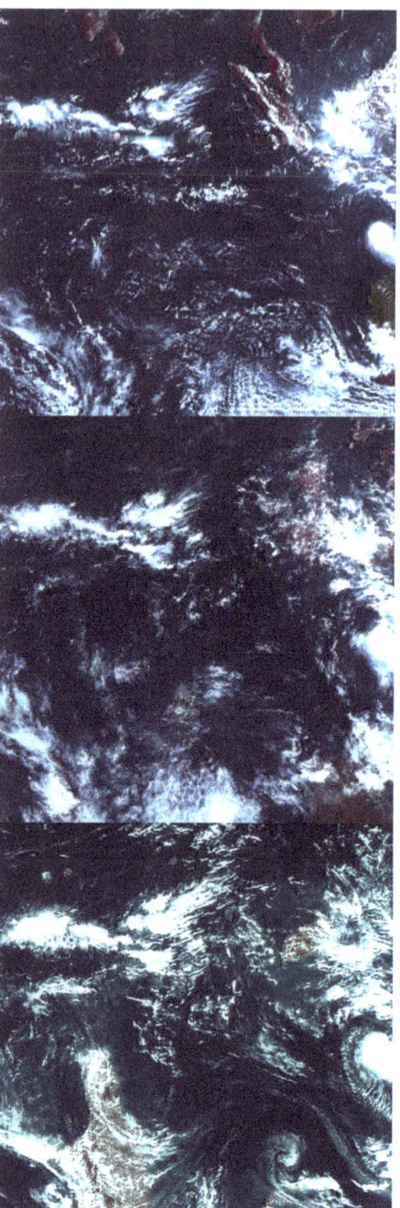

Figure 4. *Cont.*

UNIT

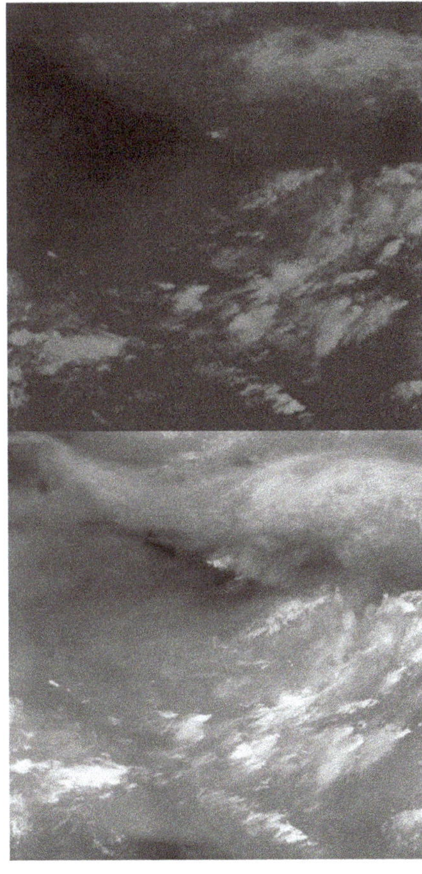

(a)

(b)

Figure 4. Visual comparison of different models for translation from infrared spectrum domain to visible spectrum domain. Subfigures (**a**) and subfigures (**b**) are two independent test cases.

SDTGAN
with
Surface
Label

SDTGAN
with
Surface and
Cloud Label

Figure 5. *Cont.*

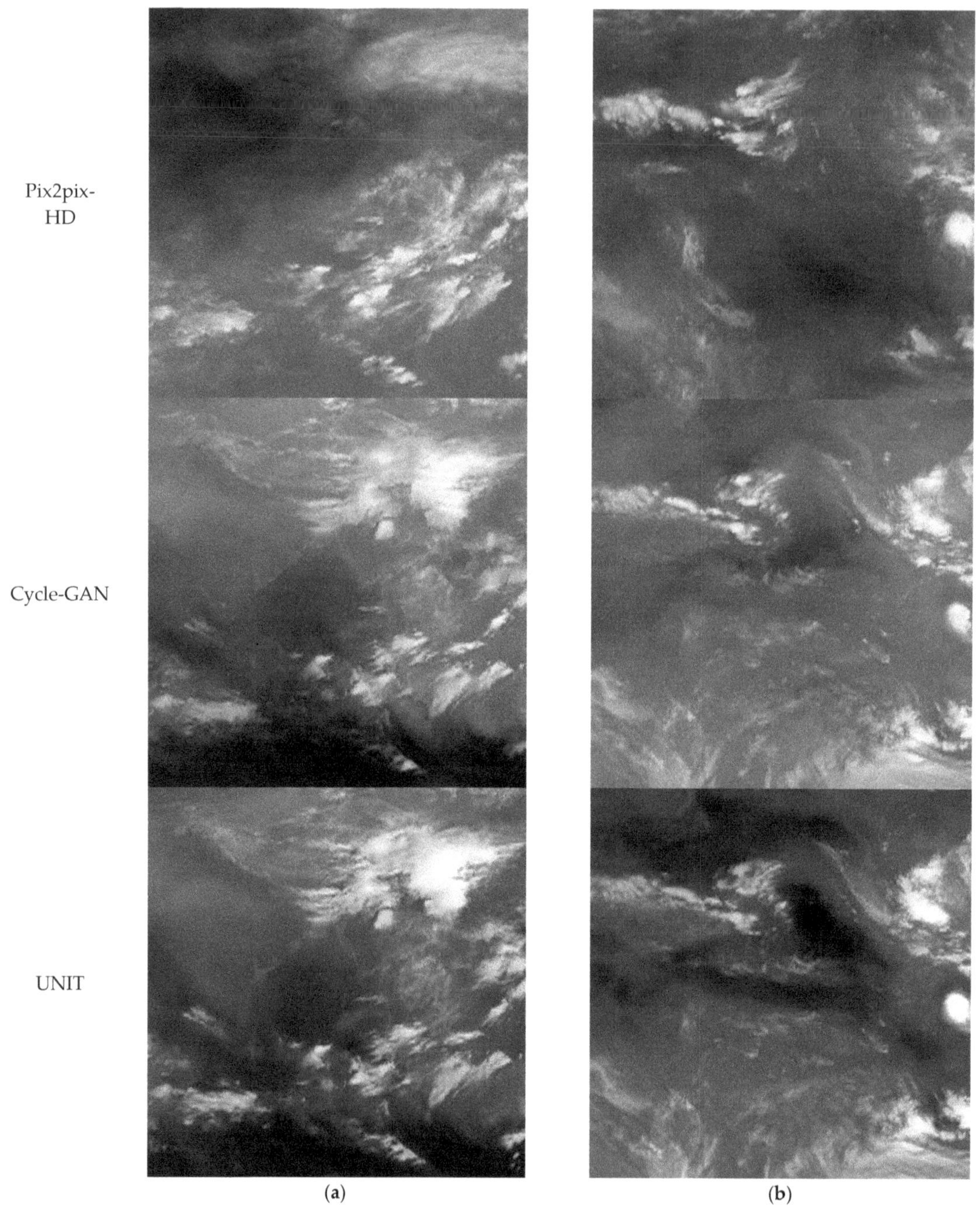

Figure 5. Visual comparison of different models for translation from visible spectrum domain to infrared spectrum domain. Subfigures (**a**) and subfigures (**b**) are two independent test cases.

3.4. Digital Comparison

To quantitatively measure the proposed method, three image quality metrics, i.e., mean-square-error (MSE), peak signal-to-noise ratio (PSNR), and structural similarity index (SSIM) [40], are selected to evaluate the translation effectiveness. MSE measures the difference between the real and simulated values. The smaller the MSE is, the more similar the two images are. PSNR is a traditional image quality evaluation index. A higher PSNR generally indicates a higher image quality. SSIM measures the structural similarity between the real image and the simulated image.

The test dataset used in the visual comparison is also taken for quantitative comparison. The dataset includes 500 remote sensing images. The average values of the evaluation metrics are calculated, and the results are listed in Tables 4 and 5, where optimal values are highlighted in bold.

Table 4. Quality result of the translation from infrared spectrum domain to visible spectrum domain.

Method	MSE	PSNR	SSIM
CycleGAN	0.0979	10.1333	0.347
UNIT	0.0931	10.3951	0.3841
Pix2pixHD	0.0663	11.8969	0.4846
SDTGAN with Surface Label	0.0361	14.5794	0.6246
SDTGAN with Surface and Cloud Label	0.0237	16.4055	0.7018

Table 5. Quality result of the translation from visible spectrum domain to infrared spectrum domain.

Method	MSE	PSNR	SSIM
CycleGAN	0.0521	13.014	0.7148
UNIT	0.1592	8.1298	0.5055
Pix2pixHD	0.0105	19.9687	0.775
SDTGAN with Surface Label	0.0017	27.9227	0.8695
SDTGAN with Surface and Cloud Label	0.0019	27.6883	0.9031

The results of Tables 4 and 5 show that the proposed SDTGAN method is superior to other comparative methods. It achieves better image recognizable structure and data authenticity in the spectral domain translation from infrared spectrum domain to visible spectrum domain and vice versa.

3.5. Ablation Study

The against ablations of within domain reconstruction loss, cross domain reconstruction loss, and latent matching loss are compared. In this case, the basic model contains only adversarial loss. As shown in Tables 6 and 7, adding reconstruction loss and latent matching loss alone can effectively improve the evaluation metrics of the images, and the model including all the loss functions achieves the highest evaluation score.

Table 6. Ablation study of the translation from infrared spectrum domain to visible spectrum domain.

Method	MSE	PSNR	SSIM
Basic	0.0393	14.1868	0.5986
Basic + within Domain Reconstruction Loss	0.0382	14.2955	0.5867
Basic + cross Domain Reconstruction Loss	0.0251	16.0984	0.6801
Basic + Latent Matching Loss	0.0326	14.9459	0.6195

Table 7. Ablation study of the translation from visible spectrum domain to infrared spectrum domain.

Method	MSE	PSNR	SSIM
Basic	0.0037	24.4947	0.5256
Basic + within Domain Reconstruction Loss	0.0040	23.3079	0.8383
Basic + cross Domain Reconstruction Loss	0.0019	27.6244	0.898
Basic + Latent Matching Loss	0.0032	25.083	0.56

3.6. Limitation

The spectral translation task involves the transformation of energy intensity and graphic texture. In most cases, the model can generate plausible cloud and continental shapes. However, there are still many cases in which the model causes loss of details and distortions. As shown in Figure 6a, in the real image, the ocean is covered with large areas of thin clouds, yet. The generated image has difficulty in inferring the random phenomenal changes over a large area, thus yielding an erroneous result. As shown in Figure 6b, in continental terrain, a certain distortion and blurring is produced for the boundary areas with variable shapes.

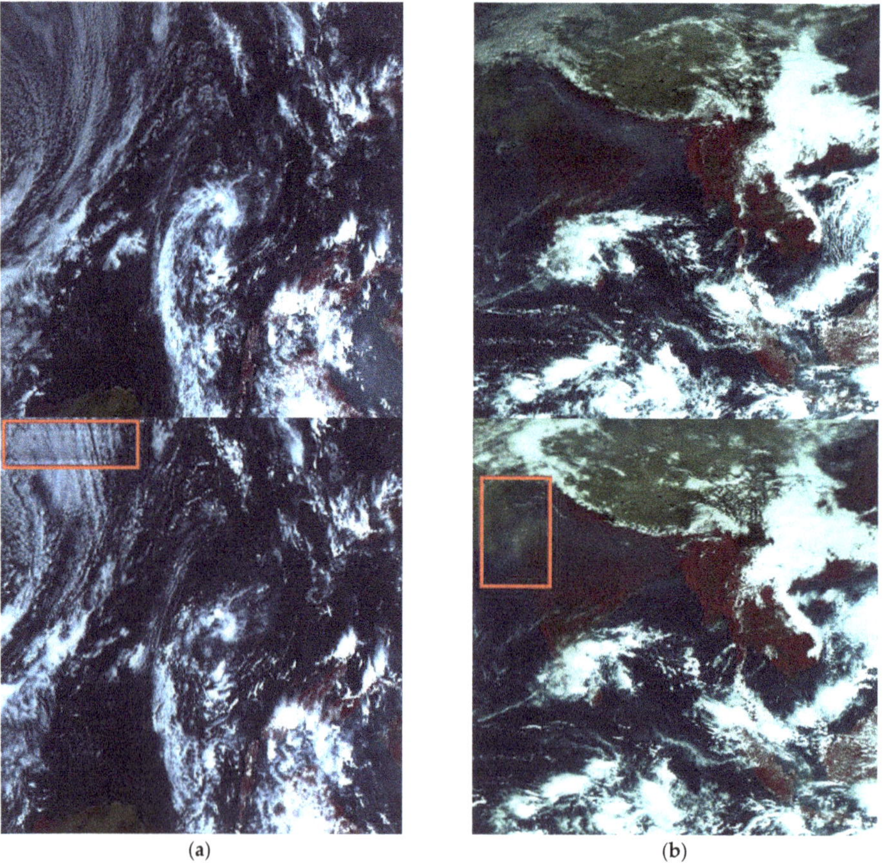

Figure 6. Example results of detail distortion on visible spectrum image. Subfigures (**a**) and subfigures (**b**) are two independent test cases.

These detail distortions may be due to the limitations in the structure of the generative network or feature selection. In future work, we will continue to explore these issues. For example, by selecting the feature data that can represent cloud height, and surface temperature, and enhancing the generated details of textures by better network structures.

4. Conclusions and Future Work

In this paper, a multi-spectral domain translation model based on conditional GAN architecture is proposed for remote sensing images of the earth background. To achieve multi-spectral domain adaptation, the model introduces feature maps of earth background and shared latent domain. In addition to adversarial loss, within domain reconstruction loss, cross domain reconstruction loss and latent matching loss are added to train the network. Besides, multi-spectral remote sensing images taken from a FY satellite are used as a dataset to test the effect of bidirectional translation between infrared band and visible band images. Compared with models such as pix2pix and cycleGAN, SDTGAN achieves more stable and accurate performance in translating spectral images at the pixel level, and simulating the surface structure and texture of clouds. In future work, we will explore a better structure for extraction, construction, and utilization of shared latent domain for spectral-domain translation, and extend it to other band combinations.

Author Contributions: Conceptualization, B.W. and J.W.; data curation, L.Z. and X.G., formal analysis, L.Z. and B.W.; funding acquisition, J.W. and X.W.; investigation, X.G., L.Z. and B.W.; methodology, J.W. and L.Z.; project administration, J.W. and X.W.; resources, J.W. and X.W.; software, L.Z. and X.G.; supervision, J.W.; validation, B.W. and L.Z.; visualization, L.Z. and X.G.; writing— original draft, B.W.; writing—review and editing, L.Z. and X.G. All authors have read and agreed to the published version of the manuscript.

Funding: This work is supported by National Natural Science Foundation of China under Grant 62005205 and Natural Science Basic Research Program of Shaanxi (Program No. 2020JQ-331).

Institutional Review Board Statement: Not applicable.

Informed Consent Statement: Not applicable.

Data Availability Statement: No new data were created or analyzed in this study. Data sharing is not applicable to this article.

Conflicts of Interest: The authors declare no conflict of interest.

References

1. Srivastava, A.; Oza, N.; Stroeve, J. Virtual sensors: Using data mining techniques to efficiently estimate remote sensing spectra. *IEEE Trans. Geosci. Remote Sens.* **2005**, *43*, 590–600. [CrossRef]
2. Miller, S.W.; Bergen, W.R.; Huang, H.; Bloom, H.J. End-to-end simulation for support of remote sensing systems design. *Proc. SPIE-Int. Soc. Opt. Eng.* **2004**, *5548*, 380–390.
3. Börner, A.; Wiest, L.; Keller, P.; Reulke, R.; Richter, R.; Schaepman, M.; Schläpfer, D. SENSOR: A tool for the simulation of hyperspectral remote sensing systems. *ISPRS J. Photogramm. Remote Sens.* **2001**, *55*, 299–312. [CrossRef]
4. Gastellu-Etchegorry, J.P.; Martin, E.; Gascon, F. DART: A 3D model for simulating satellite images and studying surface radiation budget. *Int. J. Remote Sens.* **2004**, *25*, 73–96. [CrossRef]
5. Gascon, F.; Gastellu-Etchegorry, J.P.; Lefevre, M.-J. Radiative transfer model for simulating high-resolution satellite images. *IEEE Trans. Geosci. Remote Sens.* **2001**, *39*, 1922–1926. [CrossRef]
6. Ambeau, B.L.; Gerace, A.D.; Montanaro, M.; McCorkel, J. The characterization of a DIRSIG simulation environment to support the inter-calibration of spaceborne sensors. In Proceedings of the Earth Observing Systems XXI, San Diego, CA, USA, 19 September 2016; p. 99720M.
7. Tiwari, V.; Kumar, V.; Pandey, K.; Ranade, R.; Agrawal, S. Simulation of the hyperspectral data using Multispectral data. In Proceedings of the 2016 IEEE International Geoscience and Remote Sensing Symposium (IGARSS), Beijing, China, 10–15 July 2016; pp. 6157–6160.
8. Rengarajan, R.; Goodenough, A.A.; Schott, J.R. Simulating the directional, spectral and textural properties of a large-scale scene at high resolution using a MODIS BRDF product. In Proceedings of the Sensors, Systems, and Next-Generation Satellites XX, Edinburgh, UK, 19 October 2016; p. 100000Y.
9. Cheng, X.; Shen, Z.-F.; Luo, J.-C.; Shen, J.-X.; Hu, X.-D.; Zhu, C.-M. Method on simulating remote sensing image band by using ground-object spectral features study. *J. Infrared Millim. WAVES* **2010**, *29*, 45–48. [CrossRef]

10. Geng, Y.; Mei, S.; Tian, J.; Zhang, Y.; Du, Q. Spatial Constrained Hyperspectral Reconstruction from RGB Inputs Using Dictionary Representation. In Proceedings of the IGARSS 2019 IEEE International Geoscience and Remote Sensing Symposium, Yokohama, Japan, 28 July–2 August 2019; pp. 3169–3172.
11. Han, X.; Yu, J.; Luo, J.; Sun, W. Reconstruction from Multispectral to Hyperspectral Image Using Spectral Library-Based Dictionary Learning. *IEEE Trans. Geosci. Remote Sens.* **2018**, *57*, 1325–1335. [CrossRef]
12. Goodfellow, I.; Pouget-Abadie, J.; Mirza, M.; Xu, B.; Warde-Farley, D.; Ozair, S.; Courville, A.; Bengio, Y. *Generative Adversarial Nets*; MIT Press: Cambridge, MA, USA, 2014.
13. Isola, P.; Zhu, J.-Y.; Zhou, T.; Efros, A.A. Image-to-Image Translation with Conditional Adversarial Networks. In Proceedings of the 2017 IEEE Conference on Computer Vision and Pattern Recognition (CVPR), Honolulu, HI, USA, 21–26 July 2017; pp. 5967–5976. [CrossRef]
14. Wang, T.-C.; Liu, M.-Y.; Zhu, J.-Y.; Tao, A.; Kautz, J.; Catanzaro, B. High-Resolution Image Synthesis and Semantic Manipulation with Conditional GANs. In Proceedings of the IEEE Conference on Computer Vision and Pattern Recognition, Salt Lake City, UT, USA, 18–23 June 2018; pp. 8798–8807.
15. Xiong, F.; Wang, Q.; Gao, Q. Consistent Embedded GAN for Image-to-Image Translation. *IEEE Access* **2019**, *7*, 126651–126661. [CrossRef]
16. Yi, Z.; Zhang, H.; Tan, P.; Gong, M. Dualgan: Unsupervised dual learning for image-to-image translation. In Proceedings of the IEEE International Conference on Computer Vision, Cambridge, MA, USA, 20–23 June 2017; pp. 2849–2857.
17. Zhu, J.-Y.; Park, T.; Isola, P.; Efros, A.A. Unpaired image-to-image translation using cycle-consistent adversarial networks. In Proceedings of the IEEE International Conference on Computer Vision, Venice, Italy, 22–29 October 2017; pp. 2223–2232.
18. Hoffman, J.; Tzeng, E.; Park, T.; Zhu, J.-Y.; Isola, P.; Saenko, K.; Efros, A.A.; Darrell, T. CyCADA: Cycle-Consistent Adversarial Domain Adaptation. In Proceedings of the ICML, Stockholm, Sweden, 10–15 July 2018.
19. Chen, S.; Liao, D.; Qian, Y. Spectral Image Visualization Using Generative Adversarial Networks. In Proceedings of the Swarm, Evolutionary, and Memetic Computing; Springer Science and Business Media LLC: Secaucus, NJ, USA, 2018; pp. 388–401.
20. Shi, Z.; Chen, C.; Xiong, Z.; Liu, D.; Wu, F. HSCNN+: Advanced CNN-Based Hyperspectral Recovery from RGB Images. In Proceedings of the IEEE Conference on Computer Vision and Pattern Recognition (CVPR) Workshops, Salt Lake City, UT, USA, 18–22 June 2018; pp. 939–947. [CrossRef]
21. Wu, J.; Aeschbacher, J.; Timofte, R. In Defense of Shallow Learned Spectral Reconstruction from RGB Images. In Proceedings of the 2017 IEEE International Conference on Computer Vision Workshops (ICCVW), Venice, Italy, 22–29 October 2017; pp. 471–479.
22. Zhao, Y.; Fu, G.; Wang, H.; Zhang, S. The Fusion of Unmatched Infrared and Visible Images Based on Generative Adversarial Networks. *Math. Probl. Eng.* **2020**, *2020*, 1–12. [CrossRef]
23. Ma, J.; Ma, Y.; Li, C. Infrared and visible image fusion methods and applications: A survey. *Inf. Fusion* **2019**, *45*, 153–178. [CrossRef]
24. Tang, R.; Liu, H.; Wei, J. Visualizing Near Infrared Hyperspectral Images with Generative Adversarial Networks. *Remote Sens.* **2020**, *12*, 3848. [CrossRef]
25. Cheng, W. Creating synthetic meteorology satellite visible light images during night based on GAN method. *arXiv* **2021**, arXiv:2108.04330.
26. Ma, J.; Yu, W.; Liang, P.; Li, C.; Jiang, J. FusionGAN: A generative adversarial network for infrared and visible image fusion. *Inf. Fusion* **2019**, *48*, 11–26. [CrossRef]
27. Arad, B.; Ben-Shahar, O. Sparse Recovery of Hyperspectral Signal from Natural RGB Images. In Proceedings of the European Conference on Computer Vision, Amsterdam, The Netherlands, 11–14 October 2016; pp. 11–14.
28. Liu, P.; Zhao, H. Adversarial Networks for Scale Feature-Attention Spectral Image Reconstruction from a Single RGB. *Sensors* **2020**, *20*, 2426. [CrossRef] [PubMed]
29. Huang, X.; Liu, M.-Y.; Belongie, S.; Kautz, J. *Multimodal Unsupervised Image-to-Image Translation*; Springer Science and Business Media LLC: Secaucus, NJ, USA, 2018; pp. 179–196.
30. Johnson, J.; Alahi, A.; Fei-Fei, L. Perceptual losses for real-time style transfer and super-resolution. In Proceedings of the European Conference on Computer Vision, Amsterdam, The Netherland, 8–16 October 2016; Springer: Berlin/Heidelberg, Germany, 2016; pp. 694–711.
31. Shelhamer, E.; Long, J.; Darrell, T. Fully Convolutional Networks for Semantic Segmentation. Available online: https://arxiv.org/abs/1605.06211 (accessed on 1 March 2022).
32. Durugkar, I.; Gemp, I.M.; Mahadevan, S. Generative Multi-Adversarial Networks. *arXiv* **2017**, arXiv:1611.01673.
33. Rosca, M.; Lakshminarayanan, B.; Warde-Farley, D.; Mohamed, S. Variational Approaches for Auto-Encoding Generative Adversarial Networks. *arXiv* **2017**, arXiv:1706.04987.
34. Zhang, P.; Zhu, L.; Tang, S.; Gao, L.; Chen, L.; Zheng, W.; Han, X.; Chen, J.; Shao, J. General Comparison of FY-4A/AGRI with Other GEO/LEO Instruments and Its Potential and Challenges in Non-meteorological Applications. *Front. Earth Sci.* **2019**, *6*, 6. [CrossRef]
35. Zhang, P.; Lu, Q.; Hu, X.; Gu, S. Latest Progress of the Chinese Meteorological Satellite Program and Core Data Processing Technologies. *Adv. Atmos. Sci.* **2019**, *36*, 1027–1045. [CrossRef]
36. Congalton, R.G.; Gu, J.; Yadav, K.; Thenkabail, P.S.; Ozdogan, M. Global Land Cover Mapping: A Review and Uncertainty Analysis. *Remote Sens.* **2014**, *6*, 12070–12093. [CrossRef]

37. He, K.; Zhang, X.; Ren, S.; Sun, J. Delving Deep into Rectifiers: Surpassing Human-Level Performance on ImageNet Classification. In Proceedings of the IEEE International Conference on Computer Vision, Santiago, Chile, 7–13 December 2015; pp. 1026–1034. [CrossRef]
38. Kingma, D.P.; Ba, J. Adam: A method for stochastic optimization. *arXiv* **2014**, arXiv:1412.6980.
39. Ulyanov, D.; Vedaldi, A.; Lempitsky, V.S. Instance Normalization: The Missing Ingredient for Fast Stylization. *arXiv* **2016**, arXiv:1607.08022.
40. Setiadi, D.R.I.M. PSNR vs. SSIM: Imperceptibility quality assessment for image steganography. *Multimedia Tools Appl.* **2021**, *80*, 8423–8444. [CrossRef]

remote sensing

Article

GCBANet: A Global Context Boundary-Aware Network for SAR Ship Instance Segmentation

Xiao Ke, Xiaoling Zhang * and Tianwen Zhang

School of Information and Communication Engineering, University of Electronic Science and Technology of China, Chengdu 611731, China; xke@std.uestc.edu.cn (X.K.); twzhang@std.uestc.edu.cn (T.Z.)
* Correspondence: xlzhang@uestc.edu.cn

Abstract: Synthetic aperture radar (SAR) is an advanced microwave sensor, which has been widely used in ocean surveillance, and its operation is not affected by light and weather. SAR ship instance segmentation can provide not only the box-level ship location but also the pixel-level ship contour, which plays an important role in ocean surveillance. However, most existing methods are provided with limited box positioning ability, hence hindering further accuracy improvement of instance segmentation. To solve the problem, we propose a global context boundary-aware network (GCBANet) for better SAR ship instance segmentation. Specifically, we propose two novel blocks to guarantee GCBANet's excellent performance, i.e., a global context information modeling block (GCIM-Block) which is used to capture spatial global long-range dependences of ship contextual surroundings, enabling larger receptive fields, and a boundary-aware box prediction block (BABP-Block) which is used to estimate ship boundaries, achieving better cross-scale box prediction. We conduct ablation studies to confirm each block's effectiveness. Ultimately, on two public SSDD and HRSID datasets, GCBANet outperforms the other nine competitive models. On SSDD, it achieves 2.8% higher box average precision (AP) and 3.5% higher mask AP than the existing best model; on HRSID, they are 2.7% and 1.9%, respectively.

Keywords: synthetic aperture radar; ship instance segmentation; global context modeling; boundary-aware box prediction

Citation: Ke, X.; Zhang, X.; Zhang, T. GCBANet: A Global Context Boundary-Aware Network for SAR Ship Instance Segmentation. *Remote Sens.* **2022**, *14*, 2165. https://doi.org/10.3390/rs14092165

Academic Editor: Gwanggil Jeon

Received: 1 April 2022
Accepted: 20 April 2022
Published: 30 April 2022

Publisher's Note: MDPI stays neutral with regard to jurisdictional claims in published maps and institutional affiliations.

1. Introduction

Synthetic aperture radar (SAR) is an outstanding microwave sensor. It can provide high-resolution observation images via measuring objects' radar scattering characteristics, free from both light and weather [1–5], which is extensively used in the measurement [6,7], transportation [8], ocean [9,10], and remote sensing [11,12] communities. Ship surveillance is a research highlight at present, because it is conducive to disaster reliefs traffic control, and fishery monitoring [13]. Compared with optical [14], infrared [15], and hyperspectral [16] sensors, SAR is more suitable for ocean ship surveillance because of its stronger adaptability to marine environments with changeable climate. Consequently, ship surveillance using SAR is receiving more attention [17–24].

Traditional methods [17,25–27] generally rely on hand-crafted features via expert experience, which are laborious and time-consuming, limiting broader generalization. Now, convolutional neural networks (CNNs) are offering many elegant schemes with high-efficiency and high-accuracy superiority. For example, LeCun et al. [28] proposed LeNet5 for handwritten character recognition. Krizhevsky et al. [29] proposed AlexNet, which showed great performance in 2012 ImageNet Competition. Simonyan et al. [30] deepened the layers of networks to extract more discriminative features and proposed VGG for image classification. Girshick [31] used deep convolutional networks to build Fast R-CNN for object detection. Ren et al. [32] proposed Faster R-CNN which achieved state-of-the-art object detection accuracy on PASCAL VOC datasets. Therefore, more efforts are made by an increasing number of scholars for CNN-based SAR ship detection [19–24]. For example, Cui et al. [19] proposed a dense attention pyramid network to detect multi-scale SAR ships. Zhang et al. [20] proposed a balance scene learning mechanism to improve the performance of complex inshore ships. Sun et al. [21] applied the anchor-free method for SAR ship detection. Zhang et al. [22] designed a depthwise separable convolution neural network for faster detection speed. Song et al. [24] developed an automatic methodology to generate robust training data for ship detection. However, according to the investigation in [33], most existing reports focused on detecting ships at the box level, i.e., SAR ship box detection. Regrettably, only a few reports detected ships at the box level and pixel level simultaneously, i.e., SAR ship instance pixel segmentation.

Some works [34–37] have studied SAR ship instance segmentation. Wei et al. [34] released a HRSID dataset and offered some common research baselines, but they did not offer methodological contributions. Su et al. [35] applied CNN-based models for remote sensing image instance segmentation, but the characteristics of SAR ships were not considered, which hinders further accuracy improvement. Gao et al. [36] proposed an anchor-free model, but the model cannot handle complex scenes and cases [38]. Zhao et al. [37] proposed a synergistic attention for SAR ship instance segmentation, but their method still missed many small ships and inshore ones. These existing models mostly have limited box positioning ability, hindering the further accuracy improvements of segmentation.

Thus, we propose a global context boundary-aware network (GCBANet) to solve this problem for better SAR ship instance segmentation. We designed a global context information modeling block (GCIM-Block) to capture spatial long-range dependences of ship surroundings, resulting in larger receptive fields; thus, the background interferences can be mitigated. We also designed a boundary-aware box prediction block (BABP-Block) to estimate the ship box boundary, rather than the ship box center and width-height. This can enable better cross-scale prediction, because aligning each side of the box to the target boundary is much easier than moving the box as a whole while tuning the size, especially for cross-scale targets. Here, cross-scale means that targets exhibit a large pixel-scale difference [39]. A large scale-difference is usually from the large resolution difference [40]. SAR ships have the cross-scale characteristic, i.e., small ships are extremely small and large ones are extremely large [39]. Such huge scale difference increases instance segmentation difficulty. BABP-Block tackles this problem.

We conducted ablation studies to confirm the effectiveness of GCIM-Block and BABP-Block. Combined with them, GCBANet surpasses the other nine competitive models significantly on the two public SSDD [41] and HRSID [34] datasets. Specifically, on SSDD, it achieves 2.8% higher box AP and 3.5% higher mask AP than the existing best model; on HRSID, they are 2.7% and 1.9%. The source code and the result are available online on our website [42].

The main contributions of this article are as follows:

1. GCBANet is proposed for better SAR ship instance segmentation.
2. GCIM-Block and BABP-Block are proposed to ensure GCBANet's good performance.
3. GCBANet significantly outperforms the other nine competitive models.

The rest of the materials of this article are arranged as follows. Section 2 introduces the methodology of GCBANet. Section 3 introduces the experiments. Results are shown in Section 4. Ablation studies are described in Section 5. Finally, a summary of this article is made in Section 6.

2. Methodology

Figure 1 shows the architecture of the proposed GCBANet. GCBANet follows the state-of-the-art cascade structure [43,44] for high-quality SAR ship instance segmentation, which sets three stages to refine box (B1, B2, and B3) prediction and mask (M1, M2, and M3) prediction progressively. This paradigm was demonstrated by the optimal instance segmentation performance [45].

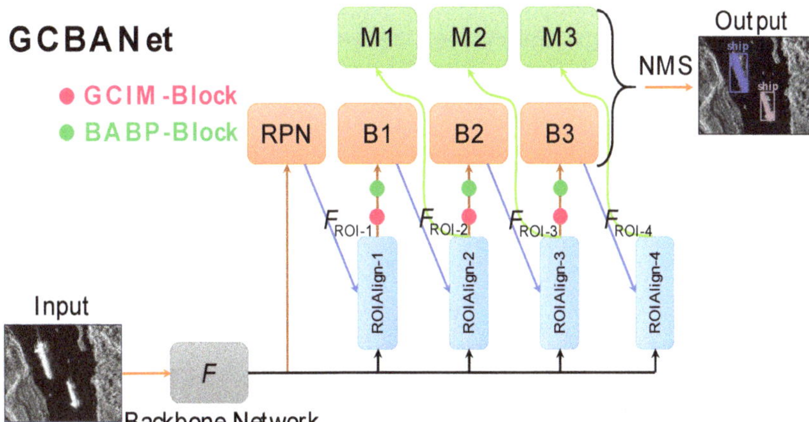

Figure 1. The architecture of the global context boundary-aware network (GCBANet). F denotes the feature maps of the backbone network. RPN denotes the region proposal network. FROI-*i* denotes the pooled ROI features in the *i*-th stage. B*i* denotes the box prediction in the *i*-th stage. M*i* denotes the mask prediction in the *i*-th stage. GCBANet adopts a cascade structure which sets three stages to refine box and mask prediction. GCIM-Block denotes the global context modeling block. BABP-Block denotes the boundary-aware box prediction block. NMS denotes non-maximum suppression.

The backbone network is used to extract SAR ship features. Without losing generality, the common ResNet-101 [46] is selected as GCBANet's backbone network. The region proposal network (RPN) [32] is used to generate some initial region candidates, i.e., regions of interests (ROIs). ROIAlign [47] is used to extract feature subsets of ROIs among the backbone network's feature maps F for the subsequent box-mask refined prediction. ROIAlign's input parameters are determined by the previous box prediction, i.e., RPN→ROIAlign-1, B1→ROIAlign-2, B2→ROIAlign-3, and B3→ROIAlign-4. The resulting feature subset is denoted by F_{ROI-i}. The box prediction in the *i*-stage is conducted by learning on F_{ROI-i} whose more refined location regression is then inputted into the next stage. The mask prediction in the *i*-stage is implemented by learning on the achieved next stage feature subset $F_{ROI-i+1}$. The final results of the box prediction B3 and mask prediction M3 are post-processed by a non-maximum suppression (NMS) [48] to delete duplicate detections.

We observe that the mask prediction mainly relies on the previous stage box prediction from the information flow direction (B1→M1, B2→M2, and B3→M3). Therefore, if one wants to further improve the segmentation performance of the mask prediction, then they should first improve the detection performance of the box prediction. In this way, the overall instance segmentation can be improved (the instance segmentation contains the box detection and the mask segmentation). This is also a direct scheme to boost the two-stage instance segmentation models' performance [49]. Thus, considering the task characteristics of SAR ships, we design two blocks, a GCIM-Block (marked by a green circle) and BABP-Block (marked by a magenta circle), to reach this goal. Their resulting benefits will be transmitted to the final box prediction B3 and mask prediction M3 for better performance.

Next, we will introduce the GCIM-Block and the BABP-Block in detail in the following two sub-sections.

2.1. Global Context Information Modeling Block (GCIM-Block)

Ships in SAR images have various surroundings, as in Figure 2, e.g., river courses, islands, inshore facilities, harbors, and wakes. Moreover, because of the special imaging mechanisms of SAR, ships are also accompanied with cross-shape sidelobes, speckle noise, and granular pixel distribution [50]. These various surroundings pose differential effects to ship instance segmentation. It is very necessary to take them into consideration for better background discrimination ability in box prediction. Therefore, we design a global context information modeling block (GCIM-Block) to model global background context information, which can capture the spatial long-range dependences of ships to decrease false alarms and missed detections. GCIM-Block offers three main design concepts, i.e., (1) content-aware feature reassembly (CAFR), (2) multi receptive-field feature response (MRFFR), and (3) global feature self-attention (GFSA). Its workflow is shown in Figure 3. The input is F_{ROI} and the output is $F_{\text{GCIM-Block}}$.

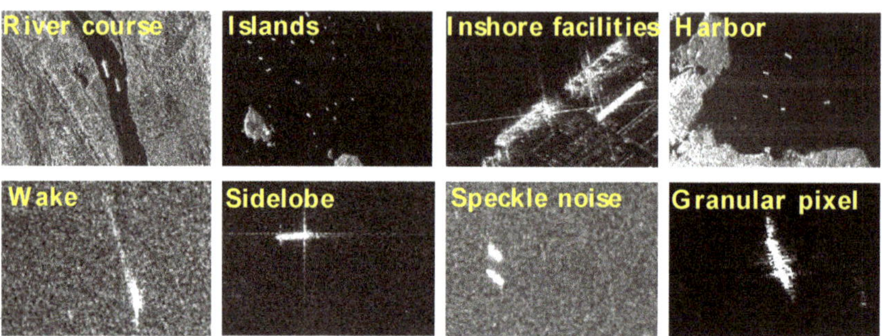

Figure 2. Various surroundings of ships in SAR images.

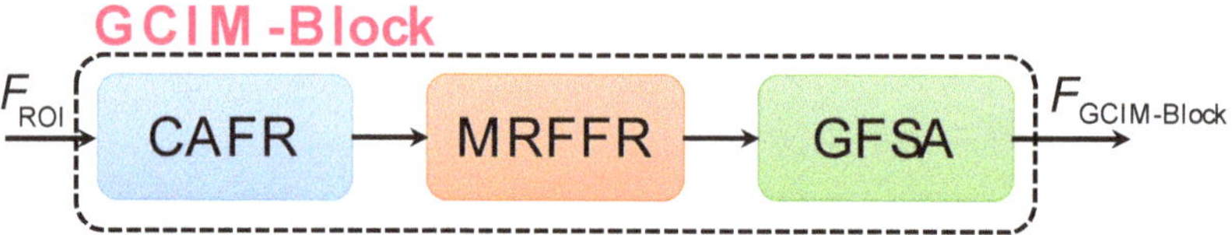

Figure 3. Workflow of the global context information modeling block (GCIM-Block). Here, CAFR denotes the content-aware feature reassembly. MRFFR denotes the multi receptive-field feature response. GFSA denotes the global feature self-attention.

2.1.1. Content-Aware Feature Reassembly (CAFR)

The standard ROI pooling size of the box prediction is 7×7 while that of the mask prediction is 14×14 [47]. Therefore, to maintain feature consistency between box and mask, before the global context modeling, we propose CAFR to up-sample the raw box feature maps from 7×7 to 14×14, which can also offer better modeling benefits in a larger feature space. We observe that this practice can offer a notable accuracy gain although the speed is sacrificed (see Section 5.1). Note that we abandon the common nearest neighbor or bilinear interpolation to reach this goal because they merely consider sub-pixel neighborhood, failing to capture the rich global context semantic information required by the dense prediction task. We also do not use the deconvolution because it applies the same kernel across the entire space, without considering the underlying global context content, limited by a limited field of view.

Differently, our proposed CAFR can enable the instance-specific content-aware handling while considering global context information, resulting in adaptive up-sampling kernels. Such a content-aware paradigm is also suggested by Wang et al. [51]. Figure 4 shows the implementation of CAFR. CAFR contains two processes—(i) content-aware kernel prediction and (ii) feature reassembly operation.

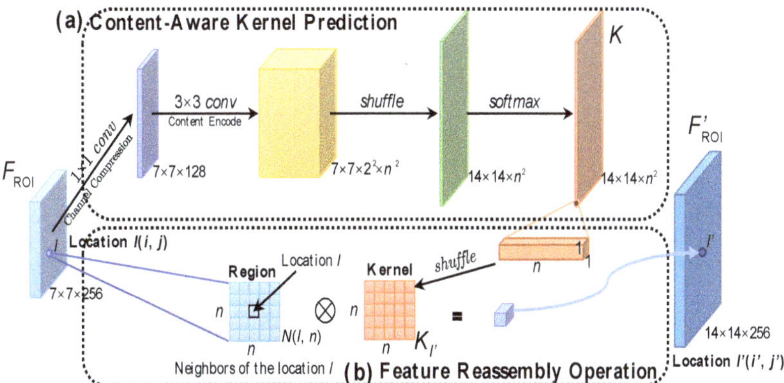

Figure 4. Implementation of the content-aware feature reassembly (CAFR) in the GCIM-Block.

The former is used to encode contents so as to predict the up-sampling kernel K. The input is the ROI's pooled feature maps denoted by F_{ROI}. To reduce the computational burden, we first adopt a 1×1 conv for channel compression where the compression is set to 0.5, i.e., from the raw 256 to the current 128, in consideration of the accuracy-speed trade-off. Then, a 3×3 conv is used to encode the entire content whose kernel number is $2^2 \times n^2$. Here, 2 denotes the up-sampling ratio, and n denotes the interpolation neighborhood scope to be considered, which is set to 5 empirically. In order to achieve the up-sampling kernel K across the entire 14×14 feature space, the previous encoded content feature maps are shuffled in space, leading to the tensor with a $14 \times 14 \times n^2$ dimension.

Finally, it is normalized via a softmax calculation function defined by $e^{x_i} / \sum_j e^{x_j}$, leading to the final up-sampling kernel K. The $1 \times 1 \times n^2$ tensor alongside the depth direction represents the corresponding kernel for a single up-sampling operation from the raw location $l(i, j)$ to the required location $l'(i', j')$. Briefly, the above is described by

$$K = \text{softmax}\{\text{shuffle}[\text{conv}_{3\times3}(\text{conv}_{1\times1}(F_{ROI}))]\} \tag{1}$$

The latter is to implement the feature reassembly, i.e., a convolution operation between the $n \times n$ neighbors of the location l denoted by $N(l, n)$ and the predicted kernel $K_{l'} \in K$ corresponding to the required location l'. The above is described by

$$F'_{ROI} = \bigcup_{l \in F_{ROI}, \, l' \in F'_{ROI}} N(l, n) \otimes K_{l'} \tag{2}$$

where F'_{ROI} denotes the output feature maps of CAFR, and $\otimes$ denotes the convolution operator.

2.1.2. Multi Receptive-Field Feature Response (MRFFR)

Inspired by the idea of the multi resolution analysis (MRA) [52] widely used in the wavelet transform community, we propose MRFFR to analyze ships in resolution from fine to coarse, which can improve the richness of global context information, i.e., from single-scale context to multi-scale contexts. Specifically, we adopt multi dilated convolutions [53] with different dilated rates r to reach this aim as shown in Figure 5, where different scale or color boxes represent different context scopes. MRFFR can not only excite feature multi-resolution responses but also capture multi-scope context information, conducive to better global context modeling.

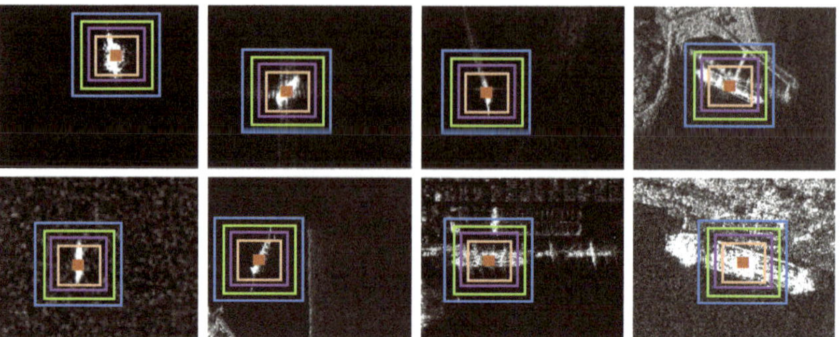

Figure 5. Multi receptive-field feature response of SAR ships.

Figure 6 depicts the implementation of MRFFR. We adopt four 3×3 convs with different dilated rates to trigger different resolution responses. More might bring better accuracy but will reduce speed. Then, the achieved four results are concatenated directly. Finally, we propose a dimension reduction squeeze-and-excitation (DRSE) to balance the contributions of different scope contexts and to achieve the channel reduction convenient for the subsequent processing. DRSE can model channel correlation to suppress useless channels and highlight valuable ones while reducing channel dimension, which reduces the risk of the training oscillation due to excessive irrelevant contextual backgrounds. We observe that only moderate contexts can enable better box and mask prediction.

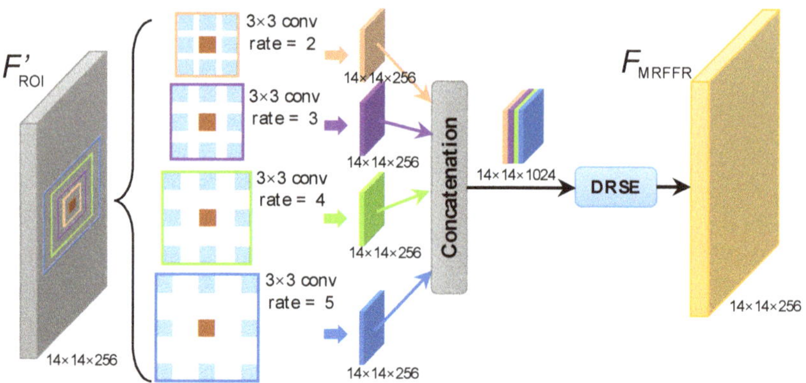

Figure 6. Implementation of the multi receptive-field feature response (MRFFR) in GCIM-Block. DRSE denotes the dimension reduction squeeze-and-excitation.

The above is described by

$$F_{\mathrm{MRFFR}} = f_{\mathrm{DRSE}} \left\{ \left[f_{3\times3}^2(F'_{\mathrm{ROI}}), f_{3\times3}^3(F'_{\mathrm{ROI}}), f_{3\times3}^4(F'_{\mathrm{ROI}}), f_{3\times3}^5(F'_{\mathrm{ROI}}) \right] \right\} \tag{3}$$

where F'_{ROI} denotes the input, F_{MRFFR} denotes the output, $f_{3\times3}^r$ denotes a 3×3 conv with a dilated rate r, and f_{DRSE} denotes the DRSE operation to reduce channels from 1024 to 256.

Figure 7 depicts the implementation of DRSE. The input is denoted by X and output is denoted by Y. In the collateral branch, a global average pooling is used to achieve global spatial information, a 1×1 conv and a sigmoid activation function are used to squeeze channels to highlight important ones. The squeeze ratio p is set to 4 ($1024 \rightarrow 256$). In the main branch, the input channel number is reduced directly by a 1×1 conv and a ReLU activation. The broadcast element-wise multiplication is used for compressed channel weighting. In this way, DRSE models the channel correlation of input feature maps in a reduced dimension space. It uses the learned weights from the reduced dimension space to pay attention to the important features of the main branch. It avoids the potential information loss of the rude dimension reduction. In short, the above is described by

$$Y = \text{ReLU}(\text{conv}_{1\times1}(X)) \odot \sigma(\text{conv}_{1\times1}(GAP(X))) \tag{4}$$

where σ denotes the sigmoid function and $\odot$ denotes the broadcast element-wise multiplication.

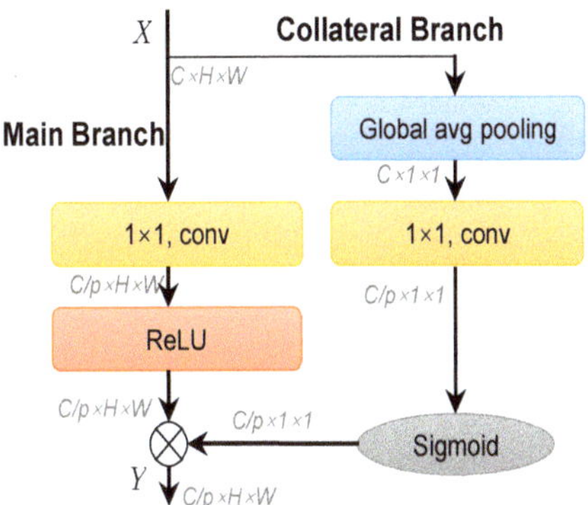

Figure 7. Implementation of dimension reduction squeeze-and-excitation (DRSE) in the MRFFR.

2.1.3. Global Feature Self-Attention (GFSA)

GFSA follows the basic idea of the non-local neural networks [54] to achieve the global context feature self-attention. It can be described by

$$\mathbf{y}_i = \frac{1}{\zeta(\mathbf{x})} \sum_{\forall j} f(\mathbf{x}_i, \mathbf{x}_j) g(\mathbf{x}_j) \tag{5}$$

where $\mathbf{x}$ denotes the input, i and j are the index position in the inputted feature maps across the whole $H \times W$ space. f is a pairwise function used to represent the spatial correlation between i and j. g is a unary function used to represent the inputted feature maps at position j. To a given i, j will enumerate the whole $H \times W$ space, resulting a sequence of spatial correlation between i and every position in the inputted feature maps. Through $\sum_{\forall j} f \times c$, the i-position's output $\mathbf{y}_i$ is related with the entire space. This means that global long-range spatial dependencies are captured. Finally, $\zeta(\mathbf{x})$ is used to normalize the response.

We instantiate Equation (5) in Figure 8. Notably, Equation (5) is only to illustrate the process of calculating a single feature vector at the j-position ($\mathbf{y}_i$) and the essence of achieving the global context feature self-attention. However, in the instantiation, the feature vectors at every position ($\mathbf{y}$) are computed in parallel through matrix calculation in consideration of computational efficiency, and we need to use existing operators such as convolution and softmax to achieve the global context feature self-attention for simplicity. Specifically, in Figure 8, features at the i-position are denoted by ϕ using a 1×1 conv W_ϕ. Features at the j-position are denoted by θ using a 1×1 conv W_θ. We model unary function g as a linear embedding which is instantiated through a 1×1 conv W_g, and embed features into $C/4$ channel space to reduce computational burdens. Moreover, pairwise function f is modeled as the Gaussian function $e^{x_i^T x_j}$ and normalization factor $\zeta(\mathbf{x})$ is modeled as $\sum_{\forall j} e^{x_i^T x_j}$. Therefore, we can instantiate f and the normalization process together through a softmax calculation function along the dimension j. Since W_ϕ and W_θ are learnable, the spatial correlation f is obtained from adaptive learning between ϕ and θ. Note that in the global self-attention process, the sizes of features need to be transposed or shift between three dimension and two dimension, which is implemented through the permute and flatten operations, respectively. The response at the i-position $\mathbf{y}_i$ is obtained by a matrix multiplication.

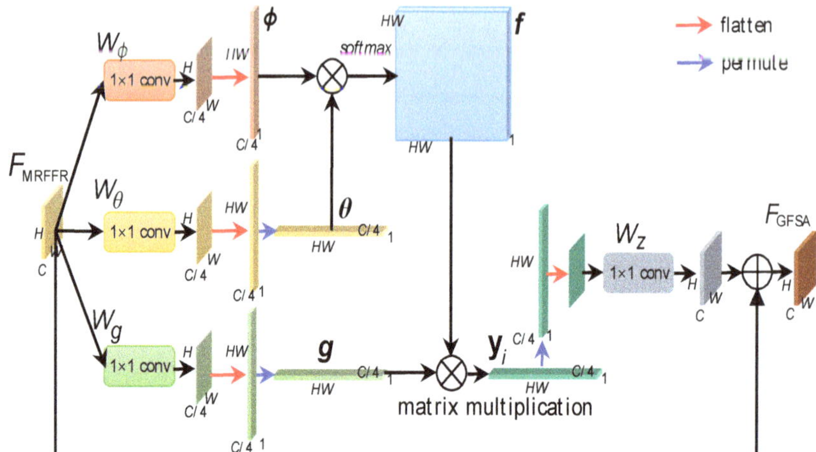

Figure 8. Implementation of the global feature self-attention (GFSA) in GCIM-Block.

Since we embed features into $C/4$ channel space to reduce computational burdens before the global self-attention process, we need to recover the channel of features after the attention process through a 1×1 conv W_z for the adding operation.

Finally, we achieve the final global feature self-attention output F_{GFSA} that will be transmitted to the subsequent boundary-aware box prediction. Here, F_{GFSA} denotes the final output of GCIM-Block $F_{\text{GCIM}-\text{Block}}$.

2.2. Boundary-Aware Box Prediction Block (BABP-Block)

The traditional box prediction is implemented via estimating the bounding box's center offset $(\Delta x, \Delta y)$ and the corresponding width and height offset $(\Delta w, \Delta h)$ with its ground truth (GT) to optimize network parameters, as shown in Figure 9a. Yet, this paradigm is not very suitable for SAR ships from the following two aspects.

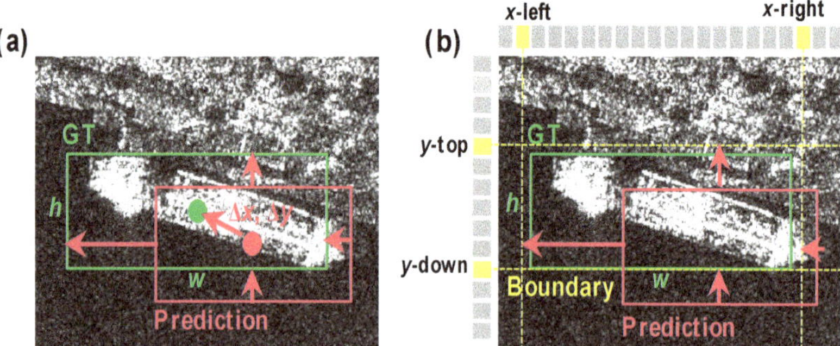

Figure 9. Different box prediction forms. (**a**) The traditional bounding box's center offset $(\Delta x, \Delta y)$ and the corresponding width and height offset $(\Delta w, \Delta h)$ estimation. (**b**) The bounding box's boundary estimation of this paper, which contains two basic steps, i.e., boundary prediction and location fine regression. The red colored box denotes prediction box. The green colored box denotes ground truth box. The green colored dot denotes the center of ground truth box. The red colored arrow denotes the trend of bounding box regression.

On the one side; as shown in Figure 10; SAR ships often exhibit a huge scale-difference due to a huge resolution difference; e.g.; 1m resolution for TerraSAR-X [55] and 20m resolution for Sentinel-1 [56]. This situation is called the cross-scale effect [39], e.g., the extremely small ships in Figure 10c vs. the extremely larger ships in Figure 10d. For example, the smallest ship in SSDD has only 28 pixels while the largest one has 62,878 pixels [57], where the scale ratio reaches 62,878/28 = 2245. For commonly-used two-stage models, presetting a series of prior anchors is required for RPN. Yet, no matter how the prior anchors are set; it is still difficult to cover such a dataset with a large scale-difference. In the dataset; since the proportion of small ships is higher than large ships; the size of the prior anchor is always closer to the small ship; but there will be a long space distance from the large ship. This will lead to the adjustment for the large ship anchors, as it becomes rather difficult if adopting the traditional scheme shown in Figure 9a. This is because it is time-consuming to adjust a small anchor to a large GT box; resulting in a great burden to the network training. As a result; the positioning accuracy of large ships will become poor

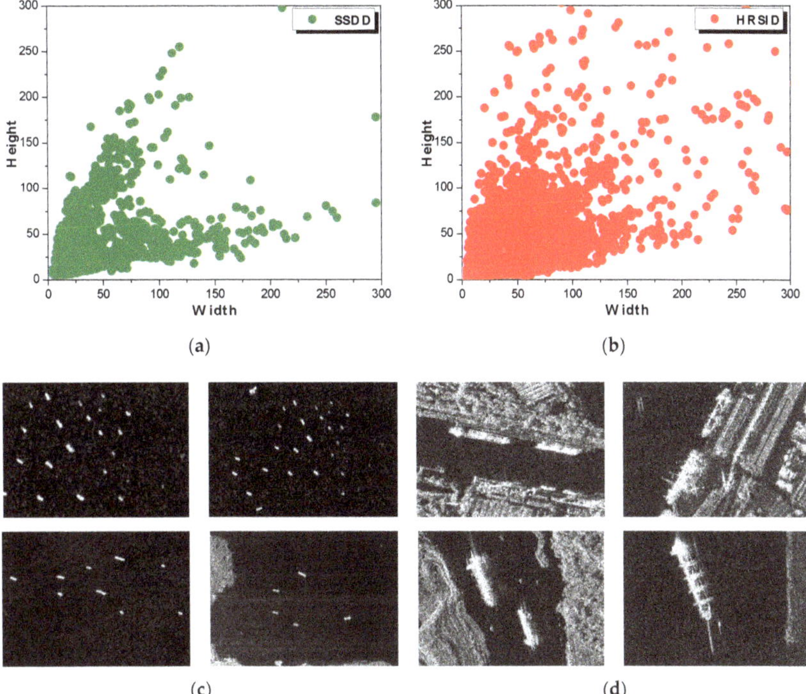

Figure 10. Some cross-scale SAR ships. (**a**) Ship size distribution in SSDD. (**b**) Ship size distribution in HRSID. (**c**) Small ships. (**d**) Large ships.

On the other side, it is rather challenging to locate the center of an SAR ship. Generally, different parts of the ship's hull have different materials, resulting in differential radar electromagnetic scatterings (i.e., radar cross section, RCS [58]). This makes the pixel brightness distribution of the ship in one SAR image extremely uneven. In many cases, the strong scattering points of the ship are not in the geometric center of the hull, but in the bow or stern. This phenomenon may directly lead to the failure

As shown above, we abandon the traditional scheme in Figure 9a, and adopt the boundary learning scheme in Figure 9b to implement the box prediction. We design a boundary-aware box prediction block (BABP-Block) to reach this goal, inspired by the gird idea from Wang et al. [59] and Lu et al. [60]. From Figure 9b, BABP-Block consists of two basic steps. (i) The first is to predict the coarse boundary of a ship marked by yellow dotted lines in the x-left, x-right, y-top and y-down (i.e., four yellow activate grids). (ii) The second is to adjust the box finely from the boundary box to the GT box. This stage is the same as the traditional scheme, but obviously it is much easier to adjust the resulting coarse boundary box to the GT box so as to achieve the final finer box. This is because the distance to be adjusted is greatly reduced. Such from coarse to fine prediction scheme divides the task into two stages where each stage is responsible for its own task, resulting in the dual-supervision of training, enabling better box prediction. Once the box prediction becomes more accurate, the mask prediction will become more accurate as well. BABP-Block offers four main design concepts, i.e., (1) boundary-aware feature extraction (BAFE), (2) boundary bucketing coarse localization (BBCL), (3) boundary regression fine localization (BRFL), and (4) boundary-guided classification rescoring (BGCR). Its workflow is depicted in Figure 11. The input is the feature maps of GCIM-Block's output $F_{\text{GCIM}-\text{Block}}$.

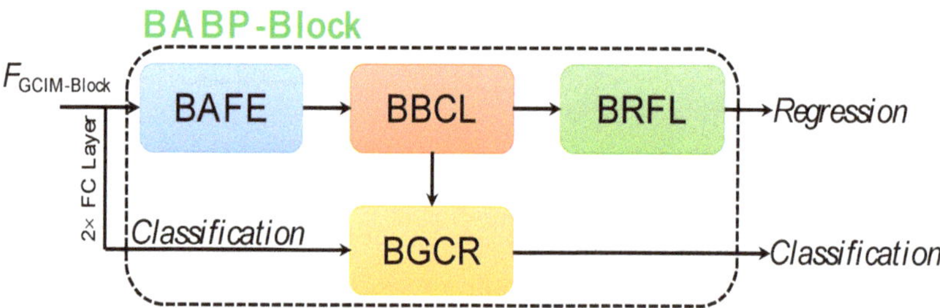

Figure 11. Workflow of the boundary-aware box prediction block (BABP-Block). Here, BAFF denotes the boundary-aware feature extraction, BBCL denotes the boundary bucketing coarse localization, BRFL denotes the boundary regression refined localization and BGCR denotes the boundary-guided classification rescoring.

2.2.1. Boundary-Aware Feature Extraction (BAFF)

The traditional feature extraction is implemented across the entire 2D space without distinguishing direction, i.e., four boundary directions including the x-left, x-right, y-top, and y-down. As a result, important boundary-sensitive features are not extracted. Thus, BAFF is arranged to solve this problem so as to ensure the subsequent boundary localization accuracy.

Figure 12 shows the implementation of BAFE. BAFE contains two parallel branches, i.e., x-boundary feature extraction and y-boundary feature extraction. Here, we take the x-boundary feature extraction as an example to introduce details. The same can be reasoned for the y-boundary feature extraction. First, we use a convolutional block attention module (CBAM) [61] to better capture direction-specific information of the ROI region. Then, a 1×1 conv with a softmax activation is used to normalize the attention map which will be weighted to the raw feature maps by the matrix element-wise multiplication. Afterwards, we sum features along the y-direction and use a 1×3 asymmetric conv to achieve the features along x-direction F_x. The above can be described by

$$F_x = \sum_y F_{\text{GCIM}-\text{Block}}(y,:) * M_x(y,:) \tag{6}$$

where M_x denotes the attention map of the x-boundary. Finally, F_x is split into two subsets evenly, i.e., $F_{x-right}$ and F_{x-left}, to represent the features of the right and left boundaries.

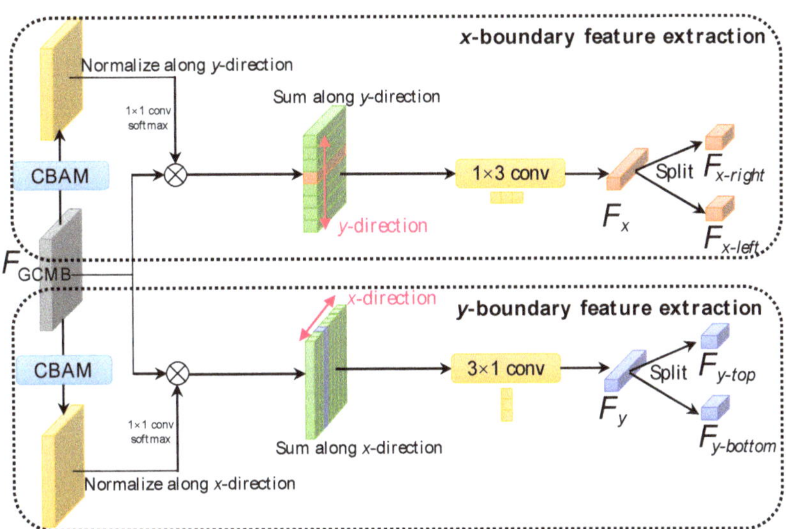

Figure 12. Implementation of the boundary-aware feature extraction (BAFE) used in BABP-Block.

Figure 13 shows the implementation of CBAM. Let its input be $F_{\text{GCIM}-\text{Block}} \in \mathbb{R}^{H \times W \times C}$ where H and W are the height and width of feature maps and C is the channel number, then the channel attention is responsible for generating a channel-dimension weight matrix $W_{CA} \in \mathbb{R}^{1 \times 1 \times C}$ to measure the important levels of C channels; the space attention is responsible for generating a space-dimension weight matrix $W_{SA} \in \mathbb{R}^{H \times W \times 1}$ to measure the important levels of space-elements across the entire $H \times W$ space. They both range from 0 to 1 by a sigmoid activation which can enrich nonlinearity of neural networks for better performance, suggested by [61]. The result of the channel attention is denoted by $F_{CA} = F \times W_{CA}$. The result of the space attention is denoted by $F_{SA} = F_{CA} \times W_{SA}$. It should be noted that here, the space attention is executed following the channel attention. It is also feasible to change their order.

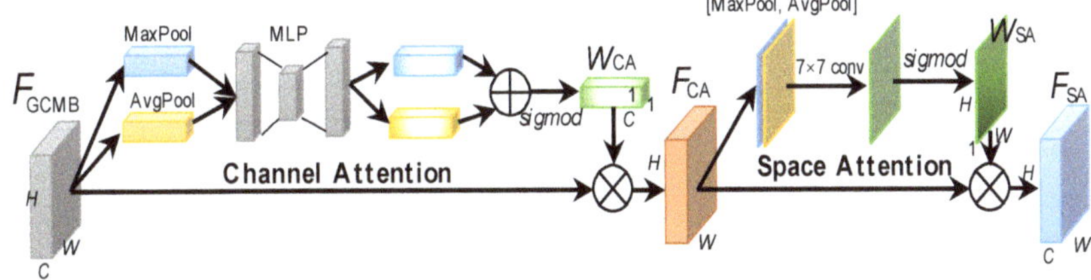

Figure 13. Implementation of the convolutional block attention module (CBAM) used in BAFE.

For the channel attention, a max-pooling (MaxPool) is used to capture its local response, and an average-pooling (AvgPool) is used to capture its global response. A multi-layer perceptron (MLP) is used to refine them for better fusion between the local and global responses. Finally, the results are normalized by a sigmoid function to obtain W_{CA}. The above is described by

$$W_{CA} = \sigma\{\text{MLP}[\text{MaxPool}(F)] + \text{MLP}[\text{AvgPool}(F)]\} \tag{7}$$

where σ denotes the sigmoid activation defined by $1/(1 + e^{-x})$.

For the space attention, MaxPool and AvgPool are also used. Still, differently, they both operate on the channel dimension to achieve 2D feature maps. Their results are concatenated directly and convolved by a common conv layer, producing the 2D spatial attention map. Finally, the results are normalized by a sigmoid activation to obtain W_{SA}. The above is described by

$$W_{SA} = \sigma\{f_{7\times7}([\text{MaxPool}(F_{CA}), \text{AvgPool}(F_{CA})])\} \tag{8}$$

where $f^{\prime\times\prime}$ is a 7×7 conv recommended by their original report [61].

2.2.2. Boundary Bucketing Coarse Localization (BBCL)

After boundary-sensitive features are achieved by the previous BAFE stage, we follow the bucketing idea [62] to predict the box boundary, referred to as BBCL. The specific implementation scheme is consistent with Wang et al. [59]. This scheme divides the target space into multiple buckets, or called discrete grid cells [60]. This coarse boundary localization is completed by searching for the correct bucket, i.e., the one in which the boundary resides. Figure 14 shows the implementation of BBCL. The candidate regions are divided into $2k$ buckets on both x-direction and y-direction, with k buckets corresponding to each boundary. Here, k is equal to 14, because the feature map's size is 14×14. From Figure 14, we adopt a fully-connected (FC) layer to serve as a binary classifier to predict whether the boundary is located in or is the closest to the bucket on each side, based on the ship boundary-aware features $F_{x-right}$, F_{x-left}, $F_{y-right}$, and F_{y-left}. We achieve the boundary probabilities of four sides, denoted by $s_{x-right}$, s_{x-left}, $s_{y-right}$, and s_{y-left}. It should be noted that the boundary probabilities of four sides, i.e., $s_{x-right}$, s_{x-left}, $s_{y-right}$, and s_{y-left} will be utilized for the final boundary-guided classification rescoring, which will be introduced in the next sections. Afterwards, the maximum activation value is then projected into the raw feature maps to achieve the corresponding index value. Finally, the four boundary positions are obtained, i.e., x-right, x-left, y-right, and y-left. In this way, the coarse boundary of a ship is predicted successfully.

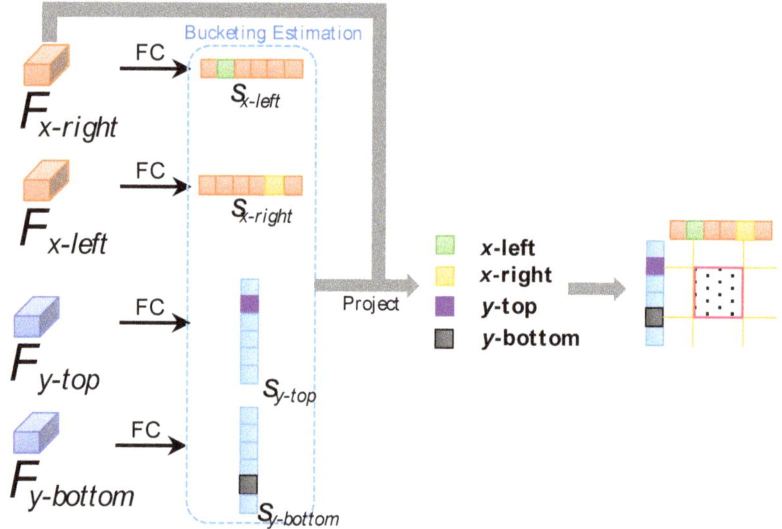

Figure 14. Implementation of the boundary bucketing coarse localization (BBCL) used in BABP-Block.

2.2.3. Boundary Regression Fine Localization (BRFL)

After the coarse boundary of a ship is obtained, we need to finely adjust the box close to the GT box in order to eliminate the boundary effects of buckets, as shown in Figure 15. This process is the same as the traditional bounding box regression scheme. Specifically, we adopt a 4-way FC layer to complete this task, i.e., the center point correction and width-height adjustment. Since this process is operated in the predicted boundary box, the distance between the initial box and the GT box becomes smaller. Consequently, such a regression task will become more relaxing to deal with the cross-scale effect, because the positioning difficulty is shared by the previous boundary prediction stage.

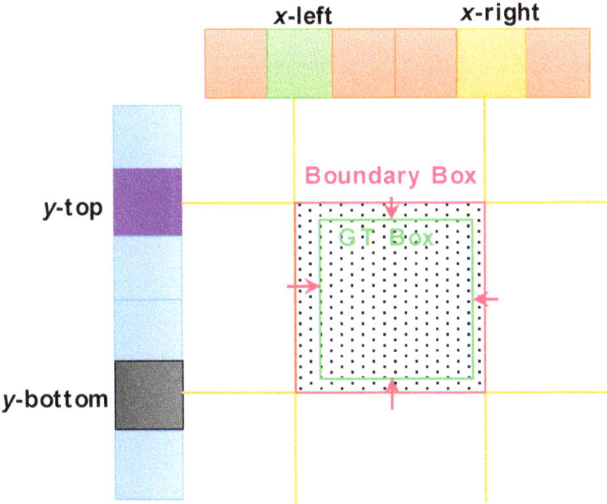

Figure 15. Implementation of the boundary regression fine localization (BRFL) used in BABP-Block.

Up to this point, we have achieved bounding box regression results by the regression branch in Figure 11.

2.2.4. Boundary-Guided Classification Rescoring (BGCR)

BBCL offers the localization reliability of the predicted boundary box, that is, the boundary probabilities of four sides $s_{x-right}$, s_{x-left}, $s_{y-right}$, and s_{y-left} as previously introduced. The idea of rescoring is also shown in FCOS [63] where the final classification score is computed by using the predicted center-ness score and the raw classification score together. And it is a direct intuition that it should be conducive to maintaining the optimum box with both high classification confidence and accurate localization if fully leveraging them. Thus, we arrange a boundary-guided classification rescoring (BGCR) strategy to reach this aim, which is described by

$$s' = \alpha \cdot s + \beta \cdot \frac{1}{4}\left(s_{x-right} + s_{x-left} + s_{y-right} + s_{y-left}\right) \tag{9}$$

where s denotes the original confidence score of the classification network (i.e., two FC layers in Figure 11), s' denotes the final confidence score, α denotes the weight coefficient of the original confidence score and β denotes the weight coefficient of the localization reliability. In our work, α and β are both set to 0.5 considering the trade-off between the spatial localization reliability and the classification reliability [64,65]. Here, in terms of the total localization reliability, we directly average the four sides' boundary probabilities, because they seem to be equally important. Finally, the resulting score s' will be inputted to the non-maximum suppression (NMS) algorithm [48] to remove repeated detections.

3. Experiments

3.1. Dataset

Two public datasets SSDD [33] and HRSID [34] are used in our work. SSDD is available online at https://github.com/TianwenZhang0825/Official-SSDD (accessed on 1 March 2022). HRSID is available online at https://github.com/chaozhong2010/HRSID, accessed on 1 April 2022. SSDD offers 1160 samples collected from RadarSat-2, TerraSAR-X, and

Sentinel-1. Polarizations are HH, VV, VH, and HV. Resolutions are from 1 m to 10 m. The test set has 232 samples with the filename suffix of 1 and 9. The remaining samples constitute the training set. HRSID offers 5604 samples collected from Sentinel-1 and TerraSAR-X. Polarizations are HH, HV, and VV. Resolutions are 0.5 m, 1 m and 3 m. The training set has 3642 samples and the test set has 1962.

3.2. Training Details

ResNet-101 [46] serves as the backbone network that is pretrained on ImageNet [66]. The FPN structure [40] is used to ensure multi-scale performance. We adopt the stochastic gradient descent (SGD) algorithm to train GCBANet and other nine comparison models by 12 epochs. The learning rate is 0.002 that is reduced by 10 times at 8-epoch and 11-epoch. The momentum is 0.9 and the weight decay is 0.0001. The batch size is 1 due to limited GPU memory. The training loss function and other hyper-parameters are same as the hybrid task cascade model [44]. The referenced source code we used for performance comparison is from MMDetection at https://github.com/open-mmlab/mmdetection (accessed on 1 March 2022).

3.3. Evaluation Criteria

The COCO metrics [67] are adopted. Its core index is the average precision (AP), i.e., the average value of precisions under ten intersection over union (IOU) thresholds from 0.50 to 0.95 with an interval of 0.05. AP_{50} denotes the AP of an IOU threshold of 0.50. AP_{75} denotes AP of an IOU threshold of 0.75. AP_S denotes AP of small ships (<322 pixels). AP_M denotes the AP of medium ships (>322 pixels and <962 pixels). AP_L denotes the AP of large ships (>962 pixels). Specifically, the IOU of the predicted mask and the ground truth mask is described by

$$IOU = \frac{P_{mask} \cap G_{mask}}{P_{mask} \cup G_{mask}} \tag{10}$$

where P_{mask} represents the predicted mask and G_{mask} represents the ground truth mask. According to a given IOU threshold and confidence threshold, the predictions of instance segmentation results can be divided into different categories, while true positive (TP), false positive (FP), and false negative (FN) represent the number of samples in each category. Then, the corresponding precision value and recall value is described by

$$\text{Precision} = \frac{TP}{TP + FP} \tag{11}$$

$$\text{Recall} = \frac{TP}{TP + FN} \tag{12}$$

With confidence threshold changes, precision and recall will be different, with the result that the precision and recall curve $P(r)$ where the recall value serves as the abscissa and precision value serves as the ordinate in Cartesian coordinate system. Then, the AP of a given IOU threshold is described by

$$AP_{IOU} = \int_0^1 P(r)dr \tag{13}$$

Then, AP is the average value of 10 AP_{IOU} whose IOU threshold ranges from 0.5 to 0.95 with the stride of 0.05., which is described by

$$AP = \frac{1}{10} \times \sum_{IOU=0.50}^{0.95} AP_{IOU} \tag{14}$$

4. Results

4.1. Quantitative Results

Tables 2 and 3 are the quantitative results on the SSDD and HRSID datasets. From Tables 2 and 3, GCBANet achieves the best accuracy, that is, on SSDD, the box AP reaches 68.4% and the mask AP reaches 63.1%; on HRSIS, the box AP reaches 69.4% and the mask AP reaches 57.3%. The mask prediction has poorer indexes than the box prediction, because the pixel-level segmentation task is more difficult than the box-level detection task. GCBANet outperforms the other nine competitive models by a significant degree. On SSDD, it achieves 2.8% higher box AP and 3.5% higher mask AP than the previous most advanced model; on HRSID, they are 2.7% and 1.9%. This fully reveals the state-of-the-art performance of GCBANet. This accuracy advantage benefits from the combined action of the proposed GCIM-Block and BABP-Block. Certainly, the speed of GCBANet does not win advantages, compared with others, thereby the speed optimization is required in the future. Moreover, although YOLACT [68] offers the fastest detection speed because it is a one-stage model, its accuracy is too poor to satisfy application requirements.

Table 1 shows the computational complexity calculations of different methods. Here, we adopt the floating point of operations (FLOPs) to measure calculations whose unit is the giga multiply add calculations (GMACs) [69]. From Table 1, the calculation amount of GCBANet is more than the others, so future model computational complexity optimization is needed.

Table 1. Computational complexity calculations of different methods. Here, we adopt the floating point of operations (FLOPs) to measure calculations whose unit is the giga multiply add calculations (GMACs) [69].

Method	Backbone	FLOPs (GMACs)
Mask R-CNN [47]	ResNet-101	121.32
Mask Scoring R-CNN [70]	ResNet-101	121.32
Cascade Mask R-CNN [43]	ResNet-101	226.31
PANet [71]	ResNet-101	127.66
YOLACT [68]	ResNet-101	67.14
GROIE [72]	ResNet-101	581.28
HQ-ISNet [35]	HRNetV2-W18	201.84
HQ-ISNet [35]	HRNetV2-W32	226.90
HQ-ISNet [35]	HRNetV2-W40	247.49
SA R-CNN [37]	ResNet-50-GCB	101.87
HTC [44]	ResNet-101	228.90
GCBANet (Ours)	ResNet-101	947.96

Table 2. Quantitative Results on SSDD. The suboptimal method is marked by underline "—".

Method	Backbone	Box (%)						Mask (%)						FPS
		AP	AP_{50}	AP_{75}	AP_S	AP_M	AP_L	AP	AP_{50}	AP_{75}	AP_S	AP_M	AP_L	
Mask R-CNN [47]	ResNet-101	62.0	91.5	75.4	62.0	64.4	19.7	57.8	88.5	72.1	57.2	60.8	27.4	11.05
Mask Scoring R-CNN [70]	ResNet-101	62.4	91.0	75.1	61.9	66.0	15.7	58.6	89.4	73.2	58.0	61.4	22.6	12.88
Cascade Mask R-CNN [43]	ResNet-101	63.0	89.6	75.2	62.4	66.0	12.0	56.6	87.5	70.5	56.3	58.8	22.6	10.55
PANet [71]	ResNet-101	63.3	93.4	75.4	63.4	65.5	40.8	59.6	91.1	74.0	59.3	61.0	52.1	13.65
YOLACT [68]	ResNet-101	54.0	90.6	61.2	56.9	48.2	12.6	48.4	88.0	52.1	47.3	53.5	40.2	15.47
GROIE [72]	ResNet-101	61.2	91.5	71.6	62.2	59.8	8.7	58.3	89.8	72.7	58.6	58.7	21.8	9.67
HQ-ISNet [35]	HRNetV2-W18	64.9	91.0	76.3	64.7	66.6	26.0	58.6	89.3	73.6	58.2	60.4	37.2	8.59
HQ-ISNet [35]	HRNetV2-W32	65.5	90.7	77.3	65.6	66.9	23.2	59.3	90.4	75.5	58.9	61.1	37.3	8.00
HQ-ISNet [35]	HRNetV2-W40	63.6	87.8	75.3	62.6	67.8	27.9	57.6	86.0	72.6	56.7	61.3	50.2	7.73
SA R-CNN [37]	ResNet-50-GCB	63.2	92.1	75.2	63.8	64.0	7.0	59.4	90.4	73.3	59.6	60.3	20.2	13.65
HTC [44]	ResNet-101	65.6	93.6	76.3	65.2	68.4	27.5	59.3	91.7	73.1	58.7	61.6	34.8	11.60
GCBANet (Ours)	ResNet-101	68.4	95.4	82.2	68.9	68.0	45.6	63.1	93.5	78.8	63.2	63.0	55.1	6.11
		+2.8	+1.8	+4.9	+3.3	+1.4	+4.8	+3.5	+1.8	+3.3	+3.6	+1.4	+3.0	

Table 3. Quantitative results on HRSID. The suboptimal method is marked by Underline "—".

Method	Backbone	Box (%)						Mask (%)						FPS
		AP	AP_{50}	AP_{75}	AP_S	AP_M	AP_L	AP	AP_{50}	AP_{75}	AP_S	AP_M	AP_L	
Mask R-CNN [17]	ResNet-101	65.1	87.7	75.5	66.1	68.4	14.1	54.8	85.7	65.2	54.3	62.5	13.3	7.07
Mask Scoring R-CNN [70]	ResNet-101	65.2	87.6	75.4	66.5	67.4	13.4	54.9	85.1	65.9	54.5	61.5	12.9	8.24
Cascade Mask R-CNN [43]	ResNet-101	65.1	85.4	74.4	66.0	69.0	17.1	52.8	83.4	62.9	52.2	62.2	17.0	6.75
PANet [71]	ResNet-101	65.4	88.0	75.7	66.5	68.2	22.1	55.1	86.0	66.2	54.7	62.8	17.8	8.74
YOLACT [68]	ResNet-101	47.9	74.4	53.3	51.7	34.9	3.3	39.6	71.1	41.9	39.5	46.1	7.3	10.02
GROIE [72]	ResNet-101	65.4	87.8	75.5	66.5	67.2	21.8	55.4	85.8	66.9	54.9	63.5	19.7	6.19
HQ-ISNet [35]	HRNetV2-W18	66.0	86.1	75.6	67.1	66.3	8.9	53.4	84.2	64.3	53.2	59.7	10.7	5.50
HQ-ISNet [35]	HRNetV2-W32	66.7	86.9	76.3	67.8	68.3	16.8	54.6	85.0	65.8	54.2	61.7	13.4	5.12
HQ-ISNet [35]	HRNetV2-W40	66.7	86.2	76.3	67.9	68.6	11.7	54.2	84.3	64.9	53.9	61.9	12.8	4.95
SA R-CNN [37]	ResNet-50-GCB	65.2	88.3	75.2	66.4	65.4	10.2	55.2	86.2	66.7	54.9	60.9	12.3	8.74
HTC [44]	ResNet-101	66.6	86.0	77.1	67.6	69.0	28.1	55.2	84.9	66.5	54.7	63.8	19.2	7.42
GCBANet (Ours)	ResNet-101	69.4	89.8	79.2	70.4	71.3	32.2	57.3	88.6	68.9	57.0	64.3	25.9	4.06
		+2.7	+1.8	+2.1	+2.5	+2.3	+4.1	+1.9	+2.4	+2.0	+2.1	+0.5	+6.7	

4.2. Qualitative Results

Figures 16 and 17 are the quantitative results on SSDD and HRSID where we only show the quantitative comparison results with the second-best mask AP.

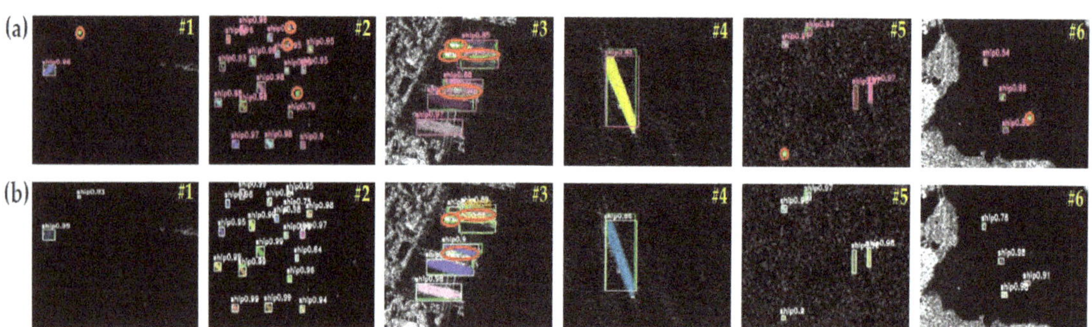

Figure 16. Qualitative results on SSDD. (**a**) PANet with the second-best mask AP. (**b**) GCBANet. Green boxes denote the ground truths. Orange boxes denote the false alarms (i.e., false positives, FP). Red circles denote the missed detections (i.e., false negatives, FN). #i denote the ith picture of results.

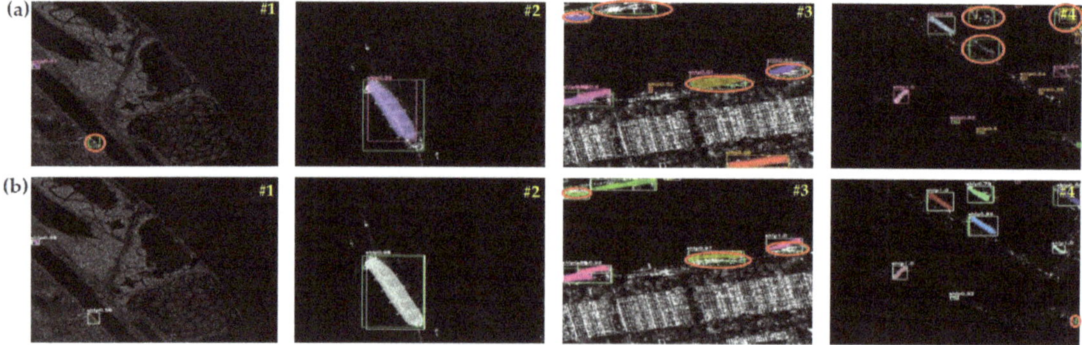

Figure 17. Qualitative results on HRSID. (**a**) GROIE with the second-best mask AP. (**b**) GCBANet. Green boxes denote the ground truths. Orange boxes denote the false alarms (i.e., false positives, FP). Red circles denote the missed detections (i.e., false negatives, FN). #i denote the ith picture of results.

From Figures 16 and 17, the following conclusions can be drawn:

1. GCBANet can detect more ships, e.g., the #1 sample in Figure 16 and the #1 sample in Figure 17. This is because GCBANet can extract more salient features, leading to better box prediction.
2. GCBANet can avoid many false alarms, e.g., the #3 sample in Figure 16 and the #4 sample in Figure 17. This is because the designed GCIM-Block can distinguish the foreground-background more effectively with the surrounding context information.
3. GCBANet can enable ship instance segmentation with higher reliability. For example, for the same ship in the #2 sample in Figure 17, the confidence score of GROIE is 0.59, which is far smaller than that of our GCBANet 0.98. This advantage benefits from the boundary-guided classification rescoring strategy in the designed BABP-Block, which can consider both high classification confidence and accurate localization.
4. GCBANet can locate large ships more accurately, e.g., the #4 sample in Figure 16 and the #2 sample in Figure 17. This advantage benefits from the designed BABP-Block, which can handle the cross-scale problem through the boundary prediction, so as to enable better box regression.

Given the above, GCBANet achieves state-of-the-art SAR ship instance segmentation performance.

5. Ablation Study

In this section, we will show the results of ablation studies to confirm the effectiveness of the proposed GCIM-Block and BABP-Block. Experiments are performed on SSDD.

5.1. Ablation Study on GCIM-Block

5.1.1. Effectiveness of GCIM-Block

Table 4 shows the quantitative results with/without GCIM-Block. The GCIM-Block offers a 1.6% box AP gain and a 1.1% mask AP gain, showing its effectiveness. It can model the global background context information so as to capture spatial long-range dependences of ships, which can enable better background discrimination ability.

Table 4. Quantitative Results with and Without GCIM-Block.

GCIM-Block	Box AP (%)	Mask AP (%)
✗	66.8	62.0
✔	68.4	63.1

5.1.2. Component Analysis in GCIM-Block

We also make a component analysis in GCIM-Block, as shown in Table 5. From Table 5, each component can offer an observable accuracy improvement, either the box AP or the mask AP. This indicates that our well-designed idea is reasonable and our theoretical analysis in Section 2.1 is correct. Moreover, we observe that GFSA does not improve the box prediction performance, but it improves the mask prediction performance further. This is because the global feature self-attention is pixel-sensitive, which can enable better pixel classification capability.

Table 5. Quantitative Results Component Analysis in GCIM-Block.

CAFR [1]	MRFFR [2]	GFSA [3]	Box AP (%)	Mask AP (%)	FPS
-	-	-	66.8	62.0	9.25
✔			67.5	62.4	7.86
✔	✔		68.4	62.8	6.72
✔	✔	✔	68.4	63.1	6.11

[1] CAFR denotes the content-aware feature reassembly. [2] MRFFR denotes the multi receptive-field feature response. [3] GFSA denotes the global feature self-attention.

5.2. Ablation Study on BABP-Block

5.2.1. Effectiveness of BABP-Block

Table 6 shows the quantitative results with and without BABP-Block. BABP-Block offers a 2.8% box AP gain and a 1.8% mask AP gain, showing its effectiveness. It can offer better box prediction by predicting the boundary so as to enable better mask prediction. Thus, this boundary estimation scheme should be more suitable for SAR ships.

Table 6. Quantitative Results with and Without BABP-Block.

BABP-Block	Box AP (%)	Mask AP (%)
✗	65.6	61.3
✔	68.4	63.1

5.2.2. Component Analysis in BABP-Block

We also make a component analysis in BABP-Block as shown in Table 7. From Table 7, each component is conducive to boosting accuracy, which shows their effectiveness. BAFE is able to extract more boundary-sensitive features so as to ensure accurate boundary bucketing coarse localization. BBCL locates four sides of the box to avoid long-distance regression, which boosts information flow. BRFL can enable more refined box regression. Finally, BGCR can leverage the boundary reliability to guide classification scores, so as to screen the detection results again, leading to more reliable predictions. As a result, the accuracy is improved progressively.

Table 7. Quantitative Results Component Analysis in BABP-Block.

BAFE [1]	BBCL [2]	BRFL [3]	BGCR [4]	Box AP (%)	Mask AP (%)	FPS
-	-	-	-	65.6	61.3	8.42
✔				66.0	61.9	7.36
✔	✔			66.7	62.2	6.58
✔	✔	✔		67.1	62.8	6.19
✔	✔	✔	✔	68.4	63.1	6.11

[1] BAFF denotes the boundary-aware feature extraction. [2] BBCL denotes the boundary bucketing coarse localization. [3] BRFL denotes the boundary regression fine localization. [4] BGCR denotes the boundary-guided classification rescoring.

6. Conclusions

In this paper, GCBANet is proposed for better SAR ship instance segmentation. In GCBANet, GCIM-Block and BABP-Block are designed to ensure its excellent performance. Specifically, GCIM-Block is used to mitigate the inferences caused by the surroundings of ships and the mechanism of SAR imaging. BABP-Block is used to locate ships more precisely. Ablation studies can confirm the effectiveness of GCIM-Block and BABP-Block. The results on two open datasets reveal the state-of-the-art performance of GCBANet compared to the other nine competitive models. On SSDD, GCBANet achieves 2.8% higher box AP and 3.5% higher mask AP than the existing best model; on HRSID, they are 2.7% and 1.9%.

Our future work is as follows:

1. We will consider optimizing the speed and the model computational complexity of GCBANet in the future.
2. We will consider simplifying GCIM-Block to further reduce the computational cost in the future.

Author Contributions: Conceptualization, T.Z.; methodology, X.K.; software, X.K.; validation, T.Z.; formal analysis, T.Z.; investigation, T.Z.; resources, T.Z.; data curation, T.Z.; writing—original draft preparation, X.K. and T.Z.; writing—review and editing, X.K. and T.Z.; visualization, T.Z.; supervision, X.K.; project administration, X.Z.; funding acquisition, X.Z. All authors have read and agreed to the published version of the manuscript.

Funding: This work was supported by the National Natural Science Foundation of China (61571099).

Data Availability Statement: Not applicable.

Acknowledgments: The authors would like to thank the editors and the four anonymous reviewers for their valuable comments that can greatly improve our manuscript.

Conflicts of Interest: The authors declare that they have no conflicts of interest.

References

1. Liu, C.; Zoughi, R. Adaptive Synthetic Aperture Radar (SAR) Imaging for Optimal Cross-Range Resolution and Image Quality in Nde Applications. *IEEE Trans. Instrum. Meas.* **2021**, *70*, 8005107. [CrossRef]
2. Zhang, T.; Zhang, X.; Shi, J.; Wei, S. Hyperli-Net: A Hyper-Light Deep Learning Network for High-Accurate and High-Speed Ship Detection from Synthetic Aperture Radar Imagery. *ISPRS J. Photogramm. Remote Sens.* **2020**, *167*, 123–153. [CrossRef]
3. Zhang, T.; Zhang, X. A Polarization Fusion Network with Geometric Feature Embedding for SAR Ship Classification. *Pattern Recognit.* **2021**, *123*, 108365. [CrossRef]
4. Jarabo-Amores, P.; Rosa-Zurera, M.; Mata-Moya, D.D.L.; Vicen-Bueno, R.; Maldonado-Bascon, S. Spatial-Range Mean-Shift Filtering and Segmentation Applied to SAR Images. *IEEE Trans. Instrum. Meas.* **2011**, *60*, 584–597. [CrossRef]
5. Gao, Y.; Qaseer, M.T.A.; Zoughi, R. Complex Permittivity Extraction from Synthetic Aperture Radar Images. *IEEE Trans. Instrum. Meas.* **2020**, *69*, 4919–4929. [CrossRef]
6. Watanabe, T.; Yamada, H. Synthetic Aperture Imaging of near-Field Scatterers Mutually Coupled with an Antenna Array. *IEEE Trans. Instrum. Meas.* **2021**, *70*, 8001218. [CrossRef]
7. Dudczyk, J.; Kawalec, A. Optimizing the minimum cost flow algorithm for the phase unwrapping process in SAR radar. *Bull. Pol. Acad. Sci. Tech.* **2014**, *62*, 511. [CrossRef]
8. Ai, J.; Luo, Q.; Yang, X.; Yin, Z.; Xu, H. Outliers-Robust CFAR Detector of Gaussian Clutter Based on the Truncated-Maximum-Likelihood- Estimator in SAR Imagery. *IEEE Trans. Intell. Transp. Syst.* **2020**, *21*, 2039–2049. [CrossRef]
9. Bentes, C.; Velotto, D.; Tings, B.O. Ship Classification in Terrasar-X Images with Convolutional Neural Networks. *IEEE J. Oceanic. Eng.* **2018**, *43*, 258–266. [CrossRef]
10. Zhang, T.; Zhang, X.; Ke, X.; Liu, C.; Xu, X.; Zhan, X.; Wang, C.; Ahmad, I.; Zhou, Y.; Wei, S.; et al. Hog-ShipCLSNet: A Novel Deep Learning Network with Hog Feature Fusion for SAR Ship Classification. *IEEE Trans. Geosci. Remote. Sens.* **2021**, *60*, 5210322. [CrossRef]
11. Koyama, C.N.; Gokon, H.; Jimbo, M.; Koshimura, S.; Sato, M. Disaster Debris Estimation Using High-Resolution Polarimetric Stereo-SAR. *ISPRS J. Photogramm. Remote Sens.* **2016**, *120*, 84–98. [CrossRef]
12. Zhang, T.; Zhang, X. Squeeze-and-Excitation Laplacian Pyramid Network with Dual-Polarization Feature Fusion for Ship Classification in SAR Images. *IEEE Geosci. Remote Sens. Lett.* **2021**, *19*, 4019905. [CrossRef]
13. Zhang, T.; Zhang, X.; Ke, X.; Zhan, X.; Shi, J.; Wei, S.; Pan, D.; Li, J.; Su, H.; Zhou, Y.; et al. Ls-Ssdd-V1.0: A Deep Learning Dataset Dedicated to Small Ship Detection from Large-Scale Sentinel-1 SAR Images. *Remote Sens.* **2020**, *12*, 2997. [CrossRef]
14. Zeng, X.; Wei, S.; Shi, J.; Zhang, X. A Lightweight Adaptive Roi Extraction Network for Precise Aerial Image Instance Segmentation. *IEEE Trans. Instrum. Meas.* **2021**, *70*, 5018617. [CrossRef]
15. Wang, B.; Dong, M.; Ren, M.; Wu, Z.; Guo, C.; Zhuang, T.; Pischler, O.; Xie, J. Automatic Fault Diagnosis of Infrared Insulator Images Based on Image Instance Segmentation and Temperature Analysis. *IEEE Trans. Instrum. Meas.* **2020**, *69*, 5345–5355. [CrossRef]
16. Tu, B.; Liao, X.; Zhou, C.; Chen, S.; He, W. Feature Extraction Using Multitask Superpixel Auxiliary Learning for Hyperspectral Classification. *IEEE Trans. Instrum. Meas.* **2021**, *70*, 1–16. [CrossRef]
17. Chen, S.W.; Cui, X.C.; Wang, X.S.; Xiao, S.P. Speckle-Free SAR Image Ship Detection. *IEEE Trans. Image Process.* **2021**, *30*, 5969–5983. [CrossRef]
18. Zhang, T.; Zhang, X. A Full-Level Context Squeeze-and-Excitation Roi Extractor for SAR Ship Instance Segmentation. *IEEE Geosci. Remote Sens. Lett.* **2022**. early access. [CrossRef]
19. Cui, Z.; Li, Q.; Cao, Z.; Liu, N. Dense Attention Pyramid Networks for Multi-Scale Ship Detection in SAR Images. *IEEE Trans. Geosci. Remote. Sens.* **2019**, *57*, 8983–8997. [CrossRef]
20. Zhang, T.; Zhang, X.; Liu, C.; Shi, J.; Wei, S.; Ahmad, I.; Su, H.; Zhan, X.; Zhou, Y.; Pan, D.; et al. Balance Learning for Ship Detection from Synthetic Aperture Radar Remote Sensing Imagery. *ISPRS J. Photogramm. Remote Sens.* **2021**, *182*, 190–207. [CrossRef]
21. Sun, Z.; Dai, M.; Leng, X.; Lei, Y.; Xiong, B.; Ji, K.; Kuang, G. An Anchor-Free Detection Method for Ship Targets in High-Resolution SAR Images. *IEEE J. Sel. Top. Appl. Earth Obs. Remote Sens.* **2021**, *14*, 7799–7816. [CrossRef]
22. Zhang, T.; Zhang, X.; Shi, J.; Wei, S. Depthwise Separable Convolution Neural Network for High-Speed SAR Ship Detection. *Remote Sens.* **2019**, *11*, 2483. [CrossRef]
23. Zhang, T.; Zhang, X. Shipdenet-20: An Only 20 Convolution Layers and <1-Mb Lightweight SAR Ship Detector. *IEEE Geosci. Remote Sens. Lett.* **2021**, *18*, 1234–1238. [CrossRef]

24. Song, J.; Kim, D.-J.; Kang, K.-M. Automated Procurement of Training Data for Machine Learning Algorithm on Ship Detection Using AIS Information. *Remote Sens.* **2020**, *12*, 1443. [CrossRef]
25. Gao, G.; Shi, G. CFAR Ship Detection in Nonhomogeneous Sea Clutter Using Polarimetric SAR Data Based on the Notch Filter. *IEEE Trans. Geosci. Remote. Sens.* **2017**, *55*, 4811–4824. [CrossRef]
26. Yang, H.; Cao, Z.; Cui, Z.; Pi, Y. Saliency Detection of Targets in Polarimetric SAR Images Based on Globally Weighted Perturbation Filters. *ISPRS J. Photogramm. Remote Sens.* **2019**, *117*, 65–79. [CrossRef]
27. Wang, X.; Li, G.; Zhang, X.P.; He, Y. A Fast Cfar Algorithm Based on Density-Censoring Operation for Ship Detection in SAR Images. *IEEE Signal Process. Lett.* **2021**, *28*, 1085–1089. [CrossRef]
28. LeCun, Y.; Bottou, L.; Bengio, Y.; Haffner, P. Gradient-based learning applied to document recognition. *Proc. IEEE* **1998**, *86*, 2278. [CrossRef]
29. Krizhevsky, A.; Sutskever, I.; Hinton, G.E. Imagenet Classification with Deep Convolutional Neural Networks. In Proceedings of the International Conference on Neural Information Processing Systems (NIPS), Lake Tahoe, NV, USA, 3–8 December 2012; pp. 1097–1105.
30. Simonyan, K.; Zisserman, A. Very Deep Convolutional Networks for Large-Scale Image Recognition. In Proceedings of the International Conference on Learning Representations (ICLR), San Diego, CA, USA, 7–9 May 2015; pp. 1–14.
31. Girshick, R. Fast R-CNN. In Proceedings of the IEEE International Conference on Computer Vision, Santiago, Chile, 7–13 December 2015; p. 1440.
32. Ren, S.; He, K.; Girshick, R.; Sun, J. Faster R-CNN: Towards Real-Time Object Detection with Region Proposal Networks. *IEEE Trans. Pattern Anal. Mach. Intell.* **2017**, *39*, 1137–1149. [CrossRef]
33. Zhang, T.; Zhang, X.; Li, J.; Xu, X.; Wang, B.; Zhan, X.; Xu, Y.; Ke, X.; Zeng, T.; Su, H.; et al. SAR Ship Detection Dataset (SSDD): Official Release and Comprehensive Data Analysis. *Remote Sens.* **2021**, *13*, 3690. [CrossRef]
34. Wei, S.; Zeng, X.; Qu, Q.; Wang, M.; Su, H.; Shi, J. HRSID: A High-Resolution SAR Images Dataset for Ship Detection and Instance Segmentation. *IEEE Access.* **2020**, *8*, 120234–120254. [CrossRef]
35. Su, H.; Wei, S.; Liu, S.; Liang, J.; Wang, C.; Shi, J.; Zhang, X. HQ-ISNet: High-Quality Instance Segmentation for Remote Sensing Imagery. *Remote Sens.* **2020**, *12*, 989. [CrossRef]
36. Gao, F.; Huo, Y.; Wang, J.; Hussain, A.; Zhou, H. Anchor-Free SAR Ship Instance Segmentation with Centroid-Distance Based Loss. *IEEE J. Sel. Top. Appl. Earth Obs. Remote Sens.* **2021**, *14*, 11352–11371. [CrossRef]
37. Zhao, D.; Zhu, C.; Qi, J.; Qi, X.; Su, Z.; Shi, Z. Synergistic Attention for Ship Instance Segmentation in SAR Images. *Remote Sens.* **2021**, *13*, 4384. [CrossRef]
38. Wang, J.; Chen, K.; Yang, S.; Loy, C.C.; Lin, D. Region Proposal by Guided Anchoring. In Proceedings of the IEEE Conference on Computer Vision and Pattern Recognition (CVPR), Long Beach, CA, USA, 16–20 June 2019; pp. 2960–2969.
39. Zhou, Z.; Guan, R.; Cui, Z.; Cao, Z.; Pi, Y.; Yang, J. Scale Expansion Pyramid Network for Cross-Scale Object Detection in SAR Images. In Proceedings of the IEEE International Geoscience and Remote Sensing Symposium (IGARSS), Brussels, Belgium, 11–16 July 2021; pp. 5291–5294.
40. Zhang, T.; Zhang, X.; Ke, X. Quad-FPN: A Novel Quad Feature Pyramid Network for SAR Ship Detection. *Remote Sens.* **2021**, *13*, 2771. [CrossRef]
41. Li, J.; Qu, C.; Shao, J. Ship Detection in SAR Images Based on an Improved Faster R-CNN. In Proceedings of the SAR in Big Data Era: Models, Methods and Applications (BIGSARDATA), Beijing, China, 13–14 November 2017; pp. 1–6.
42. Github. Available online: https://github.com/TianwenZhang0825/GCBANet (accessed on 1 March 2022).
43. Cai, Z.; Vasconcelos, N. Cascade R-CNN: High Quality Object Detection and Instance Segmentation. *IEEE Trans. Pattern Anal. Mach. Intell.* **2021**, *43*, 1483–1498. [CrossRef] [PubMed]
44. Chen, K.; Ouyang, W.; Loy, C.C.; Lin, D.; Pang, J.; Wang, J.; Xiong, L.; Li, X.; Sun, S.; Feng, W.; et al. Hybrid Task Cascade for Instance Segmentation. In Proceedings of the IEEE/CVF Conference on Computer Vision and Pattern Recognition (CVPR), Long Beach, CA, USA, 16–20 June 2019; pp. 4969–4978.
45. Wang, J.; Sun, K.; Cheng, T.; Jiang, B.; Deng, C.; Zhao, Y.; Xiao, B.; Liu, D.; Mu, Y.; Tan, M.; et al. Deep High-Resolution Representation Learning for Visual Recognition. *IEEE Trans. Pattern Anal. Mach. Intell.* **2020**, *43*, 3349–3364. [CrossRef]
46. He, K.; Zhang, X.; Ren, S.; Sun, J. Deep Residual Learning for Image Recognition. In Proceedings of the IEEE Conference on Computer Vision and Pattern Recognition (CVPR), Las Vegas, NV, USA, 26 June–1 July 2016; pp. 770–778.
47. He, K.; Gkioxari, G.; Dollar, P.; Girshick, R. Mask R-CNN. In Proceedings of the IEEE International Conference on Computer Vision (ICCV), Venice, Italy, 22–29 October 2017; pp. 2980–2988.
48. Hosang, J.; Benenson, R.; Schiele, B. Learning Non-Maximum Suppression. In Proceedings of the IEEE Conference on Computer Vision and Pattern Recognition (CVPR), Honolulu, HI, USA, 21–26 July 2017; pp. 6469–6477.
49. Hafiz, A.M.; Bhat, G.M. A Survey on Instance Segmentation: State of the Art. *Int. J. Multimed. Inf. Retr.* **2020**, *9*, 171–189. [CrossRef]
50. Zhang, T.; Zhang, X.; Shi, J.; Wei, S.; Wang, J.; Li, J.; Zhou, Y.; Su, H. Balance Scene Learning Mechanism for Offshore and Inshore Ship Detection in SAR Images. *IEEE Geosci. Remote Sens. Lett.* **2020**, *19*, 4004905. [CrossRef]
51. Wang, J.; Chen, K.; Xu, R.; Liu, Z.; Loy, C.C.; Din, D. Carafe: Content-Aware Reassembly of Features. In Proceedings of the IEEE/CVF International Conference on Computer Vision (ICCV), Seoul, Korea, 27–28 October 2019; pp. 3007–3016.

52. Benedetto, J.J.; Treiber, O.M. Wavelet Frames: Multiresolution Analysis and Extension Principles. In *Wavelet Transforms and Time-Frequency Signal Analysis*; Debnath, L., Ed.; Birkhäuser Boston: Boston, MA, USA, 2001; pp. 3–36.
53. Yu, F.; Koltun, V. Multi-scale context aggregation by dilated convolutions. In Proceedings of the 4th International Conference on Learning Representations, San Juan, PR, USA, 2–4 May 2016; pp. 1–13.
54. Wang, X.; Girshick, R.; Gupta, A.; He, K. Non-local neural networks. In Proceedings of the IEEE Conference on Computer Vision and Pattern Recognition, Salt Lake City, GA, USA, 18–22 June 2018; pp. 7794–7803.
55. Buckreuss, S.; Schättler, B.; Fritz, T.; Mittermayer, J.; Kahle, R.; Maurer, E.; Böer, J.; Bachmann, M.; Mrowka, F.; Schwarz, E.; et al. Ten Years of TerraSAR-X Operations. *Remote Sens.* **2018**, *10*, 873. [CrossRef]
56. Torres, R.; Snoeij, P.; Geudtner, D.; Bibby, D.; Davidson, M.; Attema, E.; Potin, P.; Brown, M.; Bruno, C.; Miranda, N.; et al. GMES Sentinel-1 Mission. *Remote Sens. Environ.* **2012**, *120*, 9–24. [CrossRef]
57. Zhang, T.; Zhang, X. High-Speed Ship Detection in SAR Images Based on a Grid Convolutional Neural Network. *Remote Sens.* **2019**, *11*, 1206. [CrossRef]
58. Iervolino, P.; Guida, R.; Whittaker, P. A Model for the Backscattering from a Canonical Ship in SAR Imagery. *IEEE J. Sel. Top. Appl. Earth Obs. Remote Sens.* **2016**, *9*, 1163–1175. [CrossRef]
59. Wang, J.; Zhang, W.; Cao, Y.; Chen, K.; Pang, J.; Gong, T.; Lin, D.; Shi, J.; Loy, C.C. Side-Aware Boundary Localization for More Precise Object Detection. In Proceedings of the Computer Vision—ECCV 2020, 16th European Conference, Glasgow, UK, 23–28 August 2020; pp. 403–419.
60. Lu, X.; Li, B.; Yue, Y.; Li, Q.; Yan, J. Grid R-CNN. In Proceedings of the IEEE/CVF Conference on Computer Vision and Pattern Recognition (CVPR), Long Beach, CA, USA, 16–20 June 2019; pp. 7355–7364.
61. Woo, S.; Park, J.; Lee, J.-Y.; Kweon, I.S. Cbam: Convolutional Block Attention Module. In Proceedings of the European Conference on Computer Vision (ECCV), Munich, Germany, 8–14 September 2018; pp. 3–19.
62. Tom, K.; Ondrej, B. Curriculum learning and minibatch bucketing in neural machine translation. *arXiv* **2017**, arXiv:1707.09533.
63. Tian, Z.; Shen, C.; Chen, H.; He, T. FCOS: Fully Convolutional One-Stage Object Detection. In Proceedings of the 2019 IEEE/CVF International Conference on Computer Vision (ICCV), Seoul, Korea, 27–28 October 2019; p. 9626.
64. Rybak, Ł.; Dudczyk, J. Variant of Data Particle Geometrical Divide for Imbalanced Data Sets Classification by the Example of Occupancy Detection. *Appl. Sci.* **2021**, *11*, 4970. [CrossRef]
65. Chen, H.; Zhang, F.; Tang, B.; Yin, Q.; Sun, X. Slim and Efficient Neural Network Design for Resource-Constrained SAR Target Recognition. *Remote Sens.* **2018**, *10*, 1618. [CrossRef]
66. He, K.; Girshick, R.; Dollar, P. Rethinking Imagenet Pre-Training. In Proceedings of the IEEE International Conference on Computer Vision (ICCV), Seoul, Korea, 27–28 October 2019; pp. 4917–4926.
67. Lin, T.-Y.; Maire, M.; Belongie, S.; Hays, J.; Perona, P.; Ramanan, D.; Zitnick, C.L.; Dollar, P. Microsoft Coco: Common Objects in Context. In Proceedings of the European Conference on Computer Vision (ECCV), 13th European Conference, Zurich, Switzerland, 6–12 September 2014; pp. 740–755.
68. Bolya, D.; Zhou, C.; Xiao, F.; Lee, Y.J. Yolact: Real-Time Instance Segmentation. In Proceedings of the 2019 IEEE/CVF International Conference on Computer Vision (ICCV), Seoul, Korea, 27–28 October 2019; pp. 9156–9165.
69. Eric, Q. Floating-Point Fused Multiply–Add Architectures. Ph.D. Thesis, The University of Texas at Austin, Austin, TX, USA, 2007.
70. Huang, Z.; Huang, L.; Gong, Y.; Huang, C.; Wang, X. Mask Scoring R-CNN. In Proceedings of the 2019 IEEE/CVF Conference on Computer Vision and Pattern Recognition (CVPR), Long Beach, CA, USA, 16–20 June 2019; pp. 6402–6411.
71. Liu, S.; Qi, L.; Qin, H.; Shi, J.; Jia, J. Path Aggregation Network for Instance Segmentation. In Proceedings of the IEEE Conference on Computer Vision and Pattern Recognition (CVPR), Salt Lake City, GA, USA, 18–22 June 2018; pp. 8759–8768.
72. Rossi, L.; Karimi, A.; Prati, A. A Novel Region of Interest Extraction Layer for Instance Segmentation. In Proceedings of the International Conference on Pattern Recognition (ICPR), Milan, Italy, 10–15 January 2021; pp. 2203–2209.

remote sensing

MDPI

Article

Analysing Process and Probability of Built-Up Expansion Using Machine Learning and Fuzzy Logic in English Bazar, West Bengal

Tanmoy Das [1], Shahfahad [1], Mohd Waseem Naikoo [1], Swapan Talukdar [1], Ayesha Parvez [2], Atiqur Rahman [1], Swades Pal [3], Md Sarfaraz Asgher [4], Abu Reza Md. Towfiqul Islam [5] and Amir Mosavi [6,7,8,*]

[1] Department of Geography, Faculty of Natural Sciences, Jamia Millia Islamia, New Delhi 110025, India; tanmoyblg99@gmail.com (T.D.); fahadshah921@gmail.com (S.); waseemnaik750@gmail.com (M.W.N.); swapantalukdar65@gmail.com (S.T.); arahman2@jmi.ac.in (A.R.)
[2] Henry Samueli School of Engineering, University of California, 5200 Engineering Service Rd., Irvine, CA 92617, USA; parvez.ayesha20@gmail.com
[3] Department of Geography, University of Gour Banga, Mokdumpur, Malda 732103, India; swadespal82@gmail.com
[4] Department of Geography, University of Jammu, Jammu 180006, India; sasgher20@gmail.com
[5] Department of Disaster Management, Begum Rokeya University, Rangpur 5400, Bangladesh; towfiq_dm@brur.ac.bd
[6] John von Neumann Faculty of Informatics, Obuda University, 1034 Budapest, Hungary
[7] Institute of Information Engineering, Automation and Mathematics, Slovak University of Technology in Bratislava, 81107 Bratislava, Slovakia
[8] Institute of Information Society, University of Public Service, 1083 Budapest, Hungary
* Correspondence: mosavi.amirhosein@uni-nke.hu

Citation: Das, T.; Shahfahad; Naikoo, M.W.; Talukdar, S.; Parvez, A.; Rahman, A.; Pal, S.; Asgher, M.S.; Islam, A.R.M.T.; Mosavi, A. Analysing Process and Probability of Built-Up Expansion Using Machine Learning and Fuzzy Logic in English Bazar, West Bengal. *Remote Sens.* **2022**, *14*, 2349. https://doi.org/10.3390/rs14102349

Academic Editor: Gwanggil Jeon

Received: 30 March 2022
Accepted: 10 May 2022
Published: 12 May 2022

Publisher's Note: MDPI stays neutral with regard to jurisdictional claims in published maps and institutional affiliations.

Abstract: The study sought to investigate the process of built-up expansion and the probability of built-up expansion in the English Bazar Block of West Bengal, India, using multitemporal Landsat satellite images and an integrated machine learning algorithm and fuzzy logic model. The land use and land cover (LULC) classification were prepared using a support vector machine (SVM) classifier for 2001, 2011, and 2021. The landscape fragmentation technique using the landscape fragmentation tool (extension for ArcGIS software) and frequency approach were proposed to model the process of built-up expansion. To create the built-up expansion probability model, the dominance, diversity, and connectivity index of the built-up areas for each year were created and then integrated with fuzzy logic. The results showed that, during 2001–2021, the built-up areas increased by 21.67%, while vegetation and water bodies decreased by 9.28 and 4.63%, respectively. The accuracy of the LULC maps for 2001, 2011, and 2021 was 90.05, 93.67, and 96.24%, respectively. According to the built-up expansion model, 9.62% of the new built-up areas was created in recent decades. The built-up expansion probability model predicted that 21.46% of regions would be converted into built-up areas. This study will assist decision-makers in proposing management strategies for systematic urban growth that do not damage the environment.

Keywords: land-use and land-cover; built-up expansion; probability modelling; landscape fragmentation; machine learning; support vector machine; frequency ratio; fuzzy logic; artificial intelligence; remote sensing

1. Introduction

The urban population of the world is rapidly growing, and according to a projection by the United Nations, approximately 9.8 billion people by 2050 and 11.2 billion by 2100 will be living in urban areas [1]. Due to this rapid growth in population, cities around the world are facing very fast transformation of their land use land cover (LULC) pattern [2–4]. This is creating serious problems, such as urban sprawl, unplanned expansion of impervious surfaces (built-up land), and decline in urban green spaces, as well as deterioration of the urban environmental quality [5–7]. The rate of unplanned expansion due to population growth and economic development is very high in developing countries of South and East Asia such as India, China, Bangladesh, etc. [8–10]. In India, small and medium-sized towns have experienced the very fast expansion of built-up surfaces in the last few decades [11–14], which has resulted in the unplanned expansion of these towns, as well as urban sprawling [15,16]. The rapid lateral expansion of India's small and medium-sized cities is considered a serious urban problem [17]. As of 2011, India had 3894 small and medium-sized towns [18], which are growing at a greater rate in terms of population and spatial distribution than large and metro cities [19]. Most of the studies on the urbanisation and LULC changes in medium-sized Indian cities have addressed some aspects of urban expansion and its consequences [14,20–22]. India's urban population has increased from 79 million in 1961 to 388 million in 2011, whereas its rural population doubled from 360 million to 860 million [23]. The primary cause for this continuous population rise was the development of infrastructural activities such as transport networks, internet connection, health sector development, new institutional sector establishment, etc. [7,17]. As a city's population grows, the city and countryside expand laterally to accommodate the increasing population, resulting in a lateral extension of cities beyond their formal city authority boundaries, a phenomenon known as "urban sprawl" [24]. The phenomenon of built-up expansion in the process of urban sprawl is one of the most noticed and studied phenomenon among urban researchers [12,25,26].

The expansion of built-up land leads to several negative and harmful effects on the urban environment and ecology [5,27–29]. Thus, measuring and modelling different forms of built-up land expansion might assist planners in better understanding and addressing the challenges related to the expansion of built-up areas [30–32]. Researchers have tried to analyse and quantify the urbanisation-induced built-up land expansion around the world [12,33–36]. For instance, Salem et al. [34] analysed the pattern of built-up land expansion and urban sprawl in the peri-urban area of Cairo City in the post-revolution era using remote sensing techniques coupled with logistic regression. Similarly, Talukdar et al. [37] used the transferable build-up area extraction (TBUAE) technique for assessing urban growth using high-resolution satellite datasets. Further, Fawad et al. [38] combined multispectral satellite data with synthetic aperture radar (SAR) datasets for extracting urban impervious surfaces. Xu [39] and Shahfahad et al. [12] applied an index-based built-up index (IBI) technique for urban area expansion that was based on the multiple spectral indices derived from satellite data.

Studies have been done in developing countries like South Africa [40], Pakistan [41], Nigeria [42], and India [43] for mapping and monitoring urban LULC changes and built-up area expansion using remote sensing. However, most of these studies in India have been done on large metropolitan cities like Delhi, Kolkata, Mumbai, Chennai, etc. [44–48]. On the other hand, only a few studies have been done on small and medium-sized cities in India [14,16,49]. Despite the considerable growth in the literature on urban expansion and sprawling in major cities, studies on built-up expansion in small/medium-sized blocks/districts have received little attention from scholars and urban planners in India. Therefore, to fill this literature gap, this study aims to analyse the built-up growth and LULC change in the English Bazar block, which is a small town of West Bengal. The English Bazar Municipal Corporation, which is located in the centre of the English Bazar block has witnessed very fast population growth and landscape transformation in the last few decades [49]. In this context, an analysis of the built-up growth and LULC change is a matter of concern for medium-sized towns like English Bazar. Therefore, in this work, we identified the land use features using SVM for 2001, 2011, and 2021, including built-up areas, as an essential input for further research. We introduced a model to study the process of built-up expansion using fragmentation indices and the frequency approach. Additionally, we developed a built-up probability model using fuzzy logic for exploring the future built-up expansion probable areas, which have not been studied so far.

2. Materials and Methods

2.1. Study Area

The study area English Bazar block is an important political, cultural, and administrative centre in Malda District, West Bengal, and covers an estimated area of 251.8 km^2. It is located between 24°50′–25°05′N latitude and 88°–88°10′E longitude (Figure 1). The study area is located at a height of 15–17 m above sea level. It consists of one municipality and 108 inhabited villages. This region is mainly made up of the younger alluvial plain of the Ganga River and the Mahananda River. The English Bazar, like most other parts of West Bengal, experiences highly humid and tropical weather. In May and June, temperatures can reach 42–45 °C during the day, but in December and January, they can drop to 8 °C overnight. According to the 2011 Census, the English Bazar block has a total population of 274,627, and the region is recognised as having an urban agglomeration. In the last decade (2001–2011), this region has witnessed a rapid rise in urbanisation, with the highest urbanisation rate in West Bengal at 124.81%. Various social and economic factors influenced this significant built-up expansion. The intersection of NH-34 (national highway), the state highway (SH-10), the north-eastern frontier railway, and the eastern railway all lead to the enhancement of the built-up expansion in this region. As increased urbanisation in a developing country, this region is challenged with uncontrolled built-up growth, insufficient critical infrastructure and utilities, air, noise, and water pollution, as well as bad governance.

2.2. Data Base and Its Preprocess

The LULC maps for the years 2001, 2011, and 2021 were created using the satellite images provided by different sensors of Landsat, such as Landsat 5 thematic mapper (TM) and the Landsat 8 operational land imager (OLI). Before proceeding for further research, we considered the technical issues of both sensors. However, Landsat TM started providing multispectral data in 1984. Landsat 8 launched recently (2013) with the OLI and TIRS sensors, resulting in a new orthorectified dataset (L1T) [50]. Despite the fact that Landsat 5 is no longer in operation and Landsat 7 is hampered by the failure of the Scan Line Corrector (SLC-off), the successful launch of Landsat 8 has supplied a steady stream of intermediate spatial resolution data for long-term LULC mapping and trend analysis [51]. Therefore, most research on long-term vegetation changes and LULC mapping uses different Landsat sensors together, such as TM, ETM+, and the OLI sensor. In comparison to TM and ETM+, OLI has two new spectral bands: an ultra-blue band (0.43–0.45 μm) and a cirrus band (1.36–1.39 μm). In the near-infrared (NIR), the OLI bands are generally narrower than the equivalent TM and ETM+ bands. Given these factors, it is necessary to compare Landsat 8 data to the preceding Landsat sensors before integrating them with other sensors for the trend analysis. Several researchers have examined clear sky observations taken 8 days apart for the same site to compare Landsat 5, 7, and 8 data [51,52]. These studies assumed that the phenology and land cover did not change between acquisitions and found that surface reflectance variances of about 2% and NDVI differences of about 5% exist between the two sensors [21]. However, Vinayak et al. [44] concluded that the TM, ETM+, and OLI images are comparable enough to be used as complementary data.

Therefore, based on the previous literature, we used TM and OLI sensors for our research, with two considerations. First, we did not use all bands of the TM and OLI sensors for our LULC mapping. We used only those bands that are common in both sensors and relevant for LULC research, such as red, blue, and green, while the thermal bands of both sensors, coastal and circus bands of the OLI sensors, were excluded for the analysis. Additionally, these bands did not have much influence on the LULC research. Second, before using the selected bands for further research, we did radiometric correction, like top-of-atmosphere (TOA) reflectance and/or radiance using radiometric rescaling coefficients provided in the metadata file that is delivered with the Level-1 product.

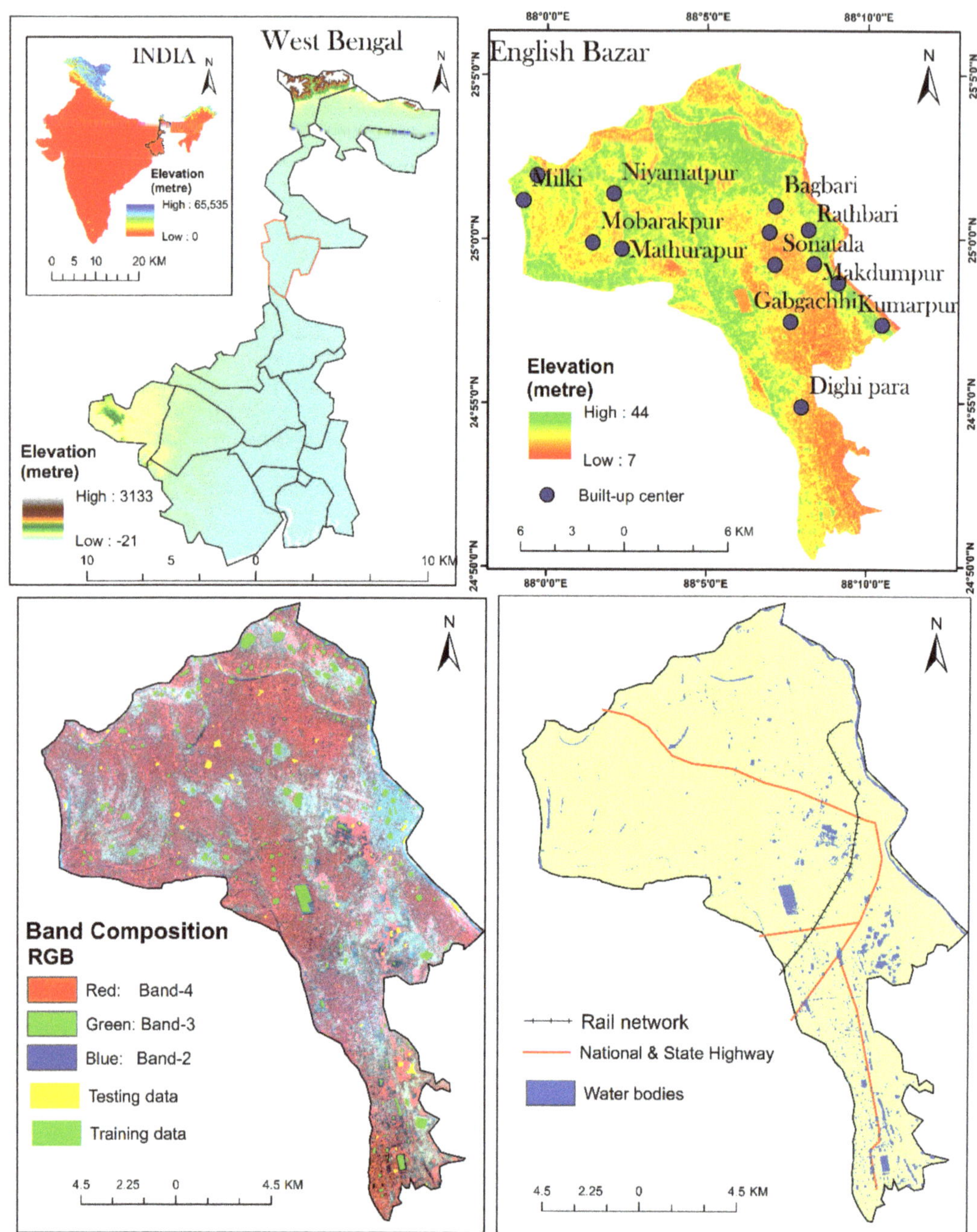

Figure 1. Location map of the study area, with the major built-up node and distribution of the training and testing datasets.

The Landsat datasets were downloaded from the Earth Explorer website (http://earthexplorer.usgs.gov/, accessed on 16 December 2021) of the United States Geological Survey. In a subtropical country like India, satellite data are generally used for the months of March and April to get cloud-free scenes [28]. However, this part of India receives significant rainfall during the pre-monsoon months (March to April) due to the nor'wester activity (locally known as Kaalbaisakhi) and, thus, experiences frequent cloud cover [12]. Therefore, in this study, the satellite data were used for the months of December to avoid any kind of errors due to local climatic effects, and we got the cloud-free images [21]. The Landsat images offer a medium spatial resolution of 30 m, which is suitable for the mapping of urban landscape pattern at different levels (Table 1). Further, for the validation of prepared LULC maps, the samples were collected through a field survey, as well as the Google Earth Pro domain.

Table 1. Details of multitemporal satellite images and their characteristics.

Satellite	Sensors	Date/Year	Path/Row	Spatial Resolution (m)	Cloud Cover (%)
Landsat 5	Thematic mapper (TM)	14 December 2001	139/43	30	0.00
Landsat 5	Thematic mapper (TM)	11 December 2011	139/43	30	0.00
Landsat 8	Operational land imager (OLI)	4 December 2021	139/43	30	0.00

2.3. Methodology

2.3.1. Method for the LULC Classification

The SVM is a binary classifier [52]; it may classify many characteristics (over two) using either the one-versus-one (OVO) or one-versus-rest (OVR) techniques. The SVM, a statistical learning approach, provides improved overfitting control and a high convergence rate and is unaffected by the local minima [53]. Many classifications issues have been successfully solved using the SVM classifier [54–56]. In this work, we did LULC classification based on three different bands of Landsat satellite data: red (R), green (G), and blue (B). We divided the pixels in the images into five categories (agricultural land, barren land, vegetation, built-up area, and water body) (Table 2).

Table 2. Description of the LULC classes identified in the study area.

S. No	LULC Type	Description
1	Water bodies	Lakes, ponds, river, canals wetlands, or other water-logged areas
2	Built-up area	Residential, commercial, and industrial areas, as well as all infrastructure facilities
3	Vegetation	Natural vegetated areas and plantation
4	Barren land/sand bar	Land without any manmade structure and vegetation cover, as well as sandy surfaces along rivers or other water-logged areas
5	Agricultural land	All types of cultivatable lands

As a result, a multiclass SVM classifier based on the RGB band composite was developed to predict the land use classes. To design and test the SVM classifier for classification into five groups, 360 datasets were collected. For training and testing the model, we partitioned the total datasets into two halves in the ratio of 80:20. In other words, 212 datasets were used to train the model, and 56 datasets were used to test it. The training dataset comprised 8619 total pixels. Out of the total pixels, we collected 1306 pixels from water bodies, 2500 from built-up, 1613 from vegetation, 2600 from agricultural land, and 600 from barren land. The SVM model classifies a typical pixel in an image as +1 if it corresponds to a specified class or −1 if it belongs to a different class [56]. This work used the OVO approach with radial basis function kernel to create a multiclass classification [57]. The SVM model was trained using training data from two different classes in the OVO approach.

The training of a new SVM model is performed until each class pair has been completed. In this scenario, picture pixels are classified into five classes, resulting in 10 [=(52 − 5)/2] combinations. Therefore, 10 SVM models are needed to train for classification into five classes. The class of a particular pixel has been predicted using the voting scheme. A typical pixel is projected as "+1" based on the most votes from all ten classifiers. Any SVM model's principal goal is to find the best separating hyperplane, w.x + b = 0, for categorising pixels into two groups. The optimal hyperplane distinguishes the pixels into +1 and −1 with the maximum margin. We used C-support vector classification to run the model. The SVM was executed with the parameters of: the penalty parameter (C) of 1, Nu of 0.5, p of 0.5, radial basis function-based kernel, coefficient 0 of 1, degree of 0.5, and gamma of 1. Based on these optimised parameters, the LULC was classified three times, for 2001, 2011, and 2021.

2.3.2. Validation of LULC Map

The most significant and last stage in the image classification process is the accuracy assessment estimation. Validation is a crucial component of robust and accurate LULC mapping, because it assists planners and administrative departments in understanding the significance, impact, and accuracy between ground reality and research work in the studied region [55,56]. Post-classification validation of the LULC maps also helps planners and administrative divisions in launching their strategies. The standard kappa coefficient, overall accuracy, producer accuracy, and user accuracy were among the accuracy measurement methodologies used in the present study. The following formula was used to determine the kappa coefficient [58]:

Kappa coefficient

$$\frac{N \sum_i^r x_{ii} - \sum_i^r (x_{i+})(x_i)}{N^2 - \sum_i^r (x_{i+})(x_i)} \tag{1}$$

where r represents the number of rows, x represents the number of observations in the rows and columns (diagonal elements), and N represents the number of total observations:

Overall accuracy

$$\text{OA} = \left(\frac{1}{N}\right) \sum_{i=1}^r n_{ii} \tag{2}$$

Producer's accuracy

$$\text{PA} = \left(\frac{n_{ii}}{n_{icol}}\right) \tag{3}$$

User's accuracy

$$\text{UA} = \left(\frac{n_{ii}}{n_{irow}}\right) \tag{4}$$

where the number of correctly classified pixels is $-ii$, the total number of pixels is N, the number of rows is r, and the column and row totals are n_{icol} and n_{irow}, respectively.

The total accuracy estimator determines how many pixels in the image have been correctly identified. For this work, we used the kappa coefficient to validate the LULC maps based on the testing samples. In the present study, we randomly chose 20% of the field data as the testing sample. Fifty-six samples were calculated as the testing sample, which comprised 2641 total pixels. Out of total pixels, we collected 286 pixels from water bodies, 780 from built-up, 690 from vegetation, 655 from agricultural land, and 230 from barren land.

2.3.3. Method for the LULC Change Detection

Using the GIS environment and a post-classification change detection approach, the changes in LULC during 2001–2011, 2011–2021, and 2001–2021 were examined (Figure 2). Post-classification change detection is one of the most common methods for analysing the details of changes in LULC patterns in a region during a time period [52]. This approach may evaluate the temporal variations in the LULC types and the level of LULC conversion caused by urbanisation. We calculated the change statistics for two time periods of LULC maps using Equation (5):

$$\text{Change (\%)} = \frac{\beta^1 - \beta^2}{\beta^1} \tag{5}$$

where β^1 = present year (2021) and β^2 = initial year (2001).

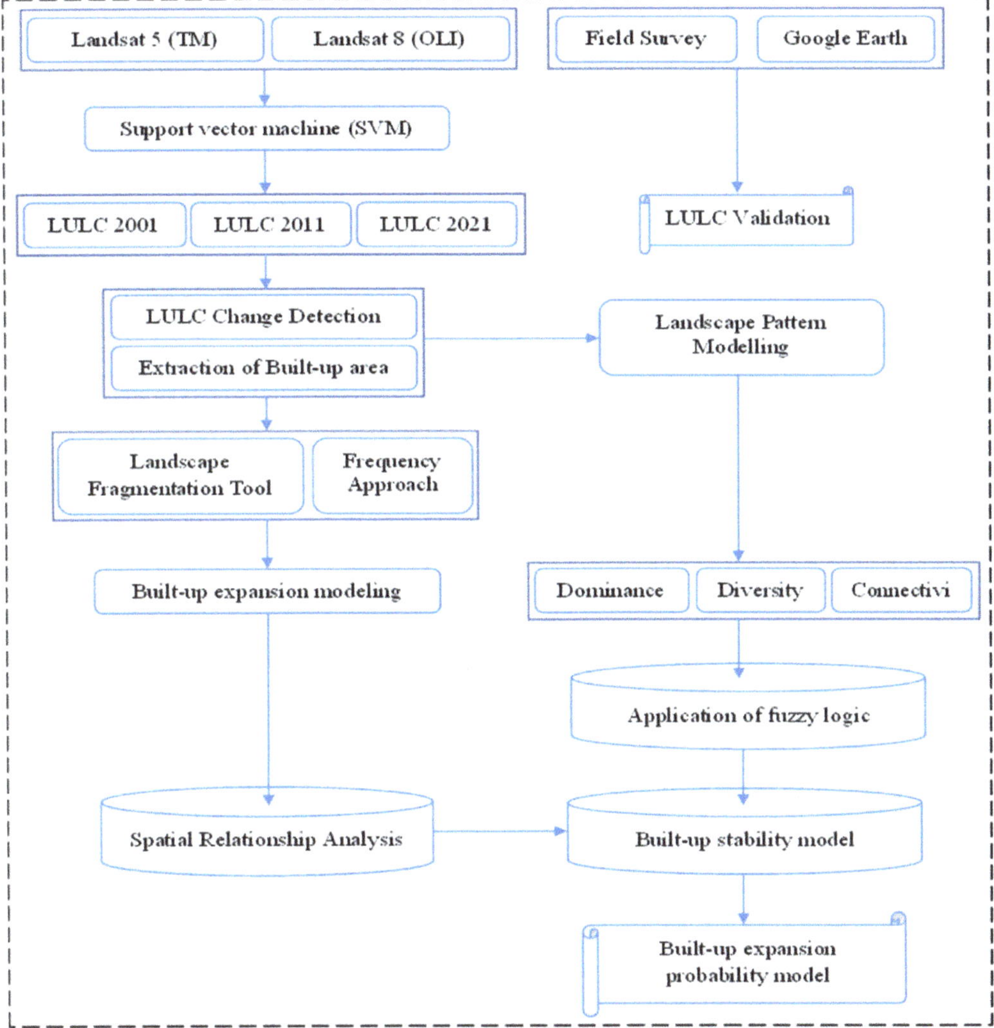

Figure 2. Hierarchical structure of the methodological framework of the entire study.

2.3.4. Methods for the Modelling of Built-Up Expansion Process

In this study, the built-up growth was represented using a frequency technique. We first categorised the LULC map, then divided the entire map into two groups, built-up and non-built up, for distinction when using the reclassification tool in GIS. The built-up area was denoted by 1 and the non-built-up area by 0. Then, we used the spatial analyst tool with the less than frequency and the reclassified raster layer as the input. This approach calculates the number of times a collection of the raster is smaller than another raster on a cell-by-cell basis. The number of instances (frequency) in which a raster in the input list has a lower value is counted for each cell position in the input value raster. Finally, a frequency map is produced. The value reflects the number of times the associated cells in the list of the raster are less than the value raster for each cell in the output raster. The final map is overlaid by a built-up area of different (three) times.

We employed landscape fragmentation tools to analyse the growth rate trend to categorise the built-up area. The LULC map was divided into two categories: built-up and non-built-up, with built-up denoted by 2 and non-built-up denoted by 1. As an input raster, we utilised this reclassified map. We followed two steps to perform this model that reclassified LULC data and analysed the fragmentation. The final map was divided into four categories: patch, edge, perforated, and core. Based on the size of the core tract, the core category is further classified into large, medium, and small cores. The primary categories are determined by a parameter called edge width. Many studies have established the deterioration of forests or grasslands along disturbance edges, but we used it in our study to demonstrate the extension of newly formed built-up areas via the edge effect.

2.3.5. Built-Up Expansion Probability Model Using Fuzzy Logic

A suitability analysis, often known as site selection, is an advanced remote sensing-based study used to discover the optimal location for anything. To generate a built-up expansion probability model, first, we created a dominance, diversity, and connectivity model in SAGA GIS, and then, we used fuzzy tools in GIS and determined three different fuzzified models with the help of suitable fuzzy membership types. We used six membership functions (linear, small, large, MS Small, MS Large, and near) for the model. The fuzzy membership tool reclassifies or converts the input data into a 0 to 1 scale, depending on the probability of being a member of a certain set [59]. We mainly used the 'large' membership function, because it has a positive relationship with built-up expansion, and 'large' indicates a high fuzzy membership value. Fuzzy membership assigns values ranging from 0 to 1, with 0 indicating that something is unlikely or inappropriate and 1 indicating that something is most probable or appropriate [60]

After that, we created a built-up stability model using fuzzy overlay methods (In a multicriteria overlay analysis, the fuzzy overlay tool may estimate the probability of a phenomena belonging to various sets. Fuzzy overlay not only determines which sets the occurrence could belong to, but it also examines the connections between the sets' membership [59]. Since the multiple classes' surfaces are compared, this phase is analogous to weighted site selection (a site selection type that allows users to rank raster cells and provide a critical relative value to each layer). Then, we utilised the fuzzy operator (And, Or, Sum, Product, and Gamma) 'And', which gave us the best results, because this operator is great for identifying places that fulfil all requirements. Finally, we created a built-up probability model using Euclidean distance from spatial analyst tools in GIS. From the centre of the source cell to the centre of each surrounding cell, the Euclidean distance was determined. Each of the distance tools calculated the actual Euclidean distance. If the shortest distance to a source is less than the maximum distance given, the value is allocated to the cell location on the output raster. Therefore, we used the built-up stability map as a raster layer to generate the final output map, signified through 0 and 1, and quickly predicted the built-up expansion probability zone in the English Bazar block. The methodology of this study is summarised in Figure 2.

3. Results

3.1. Analysis of the LULC Mapping

The LULC of the English Bazar block was classified into five categories, i.e., water bodies, built-up area, vegetation, barren land, and agricultural land (Figure 3), based on the level-I classification scheme of the National Remote Sensing Centre (NRSC). This study observed a significant shift in LULC patterns in the last 20 years. The results for the LULC map of 2001 showed that a 109 km^2 area was covered by vegetation cover, followed by agricultural land (70.1 km^2), water bodies (21.11 km^2), built-up area (15.4 km^2), and barren land (18.4 km^2). In 2011, the results showed that a 92 km^2 area was covered by agricultural land, followed by vegetation cover (80 km^2), built-up area (26.8 km^2), water bodies (17.26 km^2), and barren land (12.4 km^2). In the case of 2021, the built-up area and agricultural land increased significantly, and agricultural land covers 98 km^2, followed by the vegetation cover (72 km^2), built-up area (35.91 km^2), water bodies (19.2 km^2), and barren land (7.64 km^2) (Figure 3). The amount of built-up and agricultural land has grown, while vegetation and barren land decreased. Water bodies also increased in the south-eastern part of this study area, owing to lowlands that were flooded in 2017 [53], resulting in stagnant water bodies. Further, new farm ponds were dug under the Mahatma Gandhi National Rural Employment Guarantee Act (MGNREGA) scheme by the union, as well as the state government, to increase fish production and maintain the ecological balance, especially of aqua life. Agricultural land has primarily expanded near built-up areas because of many people involved in agricultural activity, mostly in rural areas, and to meet the growing need for food for an expanding population, both show urban growth. Barren land has been changed to agricultural land, built up, and certain areas have been converted to vegetation, particularly mango forests, since Malda District is noted for them [54]. The built-up area was expanding all over the block at a very high pace. The main growth was noticed in the surroundings part of the municipal area, which is located in the eastern part of the study area. High built-up growth was noticed in the northern part, where the expansion was mainly increased along the banks of Kalindri River. To the western part and southern part, the built-up area was increased along the State Highway (SH) 10 and National Highway (NH) 34. Maheshpur has seen a massive increase in settlements area, located on the left bank of Mahananda River, a very short distance from the municipal area (southern part). The built-up area increased by 30% during 2001–2011 and 44.28% from 2011 to 2021. The reason for the rapid urbanisation in the north-eastern part of the study area is the presence of the rapidly growing English Bazar municipality, type-I city, and decentralisation of several medium and small microenterprises (MSME), educational centres, and health centre. These governmental and private centres accelerate the process for growing urban centres across the study area rapidly, not only the north-eastern part. According to the census of 2001, the population density of this municipality was 881 persons/km^2, which was increased by 1100 persons/km^2. Therefore, it can be stated that the urban centres have been growing rapidly over time.

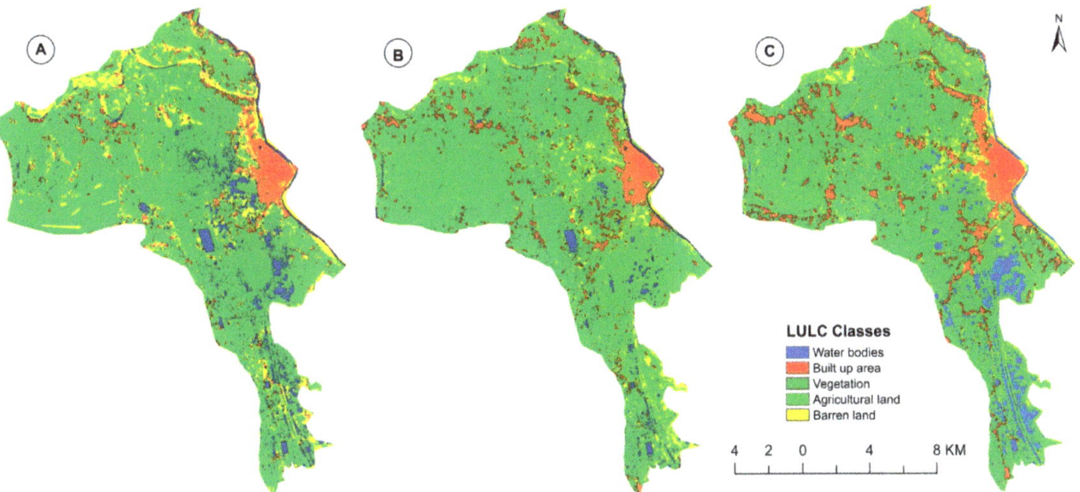

Figure 3. LULC mapping using SVM for the years of (**A**) 2001, (**B**) 2011, and (**C**) 2021 of the English Bazar block.

3.2. Validation of the LULC Classification

The accuracy of the prediction of a certain class is shown by the metric of the producer accuracy. The consistency of the group to the user is revealed by the user's accuracy. The LULC maps of 2001 had an overall average of 90.05%, 93.67% in 2011, and 96.24% in the 2021 (Table 3). These results indicate that the performance of the SVM model for classifying the LULC maps was highly satisfactory. Previous studies reported that the overall accuracy of greater than 80% could be considered satisfactory [55,56]. The results of the LULC maps of three periods showed that the LULC of 2021 had the highest user accuracy of 96.24%, followed by 93.67% in 2011, and 90.05% in 2001. The built-up area was found to be more reliable because of the 98% user accuracy for the LULC of 2021, followed by vegetation (96%), and agricultural land (96%). In the case of the LULC of 2011, vegetation had the highest user accuracy of 95%, followed by built-up area (94%) and water bodies (93%). The accuracy assessment of the LULC of 2001 showed that built-up areas achieved the second-highest user accuracy of 88%. Therefore, it can be stated that the built-up area achieved the highest accuracy compared to the other four LULC classes.

Table 3. Accuracy assessment using the Kappa statistics of the classified LULC maps for three periods.

LULC Class	2001			2011			2021		
	Producers Accuracy (%)	Users Accuracy (%)	Kappa Coefficient	Producers Accuracy (%)	Users Accuracy (%)	Kappa Coefficient	Producers Accuracy (%)	Users Accuracy (%)	Kappa Coefficient
Water bodies	90.00	89.00	0.87	94.56	93.00	0.92	96.87	95.00	0.94
Built-up area	93.25	90.00	0.88	96.82	94.00	0.94	98.00	98.00	0.97
Vegetation	91.55	87.00	0.87	95.60	95.00	0.93	97.38	96.00	0.96
Barren land	88.00	91.00	0.91	90.33	91.00	0.91	95.80	95.00	0.93
Agricultural land	91.50	87.00	0.86	89.45	90.00	0.91	95.00	96.00	0.94
Overall Classification Accuracy		90.05			93.67			96.24	
Kappa Statistics		0.88			0.92			0.95	

3.3. Analysis of the LULC Dynamics

During 2001–2021, urbanisation and development activities had the most significant impact on restructuring the land use and land cover in the English Bazar region. Figure 4 depicts how the urbanisation process, also known as built-up expansion, changed the land use at the expense of vegetation, agricultural land, and barren land, as seen in the Figure 4. From 2001 to 2011, roughly 6.2 km^2 of agricultural land and 4.6 km^2 of vegetation cover were transformed into built-up areas, while between 2001 and 2011, almost 12.5 km^2 of vegetation cover was converted into agricultural land, and between 2011 and 2021, a nearly 15.6 km^2 vegetation cover was converted to agricultural land. From 2001 to 2021, a considerable shift in the barren land was observed, with a total area of 12.63 km^2 being transformed into built-up areas, agricultural land, and vegetative cover. During the investigation, water bodies were converted into agricultural land and built-up land. The total built-up area was increased to about 35.8 km^2 between 2011 and 2021, which covered about 14.4% of the total area of the English Bazar block. It was also noticed that a large area was converting from water bodies to agricultural land by almost 15 km^2. The vegetation cover also decreased over time, which was transformed into agricultural land and barren land in 28 and 5.2 km^2. A significant change in barren land was noticed in a 3.2 km^2 area, which was transformed into a built-up area, while a 6.3 km^2 area was converted into agricultural land during 2001–2021. The area under water bodies has gained from the area under agricultural land through the conversion of 3.5 km^2.

During 2001–2021, the pattern of built-up expansion is shown in Figure 5, which illustrates how the built-up area expanded in the region. Here, we observed four classes of built-up changes, with non-built-up to built-up showing an increasing trend. The term unchanged built-up refers to a region that was built up during a preceding time. In the

English Bazar block built-up area expanding all sides of the block, some parts increased quickly and some parts slowly. In Figure 5, the changing pattern from 2001 to 2021 showed a huge increase of the built-up area; this expansion was mainly done along the road and riverside. The built-up area witnessed linear expansion along the SH-10 in the western part and NH-34 in the south-central part, as well as beside the Mahananda in the eastern part and Kalindri River in the northern part of the English Bazar block. Although, we observed few unusual conversions in the present study. That is, few small amounts of areas were converted from built-up to non-built-up (other LULC classes) because of frequent flooding and the eviction of unlawful settlements, such as slums, especially from the riverbank of Mahananda River, as reported by the local authorities.

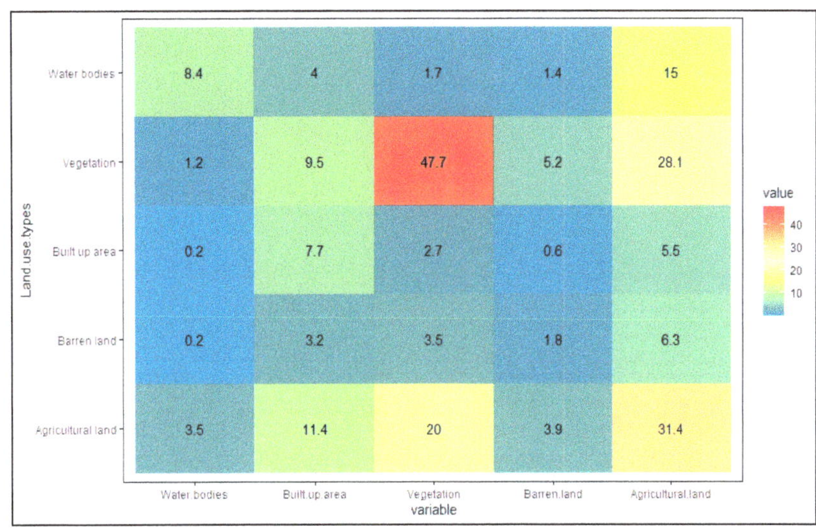

Figure 4. The LULC transition matrix using a heatmap between the periods of 2001 and 2021.

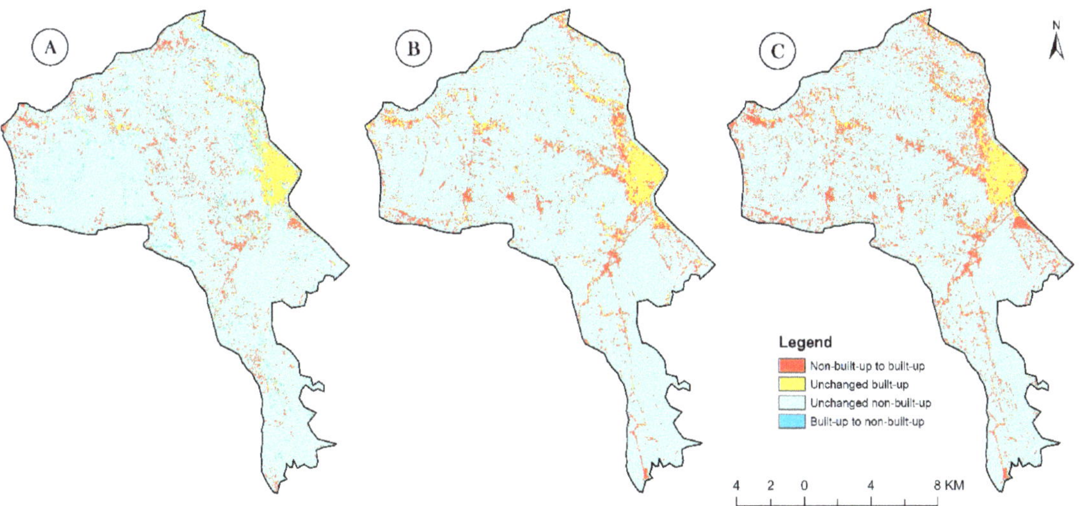

Figure 5. Built-up change maps between the years of (**A**) 2001 and 2011, (**B**) 2011 and 2021, and (**C**) 2001 and 2021 of English Bazar.

3.4. Analysis of the Built-Up Expansion Process

We employed a landscape fragmentation and frequency approach model to demonstrate the process of built-up expansion in the English Bazar. Using these models, we could easily explore the process of built-up expansion growth, its trend, and pattern. The process of built-up expansion was modelled using fragmentation indices, and it classified the built-up area into six fragmentation indices, such as large core, medium core, small core, perforated, edge, and patch (Figure 6). The large core indicates a built-up area of more than 500 acres and can be considered as the most stable and permanent built-up area. In contrast, the small core indicates an area of less than 250 acres and can be considered as the concentration of a newly formed built-up that has been expanding outwards from the main core. The results showed that the large core of the built-up area covered 2.92 km^2 in 2001, which then increased to 3.8 km^2 and was amplified by 7.42 km^2. This scenario shows that the large core of the built-up area, identified in 2001, was fixed three times. However, the medium core and small core during 2001 and 2011 were concentrated, along with the large core areas, and, ultimately, converted into large core areas in 2021. This scenario reflects increasing the large core of the built-up area through the process in the study area. Additionally, the small and medium cores in the study area observed increasing areas; for instance, in 2001, the small and medium cores were 1.4 and 0.94 km^2, which increased to 4.4 and 5.36 km^2. The small and medium cores observed significant growth because of the fresh and isolated built-up node conversion. In 2001, the core area was mainly concentrated in English Bazar City, but as the population grew over time, the municipality grew proportionately. The core area expanded in the northern and south-western parts, primarily along national highway-34 and state highway-10. Bagbari, Daulatpur, Milki, Makdumpur, Sonatala, Uttarjadupur, Maheshpur, Uttarramchandrapur, and other regions were identified as small and medium cores in the extreme northern, southern, and western parts of this study area in 2021. The edge and perforated areas mainly showed surrounding English Bazar City, particularly fresh and isolated urban nodes, as well as new patches, which can be found sparsely in the western, southern, and northern parts, namely Sultanpur, Shyampur, Niamatpur, and Kirki. The results showed that the areas under the perforated and patch categories were 6.24 km^2 and 13.42 km^2 in 2001, which increased to 13.23 km^2 and 17.86 km^2 in 2021 (Figure 6). These fresh and isolated built-up nodes were once a rural neighbourhood, but the economic and infrastructural growths triggered the urbanisation process to take place over there. Therefore, some of the perforated and patch areas progressed outwards and were connected with small and medium cores over time. This is how the small, medium, and large cores have increased tendencies. During this research, it was discovered that substantial expansion has occurred along the transportation network, primarily the road and railway [61].

The frequency approach was used in this study to illustrate the built-up expansion of the English Bazar block, and the results were promising. The probability of an occurrence, according to the frequency theory, is the upper limit of the relative frequency with which the event occurs in repeated trials under essentially identical conditions. Thus, using this model, we can immediately grasp the newly developed extended area that has been constructed. The built-up regions for three different eras were layered in this model, and a final map was created that depicted three different periods: three times, two times, and one time, respectively. Consequently, we could quantify the built-up expansion over time rapidly. The results demonstrated that the number three represents a permanent or stable built-up area, which reflects that built-up areas were common in 2001, 2011, and 2021 in those particular places, as depicted by the sky blue hue (Figure 7). While value two in the frequency approach shows that the region has observed a built-up area two times in particular places (2001 and 2011), as represented by the deep blue colour. Value one shows that the region has witnessed one instance of a newly constructed built-up area, shown by the red colour. Therefore, the findings show that the fresh and isolated built-up nodes covered a 10.98 km^2 area, indicated by value one, while the built-up transition area, indicated by value two, covered a 10.24 km^2 area (Figure 7). The transition zone is the most unstable area, which gains areas from fresh and isolated built-up areas and loses to stable or permanent built-up areas. The most economical and infrastructural growth can be observed in this zone recently. In the case of new and isolated built-up nodes, medium and small enterprises, as well as health and educational centres, help them grow into urban areas.

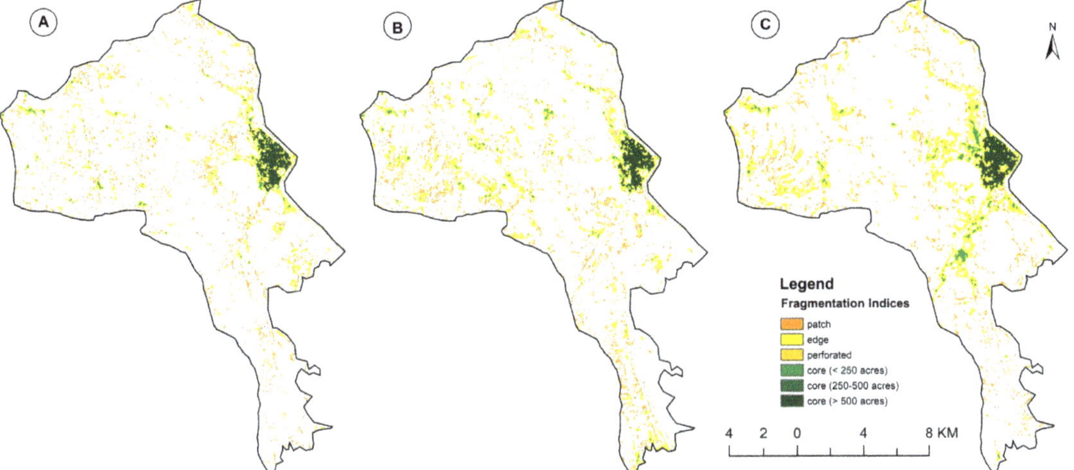

Figure 6. Built-up expansion process model using landscape fragmentation index for (**A**) 2001, (**B**) 2011, and (**C**) 2021 in English Bazar.

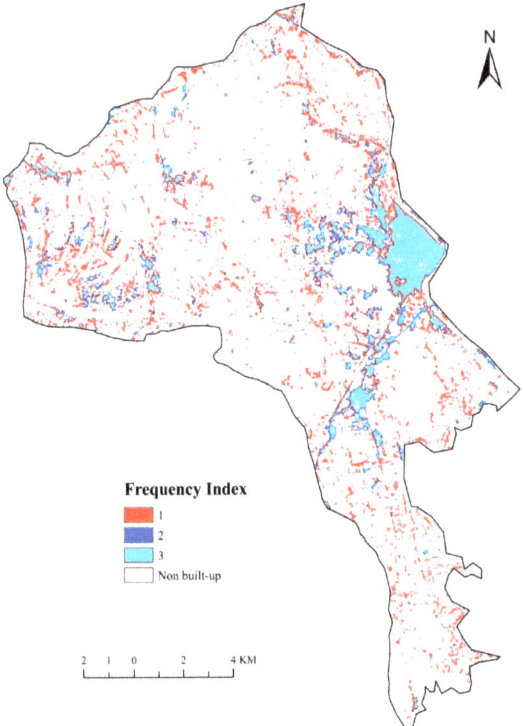

Figure 7. Frequency approach model to show the expanded built-up area over time.

3.5. Analysis of Built-Up Expansion Probability

Using SAGA GIS, we first created the dominance, diversity, and connectivity models as parameters for analysing the structural pattern of the built-up area. It used to understand the present situation and probability of built-up expansion. In this context, the parameter, like urban dominance, refers to the web of influences that particular cities sustain within a system of cities as the new orthodoxy in urban planning and development; diversity makes cities more cosmopolitan and economically productive, with different economic activity and opportunities. Diverse roles in the municipality in the urban hierarchy attract a vast population. Connectivity refers to how easily passengers or freights may move from one node to another, either directly (directly) or indirectly (through another node or a series of nodes). It is a crucial indicator of built-up growth. The values in these three models varied from 0 to 1, with 0 showing a low built-up concentration and 1 showing a high one. In this three-model analysis, it was discovered that a substantial concentration zone existed in the municipality and its surroundings and several areas next to the NH-34. The municipal area, particularly *Rathbari*, is a key transit hub for the Malda and North Bengal regions. As a result, the development grew from the municipal territory to both sides of the road network. After that, we used the fuzzy membership tool in arc GIS to unidirectionally show all the parameters, resulting in a fuzzified dominance, diversity, and connectivity model that varied from 0 to 1, showing the same outcome as before. Then, with the help of the fuzzy overlay tool and the and/or gamma operators, we utilised these three fuzzified models to build the final built-up expansion model. Its values ranged from 0 to 1, and the final map depicts the future built-up expansion probability of the English Bazar block (Figure 8).

Based on the parameters (dominance, diversity, and connectivity), the final output using a fuzzy logic model ranged between 0 and 1, indicating high to low built-up expansion probability in stretch format. Then, we used a natural break algorithm to classify the stretch built-up probability expansion index into three classes, such as high, moderate, and low built-up expansion probability zones. The results showed that a 103.2 km^2 area has been predicted as a high built-up expansion probability zone, followed by moderate (89.01 km^2) and low built-up expansion probability zones (59.62 km^2) (Figure 9A). All the parameters positively influenced the built-up expansion, which indicate that where the value of the parameters is higher, the probability of built-up expansion will also be higher. As the higher value of the dominance, diversity, and connectivity parameters were seen in the eastern part of the study area, therefore, these parts have experienced a higher built-up expansion probability. A higher concentration of all the parameters means a higher accessibility rate of the region, which will also experience higher built-up growth in the future. It appears that the surrounding rural areas of the municipality will have a greater chance to convert into urban areas, primarily the area adjacent to NH-34, so its expansion will be concentrated in the southwest and northern parts of the study area. Furthermore, while being a rural residential area, its western half has a good chance of growing in population. *Barbara, Sonatala,* and *Milik* were designated as census towns in 2011, and our research revealed that this area is in a high built-up probability expansion zone, implying that these areas have a high proclivity to expand their built-up areas and that this area has also been transformed into an urban area because of urban expansion (Figure 9). *Maheshpur, Makdumpur, Pirozpur, Lalapur, Mathurapur,* and *Uttarramchandrapur* are among the 11-g panchayats in the English Bazar block, with the majority of them having moderate built-up expansion likelihood. As a result of the spread effect of the English Bazar municipal area, its fringe areas such as *Makdumpur, Barbara, Uttarramchandrapur,* and *Maheshpur* have seen significant changes in terms of built-up expansion, and they may be converted into urban areas and merged with the municipality of English Bazar block. The areas with the lowest built-up expansion potential were found in some parts of the study area's north, south, west, and central regions, because this area either has the mango forest or water bodies, and parallel to this, these parts also suffered from lousy connectivity. Though the municipality of the English Bazar block is a type-I city, according to the 2011 Census with a population of 274,627 and a population density of 1100/km^2 and, also, the municipality had a 2.16 lakhs population, it has a high chance or probability of expanding their built-up area with a much higher rate in the future, and our study also revealed the probable expansion zone. A significant reason for this high expansion is its nodal connectivity location, fertile soil, which enhanced the agricultural activities, primarily mango groves and some food processing industries, and its growing business capacity.

The field survey was conducted to see how the built-up expansion has been taking place in the study area, such as nursing homes, two/four-wheeler showrooms, residences, apartment houses, and retail malls, and soon, which have been developed on open land. One of the most striking observations we made was that most construction activity has been taking place along the roadside, mostly alongside NH-34. Here are some photographs from the field that we took during the survey (Figure 10).

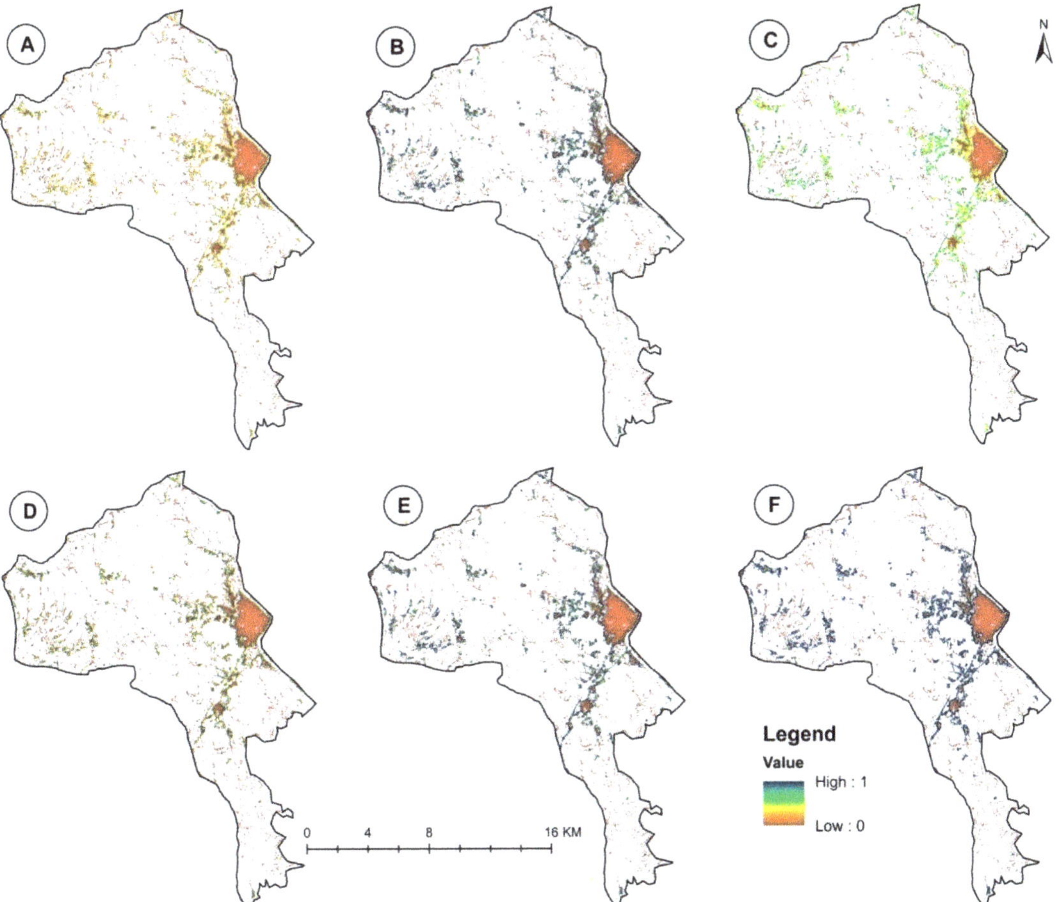

Figure 8. Triggering factors for built-up expansion probability modelling using a fuzzy logic model, such as (**A**) a dominance model, (**B**) a diversity model, and (**C**) a connectivity model. The fuzzification of the parameters using the membership function, such as (**D**) the fuzzified dominance model, (**E**) fuzzified diversity model, and (**F**) fuzzified connectivity model.

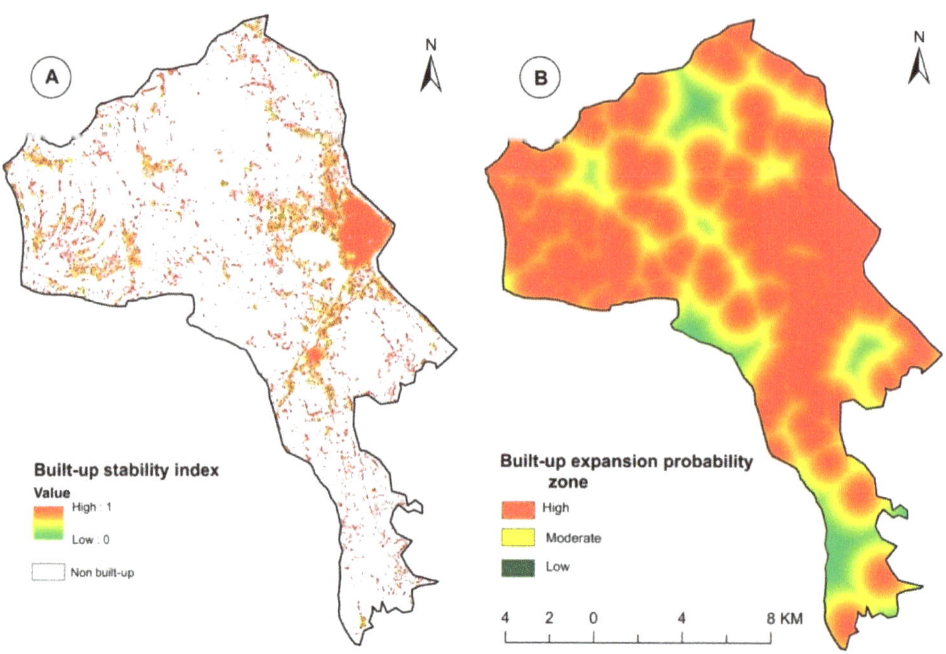

Figure 9. Fuzzy logic-based (**A**) built-up stability model, and (**B**) built-up expansion probability model for English Bazar.

Figure 10. *Cont.*

Figure 10. Under construction built-up areas on national highway-34 beyond the administrative boundary of the municipal areas, such as (**a**) a housing lodge at *Gabgachi* (1 km away from the University of Gour Banga); (**b**) a car showroom at the nearby area of *Gabgachi*, which can help to emerge other allied services in very soon upcoming days; (**c**) nursing home towards *Gazole* (another emerging site for the built-up area), which will help other allied services, hotels, restaurant, and others in the upcoming days; and (**d**) a shopping mall at *Madhabnagar* (southwards from the municipal areas), which will help to grow new built-up areas, like a garage, petrol stations, small shops, and others.

4. Discussion

This paper presents an easy and sophisticated method for exploring the process of built-up expansion mapping and modelling the built-up expansion probability. Our primary aim was to model the process of built-up expansion and delineate the probability for the English Bazar block. In the present study, we divided the whole Discussion section into several parts based on the objectives and analyses.

4.1. LULC Mapping and Dynamics

The present study generated the LULC maps for three periods using the SVM algorithm. We used this machine learning algorithm, because many studies have already been conducted successfully with this and obtained highly accurate results. Thanh Noi and Kappas [54] applied SVM, RF, and K-nearest neighbour for LULC classification and found that SVM obtained the highest overall accuracy compared to the other two models. Similarly, Singh et al. [55] used SVM to classify the LULC for Pichavaram forest on the southeast coast of India. They found that 89–94% of kappa statistics for the LULC maps of 1991, 2000, and 2009 show the higher performances of SVM classifiers. On the other hand, Rana and Suryanarayana [56] utilised the maximum likelihood, RF, and SVM algorithms to classify LULC for the Vishwamitri watershed in Vadodara, India, found that SVM outperformed other models. Therefore, it can be stated that many previous studies obtained highly accurate results for LULC classification using SVM. While, in the present study, we also obtained 90–94% accuracy for the LULC of 2001, 2011, and 2021, this suggests that the generated LULC maps are highly accurate and reliable. The results of the LULC dynamics show that the LULC transformation in English Bazar occurred rapidly from 2001 to 2021. The predominance of the built-up area replacing agricultural land characterised these transformations. Other studies have reported the increasing percentage of the built-up area in the English Bazar and West Bengal state [11,46,59]. The water bodies, especially in the southern part of the study area, created a cascading pattern (shape) from 2001 to 2021. The area under it decreased from 2001 to 2011 and then increased again from 2011 to 2021. The area under water bodies declined from 2001 to 2011 because of its conversion to agricultural land by filling up water bodies [56]. From 2001 to 2021, the area under the water bodies increased because of the widespread floods of 2017, which converted many low-lying areas into water bodies because of permanent water-logging conditions [56]. The area under barren land has decreased significantly. It has been converted into agriculture and built up because of the paucity of space for increasing the population and mango farming, which is the mainstay of the economy for Malda District [46]. The findings show that changes in the agricultural land, vegetation, water bodies, and the built-up area had the most significant influence on the landscape heterogeneities [60]. Land use change aided the transformation of the landscape, especially for the rise of the built-up area. Additionally, the conversion of vegetation and agricultural land to a built-up area, and barren land converted into agricultural land, implies that the intensity of human activities affects land use change, landscape fragmentation, and ecological change [62].

4.2. Process of Built-Up Expansion

In the present study, we introduced the process of built-up expansion through the fragmentation and frequency approach. The steps of built-up expansion over time have been identified using the six indices of the landscape fragmentation approach. The results showed that the perforated and patch category areas were 6.24 km^2 and 13.42 km^2 in 2001, respectively, and have risen to 13.23 km^2 and 17.86 km^2 in 2021 (Figure 6). These new and isolated built-up nodes were originally rural neighbourhoods. The study area has experienced an economic and infrastructure progression because of the rise of MSME and governmental schemes that has accelerated the urbanisation process. As a result, some of the perforated and patched portions have expanded outwards through time, eventually connecting with small and medium cores. Replacing the edges and patches in the study period has increased the core categories. The increase in the percentage of the core can be attributed to the increasing built-up density and the conversion of intervening spaces into the built-up and the emergence of a new infrastructure [46]. The emergence of new patches and edge is widespread, signifying the spatial expansion of built-up areas by replacing agricultural land. Additionally, the results showed that, in 2001, the large core of the built-up area encompassed 2.92 km^2, which was amplified by 7.42 km^2 in 2021. As shown in this scenario, the large core of the built-up area, defined in 2001, was fixed three times but gained extra area in 2011 and 2021. Many studies reported that agricultural land is continuously declining in English Bazar over time [11,14,46]. Additionally, the frequency approach mimics fragmentation results, in which the permanent built-up area (frequency 3) is present in the core areas. In contrast, frequency 1 was fresh and isolated built-up nodes, while frequency 2 was the built-up transition area of small and medium cores. In this way, we showed the process of built-up expansion at the spatial scale. However, the research on the process of built-up expansion has not been explored yet using remote sensing as per the authors' knowledge; most of the research has concentrated on the overall built-up expansion mapping and pattern analysis [63,64]. As shown by the current study, the area under build up has been increasing gradually over time, which can result in several socioecological problems, such as UHI [65], pollution [66], urban flooding, etc., in English Bazar and its contiguous areas in recent decades.

4.3. Built-Up Probability Modelling

In the current study, the built-up probability modelling was done using three parameters, such as dominance, diversity, and connectivity models. The results showed that the values of all three models (dominance, diversity, and connectivity) for English Bazar were very high in the urban part and its surrounding area. We constructed a built-up expansion probability model by integrating the dominance, diversity, and connectivity models using fuzzy logic. It is predicted that considerable expansion will occur around the main city centre, NH-34 and SH-10, and some western regions. Increased populations, migration, small agro-based industrial sector, and transportation nodal location of the English Bazar block are all significant factors that have led to a transformation in the land use pattern [11,46,67,68]. However, one of the significant factors of unplanned built-up expansion was local and national government plans and policies. The rise of small-scale industrial activity in industrial axes and cities encourages investors to invest and develop an infrastructure. This acts as a pull factor for the population of its surrounding rural area or neighbouring district. Therefore, the new residential area was developed for the migrants.

Similarly, the rapid development of small industrial towns, health centres, educational institutes, and transportation networks has attracted the population to come and settle over there, which significantly influences land use change. The scattered settlements in rural areas were also expanded because of many central and state government-sponsored projects, such as 'Pradhan Mantri Gram Sadak Yojana' (PMGSY) of the central government and 'Pathashree Abhijan' of the state government. These schemes boosted the process of the rapid construction of the road throughout the country (which enhanced the connectivity and linked it with the urban area). Additionally, other schemes, for *pucca* (concrete) houses under the 'Pradhan Mantri Awas Yojana' (PMAJ), accelerate the process of built-up expansion. The increase in agricultural activities may be another factor of enormous expansion in the rural built-up sector [56,59]. Due to these changes, vegetation, agricultural land, water bodies, and barren land were converted into built-up areas. Thus, with built-up growth and expansion during this period, the dominant landscape was gradually replaced by an urban landscape, resulting in a shift in the region's natural ecosystem and the formation of a more fragmented landscape pattern of natural resources [69,70]. Additionally, the distance from roads negatively affects built-up developments. Hence, built-up transitions are typically seen near roads [61,71,72]. In the present study, we delineated the future built-up expansion probability based on these factors.

We verified the results using Google Earth images (Figure 11), and it also shows the rapid built-up growth over time, similar to our models. The built-up growth can be seen outwards from the centre, which signifies the probable built-up expansion area.

4.4. Policy Recommendation for Urban Planning and Future Research Scope

To the best of the authors' knowledge, a study on built-up probability has not been conducted so far. Many studies have investigated the effect of urbanisation on land use change [11–15,30,43,68,73]. However, most of the studies were focused mainly on the urban area. However, we considered rural and urban areas in this research paper and showed their future built-up growth probability. Built-up growth centres are primarily in the southwestern and north-western regions of the municipality area, and some of them are in the western part of the English Bazar block. It also shows that not every village is adjacent to a dense region, which is a potential growth hub in the future. The outcome of the present study showed that the municipal area witnessed an urban sprawl beyond its administrative boundary after 2011. This research identified about 20 villages as high-probability future growth hubs and 38 villages as moderate-probability future growth centres based on the dominance, diversity, and connectivity parameters. The significant probability will be surrounded by the city and along the NH-34. Therefore, it is suggested that these should be immediately stopped to maintain the remaining natural resources and, if feasible, restore them. If we inspect the distribution of trees in the urban region, we can see that trees are scarce in the urban area, altering the microclimate and forming an urban heat island (UHI).

Due to this, the present study suggests that more green space be created in the English Bazar urban region, such as along the Badh road, Subhankar sisu udhyan, Chatra wetland, University of Gour Banga, etc. A similar suggestion can be followed in other built-up nodes such as Samsi, Pukhuria, Pakuahat, etc. English Bazar City has suffered from the UHI effect [61]; therefore, generating more green spaces will help reduce this UHI effect. It will also improve the air quality of the study area and provide comfort to the citizens.

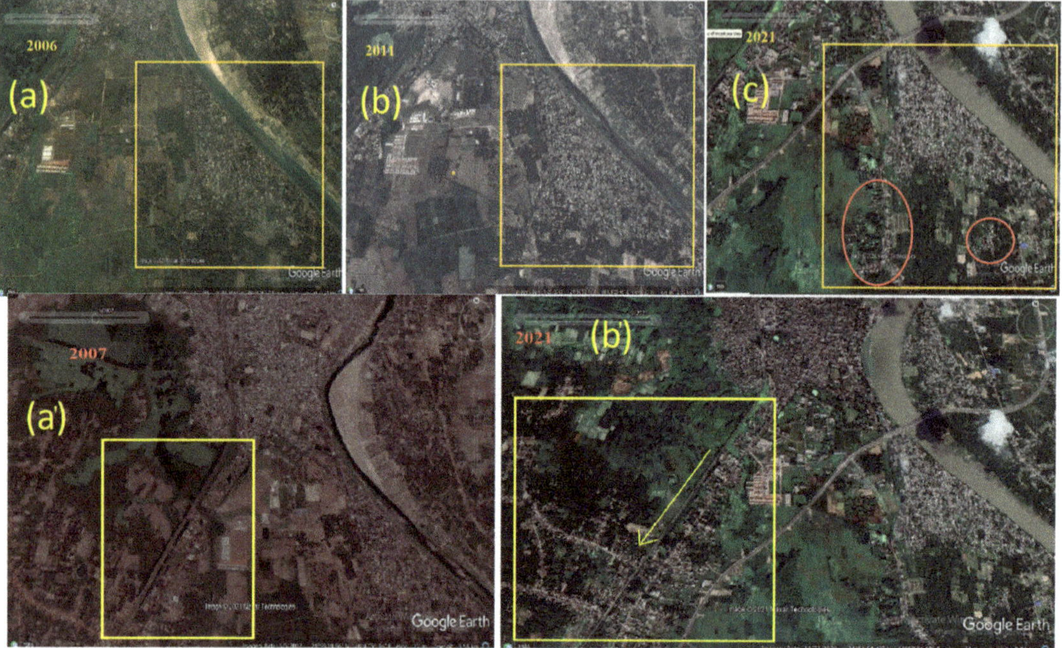

Figure 11. Present scenario of the urban landscape in the English Bazar block based on high-resolution satellite images (QuickBird, resolution 1 m) from Google Earth. (**a–c**) in upper part of figure shows South-eastern built-up expansion during 2006, 2014, and 2021, away from the municipal area. (**a',b'**) in lower part of figure shows South-west built-up growth besides NH-34 during 2007–2021). As previously stated, a significant changing pattern was observed in the southern part of the study area, with water bodies in that area changing over time from 2001 to 2011. Some of it was converted into agricultural land, but from 2011 to 2021, the water bodies area increased again for the reasons previously stated. As a result, the Google Earth images clearly indicate that the water bodies are changing into built-up lands.

Systematic and uniformly distributed urbanisation or built-up expansion has not been seen in small and medium-sized cities. It is noticed that urban growth can be observed in some regions of small and medium-sized cities. Therefore, with the help of the fragmentation model, it is possible to identify the place of growth with the direction and magnitude. If the interval is reduced to the three-five-year interval, the direction of growth, and the actual reason behind the growth, can be identified. Therefore, unwanted rapid urban growth can be prevented in those areas.

5. Conclusions

The primary purpose of this research was to explore the process of built-up expansion and model the future built-up probability zones based on the datasets from 2001 to 2021. Our findings showed that the study area observed a sharp rise in built-up areas, accompanied by a decrease in agricultural and vegetation cover. The built-up fragmentation model identified the process of built-up expansion in the form of permanent, isolated, and newly formed built-up areas. Then, using the frequency approach model, the process of built-up expansion over time was created on a single map, showing it during three separate periods. Then, the dominance, diversity, and connection models were employed as parameters for the built-up probability model. Finally, we used the fuzzy logic-based built-up stability and built-up probability model to predict future built-up growths and trends. The results showed that the built-up area in the study area increased in tandem with the substantial economic expansion. In the previous 20 years (2001–2021), the built-up area has grown by nearly 2.5 times what it was, with a 36 km^2 net increase. Between 2001 and 2021, the LULC was altered, demonstrating a rise in a built-up area (almost by 6%) but a decrease in vegetation cover (by 4.54%) and agricultural land (by 3.78%). Moreover, from 2001 to 2021, barren land was also converted into agricultural land, built-up, and vegetation cover because of the increasing population, the necessity for food, and mango farming, a dominant orchard in Malda District. The expansion and growth of the built-up regions and the elimination of agriculture and vegetation across the English Bazar increased its population and subsequent economic development. According to this study, the English Bazar municipality has established itself as a permanent and stable location for built-up areas that span time and geography (beyond its administrative boundary). The study has some limitations, although we developed a model for the urban growth process and the likelihood of future urban expansion, which can minimise urban sprawling and foster compact green towns. We employed a satellite image with a moderate resolution, which has certain limitations in recognising built-up areas and traditional machine learning methods like SVM. High-resolution satellite imagery and deep learning algorithms can solve these problems. We can finely detect the expansion of urbanisation with minor errors using high-resolution satellite images such as Sentinel, LISS-IV, Worldview, QuickBird, and others. After tackling the drawbacks, these approaches can be implemented in small and medium-sized cities for proper management. The MODIS and night-time images can be used in future research with the proposed models for exploring the urbanisation process and the probability of large and megacities. The U-Net model (deep learning model) can be used in future research to analyse and predict urbanisation expansion, providing pixel-level information with great accuracy.

Author Contributions: Conceptualisation, T.D., A.R.M.T.I., A.M. and A.R.; software, T.D., S.T. and S.; validation, M.W.N., A.P., M.S.A. and S.P.; formal analysis, T.D., S.T. and A.M.; investigation, M.W.N., S.T. and A.R.M.T.I.; resources, A.R., M.S.A. and A.M.; data curation, T.D., S.T. and S.; writing—original draft preparation, T.D., A.P., S.T., and S.; writing—review and editing, A.R. and S.P.; visualisation, S.T. and A.M.; supervision, A.R., A.M. and S.P.; project administration, A.R. and A.M.; and funding acquisition, A.M. All authors have read and agreed to the published version of the manuscript.

Funding: This research received no external funding.

Data Availability Statement: Not applicable.

Acknowledgments: All the authors are thankful to the USGS for making the Landsat data freely available.

Conflicts of Interest: The authors declare no conflict of interest.

References

1. United Nations Department of Economic and Social Affairs. World Urbanization Prospects: The 2018 Revision. Available online: https://population.un.org/wup/ (accessed on 5 March 2022).
2. Seto, K.C.; Fragkias, M.; Guneralp, B.; Reilly, M.K. A meta-analysis of global urban land expansion. *PLoS ONE* **2011**, *6*, e23777. [CrossRef] [PubMed]
3. Seto, K.C.; Güneralp, B.; Hutyra, L.R. Global forecasts of urban expansion to 2030 and direct impacts on biodiversity and carbon pools. *Proc. Natl. Acad. Sci. USA* **2012**, *109*, 16083–16088. [CrossRef] [PubMed]
4. Chakraborty, S.; Maity, I.; Dadashpoor, H.; Novotný, J.; Banerji, S. Building in or out? Examining urban expansion patterns and land use efficiency across the global sample of 466 cities with million+ inhabitants. *Habitat Int.* **2022**, *120*, 102503. [CrossRef]
5. Rahman, A.; Kumar, Y.; Fazal, S.; Bhaskaran, S. Urbanization and quality of urban environment using remote sensing and GIS techniques in East Delhi-India. *J. Geogr. Inf. Syst.* **2011**, *3*, 62–84. [CrossRef]

6. Wolch, J.R.; Byrne, J.; Newell, J.P. Urban green space, public health, and environmental justice: The challenge of making cities "just green enough". *Landsc. Urban Plan.* **2014**, *125*, 234–244. [CrossRef]
7. Hennig, E.I.; Schwick, C.; Soukup, T.; Orlitová, E.; Kienast, F.; Jaeger, J.A.G. Multi-scale analysis of urban sprawl in Europe: Towards a European de-sprawling strategy. *Land Use Policy* **2015**, *49*, 483–498. [CrossRef]
8. Chadchan, J.; Shankar, R. An analysis of urban growth trends in the post-economic reforms period in India. *Int. J. Sustain. Built Environ.* **2012**, *1*, 36–49. [CrossRef]
9. Dewan, A.M.; Yamaguchi, Y. Land use and land cover change in greater Dhaka, Bangladesh: Using remote sensing to promote sustainable urbanization. *Appl. Geogr.* **2009**, *29*, 390–401. [CrossRef]
10. Ranagalage, M.; Morimoto, T.; Simwanda, M.; Murayama, Y. Spatial Analysis of Urbanization Patterns in Four Rapidly Growing South Asian Cities Using Sentinel-2 Data. *Remote Sens.* **2021**, *13*, 1531. [CrossRef]
11. Shaw, R.; Das, A. Identifying peri-urban growth in small and medium towns using GIS and remote sensing technique: A case study of English Bazar Urban Agglomeration, West Bengal, India. *Egypt. J. Remote Sens. Space Sci.* **2018**, *21*, 159–172. [CrossRef]
12. Shahfahad; Mourya, M.; Kumari, B.; Tayyab, M.; Paarcha, A.; Asif; Rahman, A. Indices based assessment of built-up density and urban expansion of fast-growing Surat city using multi-temporal Landsat data sets. *GeoJournal* **2020**, *86*, 1607–1623. [CrossRef]
13. Chettry, V.; Surawar, M. Assessment of urban sprawl characteristics in Indian cities using remote sensing: Case studies of Patna, Ranchi, and Srinagar. *Environ. Dev. Sustain.* **2021**, *23*, 11913–11935. [CrossRef]
14. Das, M.; Das, A. Dynamics of Urbanization and its impact on Urban Ecosystem Services (UESs): A study of a medium size town of West Bengal, Eastern India. *J. Urban Manag.* **2019**, *8*, 420–434. [CrossRef]
15. Nengroo, Z.A.; BhatNissar, M.S.; Kuchay, A. Measuring urban sprawl of Srinagar city, Jammu and Kashmir, India. *J. Urban Manag.* **2017**, *6*, 45–55. [CrossRef]
16. Pawe, C.K.; Saikia, A. Decumbent development: Urban sprawl in the Guwahati Metropolitan Area, India. *Singap. J. Trop. Geogr.* **2020**, *41*, 226–247. [CrossRef]
17. Shaban, A.; Kourtit, K.; Nijkamp, P. India's urban system: Sustainability and imbalanced growth of cities. *Sustainability* **2020**, *12*, 2941. [CrossRef]
18. Census of India. Provisional Population Totals Urban Agglomerations and Cities 2011. Available online: https://censusindia.gov.in/2011-prov-results/paper2/data_files/india2/1.%20data%20highlight.pdf (accessed on 6 March 2022).
19. Raman, B.; Prasad-Aleyamma, M.; de Bercegol, R.; Denis, E.; Zérah, M.-H. Selected Readings on Small Town Dynamics in India. USR 3330 "Savoirs et MondesIndiens" Working Papers Series No. 7. No. 2. SUBURBIN Working Papers Series. 2015. Available online: https://hal.archives-ouvertes.fr/hal-01139006/document (accessed on 5 March 2022).
20. Patra, S.; Sahoo, S.; Mishra, P.; Mahapatra, S.C. Impacts of urbanization on land use/cover changes and its probable implications on local climate and groundwater level. *J. Urban Manag.* **2018**, *7*, 70–84. [CrossRef]
21. Dutta, D.; Rahman, A.; Kundu, A. Growth of Dehradun city: An application of linear spectral unmixing (LSU) technique using multi-temporal landsat satellite data sets. *Remote Sens. Appl. Soc. Environ.* **2015**, *1*, 98–111. [CrossRef]
22. Mohanta, K.; Sharma, L.K. Assessing the impacts of urbanization on the thermal environment of Ranchi City (India) using geospatial technology. *Remote Sens. Appl. Soc. Environ.* **2017**, *8*, 54–63. [CrossRef]
23. Bhagat, R.B. Emerging pattern of urbanisation in India. *Econ. Political Wkly.* **2011**, *34*, 10–12.
24. Nechyba, T.J.; Walsh, R.P. Urban sprawl. *J. Econ. Perspect.* **2004**, *18*, 177–200. [CrossRef]
25. Wang, X.; Shi, R.; Zhou, Y. Dynamics of urban sprawl and sustainable development in China. *Socio Econ. Plan. Sci.* **2019**, *70*, 100736. [CrossRef]
26. Ghazaryan, G.; Rienow, A.; Oldenburg, C.; Thonfeld, F.; Trampnau, B.; Sticksel, S.; Jürgens, C. Monitoring of Urban Sprawl and Densification Processes in Western Germany in the Light of SDG Indicator 11.3.1 Based on an Automated Retrospective Classification Approach. *Remote Sens.* **2021**, *13*, 1694. [CrossRef]
27. Bai, X.; McPhearson, T.; Cleugh, H.; Nagendra, H.; Tong, X.; Zhu, T.; Zhu, Y.-G. Linking Urbanization and the Environment: Conceptual and Empirical Advances. *Annu. Rev. Environ. Resour.* **2017**, *42*, 215–240. [CrossRef]
28. Talukdar, S.; Rihan, M.; Hang, H.T.; Bhaskaran, S.; Rahman, A. Modelling Urban Heat Island (UHI) and Thermal Field Variation and Their Relationship with Land Use Indices over Delhi and Mumbai Metro Cities. *Environ. Dev. Sustain.* **2021**, *24*, 3762–3790.
29. Aram, F.; Solgi, E.; HiguerasGarcía, E.; Mosavi, A.; Várkonyi-Kóczy, A.R. The Cooling Effect of Large-Scale Urban Parks on Surrounding Area Thermal Comfort. *Energies* **2019**, *12*, 3904. [CrossRef]
30. Wentz, E.A.; Anderson, S.; Fragkias, M.; Netzband, M.; Mesev, V.; Myint, S.W.; Quattrochi, D.A.; Rahman, A.; Seto, K.C. Supporting global environmental change research: A review of trends and knowledge gaps in urban remote sensing. *Remote Sens.* **2014**, *6*, 3879–3905. [CrossRef]
31. Yan, X.; Wang, J. Dynamic monitoring of urban built-up object expansion trajectories in Karachi, Pakistan with time series images and the LandTrendr algorithm. *Sci. Rep.* **2021**, *11*, 23118. [CrossRef]
32. Güneralp, B.; Reba, M.; Hales, B.; Wentz, E.; Seto, K. Trends in urban land expansion, density, and land transitions from 1970 to 2010: A global synthesis. *Environ. Res. Lett.* **2020**, *15*, 044015. [CrossRef]
33. As-syakur, A.R.; Adnyana, I.W.S.; Arthana, I.W.; Nuarsa, I.W. Enhanced Built-Up and Bareness Index (EBBI) for Mapping Built-Up and Bare Land in an Urban Area. *Remote Sens.* **2012**, *4*, 2957–2970. [CrossRef]
34. Zhang, J.; Li, P.; Wang, J. Urban Built-Up Area Extraction from Landsat TM/ETM+ Images Using Spectral Information and Multivariate Texture. *Remote Sens.* **2014**, *6*, 7339–7359. [CrossRef]

35. Salem, M.; Tsurusaki, N.; Divigalpitiya, P. Land use/land cover change detection and urban sprawl in the peri-urban area of greater Cairo since the Egyptian revolution of 2011. *J. Land Use Sci.* **2020**, *15*, 592–606. [CrossRef]
36. Zheng, Y.; Tang, L.; Wang, H. An improved approach for monitoring urban built-up areas by combining NPP-VIIRS nighttime light, NDVI, NDWI, and NDBI. *J. Clean. Prod.* **2021**, *328*, 129488. [CrossRef]
37. Talukdar, S.; Singha, P.; Mahato, S.; Pal, S.; Liou, Y.A.; Rahman, A. Land-use land-cover classification by machine learning classifiers for satellite observations—A review. *Remote Sens.* **2020**, *12*, 1135. [CrossRef]
38. Fawad, J.A.; Khan, M.N.; Minallah, N. Impact of Machine Learning Techniques for Lulc Classification of Peshawar Pakistan. *Int. Res. J. Mod. Eng. Technol. Sci.* **2021**, *3*, 478–486.
39. Xu, H.Q. A new index for delineating built-up land features in satellite imagery. *Int. J. Remote Sens.* **2008**, *29*, 4269–4276. [CrossRef]
40. Omer, G.; Mutanga, O.; Abdel-Rahman, E.M.; Adam, E. Exploring the utility of the additional WorldView-2 bands and support vector machines in mapping land use/land cover in a fragmented ecosystem, South Africa. *S. Afr. J. Geomat.* **2015**, *4*, 414–433. [CrossRef]
41. Ul Din, S.; Mak, H.W.L. Retrieval of Land-Use/Land Cover Change (LUCC) Maps and Urban Expansion Dynamics of Hyderabad, Pakistan via Landsat Datasets and Support Vector Machine Framework. *Remote Sens.* **2021**, *13*, 3337. [CrossRef]
42. Schulz, D.; Yin, H.; Tischbein, B.; Verleysdonk, S.; Adamou, R.; Kumar, N. Land use mapping using Sentinel-1 and Sentinel-2 time series in a heterogeneous landscape in Niger, Sahel. *J. Photogramm. Remote Sens.* **2021**, *178*, 71–111. [CrossRef]
43. Naikoo, M.W.; Rihan, M.; Ishtiaque, M. Analyses of land use land cover (LULC) change and built-up expansion in the suburb of a metropolitan city: Spatio-temporal analysis of Delhi NCR using landsat datasets. *J. Urban Manag.* **2020**, *9*, 347–359. [CrossRef]
44. Shahfahad; Rihan, M.; Naikoo, M.W.; Ali, M.A.; Usmani, T.M.; Rahman, A. Urban Heat Island Dynamics in Response to Land-Use/Land-Cover Change in the Coastal City of Mumbai. *J. Indian Soc. Remote Sens.* **2021**, *49*, 2227–2247. [CrossRef]
45. Shahfahad; Naikoo, M.W.; Islam, A.R.M.T.; Mallick, J.; Rahman, A. LULC change and its impact on surface urban heat island and urban thermal comfort in a metropolitan city. *Urban Clim.* **2022**, *41*, 101052. [CrossRef]
46. Naikoo, M.W.; Rihan, M.; Peer, A.H.; Talukdar, S.; Mallick, J.; Ishtiaq, M.; Rahman, A. Analysis of peri-urban land use/land cover change and its drivers using geospatial techniques and geographically weighted regression. *Environ. Sci. Pollut. Res.* **2022**, 1–19. [CrossRef] [PubMed]
47. Vinayak, B.; Lee, H.S.; Gedem, S.; Latha, R. Impacts of future urbanization on urban microclimate and thermal comfort over the Mumbai Metropolitan Region, India. *Sustain. Cities Soc.* **2022**, *79*, 103703. [CrossRef]
48. Dinda, S.; Chatterjee, N.D.; Ghosh, S. An integrated simulation approach to the assessment of urban growth pattern and loss in urban green space in Kolkata, India: A GIS-based analysis. *Ecol. Indic.* **2021**, *121*, 107178. [CrossRef]
49. Dutta, I.; Das, A. Application of Geo-Spatial Indices for Detection of Growth Dynamics and Forms of Expansion in English Bazar Urban Agglomeration, West Bengal. *J. Urban Manag.* **2019**, *8*, 288–302. [CrossRef]
50. Roy, D.P.; Wulder, M.A.; Loveland, T.R.; Woodcock, C.E.; Allen, R.G.; Anderson, M.C.; Helder, D.; Irons, J.R.; Johnson, D.M.; Kennedy, R.; et al. Landsat-8: Science and product vision for terrestrial global change research. *Remote Sens. Environ.* **2014**, *145*, 154–172. [CrossRef]
51. Flood, N. Continuity of reflectance data between landsat-7 ETM+ and landsat-8 OLI, for both top-of-atmosphere and surface reflectance: A study in the australian landscape. *Remote Sens.* **2014**, *6*, 7952–7970. [CrossRef]
52. Guo, Y.M.; Guo, L.B.; Hao, Z.Q.; Tang, Y.; Ma, S.X.; Zeng, Q.D.; Tang, S.S.; Li, X.Y.; Lu, Y.F.; Zeng, X.Y. Accuracy improvement of iron ore analysis using laser-induced breakdown spectroscopy with a hybrid sparse partial least squares and least-squares support vector machine model. *J. Anal. At. Spectrom.* **2018**, *33*, 1330–1335. [CrossRef]
53. Wu, C.; Du, B.; Cui, X.; Zhang, L. A post-classification change detection method based on iterative slow feature analysis and Bayesian soft fusion. *Remote Sens. Environ.* **2017**, *199*, 241–255. [CrossRef]
54. Thanh Noi, P.; Kappas, M. Comparison of random forest, k-nearest neighbor, and support vector machine classifiers for land cover classification using Sentinel-2 imagery. *Sensors* **2017**, *18*, 18. [CrossRef] [PubMed]
55. Singh, S.K.; Srivastava, P.K.; Gupta, M.; Thakur, J.K.; Mukherjee, S. Appraisal of land use/land cover of mangrove forest ecosystem using support vector machine. *Environ. Earth Sci.* **2014**, *71*, 2245–2255. [CrossRef]
56. Rana, V.K.; Suryanarayana, T.M.V. Performance evaluation of MLE, RF and SVM classification algorithms for watershed scale land use/land cover mapping using sentinel 2 bands. *Remote Sens. Appl. Soc. Environ.* **2020**, *19*, 100351. [CrossRef]
57. Galgamuwa, G.A.P.; Barden, C.J.; Hartman, J.; Rhodes, T.; Bloedow, N.; Osōrio, R. Ecological Restoration of an Oak Woodland within the Forest-Prairie Ecotone of Kansas. *For. Sci.* **2018**, *65*, 48–58. [CrossRef]
58. Amato, F.; Maimone, B.A.; Martellozzo, F.; Nolè, G.; Murgante, B. The Effects of Urban Policies on the Development of Urban Areas. *Sustainability* **2016**, *8*, 297. [CrossRef]
59. Talukdar, S.; Naikoo, M.W.; Mallick, J.; Praveen, B.; Sharma, P.; Islam, A.R.M.T.; Pal, S.; Rahman, A. Coupling geographic information system integrated fuzzy logic-analytical hierarchy process with global and machine learning based sensitivity analysis for agricultural suitability mapping. *Agric. Syst.* **2022**, *196*, 103343. [CrossRef]
60. Kowe, P.; Mutanga, O.; Dube, T. Advancements in the remote sensing of landscape pattern of urban green spaces and vegetation fragmentation. *Int. J. Remote Sens.* **2021**, *42*, 3797–3832. [CrossRef]
61. Ziaul, S.; Pal, S. Analyzing control of respiratory particulate matter on Land Surface Temperature in local climatic zones of English Bazar Municipality and Surroundings. *Urban Clim.* **2018**, *24*, 34–50. [CrossRef]

62. Saadat, M.N.; Gupta, K. Mango cultivation in Malda District, West Bengal: A historical perspective. *Asian Agri Hist.* **2017**, *21*, 309–318.
63. Thapa, R.B.; Murayama, Y. Examining spatiotemporal urbanization patterns in Kathmandu Valley, Nepal: Remote sensing and spatial metrics approaches. *Remote Sens.* **2009**, *1*, 534–556. [CrossRef]
64. Lopez, J.M.R.; Heider, K.; Scheffran, J. Frontiers of urbanization: Identifying and explaining urbanization hot spots in the south of Mexico City using human and remote sensing. *Appl. Geogr.* **2017**, *79*, 1–10. [CrossRef]
65. Basu, T.; Das, A.; Pham, Q.B.; Al-Ansari, N.; Linh, N.T.T.; Lagerwall, G. Development of an integrated peri-urban wetland degradation assessment approach for the Chatra Wetland in eastern India. *Sci. Rep.* **2021**, *11*, 4470. [CrossRef] [PubMed]
66. Das, S.; Bhunia, G.S.; Bera, B.; Shit, P.K. Evaluation of wetland ecosystem health using geospatial technology: Evidence from the lower Gangetic flood plain in India. *Environ. Sci. Pollut. Res.* **2022**, *29*, 1858–1874. [CrossRef] [PubMed]
67. Yang, Y.; Chen, J.; Lan, Y.; Zhou, G.; You, H.; Han, X.; Wang, Y.; Shi, X. Landscape Pattern and Ecological Risk Assessment in Guangxi Based on Land Use Change. *Int. J. Environ. Res. Public Health* **2022**, *19*, 1595. [CrossRef]
68. Byun, G.; Kim, Y. A street-view-based method to detect urban growth and decline: A case study of Midtown in Detroit, Michigan, USA. *PLoS ONE* **2022**, *17*, e0263775. [CrossRef]
69. Rwanga, S.S.; Ndambuki, J.M. Accuracy assessment of land use/land cover classification using remote sensing and GIS. *Int. J. Geosci.* **2017**, *8*, 611. [CrossRef]
70. Dutta, S.; Dutta, I.; Das, A.; Guchhait, S.K. Quantification and mapping of fragmented forest landscape in dry deciduous forest of Burdwan Forest Division, West Bengal, India. *Trees For. People* **2020**, *2*, 100012. [CrossRef]
71. Wentz, E.A.; Nelson, D.; Rahman, A.; Stefanov, W.L.; Roy, S.S. Expert system classification of urban land use/cover for Delhi, India. *Int. J. Remote Sens.* **2008**, *29*, 4405–4427. [CrossRef]
72. Ghosh, D.K.; Mandal, A.C.H.; Majumder, R.; Patra, P.; Bhunia, G.S. Analysis for mapping of built-up area using remotely sensed indices-A case study of rajarhat block in Barasatsadar sub-division in west Bengal (India). *J. Landsc. Ecol. Repub.* **2018**, *11*, 67–76.
73. Li, H.; Zhou, Y.; Jia, G.; Zhao, K.; Dong, J. Quantifying the Response of Surface Urban Heat Island to Urbanization Using the Annual Temperature Cycle Model. *Geosci. Front.* **2021**, *13*, 101141. [CrossRef]

Article

A Sparse-Model-Driven Network for Efficient and High-Accuracy InSAR Phase Filtering

Nan Wang, Xiaoling Zhang *, Tianwen Zhang, Liming Pu, Xu Zhan, Xiaowo Xu, Yunqiao Hu, Jun Shi and Shunjun Wei

School of Information and Communication Engineering, University of Electronic Science and Technology of China, Chengdu 611731, China; wangnan@std.uestc.edu.cn (N.W.); twzhang@std.uestc.edu.cn (T.Z.); puliming@std.uestc.edu.cn (L.P.); zhanxu@std.uestc.edu.cn (X.Z.); xuxiaowo@std.uestc.edu.cn (X.X.); hyq@std.uestc.edu.cn (Y.H.); shijun@uestc.edu.cn (J.S.); weishunjun@uestc.edu.cn (S.W.)
* Correspondence: xlzhang@uestc.edu.cn

Abstract: Phase filtering is a vital step for interferometric synthetic aperture radar (InSAR) terrain elevation measurements. Existing phase filtering methods can be divided into two categories: traditional model-based and deep learning (DL)-based. Previous studies have shown that DL-based methods are frequently superior to traditional ones. However, most of the existing DL-based methods are purely data-driven and neglect the filtering model, so that they often need to use a large-scale complex architecture to fit the huge training sets. The issue brings a challenge to improve the accuracy of interferometric phase filtering without sacrificing speed. Therefore, we propose a sparse-model-driven network (SMD-Net) for efficient and high-accuracy InSAR phase filtering by unrolling the sparse regularization (SR) algorithm to solve the filtering model into a network. Unlike the existing DL-based filtering methods, the SMD-Net models the physical process of filtering in the network and contains fewer layers and parameters. It is thus expected to ensure the accuracy of the filtering without sacrificing speed. In addition, unlike the traditional SR algorithm setting the spare transform by handcrafting, a convolutional neural network (CNN) module was established to adaptively learn such a transform, which significantly improved the filtering performance. Extensive experimental results on the simulated and measured data demonstrated that the proposed method outperformed several advanced InSAR phase filtering methods in both accuracy and speed. In addition, to verify the filtering performance of the proposed method under small training samples, the training samples were reduced to 10%. The results show that the performance of the proposed method was comparable on the simulated data and superior on the real data compared with another DL-based method, which demonstrates that our method is not constrained by the requirement of a huge number of training samples.

Keywords: interferometric phase filtering; sparse regularization (SR); deep learning (DL); neural convolutional network (CNN)

Citation: Wang, N.; Zhang, X.; Zhang, T.; Pu, L.; Zhan, X.; Xu, X.; Hu, Y.; Shi, J.; Wei, S. A Sparse-Model-Driven Network for Efficient and High-Accuracy InSAR Phase Filtering. *Remote Sens.* **2022**, *14*, 2614. https://doi.org/10.3390/rs14112614

Academic Editor: Gwanggil Jeon

Received: 28 April 2022
Accepted: 24 May 2022
Published: 30 May 2022

1. Introduction

Due to the all-weather and all-day characteristics of the synthetic aperture radar (SAR), it plays an important role in remote sensing [1–5]. Simultaneously, the continuous development of SAR has brought more and more development prospects to interferometric SAR (InSAR). At present, InSAR has a wide range of applications such as surface deformation monitoring and terrain mapping [6–11]. The basic principle of InSAR measurement technology mainly extracts the phase difference in the primary and secondary images through the observation angle difference of the primary and secondary antennas, and finally inverts the elevation information of the observation area by using the formula between the phase difference and the height.

In the whole InSAR processing flow, noise is inevitably added to the InSAR phase, which can be divided into three categories: system noise, coherent noise, and noise introduced by signal processing [12,13]. The presence of noise severely destroys the follow-up phase unwrapping step, which reduces the accuracy of phase unwrapping and even obtains the incorrect results [14,15]. Therefore, interferometric phase filtering is a necessary processing step and has also become a very important technology in InSAR measurement.

Since the invention of InSAR technology, a large number of effective interferometric phase filtering approaches have been developed, and the traditional methods fall into four main categories (i.e., spatial domain local filters [16–20], spatial domain nonlocal (NL) filters [21], transform domain local filters [22–26], and transform domain NL filters [27,28]). The main idea of spatial domain local filters is to filter out the phase noise in the space domain using a local window with pixels. A well-known spatial domain local filter is the Lee filter [18]. Unlike spatial domain filters, the transform domain local filters denoise the interferogram in the transform domain such as the Goldstein filter [26]. However, the above two kinds of filters cannot balance the noise suppression ability and phase detail preservation ability well. In order to further enhance the phase detail preservation capability while ensuring effective noise suppression, the spatial and transform domain NL filters have been proposed, which utilize the patch-by-patch method to measure the patch similarity of the interferogram and the weighted average to restore the interferometric phase [14] such as NL-InSAR [21] and InSAR-BM3D [28]. Although NL filters can consider both noise suppression and phase detail preservation, they suffer from a huge computational cost due to abundant similar block operations. Aiming to bridge this regret, a series of newly advanced filtering algorithms have been proposed [29–34].

Over the past few years, deep learning (DL) has been successfully applied to interferogram denoising due to the powerful feature extraction and calculation ability of convolutional neural networks (CNNs) such as Phi-Net [31] and PFNet [30]. However, there are two key problems with the vast majority of existing CNNs. On one hand, the underlying structure of the purely data-driven CNN with a black-box nature is difficult to interpret, that is, it lacks interpretability. Of course, interpretability is an important feature in many fields because it relates to conceptual understanding and the development of knowledge frontiers [35]. On the other hand, most modern CNNs need to learn a large number of parameters, so they excessively depend on huge amounts of data. In other words, a vast majority of CNNs improve the accuracy at the cost of increasing the network complexity. However, in many fields such as in [36,37], the performance of the network trained with small training sets will be significantly reduced, and even inferior to the traditional methods.

In recent years, a promising technique that unrolls the SR algorithm into network architectures was developed by Gregor et al. [38]. Compared to modern CNNs, the unrolled network not only has a sufficient theoretical basis, but also contains fewer layers and parameters, which do not rely on huge training sets. Therefore, some novel networks based on the idea of unrolling the SR algorithm into CNN have been proposed such as in [39,40]. However, since SR algorithm unrolling has not been applied to InSAR phase denoising, we attempted to combine this technique into this field. Inspired by [39–42], we designed an InSAR phase filtering model and established a model-driven CNN to filter the noisy interferograms.

In this article, we propose a sparse-model-driven network (SMD-Net) for efficient and high-accuracy InSAR phase filtering. In the method, we first establish a SR model for interferometric phase filtering. Then, the SMD-Net is designed as an iteration-based CNN architecture by unrolling each iteration process of the iterative shrinkage-thresholding algorithm (ISTA) [43] to solve the phase filtering model into a block. In each block, a CNN module with a local block and global context (GC) [44] block is established to adaptively learn the sparse domain transform of each iteration in ISTA. Finally, due to dealing with complex-valued data, our method is carried out by exploiting the idea of separating the real and imaginary parts of the interferometric phase. In short, the SMD-Net models the interferometric phase filtering process, rather than relying entirely on data fitting as most networks do and its network structure is simple. It thus improves the filtering performance and computational efficiency at the same time. The experimental results on the simulated and measured data demonstrate that the proposed method outperformed the Lee filter [18], Goldstein filter [26], InSAR-BM3D filter [28], ISTA-based filtering method, and the PFNet [30] in both precision and speed. Furthermore, the filtering performance of the SMD-Net on 10% of the original training samples was also slightly better than that of the PFNet. The main contributions of our work are as follows.

(1) We first built an InSAR phase filtering model. Then, the SMD-Net was designed based on the idea of unrolling the ISTA algorithm of solving the model into a simple network architecture, which enhanced the interpretability of the network. Subsequently, the SMD-Net transformed the interferometric phase into a real matrix consisting of the real and imaginary parts of the phase to achieve a complex-valued filtering operation.

(2) Unlike the traditional ISTA algorithm setting the sparse transform by handcrafting, the SMD-Net exploits a CNN module to automatically learn the sparse basis operation, which enhances the filtering performance.

(3) Plenty of simulated and measured experiments illustrate that the proposed method achieves efficiency and high-precision filtering.

The rest of this article is organized as follows. Section 2 describes the InSAR phase noise principle and the InSAR phase SR filtering model. We introduce the design of the SMD-Net and loss function in Section 3. In Section 4, we describe a method of generating the training and testing data, experimental details, and experimental metrics. Extensive experiments on the simulated and real data are conducted in Section 5. Section 6 further discusses the performance of the proposed method under small training samples. Section 7 presents our conclusions.

2. Principle and Model

In what follows, we provide a brief review of the formulation of the interferometric phase and focus on designing the interferometric phase filtering model by analyzing the formulation. This work is to prepare for the subsequent unrolling of the InSAR phase filtering algorithm.

2.1. InSAR Phase Noise Principle

The interferogram Γ is defined as the conjugate product of a pair of single-look complex SAR images.

$$\Gamma = \mathbf{z}_1 \cdot \mathbf{z}_2^* = |\mathbf{z}_1 \cdot \mathbf{z}_2^*| e^{j(\varphi)} \tag{1}$$

where $\mathbf{z}_1$ and $\mathbf{z}_2$ are the two complex SAR images; * indicates the complex conjugate; and φ denotes the measured interferometric phase with noise. The phase noising model with additive noise is as follows.

$$\varphi = \varphi_c + \mathbf{n}_\gamma \tag{2}$$

where φ_c denotes the clear interferometric phase and $\mathbf{n}_\gamma$ is the zero-mean additive Gaussian noise associated with the coherence coefficient γ, which is expressed as:

$$\gamma = \frac{E(\mathbf{z}_1 \cdot \mathbf{z}_2^*)}{\sqrt{E\left(|\mathbf{z}_1|^2\right) \cdot E\left(|\mathbf{z}_2|^2\right)}} \tag{3}$$

The interferometric phase noise level is correlated with the coherence coefficient γ. The higher the coherence coefficient, the lower the noise level.

It is worth noting that the value range of the interferometric phase is distributed in $(-\pi, \pi]$. Hence, phase denoising is implemented in the complex field in for the subsequent phase unwrapping steps. Employing a mathematical manipulation on Equation (2), we obtain the following expression.

$$\mathbf{p} = e^{j\varphi} = e^{j(\varphi_c + \mathbf{n}_\gamma)} = \varphi_r + j\varphi_i \tag{4}$$

where $\mathbf{p}$ indicates the interferogram and φ_r and φ_i denote the real and imaginary parts of $\mathbf{p}$, respectively. According to the analysis of [45], φ_r and φ_i can be given by.

$$
\begin{aligned}
\varphi_r &= \cos(\varphi_c + \mathbf{n}_\gamma) = N_c\cos(\varphi_c) + \mathbf{n}_{\gamma_c} \\
\varphi_i &= \sin(\varphi_c + \mathbf{n}_\gamma) = N_c\sin(\varphi_c) + \mathbf{n}_{\gamma_s}
\end{aligned}
\tag{5}
$$

where $\mathbf{n}_{\gamma_c}$ and $\mathbf{n}_{\gamma_s}$ are defined as the zero-mean additive noise, and N_c is a phase quality evaluation metric related to the coherent coefficient γ. Finally, combining Equations (4) and (5), the formulation of the interferometric phase is derived as follows.

$$
\mathbf{p} = N_c \cdot \mathbf{p}_c + \mathbf{n}
\tag{6}
$$

where $\mathbf{n} = \mathbf{n}_{\gamma_c} + j\mathbf{n}_{\gamma_s}$ denotes the noise of the interferogram and $\mathbf{p}_c = e^{j\varphi_c}$ is the ideal interferogram.

2.2. InSAR Phase SR Filtering Model

The conventional SR model for recovering a single from a measurement is as follows.

$$
y = \mathbf{\Phi}x + n
\tag{7}
$$

where $y \in \mathbb{R}^M$ is the measurement; $\mathbf{\Phi} \in \mathbb{R}^{M \times N}$ is dubbed the measurement matrix; $x \in \mathbb{R}^N$ is the recovered signal; and $n \in \mathbb{R}^M$ is the noise.

According to Section 2.1, the InSAR phase filtering model can be modeled as.

$$
\mathbf{p} = \mathbf{\Phi}\mathbf{p}_c + \mathbf{n}
\tag{8}
$$

Aiming to solve Equation (8), it is transformed into the following convex optimization problem.

$$
\hat{\mathbf{p}}_c = \underset{\mathbf{p}_c}{\arg\min}\,\frac{1}{2}\|\mathbf{\Phi}\mathbf{p}_c - \mathbf{p}\|_2^2 + \lambda\|\psi\mathbf{p}_c\|_1
\tag{9}
$$

where $\psi\mathbf{p}_c$ is a sparse representation of $\mathbf{p}_c$; ψ denotes a certain transform such as Wavelet, Fourier and so on; and λ is a tunable soft threshold parameter.

The main iterative steps of solving Equation (9) by the ISTA algorithm are as follows.

$$
\mathbf{h}^{(k)} = \mathbf{p}_c^{(k-1)} - \alpha\mathbf{\Phi}^{\mathrm{H}}\left(\mathbf{\Phi}\mathbf{p}_c^{(k-1)} - \mathbf{p}\right)
\tag{10}
$$

$$
\mathbf{p}_c^{(k)} = \psi^{\mathrm{H}}\mathrm{soft}\left(\psi\mathbf{h}^{(k)}, \lambda\right) = \psi^{\mathrm{H}}\mathrm{sign}(\psi\mathbf{h}^{(k)})\max\left\{\left|\psi\mathbf{h}^{(k)}\right| - \lambda, 0\right\}
\tag{11}
$$

where α indicates the step size; $\mathrm{sign}(\cdot)$ denotes a sign function; $\mathbf{\Phi}^{\mathrm{H}}$ is the conjugate transpose of $\mathbf{\Phi}$; and $\mathbf{h}^{(k)}$ is the residual error in iteration k. However, the sparse transform ψ and parameters such as α and λ are hand-crafted, which results in the algorithm being nonadaptive. Moreover, ISTA usually suffers from a huge calculative burden due to its large number of iterative steps.

We chose the identity matrix as the measurement matrix $\mathbf{\Phi}$ by the formulation of the interferometric phase. It can be seen from the above analysis that the phase filtering is operated in the complex domain to achieve high-precision phase unwrapping. Therefore, taking advantage of the idea of separating real and imaginary parts, the noisy phase $\mathbf{p}$, the ideal phase $\mathbf{p}_c$, and the measurement matrix $\mathbf{\Phi}$ are transformed as

$$
\mathbf{p}_R = \begin{pmatrix} \varphi_r \\ \varphi_i \end{pmatrix}, \mathbf{p}_{c_R} = \begin{pmatrix} \varphi_{cr} \\ \varphi_{ci} \end{pmatrix}
\tag{12}
$$

$$
\mathbf{\Phi}_R = \begin{pmatrix} \mathbf{\Phi}_r & -\mathbf{\Phi}_i \\ \mathbf{\Phi}_i & \mathbf{\Phi}_r \end{pmatrix}
\tag{13}
$$

where φ_{cr} and φ_{ci} are the real and imaginary parts of the ideal interferogram $\mathbf{p}_c$, and $\mathbf{\Phi}_r$ and $\mathbf{\Phi}_i$ denote the real and imaginary parts of the measurement matrix $\mathbf{\Phi}$, respectively.

In the end, the filtered interferometric phase is obtained by the real part φ'_r and imaginary part φ'_i of the recovered interferometric phase as follows.

$$\varphi' = \angle\left(\varphi'_r + j\varphi'_i\right) \tag{14}$$

3. Methodology

According to the phase filtering model established in Section 2.2, which aimed at performing a fast yet accurate filtering method, we propose a sparse-model-driven network (SMD-Net) for efficient and high-accuracy InSAR phase filtering by combining the interpretability and fewer parameters of SR and the speed advantage of the CNN. Inspired by the idea of the unrolled SR algorithms, the designed network casts the phase filtering model into the unrolled network and implements the complex operation of the unrolled network. In the filtering process shown in Figure 1, first of all, the SMD-Net is trained by employing the real and imaginary parts of the training datasets obtained by Equation (12). Then, the real and imaginary parts of the noisy interferograms (testing data) are entered into the trained SMD-Net, and the filtered real and imaginary parts corresponding to the input are recombined into the estimated interferometric phase using Equation (14). Finally, the filtered phase patches are spliced together by using an image fusion algorithm. Next, we will introduce this section in detail from the design of the SMD-Net and the loss function.

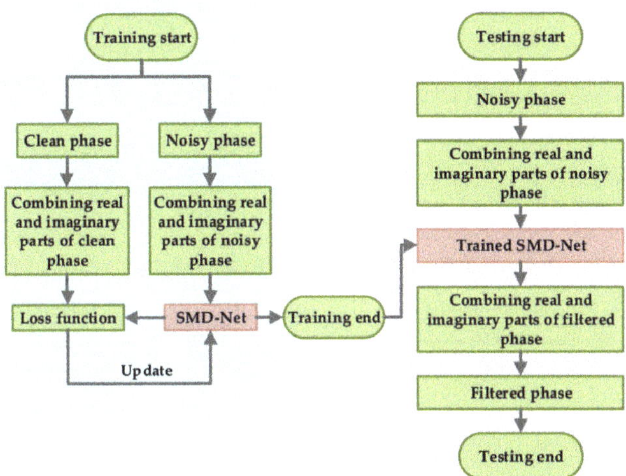

Figure 1. The flow chart of the interferometric phase filtering via the SMD-Net.

3.1. Network Architecture

3.1.1. The SMD-Net Architecture

Nowadays, some phase filtering methods [29–31] based on DL have achieved a better filtering performance than conventional filtering approaches. Nevertheless, they are purely data-driven, which means that these networks rely on a huge data volume and their underlying structures are difficult to interpret. In addition, these networks generally consist of many layers and learn a large number of parameters, which lead to great increases in the computational burden of the network. Considering that SR algorithms model the physical processes underlying the problem and a few parameters, we designed the SMD-Net by employing the idea of unrolling the ISTA algorithm. Unlike the existing based-DL phase filtering methods, the SMD-Net focuses on the interferometric phase filtering model rather than relying entirely on the data fitting and its network structure is simple. It thus is expected to improve the filtering performance and computational efficiency at the same time. The architecture of the SMD-Net is shown in Figure 2.

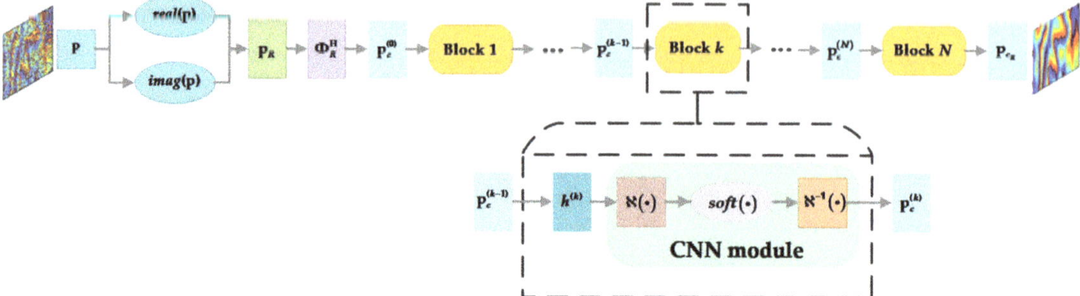

Figure 2. The architecture of the SMD-Net.

In the SMD-Net, each network block is equivalent to an iterative process in the traditional ISTA algorithm. To improve the filtering performance and computational efficiency, we exploited a CNN module that automatically learns the sparse transform instead of the artificial sparse transform in the traditional ISTA algorithm. The CNN module is shown in Figure 3.

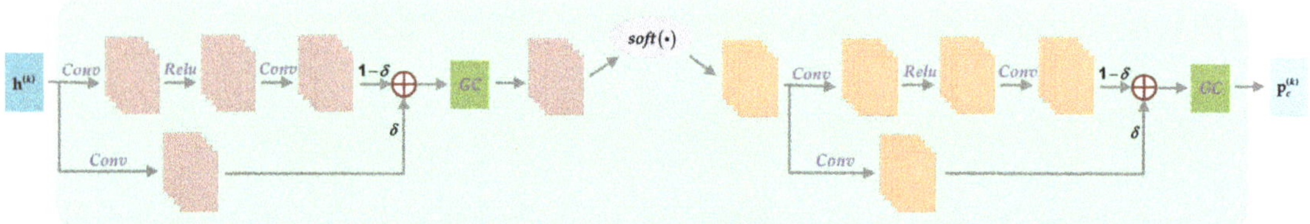

Figure 3. The CNN module in the kth block.

From each block in Figure 2, $\aleph(\cdot)$ and $\aleph^{-1}(\cdot)$ functions in the CNN module replace the sparse basis ψ and the conjugate transpose of ψ in the traditional ISTA, respectively. Thus, Equation (9) is transformed as

$$\hat{\mathbf{p}}_c = \underset{\mathbf{p}_c}{\arg\min} \frac{1}{2}\|\boldsymbol{\Phi}\mathbf{p}_c - \mathbf{p}\|_2^2 + \lambda\|\aleph(\mathbf{p}_c)\|_1 \tag{15}$$

The iterative process in ISTA converts to each block operation in the SMD-Net. Now, we can take the kth block as an example for detailed analysis.

Step 1: The SMD-Net transfers the model parameters of block $k-1$ corresponding to the ISTA algorithm parameters to block k by making use of back-propagation. Then, $\mathbf{h}^{(k)}$ is updated by

$$\mathbf{h}^{(k)} = \mathbf{p}_c^{(k-1)} - \alpha\boldsymbol{\Phi}^{\mathrm{H}}\left(\boldsymbol{\Phi}\mathbf{p}_c^{(k-1)} - \mathbf{p}\right) \tag{16}$$

where α indicates the step size; $\boldsymbol{\Phi}^{\mathrm{H}}$ is the conjugate transpose of $\boldsymbol{\Phi}$; and $\mathbf{h}^{(k)}$ is the residual error in the kth block.

Step 2: In order to satisfy the second term of Equation (15), (i.e., the sparse constraint). In the first stage of the CNN module, $\mathbf{h}^{(k)}$ is sparsely represented as $\aleph\left(\mathbf{h}^{(k)}\right)$ $\aleph(\cdot)$ is a function that automatically learns the sparse domain.

Step 3: The CNN module takes $\aleph\left(\mathbf{h}^{(k)}\right)$ and λ as inputs. The kth filtered result in the sparse domain is calculated by

$$\mathbf{s}^{(k)} = \text{soft}\left(\aleph\left(\mathbf{h}^{(k)}\right),\lambda\right) = \text{sign}(\aleph\left(\mathbf{h}^{(k)}\right))\max\left\{\left|\aleph\left(\mathbf{h}^{(k)}\right)\right| - \lambda, 0\right\} \tag{17}$$

Step 4: The $\aleph^{-1}(\cdot)$ function is designed on the constraint of $\aleph^{-1}(\cdot) \times \aleph(\cdot) = I$ to obtain the kth filtered result in the spatial domain. The result is obtained by

$$\mathbf{p}_c^{(k)} = \aleph^{-1}\left(\mathbf{s}^{(k)}\right) \tag{18}$$

where $\psi\mathbf{p}_c$ is a sparse representation of $\mathbf{p}_c$; ψ denotes a certain transform such as Wavelet, Fourier and so on; and λ is a tunable soft threshold parameter.

The SMD-Net is a combination of the merits of modern CNN and the ISTA algorithm. On one hand, the CNN can quickly process the operations between network layers, which makes up for the disadvantage of traditional ISTA algorithms relying on a large number of iterations, thus improving the computational efficiency. On the other hand, interpretability and a few parameters with specific meanings of the ISTA algorithm overcome the drawback that the CNN learns abundant parameters using a large amount of training data and improves the accuracy and the stability of the method.

3.1.2. CNN Module Architecture

To achieve the most suitable sparse representation of $\mathbf{h}^{(k)}$ and enhance the performance of the SMD-Net, we exploited a CNN module to automatically learn the sparse transform instead of the traditional hand-crafted presets one. The CNN module is shown in Figure 3, and it contains the sparse transform $\aleph(\cdot)$, soft$(\cdot)$ function, and inverse transform $\aleph^{-1}(\cdot)$. The front end of $\aleph(\cdot)$ is a local feature extraction module. In order to combine the global phase information, a GC block [44] is connected to the back end of $\aleph(\cdot)$ to extract the global feature of the phase. $\aleph(\cdot)$ can be represented by the following expression.

$$\aleph\left(\mathbf{h}^{(k)}\right) = GC\left(\delta \times \Gamma_1\mathbf{h}^{(k)} + (1-\delta) \times \Gamma_2\text{ReLU}\left(\Gamma_1\mathbf{h}^{(k)}\right)\right) \tag{19}$$

where Γ_1 represents a convolution operator with M_f filters of the size $M_s \times M_s$; Γ_2 is another convolution operator corresponding to M_f filters of the size $M_s \times M_s \times M_f$; δ denotes the weight; ReLU $(x) = \max(0, x)$; and $GC(\cdot)$ represents a GC block.

Moreover, the inverse transform function $\aleph^{-1}(\cdot)$ is the mirror-symmetrical architecture of $\aleph(\cdot)$. The constraint $\aleph^{-1}(\cdot) \times \aleph(\cdot) = I$ is required to obtain the filtered phase in the spatial domain. The two convolution operations in $\aleph^{-1}(\cdot)$ are the same as in $\aleph(\cdot)$, but the order is switched.

In the kth block of the SMD-Net, Equation (15) can be written as

$$\mathbf{p}_c^{(k)} = \underset{\mathbf{p}_c}{\text{argmin}}\frac{1}{2}\left\|\mathbf{p}_c - \mathbf{h}^{(k)}\right\|_2^2 + \lambda\|\aleph(\mathbf{p}_c)\|_1 \tag{20}$$

Finally, according to $\aleph(\cdot)$ and $\aleph^{-1}(\cdot)$, Equation (15) can be expressed in the following form.

$$\mathbf{p}_c^{(k)} = \aleph^{-1}\left(\text{soft}\left(\aleph\left(\mathbf{h}^{(k)}\right),\lambda\right)\right) \tag{21}$$

The CNN module is exploited to automatically learn the appropriate sparse basis and parameters, which not only bridges the regret of the hand-crafted setting in the conventional ISTA, but also enhances the performance of the SMD-Net for InSAR phase filtering.

3.2. Loss Function

The SMD-Net is trained with the training data $\{\mathbf{p}_i, \mathbf{p}_{ci}\}_{i=1}^{N_{tr}}$, in which $\mathbf{p}_i$ and $\mathbf{p}_{ci}$ are the measurement and the labels, respectively, and N_{tr} is the number of the training samples. Then, the loss function is defined as:

$$loss = \frac{\rho_1}{N_{tr}} \sum_{i=1}^{N_{tr}} \left\| \hat{\mathbf{p}}_{c_i} - \mathbf{p}_{c_i} \right\|_2^2 + \frac{\rho_2}{N \times N_{tr}} \sum_{k=1}^{N} \sum_{i=1}^{N_{tr}} \left\| \aleph^{-1}\left(\aleph\left(\mathbf{h}_i^{(k)} \right) \right) - \mathbf{h}_i^{(k)} \right\|_2^2 \tag{22}$$

where N denotes the total number of the SMD-Net block; ρ_1 and ρ_2 indicate the weight parameters of the two constraint items; $\hat{\mathbf{p}}_{c_i}$ and $\mathbf{p}_{c_i}$ are the ith interferogram estimated; $h^{(k)}$ the ith ideal interferogram; and $\mathbf{h}_i^{(k)}$ represents the residual error in the kth SMD-Net block.

4. Experiments

4.1. Experimental Data

It is well-known that training a deep network requires at least a few hundred or more training datasets with labels. Of course, training the SMD-Net also needs enough interferograms with noise corresponding to the ideal interferograms. In order to obtain a large number of labeled training sets, we generated abundant noisy interferograms with corresponding ideal interferograms by utilizing a digital elevation model (DEM). This is helpful in enhancing the phase detail characteristic similarity between the simulated and measured interferograms [46,47]. The simulated interferometric phase can be obtained as follows:

$$\varphi_c = \arg\left(e^{j2\pi(\mathbf{H}/h_a)} \right) \tag{23}$$

where $\mathbf{H}$ is the height value of the DEM; $\arg(\cdot)$ represents the complex argument operator; and h_a indicates the ambiguity height, which was set to 92.13 m and was consistent with the measured InSAR data used in subsequent experiments.

According to the formulation of the interferometric phase in Section 2.1, the level of noise is related to the coherence coefficient γ, and the higher the coherence value, the less the noise in the InSAR phase data. Hence, interferograms with different coherence coefficients were simulated in order to enhance the generalization ability of the network. The coherence coefficients were in the range of [0.5, 0.95] and the interval was 0.05, which is helpful in that the network adapts to the noise level of a large number of real interferograms.

The generation manner of the simulated training sets is as follows. A simulated clean interferogram, shown in Figure 4b, is generated by employing the DEM (as illustrated in Figure 4a) of the eastern part of Turkey with the size of 2048 × 2048 from the SRTM 1Sec HGT based on Equation (23). Ten noise interferograms were generated by adding different levels of noise whose coherence coefficients were [0.5, 0.95] to the clean interferograms. Noisy interferograms with coherence coefficients of 0.5 and 0.95 are shown in Figure 4c,d. In this paper, we divided the whole interferogram into interferogram patches with the size 256 × 256 with a 0.5 overlap rate to obtain enough training sets and improve the computational efficiency. Among them, each noise interferogram with a coherence coefficient was cropped into a group containing 225 patches. Hence, the total number of interferogram patches with 2250 patches contained ten groups.

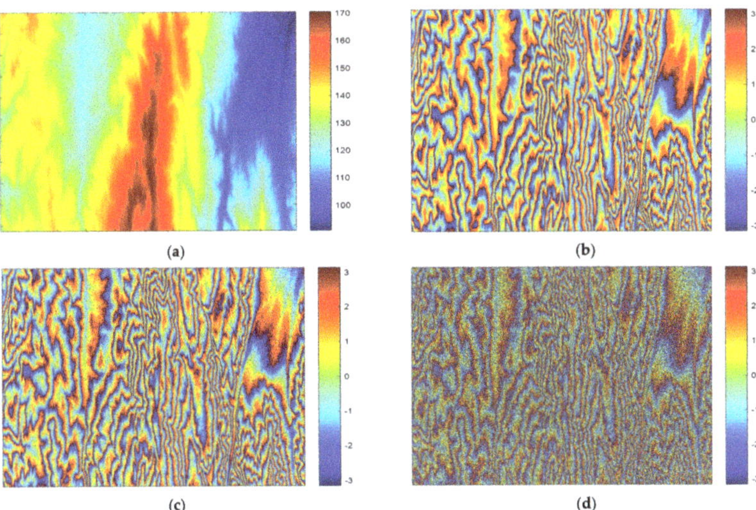

Figure 4. (**a**) The DEM used to generate the training sets; (**b**) clean interferogram generated by (**a**); (**c**) noisy interferogram with a coherence coefficient of 0.95; (**d**) noisy interferogram with a coherence coefficient of 0.5.

Figure 5a shows the DEM with the size of 1024×1024 from SRTM 1Sec HGT, which was used to generate the testing data illustrated in Figure 5b. Ten noise interferograms with different coherence coefficients were obtained by adding noise in the same way as the training data, among which the coherence coefficients of 0.5 and 0.95 are shown in Figure 5c,d. These testing interferograms with the different noise levels were also cut into 490 patches like the training data.

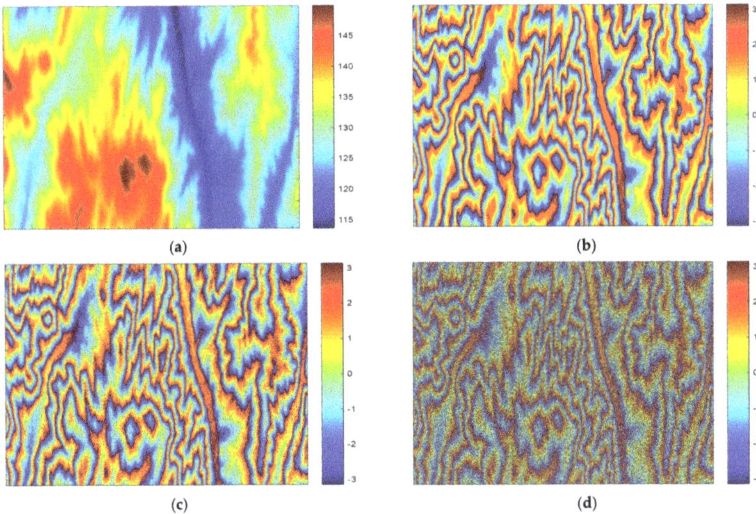

Figure 5. (**a**) The DEM used to generate the testing sets; (**b**) clean interferogram generated by (**a**); (**c**) noisy interferogram with a coherence coefficient of 0.95; (**d**) noisy interferogram with a coherence coefficient of 0.5.

4.2. Experimental Details

In this article, we analyzed the performance of the SMD-Net on the experimental results of the simulated and measured data. The proposed method outperformed the previous three widely-used methods (i.e., the Lee filter [18], Goldstein filter [26], and InSAR-BM3D filter [28]) and the two methods based on DL (Phi-Net [31] and PFNet [30]). For a fair comparison, all experiments were carried out on an Inter$^®$ Core™ i7-2790K CPU with 4 Gb random access memory (RAM) and an NVIDIA GeForce GTX 980 Ti GPU, where the previous three widely-used methods were performed in MATLAB R2016b and the proposed method was tested in Pytorch.

Inspired by the work of [39], the SMD-Net contained nine blocks (i.e., nine iterations), each of which had the same network structure as follows: Γ_1 is a convolution operator with 32 filters of the size 3×3; Γ_2 is another convolution operator corresponding to 32 filters of the size $3 \times 3 \times 32$. In our experiments, the SMD-Net was trained with the first 2000 interferograms of the training sets obtained in Section 4.1 by utilizing the Adam optimization [48] with a batch size of two. We set the initial learning rate, λ, α, and δ as 0.0001, 0.01, 0.2, and 0.1, respectively.

4.3. Evaluation Metrics

The objective of phase filtering is to suppress noise and preserve interferometric phase details as much as possible. Therefore, the precision of a filtering approach is evaluated by considering both the denoising ability and the phase detail preservation ability. Moreover, the computational complexity is also an important problem for phase filtering, so the evaluation of computational efficiency is essential. In order to compare the performance of filtering methods more intuitively, we adopted two evaluation methods based on the image and data, namely, qualitative evaluation and quantitative evaluation. Qualitative evaluation is implemented directly by the naked eye, which is highly subjective and the evaluation results are not entirely desirable. In contrast, quantitative evaluation has a certain theoretical basis, so the evaluation results are highly reliable. In this paper, the mean-square error (MSE) [49] representing the difference between the filtered interferometric phase and the corresponding ground truth, the number of residues (NOR) [50] used to reflect the denoising ability of a filtering method, the mean structural similarity index (MSSIM) [51] reflecting the phase detail feature preservation ability of a filtering method, and running time (T) were adopted to assess the experiments on the simulated data. In view of lacking the ground truth of the real InSAR data, the no-reference metric Q [52] is a quantitative assessment index of balancing between phase detail preservation and denoising, and the higher this is, the more powerful the phase detail preservation capacity.

5. Results

5.1. Results on the Simulation Data

After obtaining the trained network, in the testing stage, a noisy interferogram with a coherence coefficient of 0.5 shown in Figure 6b from the simulated testing sets was used to qualitatively assess the performance of the proposed method in this article. Figure 6a is the corresponding reference interferogram. In order to validate the proposed method, we compared it with the Lee filter [18], Goldstein filter [26], InSAR-BM3D filter [28], traditional ISTA algorithm [43], Phi-Net [31], and PFNet [30]. All filtered results are illustrated in Figure 7. Figure 7(a1–g1) represents the filtered interferograms of the seven approaches, respectively. Qualitative evaluation with the naked eye showed that the result of the proposed method was closest to the ideal interferogram shown in Figure 6a. We calculated the difference phases of the seven filtered results and the corresponding ideal interferogram, respectively, and show the difference images in Figure 7(a2–g2). From Figure 7(a2–g2), we can see that the difference image of the proposed method was closest to zero, which further verified that the filtered result obtained by the proposed method was most similar to the ideal interferogram. Moreover, the SSIM maps shown in Figure 7(a3–g3) were computed to assess the capability of the phase detail preservation, where it can be seen that the SSIM map of the proposed method (Figure 7(g3)) had the most regions whose values were close to 1 amongst all of the testing methods. The above preliminary comprehensive qualitative analysis showed that the proposed method had the best visual performance in both the suppression noise and phase edge detail preservation.

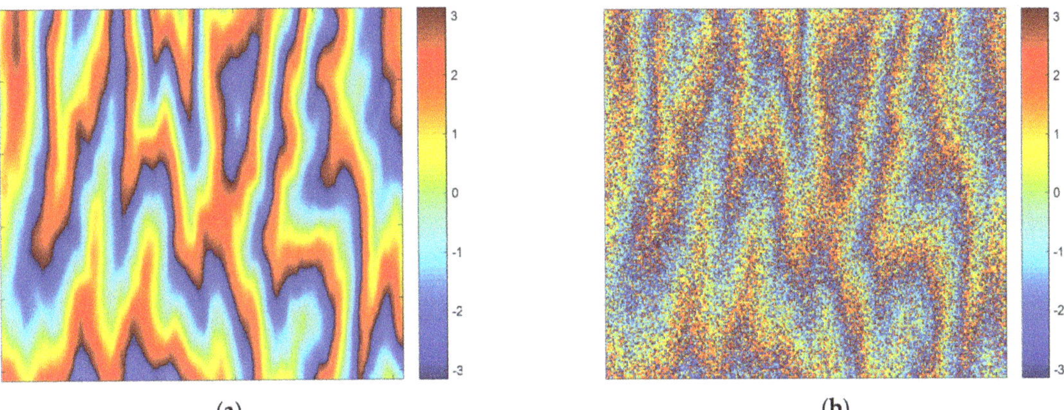

(a) **(b)**

Figure 6. A simulated testing interferogram patch: (**a**) Clean interferogram patch; (**b**) noisy interferogram patch with a coherence coefficient of 0.5.

The qualitative evaluation is too subjective and unstable, and testing only a noisy interferogram is haphazard. Therefore, we calculated the mean MSE and MSSIM of all results obtained by the experiments on all of the testing sets with the same coherence coefficients, and these results are shown in Figure 8. In Figure 8, we can intuitively see that the MSE of the proposed method was lower and the MSSIM was higher than the six reference methods. This demonstrates that the proposed method has the characteristics of the best noise suppression and phase detail feature preservation capabilities among the seven methods.

Finally, the mean values of all metrics are presented in Table 1. In Table 1, the proposed method had the highest MSSIM, lowest MSE, and NOR among the seven methods. Through a comprehensive comparison, the performance of the proposed method was optimal among the seven methods. In detail, the NORs of the proposed method, InSAR-BM3D, Phi-Net, and PFNet were equal to 0, which demonstrates that the denoising ability of the four methods is powerful. However, the proposed method could better balance between the noise reduction and phase edge feature preservation because it had the highest MSSIM and the lowest MSE. Furthermore, the running time (T) of the proposed method was the shortest. Compared to Phi-Net and PFNet, the mean T of the proposed method was 87.2% and 85.5% faster, respectively. This means that the proposed method had the most powerful computational capability among the seven methods.

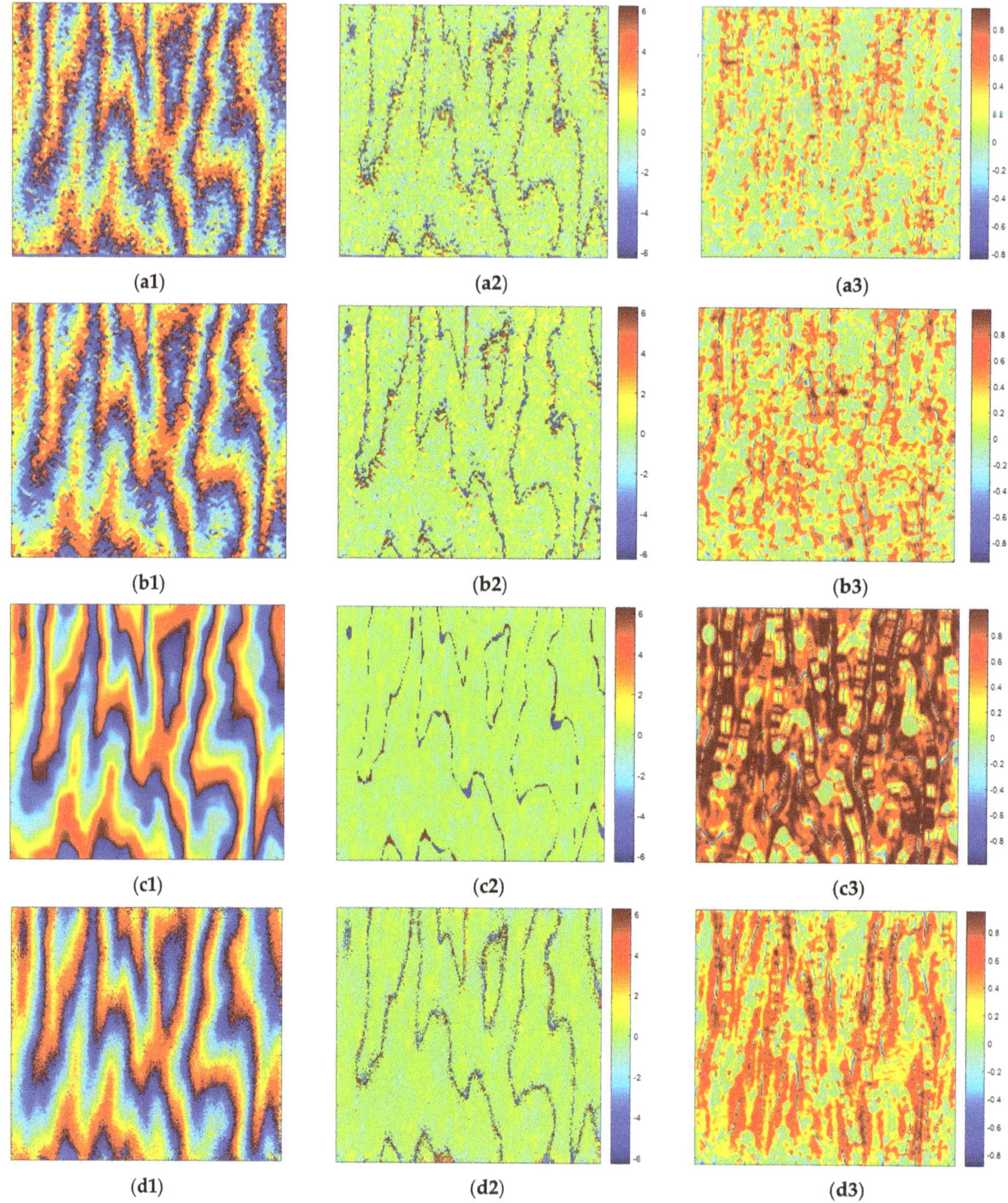

(a1)　　　　　(a2)　　　　　(a3)

(b1)　　　　　(b2)　　　　　(b3)

(c1)　　　　　(c2)　　　　　(c3)

(d1)　　　　　(d2)　　　　　(d3)

Figure 7. *Cont.*

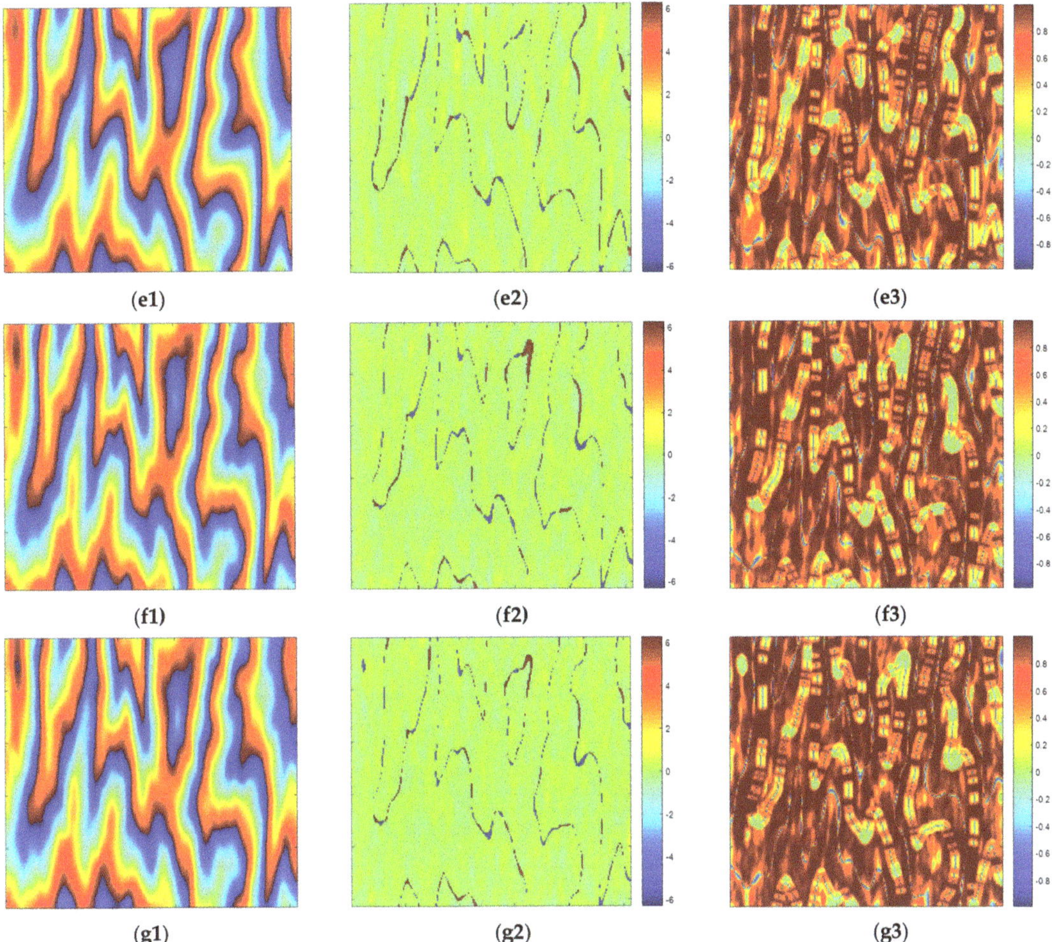

Figure 7. The analysis of the simulated interferogram patch: (**a1–g1**) Filtered interferograms of the Lee filter, Goldstein filter, InSAR-BM3D filter, ISTA, Phi-Net, PFNet, and the proposed method; (**a2–g2**) difference images of Figure 6a and (**a1–g1**) in sequence; (**a3–g3**) SSIM maps of Figure 6a and (**a1–g1**), respectively.

Table 1. The metrics of the seven methods on the simulated interferogram. MSSIM is the core accuracy index. T is the speed index.

Methods	MSE (Rad^2)	NOR	MSSIM	T (s)
Lee [18]	1.62	293.48	0.36	4.39
Goldstein [26]	1.24	91.12	0.50	5.41
InSAR-BM3D [28]	0.63	0	0.74	6.97
ISTA [43]	1.13	185.34	0.51	9.73
Phi-Net [31]	0.66	0	0.74	0.86
PFNet [30]	0.54	0	0.79	0.76
SMD-Net (Ours)	**0.50**	**0**	**0.81**	**0.11**

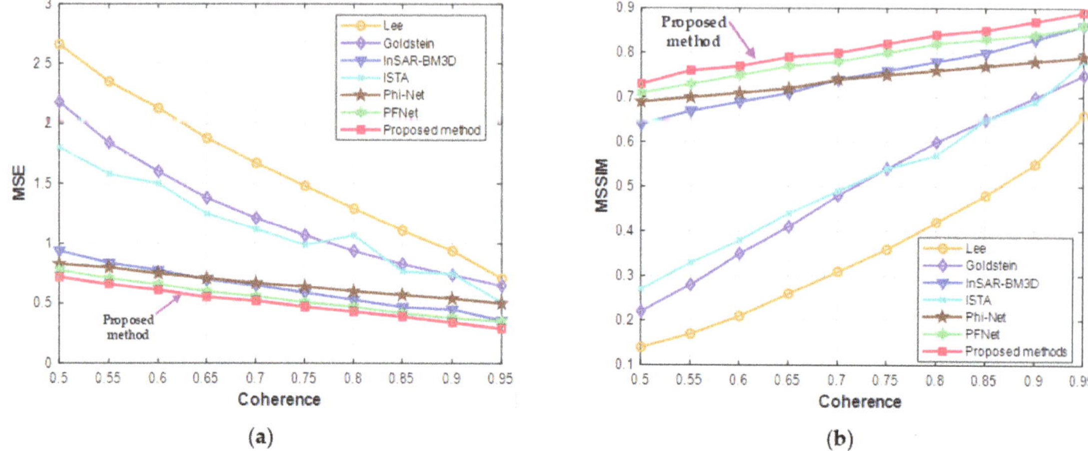

Figure 8. The metrics of the seven methods on the simulated interferograms with ten coherence coefficients: (**a**) mean MSE; (**b**) mean MSSIM.

5.2. Results on the Real Data

In order to validate the performance of the proposed method in real data, we employed two measured interferograms with the size of 512×512 pixels to perform the test experiments. As shown in Figure 9, these were provided by the Sentinel-1 SAR satellite. In order to prove the filtering performance of the proposed method in different shapes and different coherence areas, Figure 9a,b shows a high-coherence area A with a dense fringe and low-coherence area B with flatness, respectively.

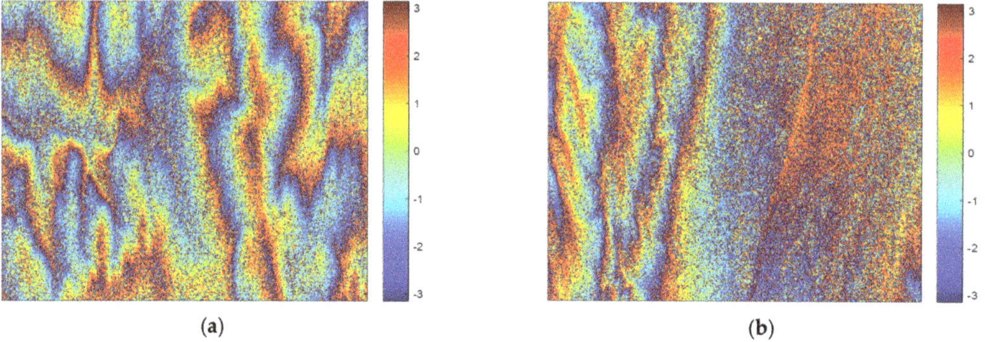

Figure 9. The measured interferograms: (**a**) Area A with high coherence; (**b**) area B with low coherence.

The seven methods were used in area A, as shown in Figure 9a and the results are shown in Figure 10a–g, respectively. From the perspective of vision, the proposed method has a more powerful noise reduction capacity than the Lee filter, Goldstein filter, InSAR-BM3D filter, and traditional ISTA. Compared to Phi-Net and PFNet, their denoising abilities were comparable, but it can be seen from the black rectangles in Figure 10e–g that the phase detail feature preservation ability of the proposed method was stronger.

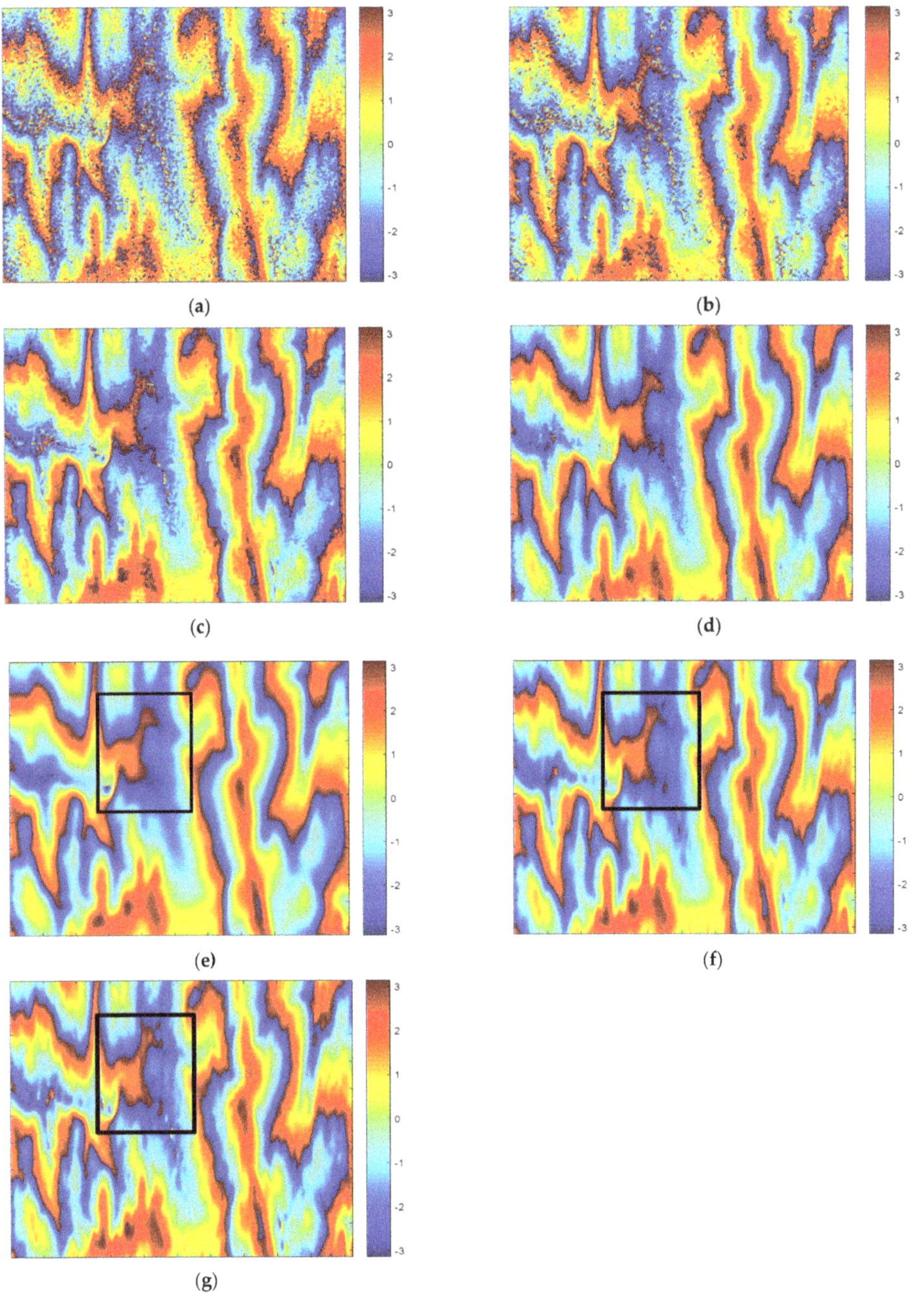

Figure 10. The filtering results obtained by processing Figure 9a using the seven methods: (**a**) filtering interferogram of the Lee filter; (**b**) filtering interferogram of the Goldstein filter; (**c**) filtering interferogram of the InSAR-BM3D filter; (**d**) the filtering interferogram of ISTA; (**e**) the filtering interferogram of Phi-Net; (**f**) the filtering interferogram of PFNet; (**g**) the filtering interferogram of the proposed method.

Next, we computed the number of residues (NOR), the no-reference metric Q, and the running time (T), listed in Table 2, to assess the performance of the proposed method more accurately. From Table 2, it can be seen that the NOR of the proposed method was lower than that of the Lee filter, Goldstein filter, InSAR-BM3D filter, and ISTA algorithm, and its metric Q was higher than that of the four methods. This strongly proves that our method was superior to the four reference methods in both the noise suppression and preservation of edge detail. Furthermore, compared with the Phi-Net and PFNet, the NOR of the proposed method was higher, but its metric Q was 9.9% and 7.8% higher, respectively, which demonstrates that the proposed method was superior to the Phi-Net and PFNet in the phase edge detail preservation (i.e., it provided a well-balanced noise reduction and fringe detail preservation). From the perspective of processing efficiency, the running time of the proposed method was the shortest with only 1.43 s among the seven methods and was 51.9% faster than the PFNet. In conclusion, the proposed method performed better than the six reference approaches in both the filtering performance and speed.

Table 2. The metrics of the seven methods on the measured interferogram of area A. Metric Q is the core accuracy index. T is the speed index.

Methods	NOR	Metric Q	T (s)
Lee [18]	2506	30.86	16.70
Goldstein [26]	1404	48.73	22.00
InSAR-BM3D [28]	66	71.38	31.26
ISTA [43]	12	54.21	81.67
Phi-Net [31]	2	82.30	11.90
PFNet [30]	0	83.87	2.97
SMD-Net (Ours)	6	**90.43**	**1.43**

Area A is a high-coherence area with dense fringe. Therefore, in order to prove the adaptability of the proposed method to different terrain regions with different levels of noise, we selected a flat area (area B shown in Figure 9b) of low coherence to experiment further. The filtering results of the seven methods are shown in Figure 11a–g. From Figure 11a–g, we can see that our method offers the best-balanced noise suppression and the preservation of the phase edge texture. In more detail, the denoising ability of the proposed method was obviously stronger than the three widely-used methods. The traditional ISTA algorithm lacks a denoising ability, but also suffers from a serious loss of phase details. As shown in the black rectangles in Figure 11e–g, compared with the Phi-Net and PFNet, it was obvious that the proposed method preserved the phase detail features more completely, while the PFNet filtering excessively resulted in a serious loss of phase details. Similarly, the indicators of all of the results obtained by the seven methods were calculated for a quantitative assessment and are listed in Table 3. The performance of the proposed method was obviously better than that of the three widely-used methods, the ISTA algorithm and Phi-Net. Then, the proposed method has a 19.8% higher metric Q compared with PFNet. Combined with the quantitative indexes of area A and area B as shown in Tables 2 and 3, we can see that the performance reduction in the PFNet was significantly higher than that of the proposed method. This proves that the proposed method had better generalization.

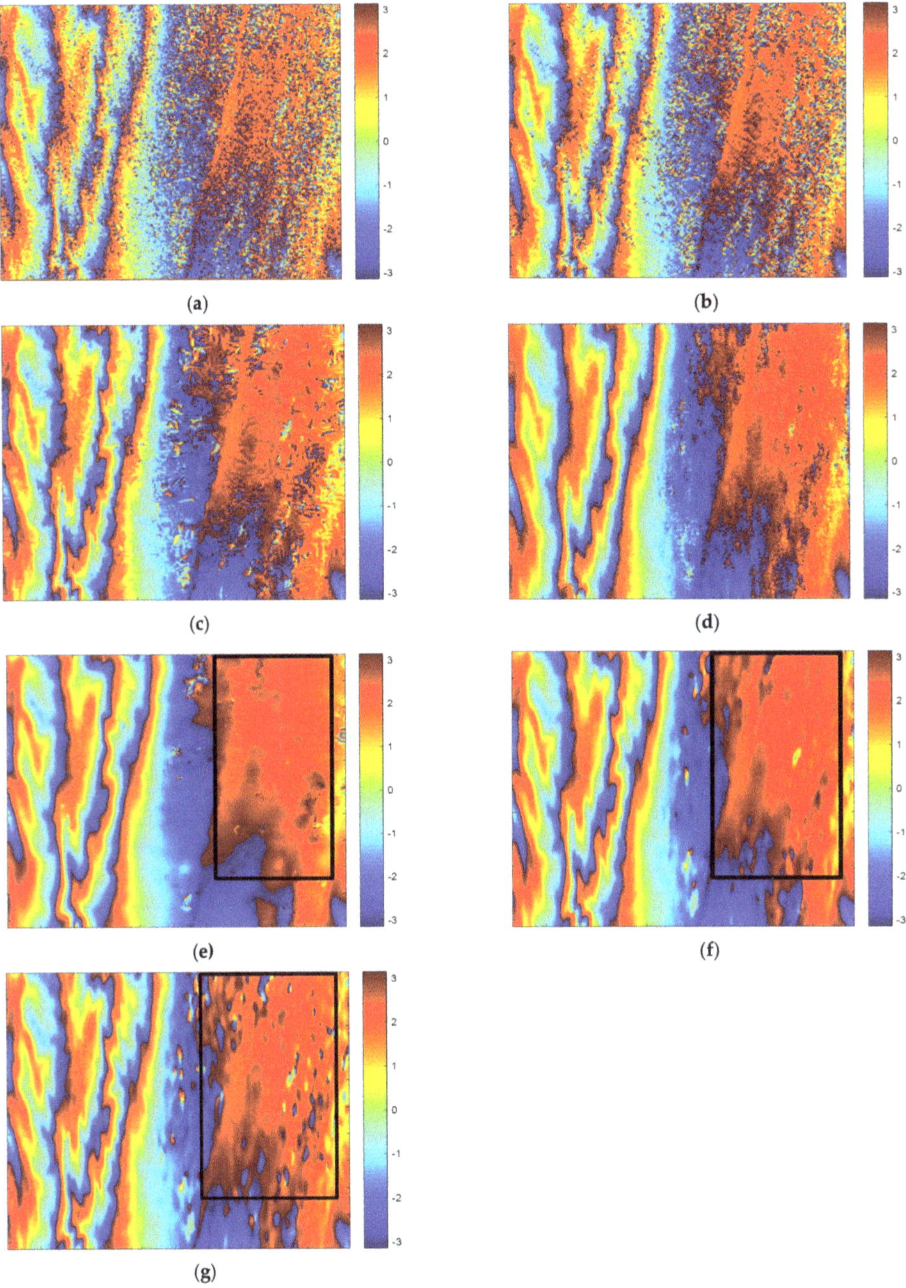

Figure 11. The filtering results obtained by processing Figure 9b using the seven methods: (**a**) the filtering interferogram of the Lee filter; (**b**) the filtering interferogram of the Goldstein filter; (**c**) the filtering interferogram of the InSAR-BM3D filter; (**d**) the filtering interferogram of ISTA; (**e**) the filtering interferogram of the Phi-Net; (**f**) the filtering interferogram of the PFNet; (**g**) filtering interferogram of the proposed method.

Table 3. The metrics of the seven methods on the measured interferogram of area B. Metric Q is the core accuracy index. T is the speed index.

Methods	NOR	Metric Q	T (s)
Lee [18]	5980	25.67	18.36
Goldstein [26]	4430	44.55	22.97
InSAR-BM3D [28]	444	65.82	29.45
ISTA [43]	203	41.08	82.85
Phi-Net [31]	70	63.96	6.79
PFNet [30]	10	74.10	2.87
SMD-Net (Ours)	81	**88.76**	**1.77**

6. Discussion

In order to further analyze the performance of the SMD-Net under the small training samples, in the 10 groups of interferogram patches in Section 4.1, 200 interferogram patches were selected starting from the first interferogram patch in each group at an interval of 10 interferograms as new training sets. Then, the SMD-Net was retrained with the training sets. The indicators of the testing results on the simulated data are listed in Table 4. As can be seen in Table 4, the MSE of the proposed method was 9.3% higher, but the MSSIM of the proposed method was equal to that of PFNet and its T was 85.5% faster. Therefore, it can be seen that the performance of the SMD-Net trained with 200 training samples was comparable to the PFNet trained with 2250 training samples. Unlike the PFNet, the performance of the SMD-Net was not constrained by the requirement of the data volume.

Table 4. The metrics of the PFNet trained with 2250 samples and the SMD-Net trained with 200 samples on the simulated data. MSSIM is the core accuracy index. T is the speed index.

Method	Samples	MSE (Rad^2)	NOR	MSSIM	T (s)
PFNet [30]	2250	0.54	0	0.79	0.76
SMD-Net (Ours)	200	0.59	0	0.79	0.11

Like the simulated data, we processed the measured data utilizing the SMD-Net trained with 200 training samples to analyze the filtering performance of the proposed method. The filtering result of area A (Figure 9a) is shown in Figure 12b, and we can see intuitively from the black rectangles in Figure 12a,b that the phase detail features of the result obtained by our method were better preserved. Next, a flat and low-coherence area B (Figure 9b) was processed to prove the generalization of the proposed method. The black rectangles in Figure 12c,d also showed that the proposed method had a stronger phase edge texture preservation capability.

Furthermore, the quantitative indicators of the two areas were calculated and are listed in Tables 5 and 6. Tables 5 and 6, compared with the PFNet, it could be observed that the NORs of the proposed method were higher in both areas, but their metric Qs were higher. This indicates that the PFNet caused the serious loss of phase fringe detail information due to over filtering and its phase detail feature preservation capability was inferior to the proposed method. In addition, we could calculate that the metric Qs of the results obtained by processing area A and area B with the proposed method were 5.6% and 17.1% higher than that of the PFNet, respectively. To sum up, it can be seen that the filtering performance of the SMD-NET trained with 200 training samples outperformed that of the PFNet trained with 2250 training samples.

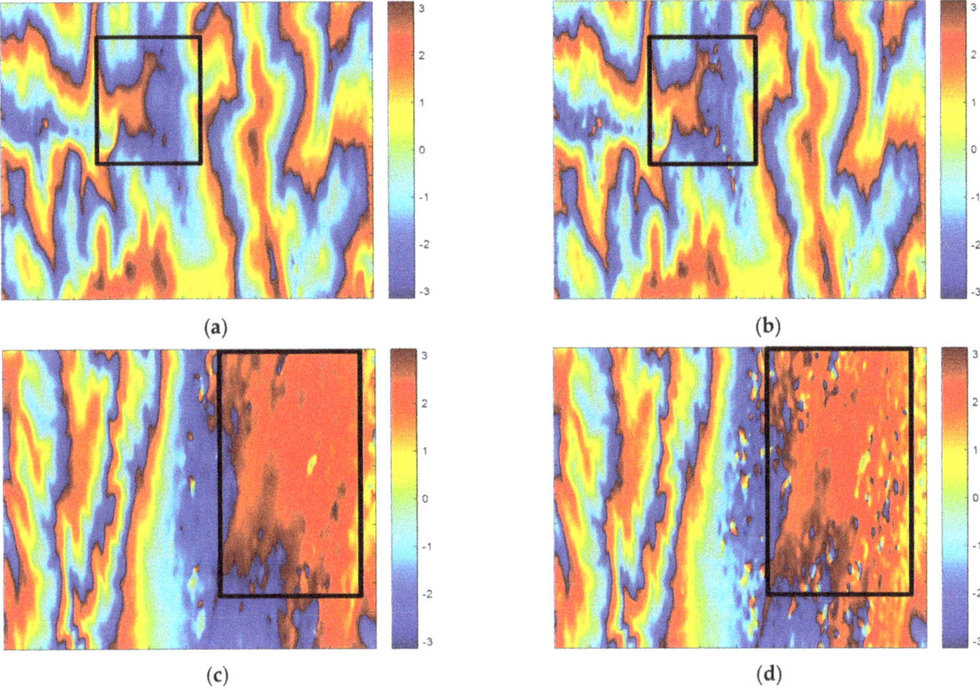

Figure 12. The filtered results obtained by processing area A and area B: (**a**) Filtered result of area A utilizing the PFNet trained with 2250 training samples; (**b**) the filtered result of area A utilizing the proposed method trained with 200 training samples; (**c**) the filtered result of area B utilizing the PFNet trained with 2250 training samples; (**d**) the filtered result of area B utilizing the proposed method trained with 200 training samples.

Table 5. The metrics of the PFNet trained with 2250 samples and the SMD-Net trained with 200 samples on area A. Metric Q is the core accuracy index. T is the speed index.

Methods	Samples	NOR	Metric Q	T (s)
PFNet [30]	2250	0	83.87	2.97
SMD-Net (Ours)	**200**	14	**88.60**	**1.52**

Table 6. The metrics of the PFNet trained with 2250 samples and the SMD-Net trained with 200 samples on area B. Metric Q is the core accuracy index. T is the speed index.

Methods	Samples	NOR	Metric Q	T (s)
PFNet [30]	2250	10	74.10	2.87
SMD-Net (Ours)	**200**	126	**86.78**	**1.45**

7. Conclusions

In this article, we propose a sparse-model-driven network (SMD-Net) for efficient and high-accuracy InSAR phase filtering. The SMD-Net was designed by casting the mathematical derivation steps of the traditional ISTA algorithm into the network structure. Unlike the ISTA algorithm, in each block of the SMD-Net, a CNN module was established to adaptively learn the sparse transform instead of the hand-crafted setting. The SMD-Net not only significantly reduced the network complexity, but was also combined with the merit of automatically learning the parameters and sparse transform of CNN. It can thus improve the filtering performance and speed at the same time. Finally, plenty of experiments were performed to validate the proposed method.

We assessed the proposed method qualitatively and quantitatively on the simulated and measured InSAR data. The experimental results on the simulated and measured data demonstrated that the proposed method could better balance the abilities of the noise suppression and phase fringe texture preservation than the several reference filtering methods. In addition, the speed of the proposed method was very fast. Compared with the PFNet, the SMD-Net was 85.5% and 51.9% faster on the simulated and measured data, respectively. Aiming to validate the performance of the proposed method was not limited by the requirement of the number of training samples, so the experiments were carried out again when the number of training samples was decreased to 10%. Compared with the PFNet trained with 2250 samples, the performance of the proposed method was comparable on the simulated data. In the experiments on the real data, the Qs of the results obtained by processing high-coherence and low-coherence areas with the proposed method were 5.6% and 17.1% higher, respectively. This proves that the comprehensive performance of our method outperformed that of the six competitive approaches, even with small samples.

Author Contributions: Conceptualization, N.W. and T.Z.; Methodology, N.W., X.Z. (Xiaoling Zhang) and X.Z. (Xu Zhan); Software, N.W., T.Z. and J.S.; Validation, N.W., L.P. and S.W.; Formal analysis, N.W., X.Z. (Xu Zhan) and Y.H.; Investigation, N.W., L.P. and J.S.; Resources, X.Z. (Xiaoling Zhang) and J.S.; Data curation, N.W. and L.P.; Writing—original draft preparation, N.W., X.Z. (Xu Zhan) and S.W.; Writing—review and editing, N.W., T.Z., X.X. and X.Z. (Xu Zhan); Visualization, N.W., X.Z. (Xiaoling Zhang) and L.P.; Supervision, T.Z., X.Z. (Xu Zhan), Y.H. and X.X.; Project administration, X.Z. (Xiaoling Zhang); Funding acquisition, X.Z. (Xiaoling Zhang). All authors have read and agreed to the published version of the manuscript.

Funding: This research was supported in part by the National Natural Science Foundation of China under grant 61571099.

Data Availability Statement: Not applicable.

Acknowledgments: We would like to thank all of the editors and reviewers for their valuable comments in improving this manuscript.

Conflicts of Interest: The authors declare no conflict of interest.

References

1. Zhang, T.; Zhang, X. A Full-Level Context Squeeze-and-Excitation ROI Extractor for SAR Ship Instance Segmentation. *IEEE Geosci. Remote Sens. Lett.* **2022**, *19*, 4506705. [CrossRef]
2. Fischer, G.; Papathanassiou, K.P.; Hajnsek, I. Modeling and Compensation of the Penetration Bias in InSAR DEMs of Ice Sheets at Different Frequencies. *IEEE J. Sel. Top. Appl. Earth Obs. Remote Sens.* **2020**, *13*, 2698–2707. [CrossRef]
3. Zhang, T.; Zhang, X. HTC+ for SAR Ship Instance Segmentation. *Remote Sens.* **2022**, *14*, 2395. [CrossRef]
4. Xu, X.; Zhang, X.; Zhang, T. Lite-YOLOv5: A Lightweight Deep Learning Detector for On-Board Ship Detection in Large-Scene Sentinel-1 SAR Images. *Remote Sens.* **2022**, *14*, 1018. [CrossRef]
5. Zhang, T.; Zhang, X.; Ke, X.; Liu, C.; Xu, X.; Zhan, X.; Wang, C.; Ahmad, I.; Zhou, Y.; Pan, D.; et al. HOG-ShipCLSNet: A Novel Deep Learning Network with HOG Feature Fusion for SAR Ship Classification. *IEEE Trans. Geosci. Remote. Sens.* **2021**, *60*, 5210322. [CrossRef]
6. Wang, H.; Zhou, Y.; Fu, H.; Zhu, J.; Yu, Y.; Li, R.; Zhang, S.; Qu, Z.; Hu, S. Parameterized Modeling and Calibration for Orbital Error in TanDEM-X Bistatic SAR Interferometry over Complex Terrain Areas. *Remote Sens.* **2021**, *13*, 5124. [CrossRef]
7. Zhu, X.X.; Wang, Y.; Montazeri, S.; Ge, N. A Review of Ten-Year Advances of Multi-Baseline SAR Interferometry Using TerraSAR-X Data. *Remote Sens.* **2018**, *10*, 1374. [CrossRef]
8. Wang, H.; Fu, H.; Zhu, J.; Liu, Z.; Zhang, B.; Wang, C.; Li, Z.; Hu, J.; Yu, Y. Estimation of subcanopy topography based on single-baseline TanDEM-X InSAR data. *J. Geod.* **2021**, *95*, 84. [CrossRef]
9. Hu, J.; Ge, Q.; Liu, J.; Yang, W.; Du, Z.; He, L. Constructing Adaptive Deformation Models for Estimating DEM Error in SBAS-InSAR Based on Hypothesis Testing. *Remote Sens.* **2021**, *13*, 2006. [CrossRef]
10. Richter, N.; Froger, J.L. The role of Interferometric Synthetic Aperture Radar in detecting, mapping, monitoring, and modelling the volcanic activity of Piton de la Fournaise, La Réunion: A review. *Remote Sens.* **2020**, *12*, 1019. [CrossRef]
11. Moreira, J.; Schwabisch, M.; Fornaro, G.; Lanari, R.; Bamler, R.; Just, D.; Steinbrecher, U.; Breit, H.; Eineder, M.; Franceschetti, G.; et al. X-SAR Interferometry: First Results. *IEEE Trans. Geosci. Remote Sens.* **1995**, *33*, 950–956. [CrossRef]
12. Xu, G.; Gao, Y.; Li, J.; Xing, M. InSAR phase denoising: A review of current technologies and future directions. *IEEE Geosci. Remote Sens. Mag.* **2020**, *8*, 64–82. [CrossRef]

13. Zebker, H.A.; Villasenor, J. Decorrelation in interferometric radar echoes. IEEE Trans. Geosci. *Remote Sens.* **1992**, *30*, 950–959. [CrossRef]

14. Xia, X.G.; Wang, G.Y. Phase Unwrapping and A Robust Chinese Remainder Theorem. *IEEE Signal Process. Lett.* **2007**, *14*, 247–250. [CrossRef]

15. Suksmono, A.B.; Hirose, A. Interferometric SAR image restoration using Monte Carlo metropolis method. *IEEE Trans. Signal Process.* **2002**, *50*, 290–298. [CrossRef]

16. Wang, Q.S.; Huang, H.F.; Yu, A.X.; Dong, Z. An Efficient and Adaptive Approach for Noise Filtering of SAR Interferometric Phase Images. *IEEE Geosci. Remote Sens. Lett.* **2011**, *8*, 1140–1144. [CrossRef]

17. Abdallah, W.B.; Abdelfattah, R. A modification to the ASM filter for improving SAR interferograms. In Proceedings of the 21st European Signal Processing Conference (EUSIPCO 2013), Marrakech, Morocco, 9–13 September 2013; pp. 1–5.

18. Lee, J.S.; Papathanassiou, K.P.; Ainsworth, T.L.; Grunes, M.R.; Reigber, A. A new technique for noise filtering of SAR interferometric phase images. *IEEE Trans. Geosci. Remote Sens.* **1998**, *36*, 1456–1465.

19. Chao, C.F.; Chen, K.S.; Lee, J.S. Refined filtering of interferometric phase from InSAR data. *IEEE Trans. Geosci. Remote Sens.* **2013**, *51*, 5315–5323. [CrossRef]

20. Fu, S.H.; Long, X.J.; Yang, X.; Yu, Q.F. Directionally adaptive filter for synthetic aperture radar interferometric phase images. *IEEE Trans. Geosci. Remote Sens.* **2013**, *51*, 552–559. [CrossRef]

21. Deledalle, C.A.; Denis, L.; Tupin, F. NL-InSAR: Nonlocal interferogram estimation. *IEEE Trans. Geosci. Remote Sens.* **2011**, *4*, 1441–1452. [CrossRef]

22. Quereda, A.M.; Lopez-Sanchez, J.M.; Selva, J.; Gonzalez, P.J. An improved phase filter for differential SAR interferometry based on an iterative method. *IEEE Trans. Geosci. Remote Sens.* **2018**, *56*, 4477–4491. [CrossRef]

23. Zha, X.J.; Fu, R.S.; Dai, Z.Y.; Liu, B. Noise reduction in interferograms using the wavelet packet transform and wiener filtering. *IEEE Geosci. Remote Sens. Lett.* **2008**, *5*, 404–408.

24. Bian, Y.; Mercer, B. Interferometric SAR phase filtering in the wavelet domain using simultaneous detection and estimation. *IEEE Geosci. Remote Sens.* **2011**, *4*, 1396–1416. [CrossRef]

25. Baran, I.; Stewart, M.; Kampes, B.; Lilly, P. A modification to the Goldstein radar interferogram filter. *IEEE Geosci. Remote Sens.* **2011**, *4*, 1396–1416. [CrossRef]

26. Goldstein, R.M.; Werner, C.L. Radar interferogram filtering for geophysical applications. *Geophys. Res. Lett.* **1998**, *25*, 4035–4038. [CrossRef]

27. Xu, H.; Li, Z.; Li, S.; Liu, W.; Li, J.; Liu, A.; Li, W. A Nonlocal Noise Reduction Method Based on Fringe Frequency Compensation for SAR Interferogram. *IEEE J. Sel. Top. Appl. Earth Obs. Remote Sens.* **2021**, *14*, 9756–9767. [CrossRef]

28. Sica, F.; Cozzolino, D.; Zhu, X.X.; Verdoliva, L.; Poggi, G. INSAR-BM3D: A nonlocal filter for SAR interferometric phase restoration. *IEEE Trans. Geosci. Remote Sens.* **2018**, *56*, 3456–3467. [CrossRef]

29. Sun, X.; Zimmer, A.; Mukherjee, S.; Kottayil, N.K.; Ghuman, P.; Cheng, I. DeepInSAR—A Deep Learning Framework for SAR Interferometric Phase Restoration and Coherence Estimation. *Remote Sens.* **2020**, *12*, 2340. [CrossRef]

30. Pu, L.; Zhang, X.; Zhou, Z.; Shi, J.; Wei, S.; Zhou, Y. A Phase Filtering Method with Scale Recurrent Networks for InSAR. *Remote Sens.* **2020**, *12*, 3453. [CrossRef]

31. Sica, F.; Gobbi, G.; Rizzoli, P.; Bruzzone, L. Φ-Net: Deep Residual Learning for InSAR Parameters Estimation. *IEEE Trans. Geosci. Remote Sens.* **2021**, *59*, 3917–3941. [CrossRef]

32. Kang, J.; Hong, D.; Liu, J.; Baier, G.; Yokoya, N.; Demir, B. Learning Convolutional Sparse Coding on Complex Domain for Interferometric Phase Restoration. *IEEE Tran. Neural Netw. Learn. Syst.* **2021**, *32*, 826–840. [CrossRef] [PubMed]

33. Ojha, C.; Fusco, A.; Pintp, I.M. Interferometric SAR phase denoising using proximity-based K-SVD technique. *Sensors* **2019**, *19*, 2684. [CrossRef] [PubMed]

34. Luo, X.; Wang, X.; Wang, Y.; Zhu, S. Efficient InSAR Phase Noise Reduction via Compressive Sensing in the Complex Domain. *IEEE J. Sel. Top. Appl. Earth Obs. Remote Sens.* **2018**, *11*, 1615–1632. [CrossRef]

35. Monga, V.; Li, Y.L.; Eldar, Y.C. Algorithm Unrolling: Interpretable, Efficient Deep Learning for Signal and Image Processing. *IEEE Signal Process. Mag.* **2021**, *38*, 18–44. [CrossRef]

36. Ronneberger, O.; Fischer, P.; Brox, T. U-Net: Convolutional Networks for Biomedical Image Segmentation. In *Proceedings of the Medical Image Computing and Computer-Assisted Intervention, Munich, Germany, 5–9 October 2015*; Navab, N., Hornegger, J., Wells, W.M., Frangi, A.F., Eds.; Springer International Publishing: Cham, Switzerland, 2015; pp. 234–241.

37. Ibtehaz, N.; Rahman, M.S. MultiResUNet: Rethinking the U-Net Architecture for Multimodal Biomedical Image Segmentation. *Neural Netw.* **2019**, *121*, 74–87. [CrossRef] [PubMed]

38. Gregor, K.; LeCun, Y. Learning Fast Approximations of Sparse Coding. In Proceedings of the 27th International Conference on International Conference on Machine Learning, Madison, WI, USA, 21–24 June 2010.

39. Zhang, J.; Ghanem, B. ISTA-Net: Interpretable Optimization-Inspired Deep Network for Image Compressive Sensing. In Proceedings of the 2018 IEEE/CVF Conference on Computer Vision and Pattern Recognition, Salt Lake City, UT, USA, 18–23 June 2018; pp. 1828–1837.

40. Jin, K.H.; McCann, M.T.; Froustey, E.; Unser, M. Deep Convolutional Neural Network for Inverse Problems in Imaging. *IEEE Trans. Image Process.* **2017**, *26*, 4509–4522. [CrossRef] [PubMed]

41. Chen, Y.; Pock, T. Trainable Nonlinear Reaction Diffusion: A Flexible Framework for Fast and Effective Image Restoration. *IEEE Trans. Pattern Anal. Mach. Intell.* **2017**, *39*, 1256–1272. [CrossRef]
42. Yang, Y.; Sun, J.; Li, H.B.; Xu, Z.B. ADMM-CSNet: A Deep Learning Approach for Image Compressive Sensing. *IEEE Trans. Pattern Anal. Mach. Intell.* **2020**, *42*, 521–538. [CrossRef]
43. Beck, A.; Teboulle, M. A Fast Iterative Shrinkage-Thresholding Algorithm for Linear Inverse Problems. *SIAM J. Imaging Vis.* **2009**, *2*, 183–202. [CrossRef]
44. Cao, Y.; Xu, J.R.; Lin, S.; Wei, F.Y.; Hu, H. GCNet: Non-local Networks Meet Squeeze-Excitation Networks and Beyond. In Proceedings of the IEEE/CVF International Conference on Computer Vision Workshop (ICCVW), Seoul, Korea, 27–28 October 2019; pp. 1971–1980.
45. Lopez-Martinez, C.; Fabregas, X. Modeling and reduction of SAR interferometric phase noise in the wavelet domain. *IEEE Trans. Geosci. Remote Sens.* **2002**, *40*, 2553–2566. [CrossRef]
46. Pu, L.; Zhang, X.; Zhou, L.; Li, L.; Shi, J.; Wei, S. Nonlocal Feature Selection Encoder–Decoder Network for Accurate InSAR Phase Filtering. *Remote Sens.* **2022**, *14*, 1174. [CrossRef]
47. Zhou, L.; Yu, H.; Lan, Y. Deep Convolutional Neural Network-Based Robust Phase Gradient Estimation for Two-Dimensional Phase Unwrapping Using SAR Interferograms. *IEEE Trans. Geosci. Remote Sens.* **2020**, *58*, 4653–4665. [CrossRef]
48. Kingma, D.P.; Ba, J. Adam: A method for stochastic optimization. *arXiv* **2014**, arXiv:1412.6980.
49. Pu, L.; Zhang, X.; Zhou, Z.; Li, L.; Zhou, L.; Shi, J.; Wei, S. A Robust InSAR Phase Unwrapping Method via Phase Gradient Estimation Network. *Remote Sens.* **2021**, *13*, 4564. [CrossRef]
50. Yu, Q.; Yang, X.; Fu, S.; Liu, X.; Sun, X. An adaptive contoured window filter for interferometric synthetic aperture radar. *IEEE Geosci. Remote Sens. Lett.* **2007**, *4*, 23–26. [CrossRef]
51. Wang, Z.; Bovik, A.C.; Sheikh, H.R.; Simoncelli, E.P. Image quality assessment: From error visibility to structural similarity. *IEEE Trans. Image Process.* **2004**, *13*, 600–612. [CrossRef]
52. Zhu, X.; Milanfar, P. Automatic Parameter Selection for Denoising Algorithms Using a No-Reference Measure of Image Content. *IEEE Trans. Image Process.* **2010**, *19*, 3116–3132.

 *remote sensing*

Article

Using Open Vector-Based Spatial Data to Create Semantic Datasets for Building Segmentation for Raster Data

Szymon Glinka, Tomasz Owerko and Karolina Tomaszkiewicz *

Faculty of Geo-Data Science, Geodesy, and Environmental Engineering, AGH University of Science and Technology, al. Mickiewicza 30, 30-059 Krakow, Poland; glinka@agh.edu.pl (S.G.); owerko@agh.edu.pl (T.O.)
* Correspondence: tomaszki@agh.edu.pl

Abstract: With increasing access to open spatial data, it is possible to improve the quality of analyses carried out in the preliminary stages of the investment process. The extraction of buildings from raster data is an important process, especially for urban, planning and environmental studies. It allows, after processing, to represent buildings registered on a given image, e.g., in a vector format. With an actual image it is possible to obtain current information on the location of buildings in a defined area. At the same time, in recent years, there has been huge progress in the use of machine learning algorithms for object identification purposes. In particular, the semantic segmentation algorithms of deep convolutional neural networks which are based on the extraction of features from an image by means of masking have proven themselves here. The main problem with the application of semantic segmentation is the limited availability of masks, i.e., labelled data for training the network. Creating datasets based on manual labelling of data is a tedious, time consuming and capital-intensive process. Furthermore, any errors may be reflected in later analysis results. Therefore, this paper aims to show how to automate the process of data labelling of cadastral data from open spatial databases using convolutional neural networks, and to identify and extract buildings from high resolution orthophotomaps based on this data. The conducted research has shown that automatic feature extraction using semantic ML segmentation on the basis of data from open spatial databases is possible and can provide adequate quality of results.

Keywords: semantic segmentation; open data; deep learning; building extraction; unet; deeplab

Citation: Glinka, S.; Owerko, T.; Tomaszkiewicz, K. Using Open Vector-Based Spatial Data to Create Semantic Datasets for Building Segmentation for Raster Data. *Remote Sens.* **2022**, *14*, 2745. https://doi.org/10.3390/rs14122745

Academic Editor: Gwanggil Jeon

Received: 5 May 2022
Accepted: 6 June 2022
Published: 7 June 2022

Publisher's Note: MDPI stays neutral with regard to jurisdictional claims in published maps and institutional affiliations.

1. Introduction

Increasing access to open spatial data and the development of machine learning algorithms mean that information can be extracted accurately from satellite and aerial imagery. On this basis, it is possible to determine the location of objects more precisely at the early stages of urban, planning and environmental analyses.

Information extraction can take place at different levels of complexity. The result is mainly dependent on the input data, the object of analysis and the algorithm used. Open spatial data are currently an increasingly important source of information in various areas of the economy. Their numbers are enormous and the amount of disk space they occupy is growing every day [1]. However, the use of such data requires processing it for specific applications. For several years, solutions based on deep neural networks have been increasingly popular. As a result, it is possible to classify, detect or segment objects, for example, from open raster data.

The application of semantic segmentation to geospatial data gives satisfactory results for: the extraction of objects, such as buildings [2–9]; roads [10,11]; the assessment of damage due to natural disasters [12]; or during population density assessment [13]. The problem with semantic segmentation is the small amount of publicly available labelled data that can be used to train the network. Creating datasets based on manual labelling of data is a tedious, time-consuming and capital-intensive process [14–16], and any errors can affect the results of the analysis. These problems motivate the search for solutions to automate the creation of masks for semantic segmentation from raster data (e.g., orthophotos), including those based on open vector spatial data [3]. Compared to our approach, existing works do not use mostly accurate and publicly available cadastral data or use less accurate data (raster with larger terrain pixel) as, for example, in Inria Dataset [17], and are not as flexible. In our approach, we can use data from different areas and create diverse datasets, as will be shown later in the paper.

Raster-based open spatial data can be divided into global and local (national). Global data are mainly remote sensing data acquired from satellites. The European Space Agency's Sentinel-2 mission allows the free acquisition of raster data which contains information not only on RGB channels, but also other spectral channels. The advantage of these type of data is that they are updated every few days, whereas the disadvantage is their spatial resolution. Therefore, they are most often used for macro analyses (also using deep neural networks) for segmentation, e.g., for fire impact assessment [18] and land cover analysis [19–21]. On the other hand, segmentation of individual buildings for open data is not possible—it would be necessary to use commercial data, whose spatial resolution is much better, such as in [22,23].

Raster local (national) data are the data made available by individual national institutions that operate (acquire, store or make available) geospatial data; for example, in Poland, this role is fulfilled by the Central Office of Geodesy and Cartography (GUGiK). The registers provide access to various resources: orthophotomaps with a resolution of up to 5 cm; vector layers of The Land and Building Register (EGiB); The Topographic Objects Database (BDOT10k) for a scale of 1:10,000; Digital Terrain Models and Digital Surface Models; LiDAR data, and others. The main problem of the data is the verification of their validity, as they are usually created every certain time unit (years). The LandCover dataset [24] which is used for land use segmentation on the basis of orthophotomaps was created on the basis of data that was made available by GUGiK. In various European countries, similar data are provided by institutions analogous to the GUGiK.

Similar to raster-based data, vector-based open spatial data can be divided into global and national scale. Open Street Map (OSM) is a global project that aims to create a free, editable map of the world. It is built by users and made available under an open-content licence. Segmentation using OSM has been carried out, among others, in [3,12].

Open vector data of national scale, similar to raster data, are made available by national institutions operating geospatial data. In Poland, such a resource is, for example, information on The Land and Building Register (EGiB) which is part of the cadastral database. The approach using open vector data for dataset creation was used by among others [9]. However, there the dataset is not described in detail the type of input data and what the problems of this dataset might have been are not described).

The problem of automatic labelling or using data resources that cannot be clearly labelled is not a simple one. Most often these data are not suitable to be directly labelled and must be processed through a transformation and rasterization process.

The aim of this paper is to present the results of work on verifying the possibility of using open vector spatial data as labels for the process of training convolutional neural networks and solving the task of the semantic segmentation of buildings for raster data. The paper uses fully open data that is available in the authors' country of residence—Poland— from the following databases: cadastral data of The Land and Building Register (EGiB) for a selected location in Poland and orthophotomaps taken from aerial photographs, made available by the Central Office of Geodesy and Cartography in Poland.

The motivation for the research was to verify the possibility of simplifying the tedious and time-consuming process of data labelling. The research goal was to verify the possibility of creating machine learning datasets based on the use of open spatial data. In addition, the research verified the impact of using available popular network architectures for solving semantic segmentation problems, i.e., UNET and DeepLabV3+, in order to obtain an algorithm that was characterised by the highest possible reliability. The algorithm was also verified in terms of differences in identification of buildings for different orthophoto terrain pixels.

The main novelty with respect to the other work is the verification of the use of fully open, accurate data to segment buildings from aerial photographs. This provides the opportunity to create large, diverse datasets that are flexible and contain multiple patterns. Additionally, the data used are characterised by high accuracy (low pixel resolution and high accuracy of vector data), where in the other works the data are far less accurate. In addition, the proposed algorithm allows for the creation of huge learning datasets from cadastral data, which are currently made publicly available by many European countries. The algorithms that were developed as a result of the work can be used, among others, for:

- Verification of the state of the cadastral databases in order to identify unpermitted buildings;
- Verification of the actual state of an area in the initial phase of an infrastructural investment process for a more reliable cost assessment;
- Mapping of buildings for unmapped areas;
- Verification of the validity of open building databases.

The structure of the article is as follows. The Section 1 introduces the topic and describes related works. The Section 2 describes the dataset that was used, discusses the issues related to it and the data pre-processing. The Section 3 discusses the network architectures that were used and presents the algorithm and processing strategies that produced the final result. The Section 4 presents the obtained results, which are then analysed—both statistically and visually. In addition, a discussion of the results is presented in this section. The paper concludes with a summary and conclusions of the conducted research in the Section 5.

2. Study Area and Datasets

2.1. Open Spatial Data

This paper focuses on the possibility of using open spatial data using the example of data that is available in the authors' country of residence, Poland. This section is a characterisation of open spatial data available in Poland which were used in the research, i.e., orthophotomaps and cadastral data—The Land and Building Register (hereinafter: EGiB). The use of OpenStreetMap (hereinafter: OSM) resource was also considered, but it was ultimately abandoned for reasons described in the next section.

Open spatial data in Poland are available on the basis of individual laws and European regulations concerning spatial information infrastructure, including the Inspire Directive [25]. The data are made available through the Geoportal [26] which is maintained by the Head Office of Geodesy and Cartography, or through individual local government units. The list of maintained resources is available at [27]. These units are obliged to maintain and make available free of charge (usually in an incomplete form for reasons of personal data protection and legal interests) geodetic resources, including those concerning the cadastre—EGiB. However, these resources are currently under development and are not yet available nationwide in a downloadable form.

The resources are maintained in various coordinate systems. Most often, data from the national dataset (e.g., orthophotomaps) are provided in the PL1992 system (EPSG 2180), while data from local government units (e.g., EGiB) are provided in the PL2000 system (EPSG 2176-2179—depending on the zone).

2.2. Selection of Study Areas

The following criteria were used to select the area for further analysis:

- Urban area;
- Architecture varying in terms of time of construction (historic buildings, often with more complicated architecture and contours, and modern buildings with simpler shapes);
- Architecture varying in terms of use (residential, industrial, public buildings, etc.);

- Building density and diversity;
- Availability of actual orthophotomap (max. up to one year back) with terrain pixel of max. 10 cm;
- Availability of data from the cadastral vector database: The Land and Building Register (EGiB).

In response to these criteria, the city of Bielsko-Biała in southern Poland in the Silesian Voivodeship was selected for further analysis. It is a city with diverse architecture, consisting of both older buildings and districts with modern buildings. Additionally, the city contains industrial areas with factories or large warehouses. In terms of building density and diversity, the city is characterised by a centre with compact buildings and, within a radius of about one kilometre, a less dense suburban area. Both an orthophotomap (dated 2021, with a maximum terrain pixel of 10 cm) and data from EGiB database were available for the city.

At the stage of selecting data sources, the use of two vector data resources, i.e., EGiB and OSM, was considered. The selection of the resource for further analysis was based on the verification of the actuality of these resources in relation to the orthophotomap of 2021, obtained from the Polish Geoportal [26]. Figure 1 shows the comparison between EGiB and OSM data. In green, the common parts of both resources are presented, in yellow the objects that are only in the EGiB database, while in red the elements that are only in the OSM database.

Figure 1. Comparison of data from EGIB and OSM (green—common parts of both databases, yellow—objects only in EGiB database, red—objects only in OSM database).

The analysis showed that the main problem of OSM resources is that they are outdated. Data from EGiB are more up-to-date, more accurate and complete and have no artefacts. Examples are presented in Figure 1 and—depending on the type of problem—are marked as the following areas:

- Area A—incorrectly determined outline of the building in the OSM database (the car park located next to the building was included in the building projection);
- Areas B1, B2, B3, B4, B5, B6—no buildings that actually exist in the OSM database;
- Area C1—presence in the OSM database of buildings which in fact do not exist;

- Area D—generalisation of building outline (simplification of building outline shape).

Since all the indicated problems may result in a much lower accuracy of the network and the wrong extraction of buildings, in further works it was decided to use only the EGiB database.

Before proceeding to further work, the input data from the EGiB database were analysed in relation to the orthophotomap. This comparison was aimed at identifying possible errors that could affect the results of the algorithm and, consequently, the possibility of extracting buildings.

Firstly, the obvious problem that was identified was that the mask outline of the dataset followed the wall outline, not the roof outline. This was due to the specificity of the EGiB database, which contains the vertices of the wall points. The applied roof eaves and other elements intended to protect the objects against, for example, the degrading activity of rainwater, increased the building outline. The problem is illustrated in Figure 2a. However, it was considered that the problem could be omitted given the purpose of the study, i.e., to identify the existence of objects with an approximate outline rather than to identify their exact outline.

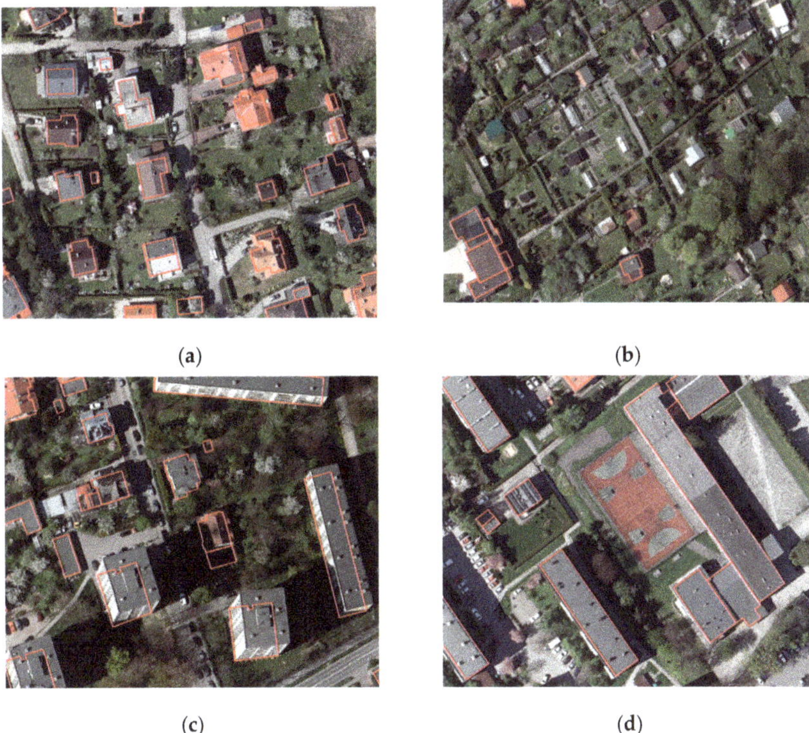

Figure 2. Identified errors for input data. Red colour marks outlines of buildings from the cadastral database. The images represent, respectively, the problems: (a) outline along the building walls—not along the roof, (b) gaps in the EGiB database, (c) radial displacement for tall buildings, (d) correct outline.

For the same reason, the problem caused by the available orthophotomap not being a true orthophotomap [28] and containing radial displacements which are particularly visible for tall objects was also omitted. A true orthophotomap is slowly being made available by the Central Office of Geodesy and Cartography. As of the writing of this article, i.e., the end of February 2022, the EGiB database was not available for the area that was subject to the creation of this type of product, and therefore it had to be excluded as a candidate for analysis. However, in each case, the building from the EGiB database is located within the outline boundaries of the objects with the orthophotomap. This problem is presented in Figure 2c.

In addition, the EGiB database does not consider objects that are not permanently connected to the ground; additional elements of buildings such as terraces, summer cottages, farmhouses, outbuildings or allotments. For this reason, the database is incomplete. Therefore, in the process of data preprocessing it was decided not to consider the objects that are not included in the EGiB database. The problem is presented in Figure 2b.

Figure 2d shows an area that is not subject to any of the problems described above. This is also the case for most of the area to be analysed, so it was decided to check the possibilities described above for the segmentation of the buildings.

2.3. Data Preprocessing

Publicly available datasets for object segmentation are most often created manually based on vector data from OSM or corrected manually based on publicly available building outlines. Datasets are also created based on commercially acquired data. However, our aim was to create a dataset completely free of charge.

The input orthophoto data included six images in. GTiff format with a ground pixel resolution of 10 cm. The areas that were selected for analysis were diverse in terms of architecture and building density. The dimensions of each image in pixels were 22,477 × 23,162, and in metres 2247.7 m × 2316.2 m. The area of analysis therefore covered an area of over 31 square kilometres. In the study area there were 21,010 buildings in vector format, available in the EGiB database.

From the input data, which consisted of orthophotomaps and vector data from EGiB, two datasets were created according to the algorithm presented in Figure 3: the first one for the input pixel with the terrain pixel of 10 cm and the second one with the terrain pixel of 50 cm. The second dataset was created by resampling data from the first, main dataset. The algorithm was programmed using the Python language and the gdal, ogr, opencv and patchify libraries. Different terrain pixels from the input data were used to compare the performance of the algorithms with respect to the size of the terrain pixels and to evaluate the possibility of segmenting buildings on these pixels.

The data were split into smaller images that were suitable for neural networks. This is a recommended action as it reduces the computing power required. Then, only those images where buildings were present were selected.

The first dataset contained 6365 images with dimensions: width—512, height—512, number of channels—3 (RGB colours) and corresponding labels in the form of binary image masks (1—buildings, 0—background) that were obtained as a result of rasterization. The data were divided into training set—80% of data, validation set—10% and testing set—10%. They contained, respectively, 5092 training images, 636 validation images and 637 test images. Similar work was carried out with the second dataset with a larger terrain pixel. The division of the large images into smaller images resulted in 1263 images with dimensions: width—256, height—256, number of channels—3 (RGB colours) and corresponding labels in the form of binary image masks (1—buildings, 0—background) that were obtained by rasterization. The data were divided in the same ratio as the first dataset and in this way 1010 training images, 126 validation images and 127 test images were obtained. A summary of both datasets is shown below in Table 1. Figure 4 shows raster data visualisations of the two datasets for visual comparison of the datasets. Clearly, more blurring is seen for the larger ground pixel, so a worse performance of the proposed architectures for this dataset is to be expected. The datasets have been made available on a repository [29] via the GitHub platform.

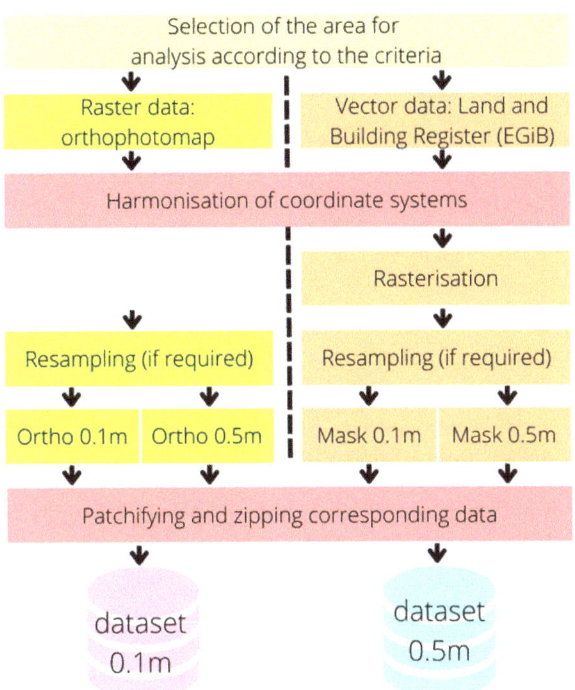

Figure 3. Algorithm for preparing datasets.

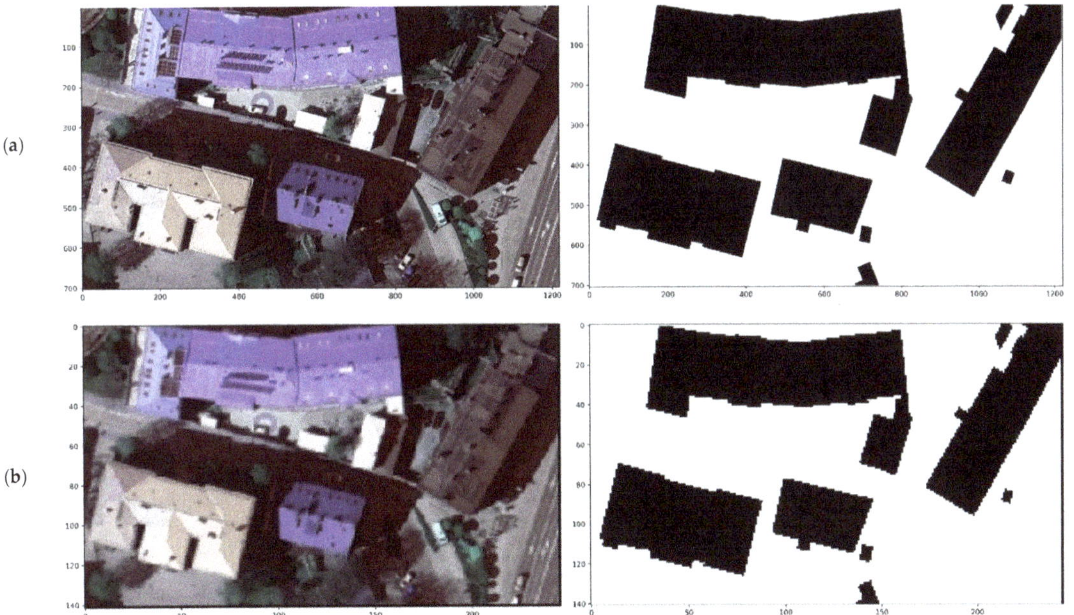

Figure 4. Comparison of datasets: (**a**) with 0.1 m pixel, (**b**) with 0.5 m pixel.

Table 1. Summary of created datasets.

Resolution [m]	Image Size [pix]	Image Size [m]	Training Images Number	Validation Images Number	Test Images Number
0.1	512 × 512 × 3	51.2 × 51.2	5092	636	637
0.5	256 × 256 × 3	128.0 × 128.0	1010	126	127

We are aware that somewhat false ground-truth data (containing the errors mentioned in the previous section) were used for testing. Testing of the algorithm with manually produced data (true ground-truth) is planned in future work. We anticipate that this may affect the accurate extraction of building edges, but as mentioned, our aim is to test the feasibility of using fully open data for building segmentation.

3. Materials and Methods

3.1. Semantic Image Segmentation Architectures

Currently, the most commonly used network architectures for image segmentation tasks are different variations of UNET and DeepLab. SegNet [30,31] and PSPNet [32] are also used, however, their use for further analyses was rejected because they are usually less efficient. Choosing the most optimal architecture was not the aim of the paper, but we would like to describe them briefly.

UNET consists of two main segments: the encoder and the decoder. The encoder at the initial stage consists of convolutional blocks, at the ends of which a pooling layer is implemented to reduce dimensionality. After moving to the dimensionality change pointbridge, dimensionality is increased by deconvolution or upsampling. Additionally, block information from the encoder is skipped and concatenated, followed by a convolution block. The operation is repeated until the input dimensions are obtained, where a predicted mask is obtained using the final convolution layer with the appropriate activation function [33]. The above description is the foundation of the network. In the years following the emergence of UNET network, various research teams have tried to modify it so that it provides even better results for different applications. Such an approach can be the use of DeepUNET [34], DeepResUNET [5,8] or combining UNET with solutions such as ASPP (Atrous Spatial Pyramid Pooling) [10]. From the point of view of information extraction from aerial images or satellite imagery, the results presented in [35,36] are particularly interesting. The obtained results allowed extraction of specific objects with varied accuracy—the mean Intersection Over Union value in most of the cited publications is around 90% in the case of building segmentation.

DeepLab, on the other hand, are network architectures based on atrous convolution in its initial version—DeepLabv1 [37], followed by the creation of atrous spatial pyramid pooling—DeepLabv2 [38]; its extension—DeepLabv3 [39]; the development of a segmentation decoder—DeepLabv3+ [40]; and the creation of networks based on NAS—Neural Architecture Search—Auto-DeepLab [41]. The use of the DeepLab architecture is particularly effective with the use of pre-trained backbones that allow for feature extraction. Part of ASPP allows for context identification by analysing links in the nearer and wider area. This approach, among others, was used in [24]. The DeepLab architecture, particularly DeepLabv3+ is also often used to segment information from satellite or aerial images e.g., [42–44].

Therefore, it was decided to test both architectures described above for solving the segmentation problem using open data resources. Part of the solution was implemented using the Python and Keras libraries, along with a Tensorflow framework as the backend. For this purpose, a publicly available library was created on GitHub [29]. A visualisation of the architectures used is shown in Figure 5.

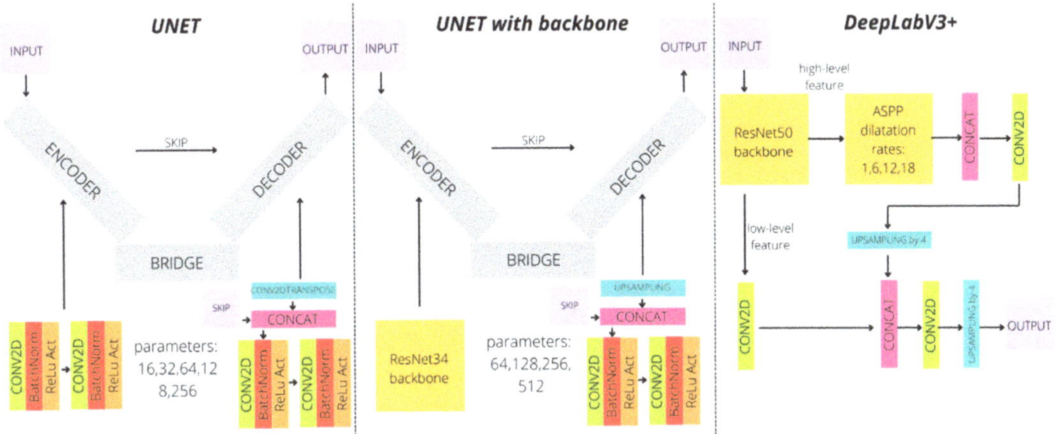

Figure 5. Used model architectures.

Various variations of UNET and DeepLabV3+ networks were implemented there, as well as the metrics and loss functions described later in this paper. A Github implementation of UNET with the ResNet34 backbone [45] was also used. The backbones were loaded with publicly available weights for the models that were obtained from the classification of the ImageNet dataset. The table below (Table 2) presents the network architectures used.

Table 2. Summary of used architectures.

Model	Description	Backbone	Number of Parameters for Input 512 × 512
UNET	Parameters: 16, 32, 64, 128, 256	does not exist	1,947,010
UNET_bb	UNET with backbone	Resnet34	24,456,299
DeepLabV3+	DeepLabV3+ with backbone	Resnet50	17,830,466

3.2. Data Augmentation

To eliminate overfitting, data augmentation was performed using the ImageDataGenerator class available in the Keras package. The rotation and flip operations were applied to the datasets. The data prepared in this way were used to check whether data augmentation improved the results that were obtained on the validation dataset. Example images that were obtained as a result of data augmentation are presented in Figure 6. The results of network training based on augmented data are presented in Section 4.

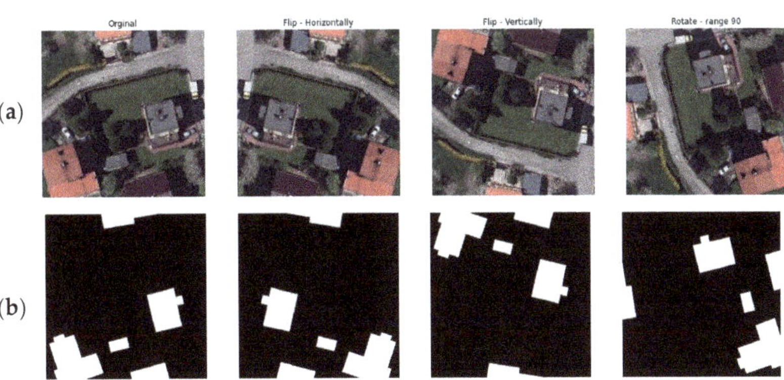

Figure 6. Example of augmented data: (**a**) images (**b**) masks.

3.3. Semantic Image Segmentation

During the development of the computational strategy for the segmentation task, the choice of network hyperparameters was considered and analysed. Special attention was paid to the selection of an appropriate loss function. The most commonly selected loss functions for the segmentation task were described in [46]. Based on the results of the analyses that were presented in the indicated publication, focal loss functions were abandoned. These functions tend to focus on difficult cases of learning patterns. Since images with problems as described in Section 2 could be classified as difficult cases, it was decided not to use this group of loss functions. To confirm the above assumption, the network was also computed using the focal loss function, which proved the above statement.

Finally, a loss function based on the Dice coefficient was chosen, which for binary segmentation is the same as the F1-score metric. This makes it possible to balance between the occurrence of False Positive and False Negative when evaluating the effects of network training. It was therefore considered possible to reduce radial displacement problems in the images in this way. The computational formula of the loss function (1) and (2) is presented below.

$$\text{Dice coefficient} = \text{F1 score} = \frac{2 \times \text{TP}}{2 \times \text{TP} + \text{FP} + \text{FN}} \times 100\% \tag{1}$$

$$L_{\text{Dice}} = 1 - \text{Dice coefficient} \tag{2}$$

The Adam optimiser was used for parameter updates [47]. As shown in the analyses conducted by the researchers, it is usually the most efficient [48], also for image segmentation tasks [49]. The optimiser parameters were used according to [47]: learning rate $\alpha = 0.001$, $\beta 1 = 0.9$, $\beta 2 = 0.999$, while $\varepsilon = 10^{-7}$.

Calculations were carried out on two standalone desktops with GPU computing capability. Computations that used 256×256 images were performed on a platform with the parameters: CPU—Intel(R) Core (TM) i7-9750H, GPU—NVIDIA GeForce GTX1650, 16 GB RAM. However, calculations for the second, larger dataset were performed on a platform with the following parameters: CPU—Intel(R) Core (TM) i7-6900K CPU @ 3.20 GHz, GPU—NVIDIA GeForce GTX1070, 62 GB RAM.

3.4. Results Evaluation

For the network evaluation, the metrics proposed in [50] were used to evaluate the quality of the results obtained in the segmentation task. The following metrics were used: precision (P); recall (R); Intersection-Over-Union (IoU, Jaccard Index); and F1 score (Dice coefficient). These are presented in Equations (3)–(6). The symbols in the formulae indicate elements of the confusion matrix, where: TP—True Positive—number of pixels correctly classified as buildings; FP—False Positive—number of background pixels classified as buildings; TN—True Negative—number of pixels correctly classified as background; and FN—False Negative—number of pixels of buildings classified as background. The average IoU value for both classes—buildings and background—was used in the presentation of the results.

$$\text{precision} = \frac{\text{TP}}{\text{TP} + \text{FP}} \tag{3}$$

$$\text{recall} = \frac{\text{TP}}{\text{TP} + \text{FN}} \tag{4}$$

$$\text{IoU} = \frac{\text{TP}}{\text{TP} + \text{FP} + \text{FN}} \tag{5}$$

$$\text{F1 score} = \frac{2 \times \text{precision} \times \text{recall}}{\text{precision} + \text{recall}} = \frac{2 \times \text{TP}}{2 \times \text{TP} + \text{FP} + \text{FN}} \tag{6}$$

4. Results

Before the learning process started, during data loading, all pixels were rescaled to values between 0 and 1. Additionally, the images were normalised. These procedures were intended to speed up the operation of the computational networks. The calculations themselves were performed according to the algorithm described in Section 3. The presented results were obtained on the basis of calculations on a test set. The validation set was used as a control during network learning.

4.1. Dataset with 0.5 m Terrain Pixel

The results that were obtained for a dataset with a 0.5 m pixel are presented in Table 3. The best results were achieved for the UNET network with the Resnet34 backbone. In contrast, the worst results were obtained for the DeepLabV3+ network, which may be due to the requirements of this architecture in relation to the number of input datasets. Part of DeepLabV3+ is ASPP (Atrous Spatial Pyramid Pooling) which uses Dilated Convolution. This is computationally demanding [51] and requires sufficient input data of satisfactory quality. The main purpose of using ASPP is to extract features of larger objects and maintain their consistency [7]. Too little input data resulted in False Positive artifacts in some parts of the predicted images. This can be seen in Figure 7, especially in example (c).

Table 3. Results for 0.5 m dataset [%].

Neural Network Architecture	Augmentation	mIoU	F1-Score	Precision	Recall
UNET	NO	90.64	95.02	94.89	95.15
UNET with backbone	NO	**92.24**	**95.91**	**95.83**	**95.99**
DeepLabV3+	NO	79.96	88.37	88.28	88.46
UNET	YES	90.33	94.85	94.57	95.13
UNET with backbone	YES	90.24	94.79	94.53	95.06
DeepLabV3+	YES	83.83	90.81	90.03	91.62

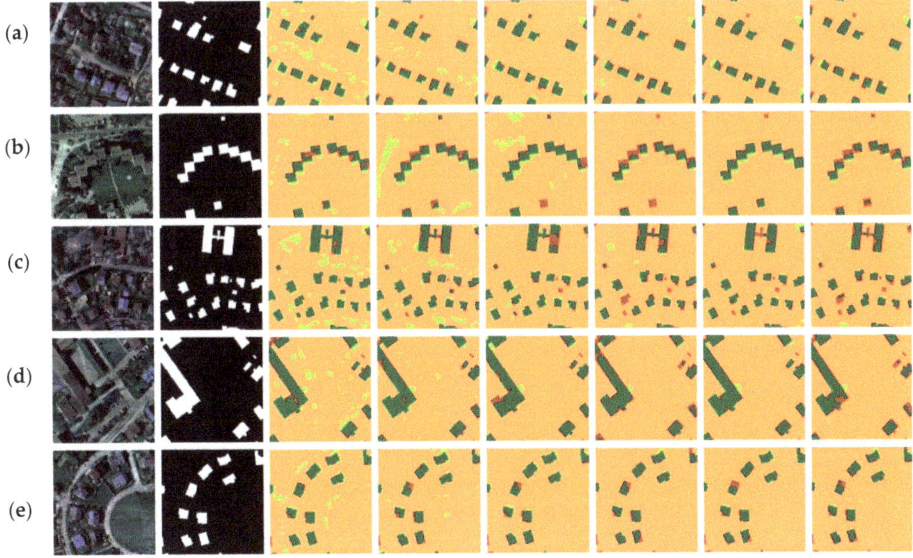

Figure 7. Visualisation of the results for the 0.5 m dataset. From left: input image, ground truth, DeepLab, DeepLab with augmentation, UNET, UNET with augmentation, UNET_bb, UNET_bb with augmentation. Colours: dark green—TP, orange—TN, light green—FP, red—FN. (**a**–**e**) various examples from the test dataset.

Table 3 also shows the results for each architecture using the data augmentation described in Section 3.2. Better results on the test set were obtained only for the DeepLabV3+ architecture, which may confirm the conjecture described in the previous paragraph about the requirements of this network. For both UNET architectures the results achieved are slightly worse. This may be due to the specifics of the dataset and the information transmitted inside the network. As the dataset is characterised by inaccuracies (described in Section 2.2), the network through augmentation may have received more bad patterns for learning. Therefore, the calculated weights changed to recognise more bad patterns. In addition, the data were highly heterogeneous, which results from a lack of spatial planning. It can therefore be concluded that data augmentation does not always produce positive results.

Figure 7 presents the predicted masks based on several images from the test set, which mirrors the results in Table 3. However, not all objects were classified. The reasons for this situation are discussed in detail in Section 4.3.

Ultimately, the UNET architecture with the Resnet34 backbone was found to perform best in terms of both metrics and visualisation. Regular UNET was slightly worse (mIOU worse by 2points). On the other hand, the worst results were obtained for the DeepLabV3+ network—mIOU which was worse than UNET by over 10 points when using the dataset without augmentation and about 7 points when using augmentation.

4.2. Dataset with 0.1 m Terrain Pixel

The results that were obtained by the individual architectures for the dataset with a 0.1 m terrain pixel are presented in Table 4. As data augmentation did not significantly improve the results that were achieved for the dataset with a 0.5 m pixel, it was decided to omit it when analysing the dataset with a smaller terrain pixel.

Table 4. Results for 0.1 m dataset [%].

Neural Network Architecture	Augmentation	mIoU	F1-Score	Precision	Recall
UNET	NO	91.08	95.31	95.19	95.43
UNET with backbone	NO	**93.00**	**96.36**	**96.33**	**96.38**
DeepLabV3+	NO	92.86	96.28	96.27	96.29

As in the case of the dataset with the 0.5 m terrain pixel, and for the dataset with the 0.1 m terrain pixel, the best performance is achieved by the UNET network with backbone, but the mIoU difference with respect to DeepLabV3+ is small at 0.14 points. This is due to the definition of an appropriate input set size for the DeepLabV3+ network. The regular UNET network performs slightly worse here, the achieved mIoU metric is worse by about 2 points compared to the other networks. However, its biggest advantage is the speed of computation, which is due to the significantly smaller number of weights to be determined.

Visualization of the results is shown in Figure 8. Considering only visual issues, it can be concluded that the best results were obtained for DeepLabV3+ architecture—the most TP and TN areas. The other architectures also give satisfactory results. Importantly, the architectures give good results for both lower density areas (Figure 8a,c–g) and higher density areas (Figure 8b,f).

The occurrence of FP-labelled pixels at the edges of buildings is largely due to the ground truth not being entirely true (errors described in Section 2.2). On the other hand, occurring FNs are often caused by obscuration caused by trees or shading from a taller building. Further, the varied shape of the roof causes the model to fail to recognise parts of the building (TN).

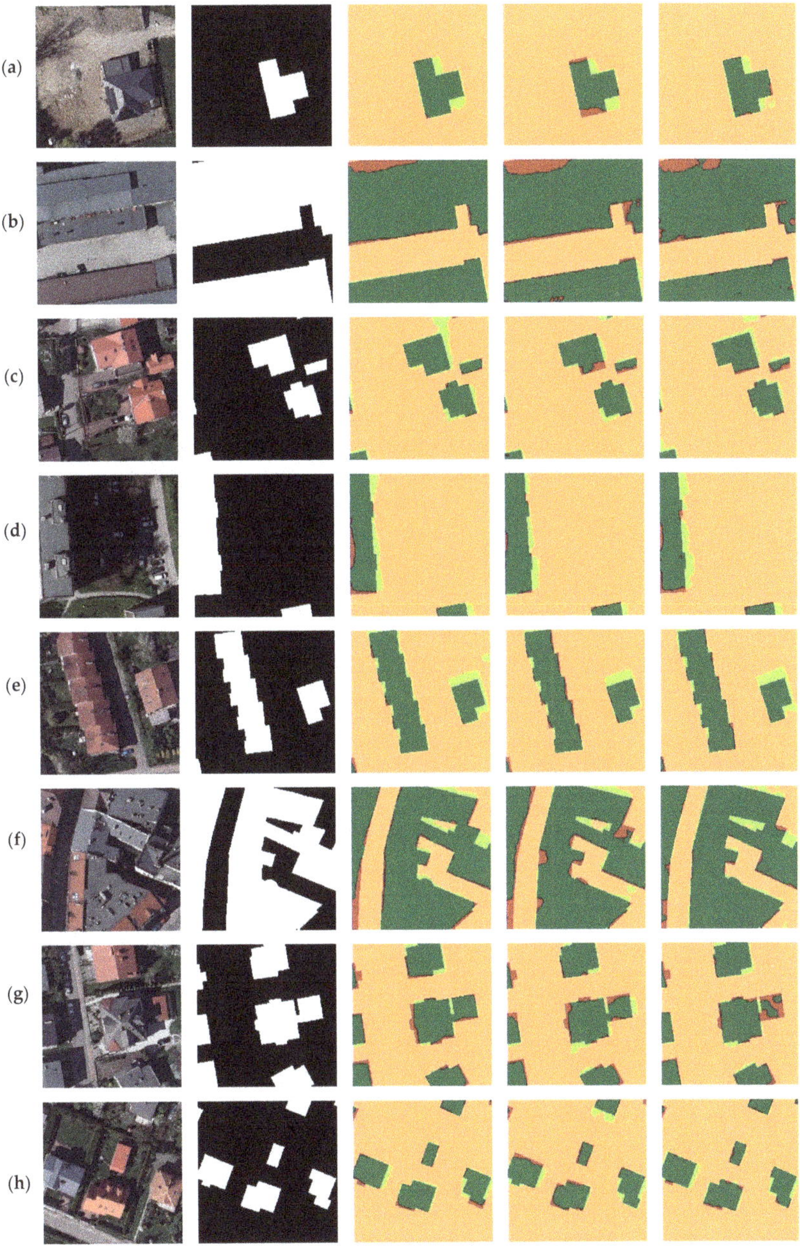

Figure 8. Visualisation of the results for the 0.1 m dataset. From left: input image, ground truth, DeepLabV3+, UNET, UNET with backbone. Colours: dark green—TP, orange—TN, light green—FP, red—FN. (**a–h**) various examples from the test dataset.

4.3. Discussion

The obtained results confirm the possibility of using open spatial data as a dataset for the task of segmenting buildings from raster images. However, it should be kept in mind that the most important issue is the requirement for data accuracy, which must be adapted to the specific task. Given this, the use of the described approach may have some limitations when the goal is to segment building outlines very accurately, where the problems discussed in Section 2.1 may have a significant impact on the results. This section will discuss the achieved results in more detail.

Firstly, comparing the results that were obtained between the two datasets, one can see slightly higher metric values for the dataset with the smaller ground pixel resolution. This difference is particularly noticeable when using the DeepLabV3+ architecture, where an mIoU score of 12.90 points better was obtained due to the specificity of this architecture. For both UNETs, the difference in value did not exceed 0.01 points, but when visually comparing the results from these networks we see that some of the buildings for the larger pixel were not classified correctly. Such a small difference in metrics is due to the size of the buildings for each dataset. For example, a building with dimensions of 10×10 m occupies 20×20 pixels for a dataset with a larger pixel and 100×100 pixels for a dataset with a smaller pixel. No segmentation of such an object or its shading/shadowing in the test image will have a much greater impact in the case of a dataset with a smaller pixel, and this is reflected during the calculation of metrics.

The visual comparison also shows that the dataset with the smaller terrain pixel is more effective. In the test images, all the objects were correctly identified, whereas the dataset with the larger pixel size showed significantly more False Negative and False Positive areas.

A noticeable problem for both datasets is shaded and wooded areas. Shadows cast by high buildings cause the objects directly below to be covered or shaded. The result is a pixel misclassification or partial segmentation of the object. This problem concerns mainly garages or extensions to the main building. Similar results are generated on images where vegetation—usually trees—covers the image. While the first problem can be eliminated by creating a true-orthophotomap, the second problem in the case of using photogrammetric digital cameras will always occur. Its solution may be the use of LiDAR and adding more analysis dimensions in addition to RGB colours.

The loss function that was used achieved its purpose. The aim of using DICE loss was to maintain a balance between FP and FN values. The results obtained, i.e., similar Recall and Precision values, show that this aim can be considered satisfied.

The mIoU values obtained for building segmentation are similar to those achieved by other researchers for public datasets. Typically, this value reaches a value of around 90%.

The limitations of the used dataset, as described in this section and Section 2.2, may be difficult to fully overcome. The solution of this may be to expand the dataset to include other areas that generate new learning patterns. The analysed dataset was also not varied by lighting, so applying the algorithm to images with different histogram characteristics may not give satisfactory results. Therefore, further tests for more areas, and more diverse areas are possible. In addition, in order to fully assess the accuracy of the obtained results, a comparison of the obtained results with fully correct, hand-made building outlines is required.

The applied models and their hyperparameters can be optimised. However, the aim of this paper was not to develop new architectures or approaches, but to verify the possibility of using open data to generate data for training sets to solve the semantic segmentation problem.

5. Conclusions

Open spatial data allow the extraction of a lot of information. Their biggest advantage is fast and free access. The increase in data volume, combined with the development of machine learning algorithms for object identification increases the ability to accurately extract information from satellite and aerial images. As a result, it is possible to identify the location of objects more accurately at early stages of urban, planning or environmental analyses. It should be remembered, however, that the use of open data depends on the accuracy requirements for the problem being analysed.

On this basis, the analyses presented in this paper conclude that open vector spatial cadastral data can be used as labels in the training process of convolutional neural networks, and solve the task of the semantic segmentation of buildings for raster data. The solution presented in this paper enables simplification of the tedious, time-consuming and capital-intensive process of data labelling. This solution also enables the minimisation of errors that may be reflected in the later results of the analyses. The results of the conducted analyses also allow a comparison of the effectiveness of available popular network architectures, i.e., UNET and DeepLabV3+, in solving semantic segmentation problems, and the influence of backbones on the accuracy of building detection.

This paper also analysed the effect of the orthophoto ground pixel resolution on the accuracy of building identification. The analyses showed that for each of the network architectures used, better results are achieved for data with a smaller terrain pixel. The use of data with a smaller terrain pixel is particularly important when using DeepLabV3+, where it allows the mIoU value to be increased by approximately 13 points.

This needs to be confirmed in separate studies, but most probably these data can be successfully applied for the purposes mentioned in the introduction, such as verifying the state of cadastral databases for the identification of unauthorised buildings; verifying the actual state of the land in the initial phase of the infrastructural investment process for a more reliable cost assessment; the mapping of buildings for unmapped areas; verifying the validity of open databases.

However, it is important to keep in mind the identified limitations of the datasets described in Section 2, such as the building outline following the walls rather than the roof or radial offsets. These limitations were omitted in this analysis as unimportant to the problem being solved. However, they may reduce the efficiency of directly implementing the presented algorithm to solve other problems, such as the accurate detection of the position of buildings.

Therefore, the authors plan to further develop the presented dataset with other, more diverse areas, which will allow to generate new learning patterns and optimise the hyper-parameters of the applied models in order to increase the accuracy of object detection.

The authors also plan to apply the algorithm presented in this paper using true-orthophoto mapping, which should become more widely available for a larger area of Poland over time. This approach will allow examination of the extent to which the reduction in radial displacement problems improves the extraction of buildings using the algorithm presented in this paper. Additionally, in order to minimise the problem of detecting objects under vegetation and shaded areas, the possibility of adding more image dimensions on the basis of available LiDAR data from Airborne Laser Scanning also made available by governmental units will be verified.

Moreover, the authors are planning to compare a dataset for which masks will be created manually with a dataset that includes masks adopted based on cadastral data. The aim of this activity will be to verify the existence of the algorithm limitations that are proposed in this publication.

Author Contributions: Conceptualization, S.G.; Data curation, S.G. and K.T.; Formal analysis, T.O.; Methodology, S.G., T.O. and K.T.; Software, S.G.; Supervision, T.O.; Validation, T.O.; Visualization, S.G. and K.T.; Writing—original draft, S.G. and K.T.; Writing—review & editing, T.O. All authors have read and agreed to the published version of the manuscript.

Funding: This research received no external funding.

Data Availability Statement: The data and algorithms used are available through the GitHub platform (references in the text).

Conflicts of Interest: The authors declare no conflict of interest.

References

1. European Commission. *Open Data Maturity Report 2020*; European Commission: Brussels, Belgium, 2020.
2. Liu, P.; Liu, X.; Liu, M.; Shi, Q.; Yang, J.; Xu, X.; Zhang, Y. Building footprint extraction from high-resolution images via spatial residual inception convolutional neural network. *Remote Sens.* **2019**, *11*, 830. [CrossRef]
3. Touzani, S.; Granderson, J. Open data and deep semantic segmentation for automated extraction of building footprints. *Remote Sens.* **2021**, *13*, 2578. [CrossRef]
4. Liu, J.; Wang, S.; Hou, X.; Song, W. A deep residual learning serial segmentation network for extracting buildings from remote sensing imagery. *Int. J. Remote Sens.* **2020**, *41*, 5573–5587. [CrossRef]
5. Li, W.; He, C.; Fang, J.; Fu, H. Semantic segmentation based building extraction method using multi-source GIS map datasets and satellite imagery. In Proceedings of the 2018 IEEE/CVF Conference on Computer Vision and Pattern Recognition Workshops (CVPRW), Salt Lake City, UT, USA, 18–22 June 2018; pp. 233–236. [CrossRef]
6. Chen, Z.; Li, D.; Fan, W.; Guan, H.; Wang, C.; Li, J. Self-attention in reconstruction bias U-net for semantic segmentation of building rooftops in optical remote sensing images. *Remote Sens.* **2021**, *13*, 2524. [CrossRef]
7. Wang, H.; Miao, F. Building extraction from remote sensing images using deep residual U-Net. *Eur. J. Remote Sens.* **2022**, *55*, 71–85. [CrossRef]
8. Bischke, B.; Helber, P.; Folz, J.; Borth, D.; Dengel, A. Multi-Task Learning for Segmentation of Building Footprints with Deep Neural Networks. In Proceedings of the 2019 IEEE International Conference on Image Processing (ICIP), Taipei, Taiwan, 22–25 September 2019; pp. 1480–1484. [CrossRef]
9. Yi, Y.; Zhang, Z.; Zhang, W.; Zhang, C.; Li, W.; Zhao, T. Semantic segmentation of urban buildings from VHR remote sensing imagery using a deep convolutional neural network. *Remote Sens.* **2019**, *11*, 1774. [CrossRef]
10. He, H.; Yang, D.; Wang, S.; Wang, S.; Li, Y. Road extraction by using atrous spatial pyramid pooling integrated encoder-decoder network and structural similarity loss. *Remote Sens.* **2019**, *11*, 1015. [CrossRef]
11. Boonpook, W.; Tan, Y.; Bai, B.; Xu, B. Road Extraction from UAV Images Using a Deep ResDCLnet Architecture. *Can. J. Remote Sens.* **2021**, *47*, 450–464. [CrossRef]
12. Gupta, A.; Watson, S.; Yin, H. Deep learning-based aerial image segmentation with open data for disaster impact assessment. *Neurocomputing* **2021**, *439*, 22–33. [CrossRef]
13. Robinson, C.; Hohman, F.; Dilkina, B. A deep learning approach for population estimation from satellite imagery. In Proceedings of the 1st ACM SIGSPATIAL Workshop on Geospatial Humanities, Online, 7 November 2017; pp. 47–54. [CrossRef]
14. Cai, L.; Xu, X.; Liew, J.H.; Sheng Foo, C. Revisiting Superpixels for Active Learning in Semantic Segmentation with Realistic Annotation Costs. In Proceedings of the 2021 IEEE/CVF Conference on Computer Vision and Pattern Recognition (CVPR), Nashville, TN, USA, 20–25 June 2021; pp. 10983–10992. [CrossRef]
15. Li, Y.; Chen, J.; Xie, X.; Ma, K.; Zheng, Y. Self-loop uncertainty: A novel pseudo-label for semi-supervised medical image segmentation. In *Medical Image Computing and Computer Assisted Intervention—MICCAI 2020*; Lecture Notes in Computer Science (Including Subseries Lecture Notes in Artificial Intelligence and Lecture Notes in Bioinformatics); Springer: Cham, Switzerland, 2020; Volume 12621, pp. 614–623. [CrossRef]
16. Sun, W.; Zhang, J.; Barnes, N. 3D Guided Weakly Supervised Semantic Segmentation. In *Computer Vision—ACCV 2020*; Lecture Notes in Computer Science (Including Subseries Lecture Notes in Artificial Intelligence and Lecture Notes in Bioinformatics); Springer: Cham, Switzerland, 2020; Volume 12622, pp. 585–602. [CrossRef]
17. Maggiori, E.; Tarabalka, Y.; Charpiat, G.; Alliez, P. Can semantic labeling methods generalize to any city? the inria aerial image labeling benchmark. In Proceedings of the 2017 IEEE International Geoscience and Remote Sensing Symposium (IGARSS), Fort Worth, TX, USA, 23–28 July 2017; pp. 3226–3229. [CrossRef]
18. Farasin, A.; Colomba, L.; Garza, P. Double-step U-Net: A deep learning-based approach for the estimation of wildfire damage severity through sentinel-2 satellite data. *Appl. Sci.* **2020**, *10*, 4332. [CrossRef]
19. Ulmas, P.; Liiv, I. Segmentation of Satellite Imagery using U-Net Models for Land Cover Classification. *arXiv* **2020**, arXiv:2003.02899.
20. Gargiulo, M.; Dell'aglio, D.A.G.; Iodice, A.; Riccio, D.; Ruello, G. Integration of sentinel-1 and sentinel-2 data for land cover mapping using w-net. *Sensors* **2020**, *20*, 2969. [CrossRef] [PubMed]
21. Karra, K.; Kontgis, C.; Statman-Weil, Z.; Mazzariello, J.C.; Mathis, M.; Brumby, S.P. Global Land Use/Land Cover with Sentinel 2 and Deep Learning. In Proceedings of the 2021 IEEE International Geoscience and Remote Sensing Symposium IGARSS, Brussels, Belgium, 11–16 July 2021; pp. 4704–4707. [CrossRef]
22. Pan, Z.; Xu, J.; Guo, Y.; Hu, Y.; Wang, G. Deep learning segmentation and classification for urban village using a worldview satellite image based on U-net. *Remote Sens.* **2020**, *12*, 1574. [CrossRef]
23. Shahi, K.; Shafri, H.Z.M.; Taherzadeh, E.; Mansor, S.; Muniandy, R. A novel spectral index to automatically extract road networks from WorldView-2 satellite imagery. *Egypt. J. Remote Sens. Sp. Sci.* **2015**, *18*, 27–33. [CrossRef]
24. Boguszewski, A.; Batorski, D.; Ziemba-Jankowska, N.; Dziedzic, T.; Zambrzycka, A. LandCover.ai: Dataset for automatic mapping of buildings, woodlands, water and roads from aerial imagery. In Proceedings of the IEEE Computer Society Conference on Computer Vision and Pattern Recognition Workshops, Nashville, TN, USA, 19–25 June 2021; pp. 1102–1110.

25. Directive 2007/2/EC of the European Parliament and of the Council of 14 March 2007 Establishing an Infrastructure for Spatial Information in the European Community (INSPIRE). Available online: https://eur-lex.europa.eu/legal-content/EN/ALL/?uri=CELEX%3A32007L0002 (accessed on 15 March 2022).

26. Geoportal Krajowy (National Geoportal). Available online: https://www.geoportal.gov.pl/ (accessed on 15 March 2022).

27. Ewidencja Zbiorów i Usług Danych Przestrzennych (Register of Spatial Data Sets and Services). Available online: https://integracja.gugik.gov.pl/eziudp/ (accessed on 15 March 2022).

28. Habib, A.F.; Kim, E.M.; Kim, C.J. New methodologies for true orthophoto generation. *Photogramm. Eng. Remote Sens.* **2007**, *73*, 25–36. [CrossRef]

29. Glinka, S. Keras Segmentation Models. Available online: https://github.com/sajmonogy/keras_segmentation_models (accessed on 1 May 2022).

30. Badrinarayanan, V.; Kendall, A.; Cipolla, R. SegNet: A Deep Convolutional Encoder-Decoder Architecture for Image Segmentation. *IEEE Trans. Pattern Anal. Mach. Intell.* **2017**, *39*, 2481–2495. [CrossRef]

31. Abdollahi, A.; Pradhan, B.; Alamri, A.M. An ensemble architecture of deep convolutional Segnet and Unet networks for building semantic segmentation from high-resolution aerial images. *Geocarto Int.* **2020**, *35*, 1856199. [CrossRef]

32. Zhao, H.; Shi, J.; Qi, X.; Wang, X.; Jia, J. Pyramid scene parsing network. In Proceedings of the 2017 IEEE Conference on Computer Vision and Pattern Recognition (CVPR), Honolulu, HI, USA, 21–26 July 2017; pp. 6230–6239. [CrossRef]

33. Weng, W.; Zhu, X. UNet: Convolutional Networks for Biomedical Image Segmentation. *IEEE Access* **2021**, *9*, 16591–16603. [CrossRef]

34. Li, R.; Liu, W.; Yang, L.; Sun, S.; Hu, W.; Zhang, F.; Li, W. DeepUNet: A Deep Fully Convolutional Network for Pixel-Level Sea-Land Segmentation. *IEEE J. Sel. Top. Appl. Earth Obs. Remote Sens.* **2018**, *11*, 3954–3962. [CrossRef]

35. Sofla, R.A.D.; Alipour-Fard, T.; Arefi, H. Road extraction from satellite and aerial image using SE-Unet. *J. Appl. Remote Sens.* **2021**, *15*, 014512. [CrossRef]

36. He, N.; Fang, L.; Plaza, A. Hybrid first and second order attention Unet for building segmentation in remote sensing images. *Sci. China Inf. Sci.* **2020**, *63*, 140305. [CrossRef]

37. Chen, L.C.; Papandreou, G.; Kokkinos, I.; Murphy, K.; Yuille, A.L. Semantic Image Segmentation with Deep Convolutional Nets and Fully Connected CRFs. *arXiv* **2015**, arXiv:1412.7062.

38. Chen, L.C.; Papandreou, G.; Kokkinos, I.; Murphy, K.; Yuille, A.L. DeepLab: Semantic Image Segmentation with Deep Convolutional Nets, Atrous Convolution, and Fully Connected CRFs. *IEEE Trans. Pattern Anal. Mach. Intell.* **2018**, *40*, 834–848. [CrossRef] [PubMed]

39. Chen, L.-C.; Papandreou, G.; Schroff, F.; Adam, H. Rethinking Atrous Convolution for Semantic Image Segmentation. *arXiv* **2017**, arXiv:1706.05587.

40. Chen, L.C.; Zhu, Y.; Papandreou, G.; Schroff, F.; Adam, H. Encoder-decoder with atrous separable convolution for semantic image segmentation. In *Computer Vision—ECCV 2018*; Lecture Notes in Computer Science (Including Subseries Lecture Notes in Artificial Intelligence and Lecture Notes in Bioinformatics); Springer: Cham, Switzerland, 2018; Volume 11211, pp. 833–851. [CrossRef]

41. Liu, C.; Chen, L.C.; Schroff, F.; Adam, H.; Hua, W.; Yuille, A.L.; Fei-Fei, L. Auto-deeplab: Hierarchical neural architecture search for semantic image segmentation. In Proceedings of the 2019 IEEE/CVF Conference on Computer Vision and Pattern Recognition (CVPR), Long Beach, CA, USA, 15–20 June 2019; pp. 82–92. [CrossRef]

42. Liu, Z.; Liu, W.; Qi, H.; Li, Y.; Zhang, G.; Zhang, T. Extracting River Illegal Buildings from UAV Image Based on Deeplabv3+. In *Geoinformatics in Sustainable Ecosystem and Society, Proceedings of the 7th International Conference, GSES 2019, and First International Conference, GeoAI 2019, Guangzhou, China, 21–25 November 2019*; Springer: Singapore, 2020; Volume 1228, pp. 259–272. [CrossRef]

43. Xiang, S.; Xie, Q.; Wang, M. Semantic Segmentation for Remote Sensing Images Based on Adaptive Feature Selection Network. In *IEEE Geoscience and Remote Sensing Letters*; IEEE: New York, NY, USA, 2022; Volume 19. [CrossRef]

44. Zhang, D.; Ding, Y.; Chen, P.; Zhang, X.; Pan, Z.; Liang, D. Automatic extraction of wheat lodging area based on transfer learning method and deeplabv3+ network. *Comput. Electron. Agric.* **2020**, *179*, 105845. [CrossRef]

45. Yakubovskiy, P. Segmentation Models. Available online: https://github.com/qubvel/segmentation_models (accessed on 5 March 2022).

46. Jadon, S. A survey of loss functions for semantic segmentation. In Proceedings of the 2020 IEEE Conference on Computational Intelligence in Bioinformatics and Computational Biology (CIBCB), Via del Mar, Chile, 27–29 October 2020. [CrossRef]

47. Kingma, D.P.; Ba, J.L. Adam: A method for stochastic optimization. In Proceedings of the 3rd International Conference for Learning Representations, San Diego, CA, USA, 7–9 May 2015; pp. 1–15.

48. Schneider, F.; Balles, L.; Hennig, P. Deepobs: A deep learning optimizer benchmark suite. In Proceedings of the 7th International Conference on Learning Representations, ICLR 2019, New Orleans, LA, USA, 6–9 May 2019; pp. 1–14.

49. Yaqub, M.; Jinchao, F.; Zia, M.S.; Arshid, K.; Jia, K.; Rehman, Z.U.; Mehmood, A. State-of-the-art CNN optimizer for brain tumor segmentation in magnetic resonance images. *Brain Sci.* **2020**, *10*, 427. [CrossRef]

50. Minaee, S.; Boykov, Y.Y.; Porikli, F.; Plaza, A.J.; Kehtarnavaz, N.; Terzopoulos, D. Image Segmentation Using Deep Learning: A Survey. *IEEE Trans. Pattern Anal. Mach. Intell.* **2021**, *14*, 3523–3542. [CrossRef]
51. Wang, Z.; Ji, S. Smoothed dilated convolutions for improved dense prediction. In Proceedings of the 24th ACM SIGKDD International Conference on Knowledge Discovery & Data Mining, KDD 2018, London, UK, 19–23 August 2018; pp. 2486–2495. [CrossRef]

Article

A Classifying-Inversion Method of Offshore Atmospheric Duct Parameters Using AIS Data Based on Artificial Intelligence

Jie Han [1,2], Jiaji Wu [1,*], Lijun Zhang [2], Hongguang Wang [2], Qinglin Zhu [2], Chao Zhang [2], Hui Zhao [2] and Shoubao Zhang [2]

[1] School of Electronic Engineering, Xidian University, Xi'an 710071, China; hanj@crirp.ac.cn
[2] China Research Institute of Radiowave Propagation, Qingdao 266107, China; zhanglj@crirp.ac.cn (L.Z.); wanghg@crirp.ac.cn (H.W.); zhuql1@crirp.ac.cn (Q.Z.); zhangc@crirp.ac.cn (C.Z.); zhaoh1@crirp.ac.cn (H.Z.); zhangsb@crirp.ac.cn (S.Z.)
* Correspondence: wujj@mail.xidian.edu.cn

Abstract: Atmospheric duct parameters inversion is an important aspect of microwave-band radar and communication system performance evaluation. AIS (Automatic Identification System) is one of the signal sources used for atmospheric duct parameters inversion. Before the inversion of atmospheric duct parameters, determining the type of atmospheric duct plays an important role in the inversion results, but the current inversion methods ignore this point. We outlined a classifying-inversion method of atmospheric duct parameters using AIS signals combined with artificial intelligence. The method consists of an atmospheric duct classification model and a parameter inversion model. The classification model judges the type of atmospheric duct, and the inversion model inverts the atmospheric duct parameters according to the type of atmospheric duct. Our findings demonstrated that the accuracy of the atmospheric duct classification model based on deep neural network (DNN) even exceeds 97%, and the atmospheric duct parameters inversion model has better inversion accuracy than that of the traditional method, thereby illustrating the effectiveness and accuracy of this novel method.

Keywords: classifying-inversion method; AIS; atmospheric duct; artificial intelligence

Citation: Han, J.; Wu, J.; Zhang, L.; Wang, H.; Zhu, Q.; Zhang, C.; Zhao, H.; Zhang, S. A Classifying-Inversion Method of Offshore Atmospheric Duct Parameters Using AIS Data Based on Artificial Intelligence. *Remote Sens.* **2022**, *14*, 3197. https://doi.org/10.3390/rs14133197

Academic Editor: Michael J. Newchurch

Received: 28 May 2022
Accepted: 30 June 2022
Published: 3 July 2022

Publisher's Note: MDPI stays neutral with regard to jurisdictional claims in published maps and institutional affiliations.

1. Introduction

The atmospheric duct is an abnormal phenomenon in the tropospheric atmosphere that includes evaporation, surface, and elevated ducts. Ducting occurs when a radio ray originating at the Earth's surface is sufficiently refracted so that it is either bent back toward the Earth's surface or travels in a path parallel to the Earth's surface. These types have different causes. Evaporation duct is caused by water surface evaporation, and it mainly appears over the ocean, with an occurrence of over 85% [1]. Surface and elevated ducts (low-altitude atmospheric ducts) are mainly caused by weather phenomena, such as radiation-inversion, sinking-inversion, and advection-inversion. The occurrence of offshore low-altitude atmospheric ducts is 20–60% [2]. The atmospheric duct has an important impact on radio wave propagation. Figure 1 illustrates the comparison of the distribution of electromagnetic wave propagation loss in a standard atmosphere and surface ducts. From the diagram, when the surface duct appears, the distribution of electromagnetic wave propagation loss changes significantly. Electromagnetic waves can propagate beyond the visual range with small propagation loss, an event called the over-the-horizon phenomenon.

The atmospheric duct will cause the radar system to produce the detection blind area, the clutter echo enhancement, the target-positioning error increase, and other adverse effects, thereby affecting its performance [3]. Therefore, it is important to obtain the atmospheric duct parameters when evaluating the radar performance.

The acquisition methods of atmospheric duct parameters include direct detection and remote-sensing inversion. The direct detection method uses radiosondes or rocketsondes to measure the atmospheric duct parameters, though it is expensive and difficult to operate. However, remote-sensing inversion has a high spatial-temporal resolution and has gained great attention recently. Ground-based Global Navigation Satellite System (GNSS) occultation signal is one of the signal sources used for atmospheric remote-sensing [4,5]. Zuffada [6] realized that the use of ground based occultation signal bending angles laid a theoretical foundation for the inversion of the atmospheric duct. Wang [7] proposed a method of retrieving atmospheric duct parameters using a ground-based GNSS occultation signal and carried out experimental verification. Due to the fixed number of GNSS, the number of occultation events received every day was limited (about 100 times) [7], which leads to atmospheric ducts often being missed.

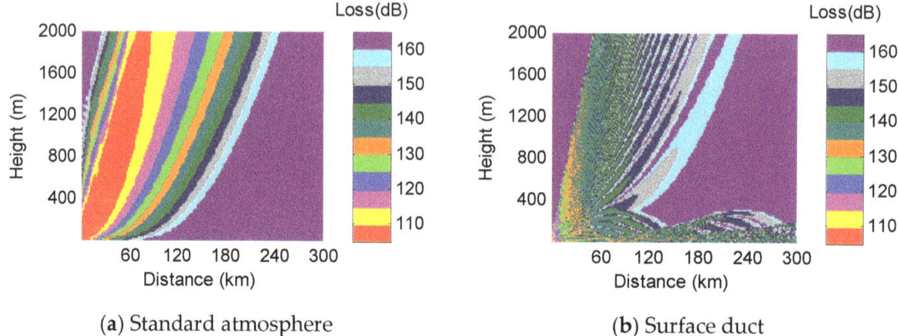

(**a**) Standard atmosphere (**b**) Surface duct

Figure 1. Distribution diagram of radio wave propagation loss.

AIS system is a navigation aid system applied in maritime safety and communication between ships and shore, and between ships [8,9]. The International Maritime Organization stipulates that AIS systems should be installed on all international sailing ships of 300 tons and above, and all non-international sailing ships of 500 gross tons and above. Therefore, in offshore waters, there is a large amount of widely-distributed AIS information. E.R. Bruin [10] analyzed the influence of different atmospheric duct environments on AIS signals. Atmospheric ducts can increase the propagation distance of AIS signals. Zhang [11] discussed the propagation characteristics of AIS signals in different atmospheric duct environments and demonstrated that the low-altitude atmospheric duct (especially the surface duct) had a significant influence on AIS signals at sea. From previous observations, the AIS signal is affected by the atmospheric duct in the process of propagation and can be used to invert atmospheric duct parameters.

The inversion algorithm is an important aspect in the field of atmospheric duct remote sensing. The common inversion algorithm used for remote-sensing of atmospheric ducts was the global optimization algorithm, such as the genetic algorithm and particle swarm optimization [12]. Gerstoft [12] proposed a method for inverting atmospheric duct parameters using sea surface echo from the genetic algorithm called refraction-from-cluster (RFC) technology. In 2007, Yardim [13] proposed a GA-MC hybrid algorithm, which can ensure the inversion accuracy and improve the inversion speed. With the development of artificial intelligence technology, the deep-learning theory has been applied to the inversion of atmospheric duct parameters. Guo [14] outlined a method of inverting atmospheric duct parameters using deep-learning network and sea clutter that greatly improved the inversion speed of atmospheric duct parameters. Han [15] illustrated a method to predict the height of the evaporation duct using a recurrent neural network. Hilesit [16] demonstrated a method to characterize the parameters of the evaporation duct in the ocean boundary layer based on an artificial neural network. Han [17] outlined a cooperative inversion model of atmospheric duct parameters using ground-based GNSS occultation signals and a deep-learning network and established a weight loss function construction method. Tepecik [18] demonstrated an atmospheric duct inversion method using a genetic algorithm and deep learning.

Wang [7] and Gerstoft [12] adopted a one-step inversion strategy and only one model is established to judge the type of atmospheric duct and invert the parameters of atmospheric duct. Hilaire [19] showed a two-step inversion strategy: the classification of atmospheric duct types, and the inversion of atmospheric duct parameters. This effectively improved the inversion accuracy of atmospheric duct parameters.

From previous findings, we adopted a classifying-inversion model of atmospheric duct parameters based on AIS signals including two parts: classification of duct type and inversion of duct parameters. Before the inversion of atmospheric duct parameters, the types of atmospheric ducts were classified and judged. This model has higher inversion accuracy than that of the traditional method.

The content of this manuscript is arranged as follows: In Section 2, using the AIS signal simulation algorithm and data of the AIS signal power, we deduced the influence of different atmospheric duct types on AIS signal power distribution. Section 3 introduces the modeling methods of the classifying-inversion model, including the modeling methods of the atmospheric duct classification model and duct parameters inversion model. Section 4 illustrates the analysis of test results. The conclusions are presented in Section 5.

2. The Effect of the Atmospheric Duct on the AIS Signal

The atmospheric duct includes evaporation, surface, and elevated ducts. We focused on the effect of elevated and surface ducts on AIS signals as the evaporation duct has almost no influence on AIS signals [20]. In this part, AIS power simulation and measurement data were used to analyze the effect of different atmospheric duct types on AIS signals, necessary for modeling the atmospheric duct classifying-inversion model.

2.1. Atmospheric Duct Model

The atmospheric duct structure is described by a modified refractive index that varies with height. When the modified refractive index has a negative gradient, the atmospheric duct phenomenon appears [21].

$$M = N + \frac{h}{r_e} \times 10^6 \tag{1}$$

$$N = \frac{77.6}{T} \times \left(P + \frac{4810e}{T}\right) \tag{2}$$

where r_e, P, T, and e are the average Earth radius, the atmospheric pressure, absolute temperature, and water vapor partial pressure at height h from the ground. The units for P, T, and e are kPa, K and kPa.

The surface duct model is a two-parameter model:

$$M(z) = \begin{cases} M_0 - \frac{M_d}{z_t}z & 0 \leq z \leq z_t \\ M_0 - \frac{M_d}{z_t}z + 0.118z & z \geq z_t \end{cases} \tag{3}$$

where z_t is surface duct height, M_d is surface duct strength, and M_0 is the modified refractive index of surface or sea surface. The elevated duct model is a four-parameter model [22] expressed in Equation (4) as shown below:

$$M(z) = M_0 + \begin{cases} kz & 0 \leq z \leq z_b \\ kz_b - \frac{M_d}{z_t}(z - z_b) & z_b \leq z \leq z_b + z_t \\ kz_b - M_d + 0.118(z - z_b - z_t) & z \geq z_b + z_t \end{cases} \tag{4}$$

where k is the foundation layer slope, $M(z)$ is the modified refractive index at height z, z_b is the trapped layer bottom height, z_t is the trapped layer thickness, and M_d is elevated duct strength. The structure diagram of the surface duct and elevated duct is shown in Figure 2.

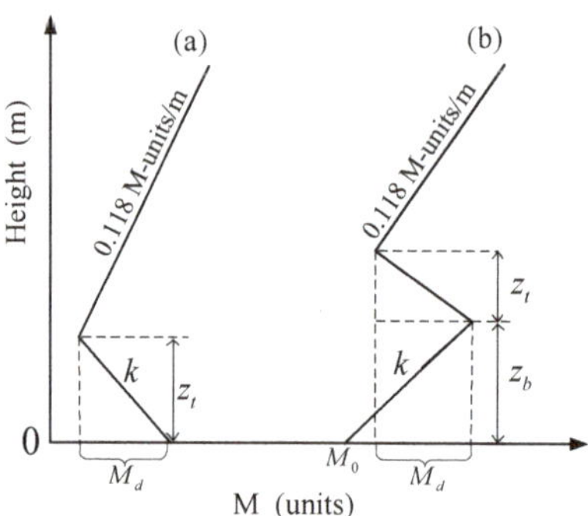

Figure 2. Elevated duct and surface duct structure diagram ((**a**) is the elevated duct structure diagram; (**b**) is the surface duct structure diagram).

2.2. AIS Signal Power Simulation

The AIS signal power calculation formula is shown in Equation (5):

$$P = P_t + G_t + G_r - L_1 - L \tag{5}$$

where P_t is the transmission power of the AIS system, which is 41 dBm. G_t is the transmit antenna gain, G_r is the receiving antenna gain, L is the propagation loss of AIS in an atmospheric environment, and L_1 is the cable transmission loss of AIS receiving equipment.

The propagation loss of the AIS signal in an atmospheric environment was obtained using the parabolic equation method [23]. Parabolic equations are divided into the narrow-angle parabolic equation and wide-angle parabolic equation. We employed the narrow-angle parabolic equation, suitable for the calculation of radio wave propagation with an elevation angle of less than 10 degrees. The expression of the narrow-angle parabolic equation is shown in Equation (6).

$$\frac{\partial^2 u(x,z)}{\partial z^2} + 2ik\frac{\partial u(x,z)}{\partial x} + k^2(n^2(x,z) - 1)u(x,z) = 0 \tag{6}$$

where $u(x,z)$ is the component of the electric or magnetic field, k_0 is the wave number, and $n(x,z)$ is the atmospheric refractive index at different distances and heights. The Split-Step Fourier Transform (SSFT) method is the main method for solving parabolic equations [24]. The SSFT solution of the narrow-angle parabolic equation is shown in Equation (7) [23].

$$u(x + \Delta x, z) = e^{\frac{ik(n^2-1)\Delta x}{2}}\Im^{-1}\left\{e^{\frac{-i\pi^2 p^2 \Delta x}{2k}}\Im u(x,z)\right\} \tag{7}$$

where α_e is the radius of the Earth, p is the transform domain variable, $\Im$ and $\Im^{-1}$ are Fourier transform and inverse transform respectively. The equation of AIS signal path propagation loss obtained from Equation (8) is:

$$L = 20\lg f + 10\lg r - 20\lg|u(x,z)| - 27.6 \tag{8}$$

where L is propagation loss, f is AIS signal frequency, and r is the propagation distance.

2.3. AIS Signal Receiving Test

In June 2020, the China Research Institute of Radiowave Propagation carried out an AIS signal receiving test in the coastal area of Nantong, China. AIS signal-receiving equipment are often used to receive AIS signals in coastal areas and to collect meteorological sounding data in the test area. The sounding data were obtained twice a day at 08:00 and 20:00 Beijing time respectively. The test area is shown in Figure 3. Parameters of AIS signal-receiving equipment are shown in Table 1.

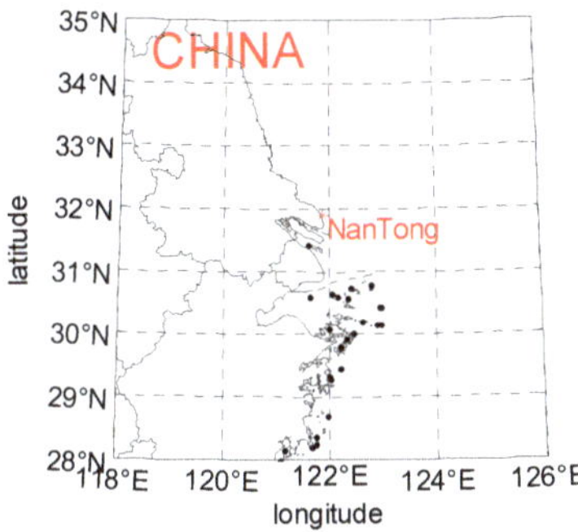

Figure 3. AIS signal receiving test position.

Table 1. Parameters of AIS signal-receiving equipment.

Parameter	Value	Unit
Antenna frequency range	118~164	MHz
Receiving antenna height	25	meter
Receiving antenna gain	2	dB
Cable loss	16	dB

We selected three typical atmospheric environments: no atmospheric duct, surface duct, and elevated duct. The corresponding atmospheric duct profile is illustrated in Figure 4.

AIS signal data were selected at the same time as sounding data, and the signal position and power distribution are shown in Figure 5. The x-axis is the longitude direction distance, the y-axis is the latitude direction distance, and the AIS signal-receiving equipment is located at point 0 of the y-axis. When there is no atmospheric duct, the AIS signal is distributed within 100 km as seen in Figure 5. When the surface duct appeared, the maximum distance of the AIS signal was over 500 km, and the signal power was strong. The signal power beyond 100 km was about −80 dBm. When the elevated duct appeared, the AIS signal was distributed within 200 km, and the signal power was weak (about −100 dBm).

Using sounding data and Equation (5), the AIS signal power variation with distance was determined in three atmospheric environments and was compared with the actual received AIS signal power, as shown in Figure 6. The red dotted line shows the sensitivity of AIS signal-receiving equipment (−112 dBm); the solid green line is the simulated AIS signal power variation curve with distance; the blue points are the distribution of measured AIS signal power with distance. Figure 6 illustrates that the AIS simulation results are in good agreement with the measured data, thereby revealing the effectiveness of the AIS signal power simulation algorithm used in our study.

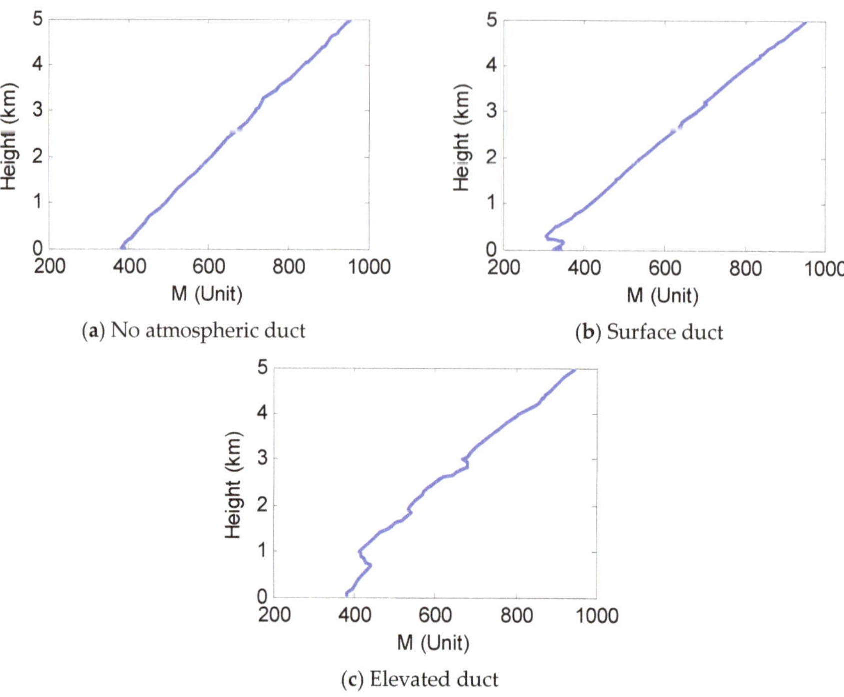

Figure 4. Atmospheric duct profile calculated by sounding data.

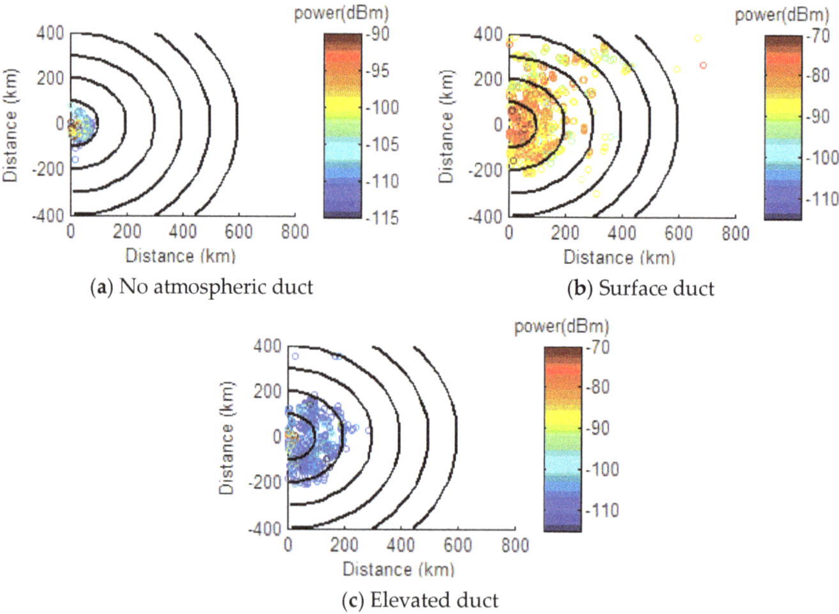

Figure 5. AIS signal distribution in different atmospheric ducts.

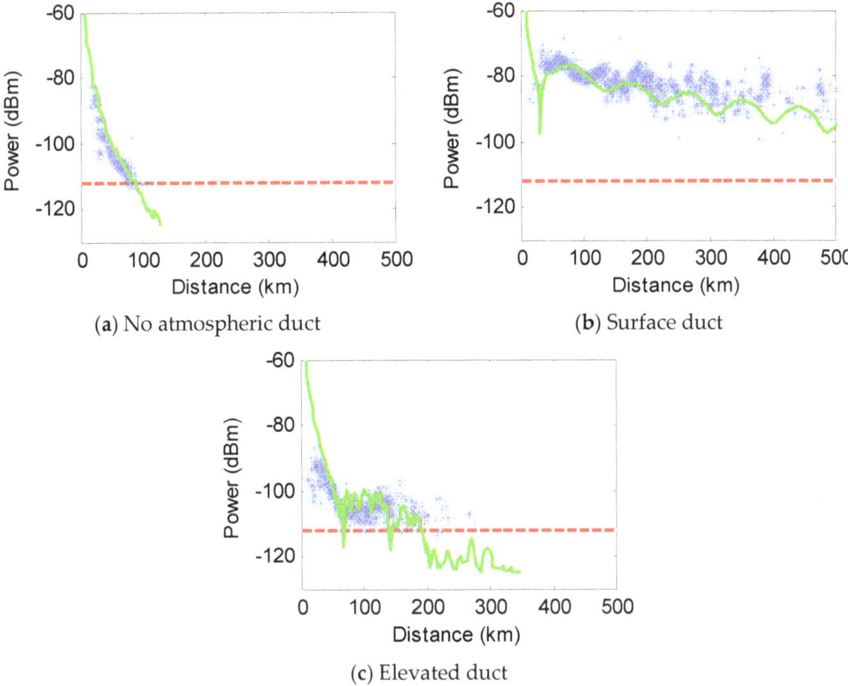

(**a**) No atmospheric duct

(**b**) Surface duct

(**c**) Elevated duct

Figure 6. Comparison between simulated AIS signal power and measured AIS signal power.

From the above analysis, we observed obvious differences in AIS signal distribution in different atmospheric environments, mainly as follows:

(1) The maximum distances of signals that can be received were different. Without the atmospheric duct, the maximum distance was about 80 km; when the surface duct appeared, AIS signals beyond 500 km were received; when the elevated duct appeared, the maximum distance was 200 km.

(2) The signal strength was different. In the surface duct environment, the signal power was strong and was approximately −80 dBm within 100 km. The signal strength in the elevated duct environment was weak and was within −110 dBm within 100 km.

These show that AIS signals can be used to invert atmospheric ducts, and the types of atmospheric ducts can be distinguished since surface and elevated duct have different influences on AIS.

3. Modeling of Duct Parameters Classifying-Inversion Model

In this section, we introduced two artificial intelligence methods: genetic algorithm (GA) and DNN, as well as the modeling process of the classifying-inversion model of atmospheric duct parameters using AIS data.

3.1. Artificial Intelligence Method for Atmospheric Duct Inversion

From previous studies, the main artificial intelligence methods used for atmospheric duct parameter inversion were GA and DNN. DNN is a deep learning network structure. GA is designed according to the evolution law of organisms in nature, and the optimal solution is searched by simulating the natural evolution process. In this algorithm, the problem-solving process is transformed into the evolutionary process of biological chromosome genes through mathematical means and computer simulation. GA has been widely used in combinatorial optimization, machine-learning, signal processing, adaptive control, and artificial life [25]. GA consists of three steps: selection, crossover, and mutation. The GA flow chart is shown in Figure 7.

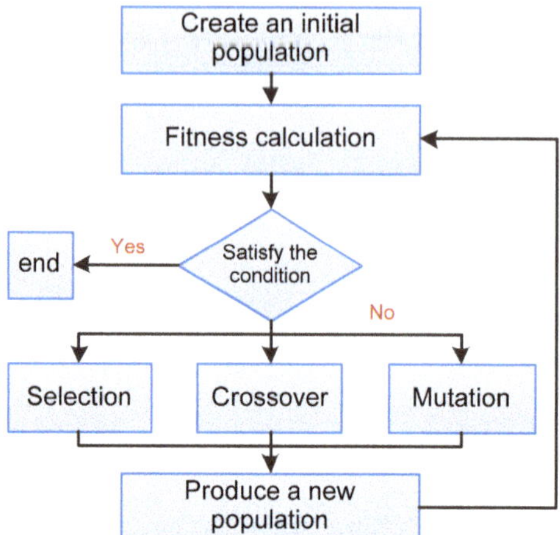

Figure 7. GA flow chart.

DNN is composed of input, hidden, and output layers. The hidden layer can have multiple layers that enhance the expression ability of DNN. The neurons in the output layer have multiple outputs that flexibly apply to classification regression, dimensionality reduction, and clustering. The schematic diagram of DNN is shown in Figure 8. The DNN layer is fully connected with the other layers, and any neuron in the i layer must be connected with a neuron in the $i+1$ layer.

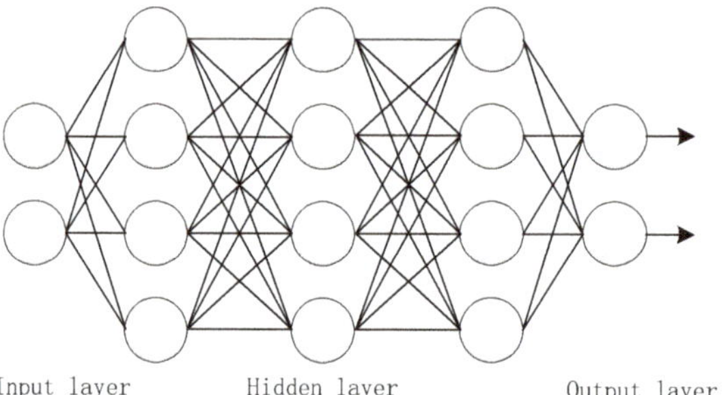

Figure 8. The schematic diagram of DNN.

3.2. Classifying-Inversion Flow of Atmospheric Duct

In this study, we employed the idea of "classification before inversion" for atmospheric duct parameters inversion. The first step was to establish a classification model of atmospheric duct, use the received AIS signal to judge the occurrence of atmospheric ducts, and distinguish the types of atmospheric duct occurrence. Secondly, the surface duct parameter inversion model and the elevated duct parameter inversion model were established respectively. The flow chart of atmospheric duct Classifying-inversion model is shown in Figure 9.

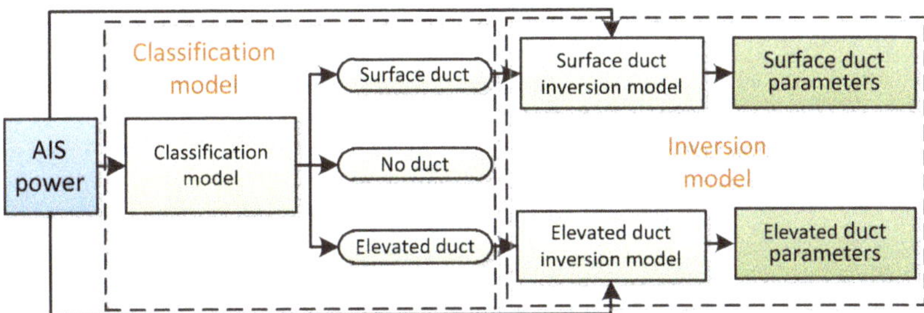

Figure 9. Flow chart of atmospheric duct Classifying-inversion model.

The atmospheric duct classification model adopted DNN, and the atmospheric duct parameters inversion model adopted GA and DNN respectively, and two classifying-inversion models were established as Model-1 and Model-2. In addition, we used GA to establish the traditional atmospheric duct inversion model (Model-3) and compared it with the aforementioned models. The model information is illustrated in Table 2.

Table 2. The model information.

Model	Type	Algorithm Combination (Classify-Inversion)
Model-1	Proposed model	DNN-DNN
Model-2	Proposed model	DNN-GA
Model-3	Traditional model	GA

3.3. Atmospheric Duct Classification Model

The atmospheric duct classification model adopted the DNN of three hidden layers, and the nodes of each hidden layer were 512, 256 and 32. The input layer data involved the AIS signal power data, and the output layer data involved the atmospheric duct type data. The type of data consisted of three numbers, and the format and value are shown in Table 3.

Table 3. Type data format.

Duct Type	First Number	Second Number	Third Number
No Duct	1	0	0
Surface Duct	0	1	0
Elevated duct	0	0	1

The hidden layer activation function is tanh, and the expression is:

$$\tanh(x) = \frac{1 + e^{-2x}}{1 - e^{-2x}} \tag{9}$$

The output layer activation function is softmax, which is one of the most common activation functions. The expression is:

$$S(x) = \frac{1}{1 + e^{-x}} \tag{10}$$

The Stochastic Gradient Descent Method was selected for optimization; it splits the dataset into batches and randomly selects a batch to calculate and update the parameters.

3.4. Atmospheric Duct Parameters Inversion Model

3.4.1. Solution Based on DNN

The atmospheric duct parameter inversion model based on DNN consists of four hidden layers, and the nodes of each hidden layer were 1024, 512, 256 and 32. The input data was the AIS signal power data, and the output data was the surface or elevated duct parameters. Surface duct parameters consisted of duct height and strength, and elevated duct parameters consisted of atmospheric duct top height, the slope of the base layer, and duct layer thickness and strength. The hidden layer activation function was Rectified Linear Unit (ReLU), and the expression is:

$$f(x) = \max(x, 0) \tag{11}$$

The adaptive moment estimation method is used for optimization, and can dynamically adjust the learning rate in the process of training to adapt to different weight parameters and achieve better optimization results.

3.4.2. Solution Based on GA

The steps of atmospheric duct parameters inversion based on GA were as follows:

(1) AIS power data processing, using the actual received AIS signal power data, and the power sequence P^{obs} obtained through median filtering.
(2) Determine the search range of atmospheric duct parameters as shown in Table 4.
(3) AIS signal power forward simulation. From Table 4, the atmospheric duct parameters are initialized, and the simulated power sequence P^{sim} corresponding to each profile was calculated using Equations (3)–(5).
(4) Objective function. The objective function was used to evaluate the coincidence between AIS measured power and AIS simulated power. It adopted the following format:

$$\phi(m) = e^T e \tag{12}$$

$$e = P^{obs} - P^{sin} - \hat{T} \tag{13}$$

$$\hat{T} = \overline{P}^{obs} - \overline{P}^{sim} \tag{14}$$

where $\overline{P}^{obs}$ and $\overline{P}^{sim}$ are the average values of P^{obs} and P^{sim}, respectively.

(5) Optimize. There is a very complicated non-linear relationship between AIS signal power and atmospheric duct parameters. Once the objective function and model parameter space are determined, the whole inversion problem is transformed into a minimum optimization problem. In this paper, GA is used for iterative optimization to find the optimal solution.

Table 4. The search range of atmospheric duct parameters.

Duct Type	Parameter	Minimum Value	Maximum Value
Elevated duct	Foundation layer slope	0.03	0.19
	Duct layer bottom height	400	2500
	Duct strength	1	80
	Duct layer thickness	50	400
Surface Duct	Duct height	100	1000
	Duct strength	1	80

4. Test and Analysis

4.1. Dataset

Three types of data were used for DNN modeling: AIS signal power data, atmospheric duct parameters data, and type data. Among them, AIS signal power data and type data were used to establish the atmospheric duct classification model. AIS signal power data and atmospheric duct parameters data were used to establish the inversion model of atmospheric duct parameters.

The data were mainly obtained from (1) AIS data receiving test introduced in Section 2.3, including the measured AIS power and atmospheric duct parameters calculated using sounding data; (2) simulation data, including the simulation data of atmospheric duct parameters and AIS power. The sample size was about 5900 groups; 800 groups of no atmospheric duct data, 1300 groups of surface duct data, and 3800 groups of elevated duct data.

We divided the dataset into training, verification, and test sets, among which the training set accounted for 80%, the verification set for 17.5%, and the test set for 2.5%.

4.2. Comparison of Atmospheric Duct Classification Results

The accuracy of the atmospheric duct classification model was compared using the test set data, and the results are shown in Figure 10. In Figure 10, the red asterisk indicated the type of atmospheric duct in the test set (0 indicates no duct; 1 indicates a surface duct; 2 indicates elevated duct), and the blue circle indicated the prediction result of the model. The prediction accuracy of the classification model attained 97%.

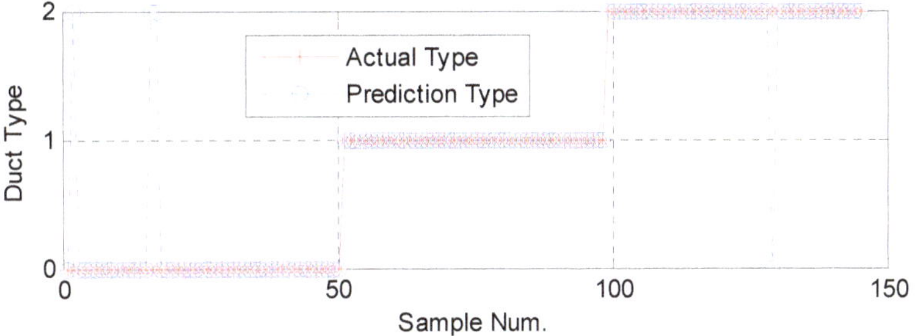

Figure 10. Comparison of atmospheric duct classification results.

From Figure 10, the classification of the surface duct was correct, because when the surface duct appeared, AIS signal power increased. The wrong samples mainly occurred when there was either no duct or an elevated duct. When the elevated duct strength was relatively small, it exerted little influence on the AIS signal, similar to that without an atmospheric duct.

4.3. Comparison of Inversion Results of Surface Duct Parameters

We selected three surface duct samples and used Model-1, Model-2, and Model-3 to invert the surface duct parameters. The inversion results are shown in Table 5. The bold figures in the table are the inversion results with the smallest error from the true value.

From Table 5, the inversion results of the atmospheric duct inversion models (Model-1 and Model-2) established using the classifying-inversion idea proposed in our study are much closer to the true values than those of the traditional inversion model (Model-3). Especially for Sample 1 and Sample 2, the inversion results of the traditional model are quite different from the true values.

Table 5. Inversion results of surface duct parameters.

Sample	Model	Duct Height (m)	Duct Strength (M)
1	True value	305	41
	Model-1	**362**	29
	Model-2	413	**32**
	Model-3	115	73
2	True value	368	28
	Model-1	**429**	**39**
	Model-2	446	46
	Model-3	469	55
3	True value	302	47
	Model-1	383	**48**
	Model-2	**281**	42
	Model-3	125	64

The bold are the closest to the true value

4.4. Comparison of Inversion Results of Elevated Duct Parameters

We selected three elevated duct samples and used Model-1, Model-2, and Model-3 to invert the parameters. The inversion results are shown in Table 6. The bold part in the table is the inversion result with the smallest error from the true value.

Table 6. Inversion results of elevated duct parameters.

Sample	Model	Foundation Layer Slope	Duct Layer Bottom Height	Duct Layer Thickness	Duct Strength
1	True value	0.055	725	113	8.8
	Model-1	**0.094**	**704**	154	**9.1**
	Model-2	0.12	560	**149**	35.0
	Model-3	0.17	279	76	45
2	True value	0.108	845	163	28.1
	Model-1	**0.105**	**949**	**109**	**32.2**
	Model-2	0.11	576	77	18.0
	Model-3	0.13	325	89	46
3	True value	0.038	632	48	9.7
	Model-1	0.091	**625**	133	**9.6**
	Model-2	**0.085**	974	**70**	44
	Model-3	0.14	152	83	42

The bold are the closest to the true value

Table 6 illustrates that the inversion results of the traditional method (Model-3) have a big deviation from the true value, similar to the surface duct inversion result. In the model established in this paper, the error between the result of Model-1 inversion and the real value was smaller. For example, consider Sample 2 with a comparison diagram of the atmospheric duct profile illustrated in Figure 11. The red line is the true value calculated by sounding data, the blue line is the inversion result by Model-1, and the green line is the inversion result by Model-2. Model-1 inversion of atmospheric duct layer height was consistent with the true value, while Model-2 inversion of atmospheric duct layer height was lower than the true value.

Figure 12 is a comparison diagram of AIS signal power. The black asterisks are the measured AIS signals' power, the red line is the AIS signal power distribution determined from the sounding data, the blue line is the AIS signal power distribution obtained using Model-1 inversion results, and the green line is the AIS signal power distribution determined using Model-2 inversion results. We observed that the AIS signal power distribution obtained using Model-1 was closer to the actual AIS signal power distribution than Model-2 in the range of 80–200 km (the range affected by atmospheric ducts).

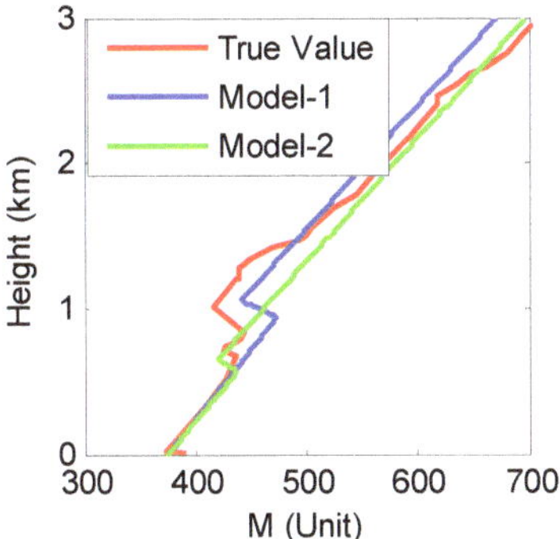

Figure 11. Comparison of atmospheric duct profiles.

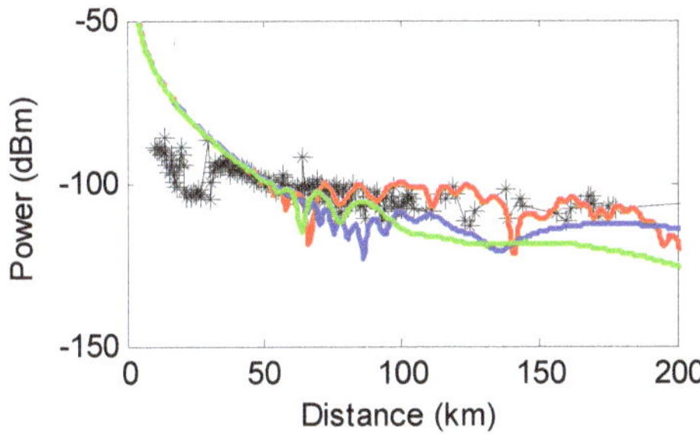

Figure 12. Comparison diagram of AIS signal power.

5. Conclusions

In the present study, we designed an inversion model of an offshore atmospheric duct using an AIS signal. The model adopted the idea of classification before inversion: Firstly, the classification model of the atmospheric duct was established to judge the type of atmospheric duct. Secondly, the inversion models of the surface and elevated duct parameters were established, and the atmospheric duct parameters were obtained. In this study, the atmospheric duct classification model was established using DNN, and the duct parameters inversion model was established using DNN and GA respectively. Through experimental comparison, the inversion result of atmospheric duct parameters based on the model in our study was better than that of the traditional model. The atmospheric duct classification model in our study attained a classification accuracy of 97%. Comparing the inversion results of the atmospheric duct parameter inversion model established by DNN and GA, we found that the DNN method is more advantageous than the GA method in inverting the elevated duct. Thus, more AIS measured power should be included in the data set, and the model parameters must be optimized in future studies, to improve the inversion accuracy of atmospheric duct parameters.

Author Contributions: Conceptualization, J.W. and J.H.; Methodology, J.H.; Software, J.H., Q.Z. and L.Z.; Validation, J.H., J.W. and H.W.; Formal analysis, J.H.; Investigation, J.H.; Resources, J.H. and H.Z.; Data curation C.Z., S.Z. and L.Z.; Writing—original draft preparation, J.H.; Writing—review and editing, J.W., H.W. and H.Z.; Visualization, H.Z.; Supervision, J.W.; Project administration, J.W.; Funding acquisition, H.W. All authors have read and agreed to the published version of the manuscript.

Funding: This research was funded by the National Natural Science Foundation of China (grant number 42076195).

Institutional Review Board Statement: Not applicable.

Informed Consent Statement: Not applicable.

Data Availability Statement: Not applicable.

Conflicts of Interest: The authors declare that they have no conflict of interest.

References

1. Babin, S.M.; Young, G.S.; Carton, J.A. A New Model of the Oceanic Evaporation Duct. *J. Appl. Meteorol.* **1997**, *36*, 193–204. [CrossRef]
2. Hao, X.J.; LI, Q.L.; Huo, L.X.; Lin, L.K. Digital Maps of Atmospheric Refractivity and Atmospheric Ducts Based on a Meteorological Observation Datasets. *IEEE Trans. Antennas Propag.* **2022**, *70*, 2873–2883. [CrossRef]
3. Skolnik, M. Electrical Engineering Series. In *Introduction to Radar Systems*; McGraw-Hill: New York, NY, USA, 1980.
4. Anderson, K.D. Tropospheric Refractivity Profiles Inferred from Low Elevation Angle Measurements of Global Positioning System (GPS) Signals. In Proceedings of the of AGARD Conference, Propagation Assessment in Coastal Environments, Bremerhaven, Germany, 19–22 September 1994; pp. 1–6.
5. Hajj, G.A.; Kursinski, E.R.; Romans, L.J.; Bertiger, W.I.; Leroy, S.S. A Technical Description of Atmospheric Sounding by GPS Occultation. *J. Atmos. Solar-Terr. Phys.* **2002**, *64*, 451–469. [CrossRef]
6. Zuffada, C.; Hajj, G.A.; Kursinski, E.R. A Novel Approach to Atmospheric Profiling with a Mountain-Based or Airborne GPS Receiver. *J. Geophys. Res.* **1999**, *104*, 24435–24447. [CrossRef]
7. Wang, H.G.; Wu, Z.S.; Kang, S.F.; Zhao, Z.W. Monitoring the Marine Atmospheric Refractivity Profiles by Ground-Based GPS Occultation. *IEEE Geosci. Remote Sens.* **2013**, *10*, 962–965. [CrossRef]
8. *ITU-R M.1371-5*; Technical Characteristics for an Automatic Identification System Using Time-Division Multiple Access in the VHF Maritime Mobile Frequency Band. International Telecommunication Union (ITU): Geneva, Switzerland, 2014.
9. *ITU-R M.2123*; Long Range Detection of Automatic Identification System (AIS) Messages Under Various Tropospheric Propagation Conditions. International Telecommunication Union (ITU): Geneva, Switzerland, 2007.
10. Bruin, E.R. Modelling the Impact of North Sea Weather Conditions on the Performance of AIS and Coastal Radar Systems. Master's Thesis, Utrecht University, Utrecht, The Netherlands, 2016.
11. Zhang, L.J.; Wang, H.G.; Li, J.R. Experimental Analysis of Low Atmospheric Duct Monitoring Based on AIS Signal. *Chin. J. Radio Sci.* **2022**. (In Chinese) [CrossRef]
12. Gerstoft, P.; Rogers, L.T.; Krolik, J.L.; Hodgkiss, W.S. Inversion for Refractivity Parameters from Radar Sea Clutter. *Radio Sci.* **2003**, *38*. [CrossRef]
13. Yardim, C.; Gerstoft, P.; Hodgkiss, W.S. Statistical Maritime Radar Duct Estimation Using a Hybrid Genetic Algorithm-Markov Chain Monte Carlo Method. *Radio Sci.* **2007**, *42*, 1–15. [CrossRef]
14. Guo, X.W.; Wu, J.J.; Zhang, J.P.; Han, J. Deep Learning for Solving Inversion Problem of Atmospheric Refractivity Estimation. *Sustain. Cities Soc.* **2018**, *43*, 524–531. [CrossRef]
15. Han, J.; Wu, J.J.; Zhu, Q.L.; Wang, H.G.; Zhou, Y.F.; Jiang, M.B.; Zhang, S.B.; Wang, B. Evaporation Duct Height Nowcasting in China's Yellow Sea Based on Deep Learning. *Remote Sens.* **2021**, *13*, 1577. [CrossRef]
16. Sit, H.; Earls, C.J. Characterizing Evaporation Ducts Within the Marine Atmospheric Boundary Layer Using Artificial Neural Networks. *Radio Sci.* **2019**, *54*, 1181–1191. [CrossRef]
17. Han, J.; Wu, J.J.; Wang, H.G.; Zhu, Q.L.; Zhang, L.J.; Zhang, C.; Wang, Q.N.; Zhao, H. Weight Loss Function for the Cooperative Inversion of Atmospheric Duct Parameters. *Atmosphere* **2022**, *13*, 338. [CrossRef]
18. Tepecik, C.; Navruz, I. A Novel Hybrid Model for Inversion Problem of Atmospheric Refractivity Estimation. *Int. J. Electron. Commun.* **2018**, *84*, 258–264. [CrossRef]
19. Hs, A.; Cjea, B. Deep Learning for Classifying and Characterizing Atmospheric Ducting Within the Maritime Setting. *Comput. Geosci.* **2021**, *157*, 104919.
20. Tang, W.C.; Wei, H.M. A Study on the Propagation Characteristics of AIS Signals in the Evaporation Duct Environment. *Appl. Comp. Electromagn. Soc. J.* **2019**, *34*, 996–1001.
21. Liu, C.G. *Research on Evaporation Duct Propagation and Its Applications*; Xidian University: Xi'an, China, 2003.
22. Karimian, A.; Yardim, C.; Gerstoft, P.; Hodgkiss, W.S.; Barrios, A.E. Refractivity Estimation from Sea Clutter: An Invited Review. *Radio Sci.* **2011**, *46*, 1–16. [CrossRef]
23. Dockery, G.D. Modeling Electromagnetic Wave Propagation in the Troposhere Using the Parabolic Equation. *IEEE Trans. Antennas Propag.* **1988**, *36*, 1464–1470. [CrossRef]
24. Levy, M.F. *Parabolic Equation Methods for Electromagnetic Wave Propagation*; The Institution of Electrical Engineers: London, UK, 2000.
25. Coley, D.A. *An Introduction to Genetic Algorithms for Scientists and Engineers*; World Scientific Publishing: Singapore, 1999.

remote sensing

MDPI

Article

RBFA-Net: A Rotated Balanced Feature-Aligned Network for Rotated SAR Ship Detection and Classification

Zikang Shao [1], Xiaoling Zhang [1,*], Tianwen Zhang [1], Xiaowo Xu [1] and Tianjiao Zeng [2]

[1] School of Information and Communication Engineering, University of Electronic Science and Technology of China, Chengdu 611731, China; zkshao@std.uestc.edu.cn (Z.S.); twzhang@std.uestc.edu.cn (T.Z.); xuxiaowo@std.uestc.edu.cn (X.X.)

[2] School of Aeronautics and Astronautics, University of Electronic Science and Technology of China, Chengdu 611731, China; tzeng@uestc.edu.cn

* Correspondence: xlzhang@uestc.edu.cn

Abstract: Ship detection with rotated bounding boxes in synthetic aperture radar (SAR) images is now a hot spot. However, there are still some obstacles, such as multi-scale ships, misalignment between rotated anchors and features, and the opposite requirements for spatial sensitivity of regression tasks and classification tasks. In order to solve these problems, we propose a rotated balanced feature-aligned network (RBFA-Net) where three targeted networks are designed. They are, respectively, a balanced attention feature pyramid network (BAFPN), an anchor-guided feature alignment network (AFAN) and a rotational detection network (RDN). BAFPN is an improved FPN, with attention module for fusing and enhancing multi-level features, by which we can decrease the negative impact of multi-scale ship feature differences. In AFAN, we adopt an alignment convolution layer to adaptively align the convolution features according to rotated anchor boxes for solving the misalignment problem. In RDN, we propose a task decoupling module (TDM) to adjust the feature maps, respectively, for solving the conflict between the regression task and classification task. In addition, we adopt a balanced L1 loss to balance the classification loss and regression loss. Based on the SAR rotation ship detection dataset, we conduct extensive ablation experiments and compare our RBFA-Net with eight other state-of-the-art rotated detection networks. The experiment results show that among the eight state-of-the-art rotated detection networks, RBFA-Net makes a 7.19% improvement with mean average precision compared to the second-best network.

Keywords: synthetic aperture radar (SAR); ship detection and classification; rotated bounding box; deep learning (DL); attention; feature alignment

Citation: Shao, Z.; Zhang, X.; Zhang, T.; Xu, X.; Zeng, T. RBFA-Net: A Rotated Balanced Feature-Aligned Network for Rotated SAR Ship Detection and Classification. *Remote Sens.* **2022**, *14*, 3345. https://doi.org/10.3390/rs14143345

Academic Editor: Gwanggil Jeon

Received: 14 May 2022
Accepted: 28 June 2022
Published: 11 July 2022

Publisher's Note: MDPI stays neutral with regard to jurisdictional claims in published maps and institutional affiliations.

1. Introduction

Synthetic aperture radar (SAR) has the ability to work all day and in all weathers, so it has a wide and important application in marine ship monitoring [1,2]. As a basic maritime task, SAR ship detection is of great significance to marine transportation department, fishery department and national defense department. In maritime traffic control, we need to accurately identify the location and category information of the target ship, so that the traffic management department can reasonably mobilize the ship route. For fisheries management, correctly identifying target ships, such as fishing ships, from SAR images is of great significance for rational management of fishery resources and combating illegal fishing.

Many traditional SAR ship detection methods mainly rely on the manual design of ship features. For example, the constant false alarm rate (CFAR) [3] estimates the statistical data of background clutter, adaptively calculates the detection threshold and maintains a constant false alarm probability. However, the determination of the detection threshold depends on the distribution of sea clutter, which is not robust enough. There are other traditional methods based on super-pixel and transform [4,5], but their algorithms are complex and not robust enough, resulting in limited migration applications. Many traditional algorithms often use limited images for theoretical analysis to define ship features However, these images are difficult for reflecting the characteristics of various ship sizes under different backgrounds. This leads to low detection accuracy under multi-scale scenes

Recently, with the development of deep learning, target detection using convolutional neural network (CNN) has been widely used in many fields. At present, the mainstream object detection methods can be divided into two types: single-stage algorithms [6–13] and two-stage algorithms [14–19]. One-stage algorithms, such as YOLO [6] and RetinaNet [13], use a single convolutional network to directly predict the bounding boxes and corresponding classes. Two-stage algorithms generate candidate regions of interests in the first stage and then perform classification in these regions in the second stage. R-CNN [14] and Faster R-CNN [16] are two typical two-stage algorithms.

SAR ship detection models based on convolutional neural network make up for the defects of traditional methods in many aspects. Compared with traditional model-driven methods, the methods based on deep learning have the advantages of full automation, high speed and strong model migration ability [20]. On the basis of a large amount of data training, the deep-learning method can mine features that cannot be mined by traditional algorithms, so as to better realize SAR ship detection. Thus, many researchers in the SAR ship detection community started to pay attention to CNN-based methods. In terms of SAR datasets, Zhang et al. released the first dataset SSDD for SAR ship detection [21]. Lei et al. released the dataset SRSDD for rotated SAR ship detection [22]. In terms of network structure, Zhang et al. [23] proposed a quad feature pyramid network to extract multiple-scale SAR ship features. Sun et al. [24] focused on reducing computation complexity and proposed a lightweight densely connected sparsely activated detector. Wang et al. [25] used RetinaNet to realize automatic ship detection. Based on Faster R-CNN, Jiao et al. [26] proposed a densely connected multiscale neural network to handle multiscale SAR ship images. So, SAR ship detection based on deep learning has broad development prospects.

Although there has been a lot of research on CNN in SAR ship detection, we still face many problems. Firstly, the horizontal bounding boxes cannot fit the oriented ships very well, which results in introducing more background interference [27]. Secondly, for dense ships in SAR images, the densely arranged horizontal bounding boxes have high intersection over union (IoU), which leads to missed inspections after non-maximum suppression (NMS). As a response to these problems, these researchers [28–31] started to use rotated bounding boxes to solve these problems. Figures 1 and 2 show the advantages of using rotated bounding boxes. Many scholars have published research applying the rotated bounding box in the field of SAR ship detection. Based on RetinaNet and rotated bounding boxes, Yang et al. [28] proposed R-RetinaNet. Pan et al. [29] proposed a multi-stage rotational region-based network in order to eliminate close false positive proposals successively. Chen et al. [30] proposed a rotated refined feature alignment detector to balance accuracy and speed.

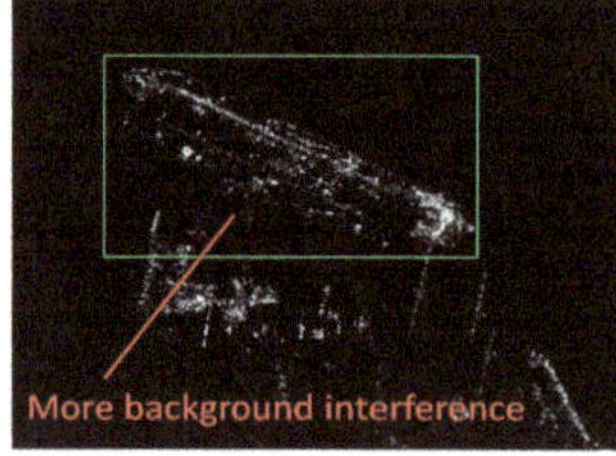

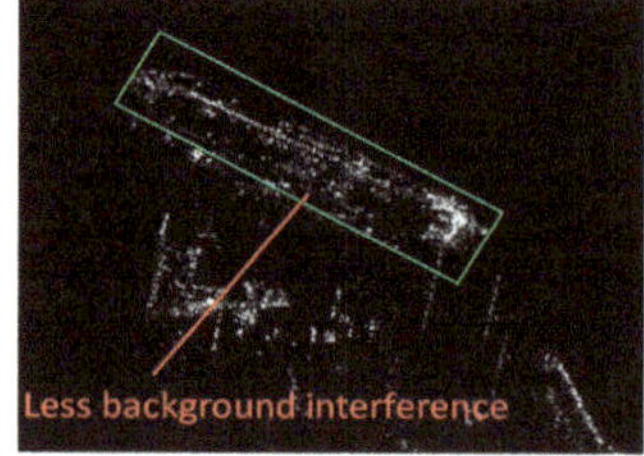

Figure 1. Comparison of horizontal bounding box and rotated bounding box. The left picture shows that horizontal bounding box contains more background interference, while rotated bounding box contains less.

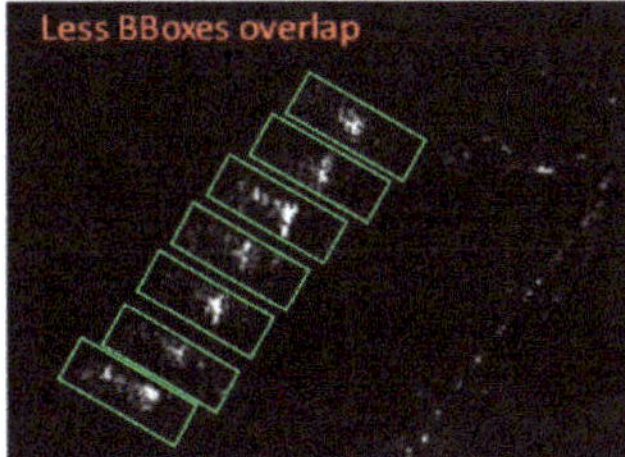

Figure 2. Comparison of horizontal bounding box and rotated bounding box. The left picture shows that horizontal bounding box leads to much overlap, while rotated bounding box does not.

Despite current research on rotated SAR ship detection, there are still some problems to be solved. Firstly, using rotated anchor boxes leads to the dislocation between the rotated anchor boxes and feature maps, which reduces the accuracy of the regression network [31]. Secondly, many researchers [32,33] pay less attention to the huge difference of ship scales in the existing SAR datasets, which is negative for the detection accuracy [23]. Thirdly, some SAR detection models [29] ignore the fact that classification tasks and localization tasks have different requirements for the spatial sensitivity of features [34]. Fourthly, few researchers focus on both SAR ship detection and SAR ship classification. For example, Zhang et al., He et al. and Zeng et al. [35–39] conducted SAR ship classification, but their networks were not able to achieve SAR ship detection.

Therefore, aiming at the above problems, we propose a rotated balanced feature-aligned network (RBFA-Net). The main goal of RBFA-Net is to accurately realize the recognition of SAR ships, that is, the detection and classification of SAR ships. Firstly, RBFA-Net uses the rotated bounding box, which greatly reduces the impact of redundant background noise. Secondly, we improve FPN into a balanced-attention FPN, which can better fuse and enhance multi-scale feature maps. Thirdly, we adopt alignment convolution in AFAM to adaptively align the convolution features according to rotated anchor boxes. Finally, in the rotational detection network (RDN), the input feature maps are adjusted, respectively, for regression task and classification task.

The main contributions are as follows:

1. Balanced rotated feature-aligned network (RBFA-Net) is proposed for SAR ship recognition.
2. A balanced-attention FPN is used for fusing multi-scale features and reducing the negative impact of the imbalance of different scale ships.
3. A rotational detection module is used for fixing the position of ships with rotated bounding boxes and classifying the categories of ships.

The rest of this paper is arranged as follows. Section 2 introduces the methodology. Experiments are described in Section 3. Results and ablation studies are shown in Section 4. Finally, a summary of this paper is put forward in Section 5.

2. Proposed Methods

RBFA-Net is designed on the basis of RetinaNet [13], which consists of FPN and detection subnets. First, we improve the detection subnets with rotated anchors where RetinaNet uses horizontal anchors. Then, we use the BAFPN instead of the FPN originally used by RetinaNet to enhance feature extraction ability. In addition, we add an anchor-guided feature alignment network after FPN to solve the misalignment between the features and rotated anchors. Finally, unlike RetinaNet directly inputting the features into the classification and regression subnets, we add TDN to solve the conflict problem before inputting the features into the classification and regression subnets.

The whole framework of RBFA-Net is divided into three parts: (1) a balanced-attention FPN (BAFPN), (2) an anchor-guided feature alignment network (AFAN) and (3) a rotational detection network (RDN). BAFPN is used for feature extraction, fusion and enhancement. AFAM is used for decreasing the dislocation between the rotated anchor boxes and feature maps. RDN is used for fixing the position of ships with rotated bounding boxes and classifying the categories of ships. The architecture of RBFA-Net is shown in Figure 3.

In this section, we will first introduce BAFPN, and then, we will explain AFAM. Finally, we will introduce RDN in detail. At the end of this section, we will introduce our loss function.

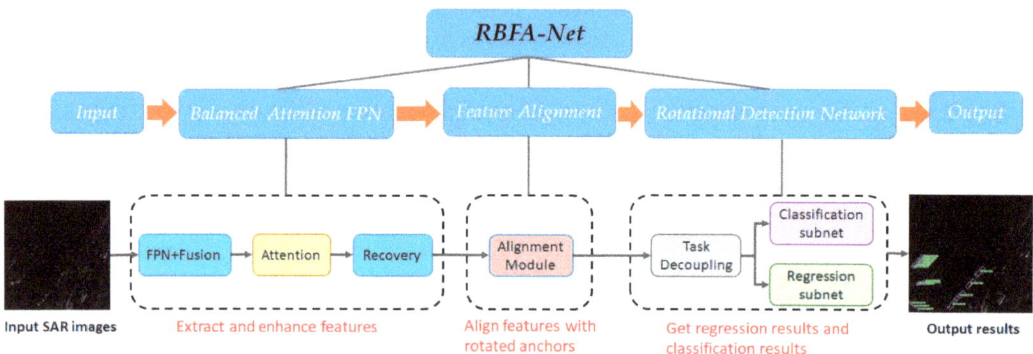

Figure 3. Architecture of RBFA-Net.

2.1. Balanced-Attention FPN

Previous works [40–44] tended to use the feature pyramid network (FPN) [45] as the backbone because the low-level feature maps with higher resolution are suitable for small-scale ship detection, while the high-level feature maps with more semantic information are very suitable for large-scale objects. However, using horizontal connection to integrate multi-level information makes the network pay more attention to the feature maps of adjacent layers, and the semantic information of non-adjacent layers will be diluted many times in the network. In SAR ship datasets [21,46,47], the scale span of the SAR ship target tends to be very large, which makes the ship features exhibit significant difference. In order to solve this problem, inspired by Ref. [48], we use a balanced-attention FPN (BAFPN) to balance and enhance the feature maps of different scales. As shown in Figure 3, BAFPN mainly consists of three parts: (1) feature extraction and fusion, (2) self-attention enhancement module, (3) feature pyramid recovery.

2.1.1. Feature Extraction and Fusion

Considering the feature dilution caused by multi-level horizontal connection structure in traditional FPN, we adopt another way for feature fusion. First, we use ResNet-50 to extract feature maps of different scales. As shown in Figure 4, the extracted feature maps of five levels are denoted by $\{C1, C2, C3, C4, C5\}$. Since $C3$ is in the middle of the pyramid, we recognize that $C3$ can synthesize top semantic information and bottom spatial information better [48]. So, we resize $\{C1, C2, C4, C5\}$ to the $C3$ resolution with up-sampling and max-pooling. The rescaled feature maps are denoted by $\{C1', C2', C3', C4', C5'\}$.

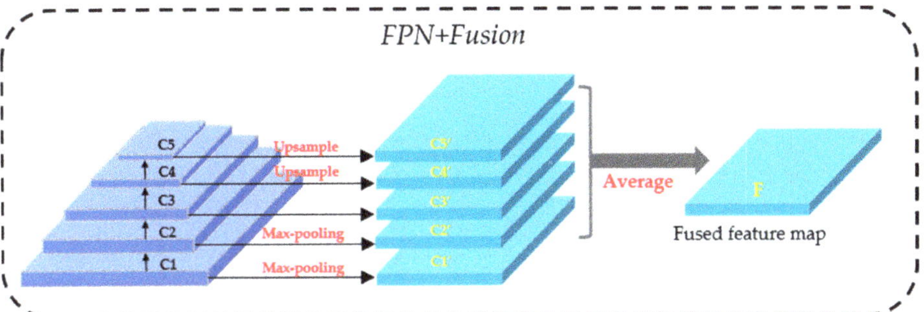

Figure 4. Architecture of feature fusion module.

Then, we fuse these rescaled feature maps by averaging, and the result is executed by

$$F = \sum_{i=1}^{5} C_i'$$ (1)

where i represents the i-th detection.

During averaging, the fused feature map obtains information from all resolutions. By using feature fusion, the impact of different ship scales on detection and classification performance can be reduced.

2.1.2. Attention Module

In order to enhance the global response ability of receptive field and avoid the more complex network structure caused by stacking convolution, we use the non-local [49] module to enhance the fused feature map. Through the module based on a self-attention mechanism, the network can pay more attention to the more important global information. The non-local module establishes the correlation between the global information and the local information without stacking the convolution kernels. This is equivalent to expanding the field of vision, so that the network can better integrate the global information of the image. The expression of the non-local attention formula is as follows:

$$H_i = \frac{1}{C(F)} \sum_{\forall j} f(F_i, F_j) g(F_j)$$ (2)

where F_i represents the i-th location of the input feature map. $C(\cdot)$ represents the normalized coefficient. Function $f(\cdot)$ is used for calculating the similarity between F_i and F_j. Function $g(\cdot)$ is used for calculating the representation of the input feature map at j-th location. Coefficient $\frac{1}{C(F)}$ is used for normalization.

In order to express it concisely, the function $g(\cdot)$ can be regarded as a linear embedding, i.e.,

$$g(F_j) = W_g F_j$$ (3)

where W_g is the weight matrix.

In this module, we use the embedded Gaussian as the function $f(\cdot)$. Embedded Gaussian is a simple extension of Gaussian and calculates the similarity of embedded space.

$$f(F_i, F_j) = e^{\theta(F_i)^T \varphi(F_j)}$$

$$\theta(F_i) = W_\theta F_i$$ (4)

$$\varphi(F_j) = W_\varphi F_j$$

where $\theta(\cdot)$ and $\varphi(\cdot)$ is the weight matrix. W_θ and W_θ are the weight matrices.

In addition, the normalized coefficient $C(\cdot)$ is

$$C(F) = \sum_{\forall j} f(F_i, F_j) \tag{5}$$

Therefore, through the above formula, we can deduce that the output expression is

$$H_i = (\sum_{\forall j} e^{\theta(F_i)^T \varphi(F_j)} \times W_g F_j) / \sum_{\forall j} f(F_i, F_j) \tag{6}$$

Figure 5 shows the architecture of the non-local module. We put the input feature maps F into three 1×1 convolutional layers at the same time to calculate φ, θ and g. Then, we flatten the H and W dimensions of φ and θ. With the flattened layers, we can calculate the similarity f by matrix multiplication. Finally, the similarity f is normalized by a soft-max function, and the result is multiplied by the flattened feature maps g. The output is also processed by a 1×1 convolution layer to make it match the size of the input feature map. Apart from that, a skip connect is added into the architecture. So, the final output refined feature map H' is

$$H' = W_H H + F \tag{7}$$

where W_H is the weight matrix.

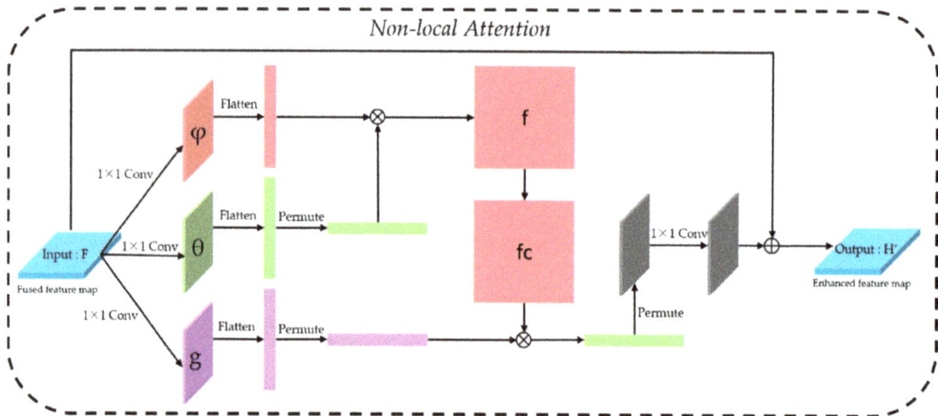

Figure 5. Architecture of non-local module.

For example, for the input feature map F, the following operations are performed in the non-local module:

1. Use 1×1 convolution for down-sampling to obtain three variants: φ, θ and g.
2. φ and θ perform channel merging and transposing, respectively, and then perform matrix multiplication to gain similarity f.
3. The similarity f is normalized by a soft-max function, and then multiply f with the channel-merged g.
4. The obtained results first restore the original size and then restore the number of channels through 1×1 convolution.
5. Finally, add it to the original input F to form a complete residual non-local module.

2.1.3. Feature Pyramid Recovery

As shown in Figure 6, the fused feature maps are restored to a feature pyramid through up-sampling and max-pooling. This part can be regarded as the reverse operation of feature fusion. As shown in Figure 6, the enhanced feature maps H' from the non-local module are restored to a feature pyramid consisting of five levels {$F1$, $F2$, $F3$, $F4$, $F5$}. Among them,

{**F1**, **F2**} are obtained from H' through max-pooling. {**F4**, **F5**} are obtained from H' through up-sampling. Feature maps H' are retained as **F3** in the recovered feature pyramid. In the recovered feature pyramid, each level contains more multi-scale features and balanced information from all resolutions. To conclude, the BAFPN enables RBFA-Net to better focus on multi-scale ship features while integrating global features.

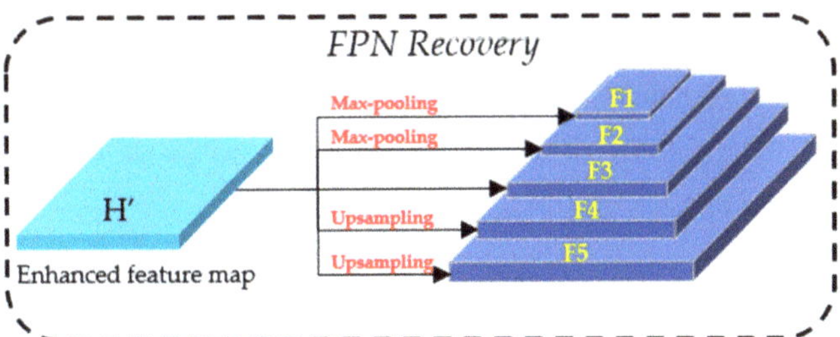

Figure 6. Architecture of feature pyramid recovery.

2.2. Anchor-Guided Feature Alignment Network

In networks using horizontal anchors, the convolution features are aligned with anchors, so the convolution features can reflect the anchor representations [30]. However, in the network using rotating anchor, solving the misalignment between rotated anchors and convolution features is an important problem. To solve this problem, we introduce an anchor-guided feature alignment network. With the guidance of rotated anchors, the feature map will be adjusted to align with rotated anchors in the next rotational detection network. In this section, we will first introduce the basic information about the rotated bounding box and then introduce the anchor-guided feature alignment network.

2.2.1. Introduction to Rotated Bounding Box

In this paper, we use the geometric definition of the rotated bounding box used by MMRotate [50]. The rotated bounding box can be represented by five parameters (x, y, w, h, θ). The two tuples (x, y) represent the location of the center point of the rotated bounding box. The two tuples (w, h) represent the length and width of the rotated bounding box. θ is the rotation angle of the rotated bounding box, and its value range is $[-45°, 135°)$. Figure 7 shows more details about the geometric definition of the rotated bounding box.

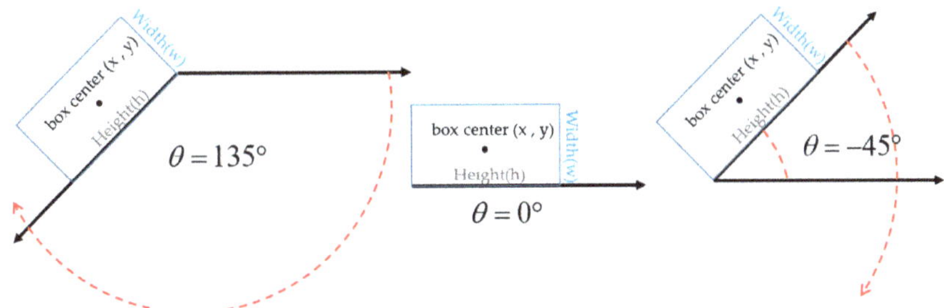

Figure 7. Geometric representation of rotated bounding boxes.

2.2.2. Anchor-Guided Feature Alignment Network

Generally, many existing networks with a rotated bounding box use heuristically defined anchors with different scales and aspect ratios. As a result, these networks tend to suffer from misalignment between the rotated anchor boxes and axis-aligned convolution features. Previous works [30,51] have proved that this misalignment will lead to the decline of detection accuracy. In order to solve this problem, we introduce the alignment convolution layer to establish the anchor-guided feature alignment network (AFAN) to extract the adaptive features according to the predicted shape of the refined anchor.

The core idea of AFAN is that we use the alignment convolution layer to reset the sampling locations according to the roughly generated rotated bounding box. As shown in Figure 8, the extracted feature maps are sent into a rough regression subnet consisting of two convolution layers. The subnet roughly generates rotated bounding boxes in this part. Then, through these roughly generated bounding boxes, we can calculate their offset from the bounding boxes. Finally, these offsets and the original feature maps are put into the alignment convolution layer to reset the sampling locations.

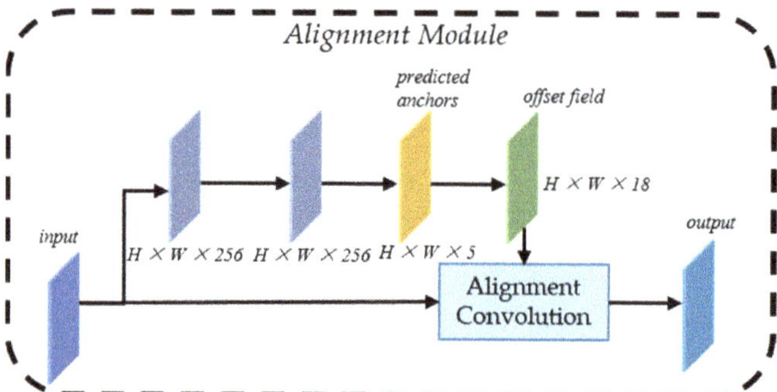

Figure 8. Architecture of anchor-guided feature alignment network (AFAN).

For the feature maps extracted through BAFPN, firstly, they will be input into a regression network to preliminarily locate the anchors and adjust their directions. From this, we obtain the feature maps of $H \times W \times 5$. For each 5-dimensional anchor box, we sample 9 points to obtain the 18-dimension offset field O. The calculation formula of the offset field O is as follows:

$$L_{p_0}^{p_k} = \frac{1}{S}((x,y) + \frac{1}{k}(w,h)p_k)R^T(\theta) \tag{8}$$

$$O = \left\{ L_{p_0}^{p_k} - p_0 - p_k \right\}_{p_k \in R} \tag{9}$$

where $L_{p_0}^{p_k}$ represents the sampling location. S represents the stride of the feature maps. k represents the kernel size. p_0 represents each location on the feature map X_A. p_k represents the p_k-th location in X_A. $R(\theta) = (cos\theta, -sin\theta; sin\theta, cos\theta)^T$ is the rotation matrix. O represents the offset field.

Finally, the offset field and feature maps extracted with BAFPN are input into the alignment convolution layer. By alignment convolution, we align the feature maps with the rotated bounding boxes. These aligned feature maps will be input into RDN for ship target detection and classification.

The alignment convolution is established by adding the offset field o to the ordinary convolution. The alignment convolution can be defined as follows:

$$X_A(p_o) = \sum_{p_k \in R, o \in O} w(p_k)X(p_o + p_k + o) \tag{10}$$

o represents the offset in the offset field O.

2.3. Rotational Detection Network

In this section, we propose a rotational detection network (RDN) to realize ship detection and classification. RDN is designed on the basis of RetinaNet [13], where the feature maps are sent to the regression subnet and classification subnet, respectively. However, the features used for classification should remain invariant, while for the regression task, the features should represent the changes of the target position, size and rotation angle. The opposite requirements bring negative impact on detection accuracy. In order to solve this conflict, we propose a task decoupling module. Recent studies [52] have shown that decoupling the classification task and regression task in the spatial dimension of the feature maps can relieve this conflict. Therefore, we add a task decoupling module consisting of two squeeze-and-excitation (SE) modules [53]. Figure 9 shows the architecture of the rotational detection network.

As shown in Figure 9, a global average pooling layer, a full connection layer and a ReLU activation function form the encoder of the task decoupling module. For the feature maps from AFAN, their dimension is $H \times W \times C$. Firstly, these feature maps are put into the global average pooling layer to extract the global feature information. Then, we send the output feature maps of the global pooling layer to a full connection layer and a ReLU activation function [54]. According to the design of SENet, the size of the output feature map is compressed to $1 \times 1 \times C/8$. Then, the output of the encoder will be input into two decoders composed of a full connection layer and a sigmoid activation function, respectively, and the size will be restored to $1 \times 1 \times C$ in this process. Finally, the feature maps are multiplied by the corresponding elements in the decoder output vector to adjust the feature maps to adapt to different learning tasks. The output feature maps are sent to the classification subnet and regression subnet, respectively.

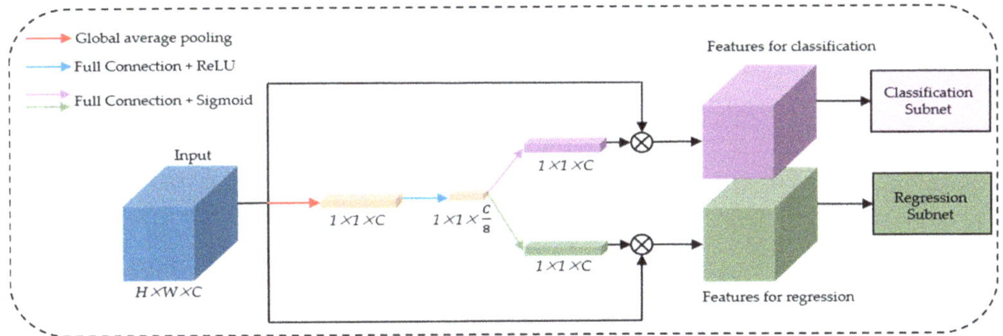

Figure 9. Architecture of rotational detection network (RDN).

2.4. Loss Function

The loss function of RBFA-Net is composed of two parts: the loss of AFAN and the loss of RDN. Each part consists of classification loss and regression loss. The loss of AFAN is similar to that of RDN. Both of them are obtained by adding regression loss and classification loss. The formulae are as follows:

$$L = \frac{1}{N_A} L_{AFAN} + \frac{\lambda}{N_D} L_{RDN}$$

$$L_{AFAN} = \sum_{i=1}^{N} L_{cls}\left(p_i^A, p_i^*\right) + \sum_{i=1}^{N} p_i^* L_{reg}\left(t_i, t_i^*\right) \tag{11}$$

$$L_{RDN} = \sum_{i=1}^{N} L_{cls}\left(p_i, p_i^*\right) + \sum_{i=1}^{N} p_i^* L_{reg}\left(t_i^D, t_i^*\right)$$

where p_i represents the predictive class probability, p_i^* represents the ground-truth category. $p_i^* = 1$ when sample I is a positive one, else $p_i^* = 0$. p_i^A indicates that the prediction probability is obtained by AFAN. λ is a hyper-parameter to balance the loss of the alignment network and that of the detection network. t_i^* represents the offset between the i-th sample and ground truth. t_i represents the offset between the i-th prediction and ground truth. t_i^D

indicates that the offset is obtained by RDN. We use focal loss [13] and balanced L1 loss [48] as the classification loss L_{cls} and regression loss L_{reg}.

The main purpose of the regression subnet is predicting the location, size and angle of the rotated bounding boxes. According to the definition of the rotated bounding box, the regression offset t_i^* and t_i can be denoted by

$$
\begin{aligned}
t_x^* &= \frac{G_x - A_x}{A_w}, \; t_y^* = \frac{G_y - A_y}{A_h} \\
t_w^* &= log\frac{G_w}{A_w}, \; t_h^* = log\frac{G_h}{A_h} \\
t_\theta^* &= \tan(G_\theta - A_\theta) \\
t_x &= \frac{B_x - A_x}{A_w}, \; t_y = \frac{B_y - A_y}{A_h} \\
t_w &= log\frac{B_w}{A_w}, \; t_h = log\frac{B_h}{A_h} \\
t_\theta &= \tan(B_\theta - A_\theta)
\end{aligned}
\tag{12}
$$

where G_i, A_i and B_i represent the five-tuple coordinate $i \in (x, y, w, h, \theta)$ of ground truth, anchor box and predicted bounding box. t_i^* represents the regression offset between the ground truth and anchor box. t_i represents the regression offset between the predicted bounding box and anchor box.

Like Refs [13,55,56], we use focal loss as the classification loss L_{cls}. However, previous work proved that the imbalance of classification loss and regression loss has negative impact on the detection accuracy. Directly adjusting the weight to increase the regression loss will make the network more sensitive to outliers, which is unfavorable for SAR images with much speckle noise. So, we use balanced L1 loss as the regression loss L_{reg} instead of the widely used smooth L1 loss.

Researchers usually balance the regression loss and classification loss by adjusting the loss weight. However, directly increasing the weight of regression loss will make the model more sensitive to the noise in the image. Moreover, SAR images are often disturbed by much background noise and speckle noise, which seriously affects the detection accuracy. Therefore, to solve this problem, we use balanced L1 loss instead of the commonly used smooth L1 loss as the regression loss. The formula of balanced L1 loss is as follows:

$$
L_b(x) = \begin{cases} \frac{a}{b}(b|x| + 1)ln(b|x| + 1) - a|x|, & if \; |x| < 1 \\ \gamma|x| + C & , \; otherwise \end{cases}
\tag{13}
$$

where a, b and γ are hyper-parameters and meet $aln(b + 1) = \gamma$. According to the configuration in Ref. [48], we set $a = 0.5$, $\gamma = 1.5$. x is the difference between the predicted value and the ground truth.

As Ref. [48] points out, compared with smooth L1 loss, balanced L1 loss can increase more gradient for accurate samples, reducing the contribution of outliers to loss. For example, compared with the contribution of outliers, the contribution of inliers to loss is only 30%, which makes the network very sensitive to outliers. Using balanced l1loss can improve the contribution of normal values to loss and reduce the sensitivity of the network to outliers. Therefore, using balanced L1 loss is helpful for ship detection, especially for ship detection in inshore scenes.

3. Experiments

Our experiments are run on a personal computer with i9-9900K CPU and RTX2080Ti GPU based on Pytorch. Our experiments are under the MMDetection toolbox [57] to ensure comparison fairness.

3.1. Experimental Datasets

One of the main reasons for the lack of research on SAR ship recognition in the past has been insufficient data. Now, thanks to the SAR rotation ship detection dataset (SRSDD) released by Lei et al. [22] in 2021, the research on SAR ship detection and classification can be further developed. The images in SRSDD all come from China's GF-3 satellite, which photographed more than 30 ports from five locations. The size of each slice is 1024 × 1024. Table 1 shows more details about SRSDD.

Table 1. The basic parameters of SRSDD.

Parameter	Value
Number of images	666
Waveband	C
Image Size	1024 × 1024
Image Mode	Spotlight Mode
Polarization	HH, VV
Resolution(m)	1
Ship Classes	6
Position	Nanjing, Hongkong, Zhoushan, Macao, Yokohama

The ships in SRSDD are marked by the rotated bounding box and are divided into six categories, including ore–oil ships, bulk cargo ships, fishing boats, law enforcement ships, dredger ships and container ships. The rotated bounding box and the category of each ship target are given by experts after checking the corresponding SAR image and corresponding optical image. This ensures their authenticity and accuracy. Table 2 shows the number of each category. It can be seen from the table that the number of bulk cargo accounts for most of the total, while the number of law enforcement is almost one tenth of that of bulk cargo.

In addition, considering the problem that the number of offshore scenes is larger than that of inshore scenes in the existing SAR dataset, such as SSDD [21,46] and Gaofen-SSDD [47], SRSDD focuses on taking nearshore scenes during sampling. In SRSDD, inshore scenes account for 63.1%, and offshore scenes account for 36.9%. In order to ensure fairness of the experimental results, our experiment is completely consistent with Ref. [22], that is, 532 pictures are used for training, and 134 pictures are used for testing.

Table 2. The number of each ship category in SRSDD.

Category	Train Number	Test Number	Total Number
Ore–oil ships	132	34	166
Bulk cargo ships	1603	450	2053
Fishing boats	206	82	288
Law enforcement	20	5	25
Dredger ships	217	46	263
Container ships	67	22	89
Total	2245	639	2884

3.2. Experimental Details

We refer to the work of Pang et al. [48] and use Resnet-50 [58] pretrained on Image-net as the backbone of RBFAN. A large number of images in Image-net can better train Resnet-50 to extract underlying features. Limited by GPU, we set the batch size to 4. Similar to Ref. [59], we use stochastic gradient descent (SGD) [60] as the optimizer, with a 0.005 learning rate, 0.9 momentum and 0.0001 weight. In addition, the learning rate is reduced from 130 epochs to 140 epochs, and each epoch is reduced by 10 times to ensure sufficient loss reduction. In addition, the intersection of union (IOU) threshold in the experiment is set to 0.5. According to the configuration in Ref. [48], we set $a = 0.5$, $\gamma = 1.5$. The hyper-parameter λ in loss function is set to 1. The hyper-parameter C in balanced L1 loss is set to 0.5.

According to Refs [13,30,48], we set the following parameters. In BAFPN, the resolution of each layer is $[512 \times 512,\ 256 \times 256,\ 128 \times 128,\ 64 \times 64,\ 32 \times 32]$. The size of the fused feature map is $128 \times 128 \times 256$. The resolution of the restored feature pyramid is $[512 \times 512,\ 256 \times 256,\ 128 \times 128,\ 64 \times 64,\ 32 \times 32]$. In AFAN, the convolution kernel size is 3×3. In RDN, the size of the used convolution kernels is 3×3 for all. The number of convolution layers stacked in the classification subnet and the regression subnet is 2.

3.3. Training Process

RBFA-Net is a single-stage network, so we refer to the training method of single-stage network in MMDetection [57] during training.

The specific training process is as follows:

1. Obtain the SAR image dataset preprocessed by SRSDD publisher.
2. Input the SAR image data into RBFA-Net for forward propagation to obtain the regression score and classification score.
3. Input the regression score and classification score and the ground truth into the loss function to obtain loss value.
4. Determine the gradient vector by back propagation.
5. Adjust each weight to make the loss value tend to zero or converge. We use the stochastic gradient descent (SGD) method for adjustment.
6. Repeat the above process until the set number of training times (training epoch) or loss value does not decrease.

3.4. Evaluation Indices

In this paper, we use mean average precision (mAP) as the evaluation index. The larger the mAP, the higher the network detection accuracy. To calculate mAP, we need to calculate recall and precision first. The recall and precision can be calculated as

$$Recall = \frac{TP}{TP + FN} \tag{14}$$

$$Precision = \frac{TP}{TP + FP} \tag{15}$$

where TP represents the number of true positives, FN represents the number of false negatives, FP represents the number of false positives. Then, we can obtain the precision–recall curve and calculate mAP.

$$AP = \int_0^1 P(R)dR \tag{16}$$

$$mAP = \frac{1}{k}\sum_{i=1}^{k} AP_i \tag{17}$$

where R represents recall, and P represents precision. $P(R)$ represents the precision–recall curve.

4. Results

4.1. Qualitative and Quantitative Analyses of Results

Table 3 shows the comparison of the detection results with the other eight rotated detectors on SRSDD. In Table 3, labels C1–C6 correspond to ore–oil ships, fishing boats, law enforcement ships, dredger ships, bulk cargo ships and container ships. The detection results of the other methods are from Ref. [22]. It can be seen from the results that the performance of our network is better than the eight state-of-the-art methods. In addition, our RBFA-Net achieves the highest mAP with a small model size, which proves the excellent performance of our RBFA-Net. Our RBFA-Net is only half the size of BBAVectors [61]. Moreover, the size of RBFA-Net is smaller than the second-best O-RCNN [62].

For each category, except for fishing boat and dredger, our model obtains optimal or suboptimal results. For fishing boat and dredger, our detection accuracy is also above the average level. For the largest number of categories (bulk cargo) and the lowest number of categories (law enforcement) in SRSDD, our detection accuracy reaches the best, which proves that our network can not only focus on small samples but also ensure the accuracy of large sample targets. It is worth noting that the detection accuracy of our network in law enforcement category is much higher than that of other models. Because law enforcement often appears in a fixed position, it is more sensitive to spatial information. With BAFPN, our network integrates global information, so it is helpful for the detection of such spatial sensitive targets.

Table 3. Quantitative evaluation comparison with the eight state-of-the-art detectors. Labels C1–C6 correspond to ore–oil ships, fishing boats, law enforcement ships, dredger ships, bulk cargo ships and container ships.

Models	C1	C2	C3	C4	C5	C6	mAP	Model Size
FR-O [16]	55.62	30.86	27.27	77.78	46.71	**85.33**	53.93	315 MB
R-RetinaNet [13]	30.37	11.47	2.07	67.71	35.79	48.94	32.73	<u>277 MB</u>
ROI [31]	<u>61.43</u>	32.89	27.27	79.41	48.89	76.41	54.38	421 MB
R3Det [55]	44.61	18.32	1.09	54.27	42.98	73.48	39.12	468 MB
BBAVectors [61]	54.33	21.03	1.09	<u>82.21</u>	34.84	78.51	45.33	829 MB
R-FCOS [40]	54.88	25.12	5.45	**83.00**	47.36	<u>81.11</u>	49.49	**244 MB**
Glid Vertex [63]	43.41	34.63	27.27	71.25	52.80	79.63	51.50	315 MB
O-RCNN [62]	**63.55**	<u>35.35</u>	<u>27.27</u>	77.50	**57.56**	76.14	<u>56.23</u>	315 MB
RBFA-Net (Ours)	59.39	**41.51**	**73.48**	77.17	<u>57.36</u>	71.62	**63.42**	302 MB

The best detector is in bold and the second best is underlined.

We also draw the confusion matrix showing the classification results of our network in more detail. The confusion matrix evaluates the classification accuracy of the network. As shown in Figure 10, the abscissa is the prediction category, and the ordinate is the real category. In the confusion matrix, the diagonal is the correct classification probability, and the others are the wrong classification probability. The confusion matrix is composed of ore–oil ships, fishing boats, law enforcement ships, dredger ships, bulk cargo ships, container ships and other class. Among them, the other class includes ships not in the dataset and other sea targets, such as oil platforms.

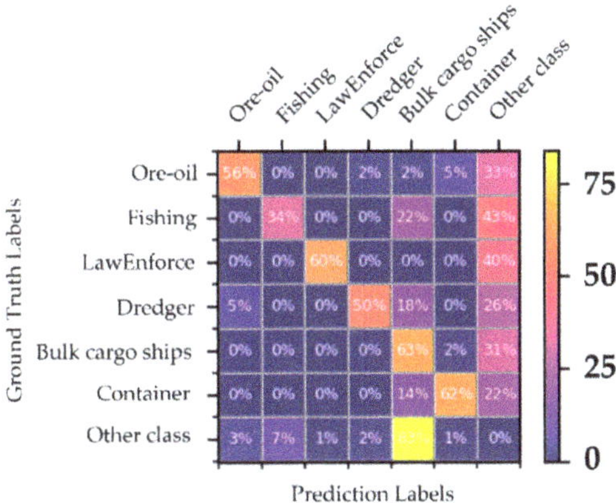

Figure 10. Confusion matrix.

Limited by space, Figure 11 shows the qualitative comparison of second-best method O-RCNN [62] and our RBFA-Net in detail. In order to observe the detection results of densely arranged ships near the shore, we compare the sliced and enlarged SAR images.

From the experimental results, we can draw the following conclusions:

1. RBFA-Net net can avoid some missing inspection of some densely arranged inshore small ships. As shown in Figure 11, RBFA-Net detects more fishing boats and bulk cargo than O-RCNN. This is because RBFA-Net uses AGFAN to align the feature map with the rotating anchor boxes, which reduces the negative impact of densely arranged ships and background interference.

2. For inshore ship detection, RBFA-Net has better detection accuracy, as shown in Figure 11. In the same SAR image, RBFA-Net can detect and correctly classify the inshore ships and the complex coastal background. This is because RBFA-Net uses FAFPN to fuse and enhance multi-scale features, which enhances the ability of focusing on global information of SAR images.

3. For the problem of difficult target ship detection, RBFA-Net has a higher detection effect, as shown in Figure 11. RBFA-Net can successfully classify the fishing boat, which has similar features with bulk cargo ships. This is because RBFA-Net uses the task decoupling module to adaptively enhance the feature map, making it better in the classification network.

4. Nevertheless, there are still some problems in our network. For example, there is still the problem of missing inspection when there are too many nearshore ships. For some ships whose characteristics are not obvious, there will also be classification errors. Finally, for some ships in specific directions, there is also the problem of dislocation of detection frame.

In addition, to better demonstrate the performance of our network, we compare the ground truth, the detection results of the third-best method RoI Transformer (ROI) [31], the detection results of the second-best method O-RCNN [62] and the detection results of RBFA-Net. Figures 12–14 show more detection results. For the ground truth, we show the correct category of each ship. For the test results, the category information and its confidence are displayed in the green label box. The higher the confidence, the greater the possibility that the test result is of this category. In order to display the SAR images more clearly, we increased the brightness of all displayed images.

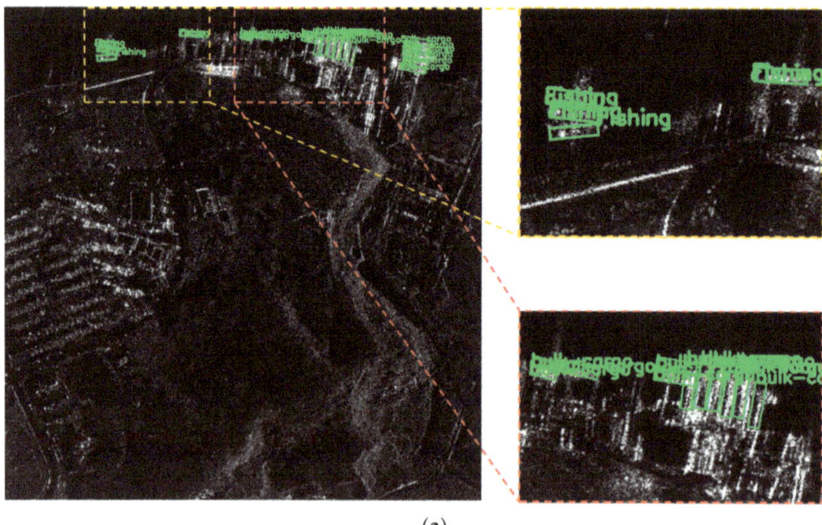

(a)

Figure 11. *Cont.*

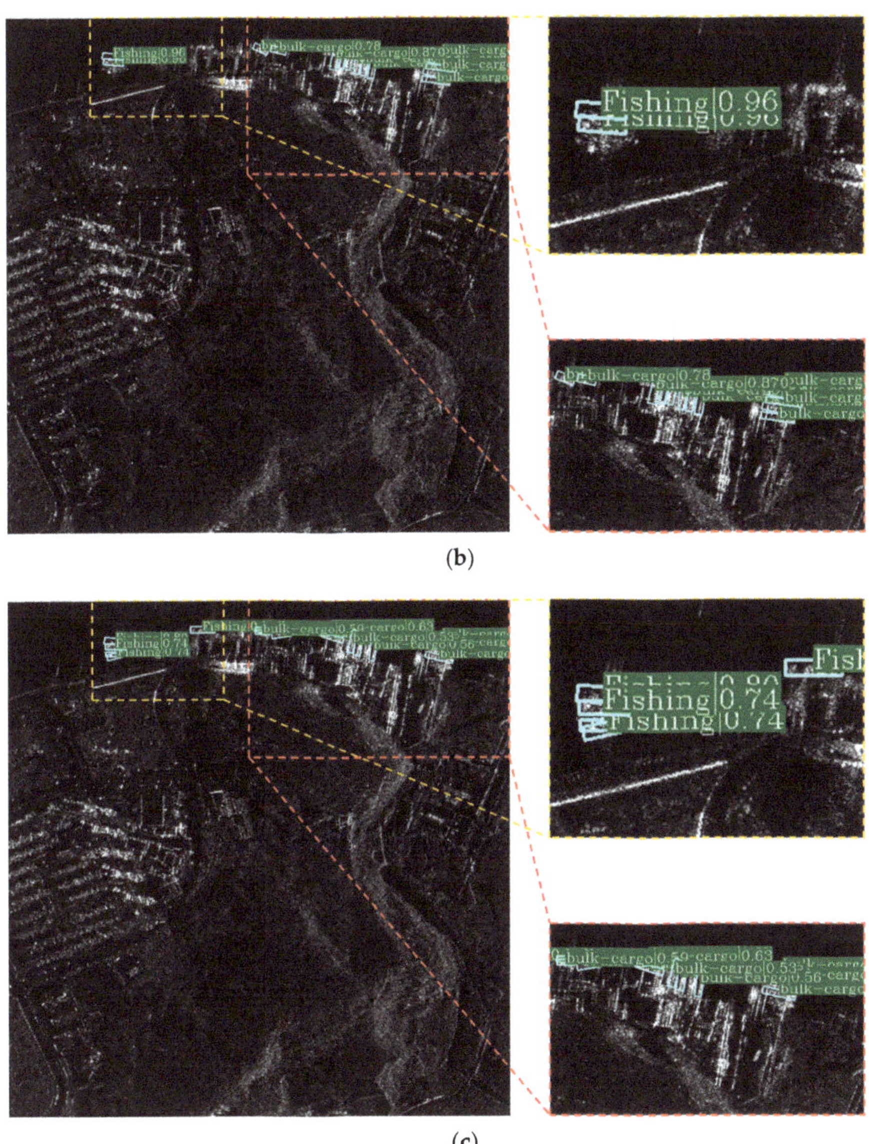

Figure 11. Detailed detection results. (**a**) Ground truth; (**b**) Result of O-RCNN; (**c**) Result of the proposed RBFA-Net.

Figure 12 shows the detection and classification results of the offshore ship. As can be seen from the SAR image, RBFA-Net successfully suppresses the false alarm in the SAR image. ROI and O-RCNN mistakenly identify the interference noise in the SAR image as a bulk cargo target. The detection result shows that RBFA-Net has better scene adaptability. Because our network uses BAFPN, it can fuse and enhance the global information to make RBFA-Net more robust, so as to reduce the impact of noise on the network detection and classification results to a certain extent.

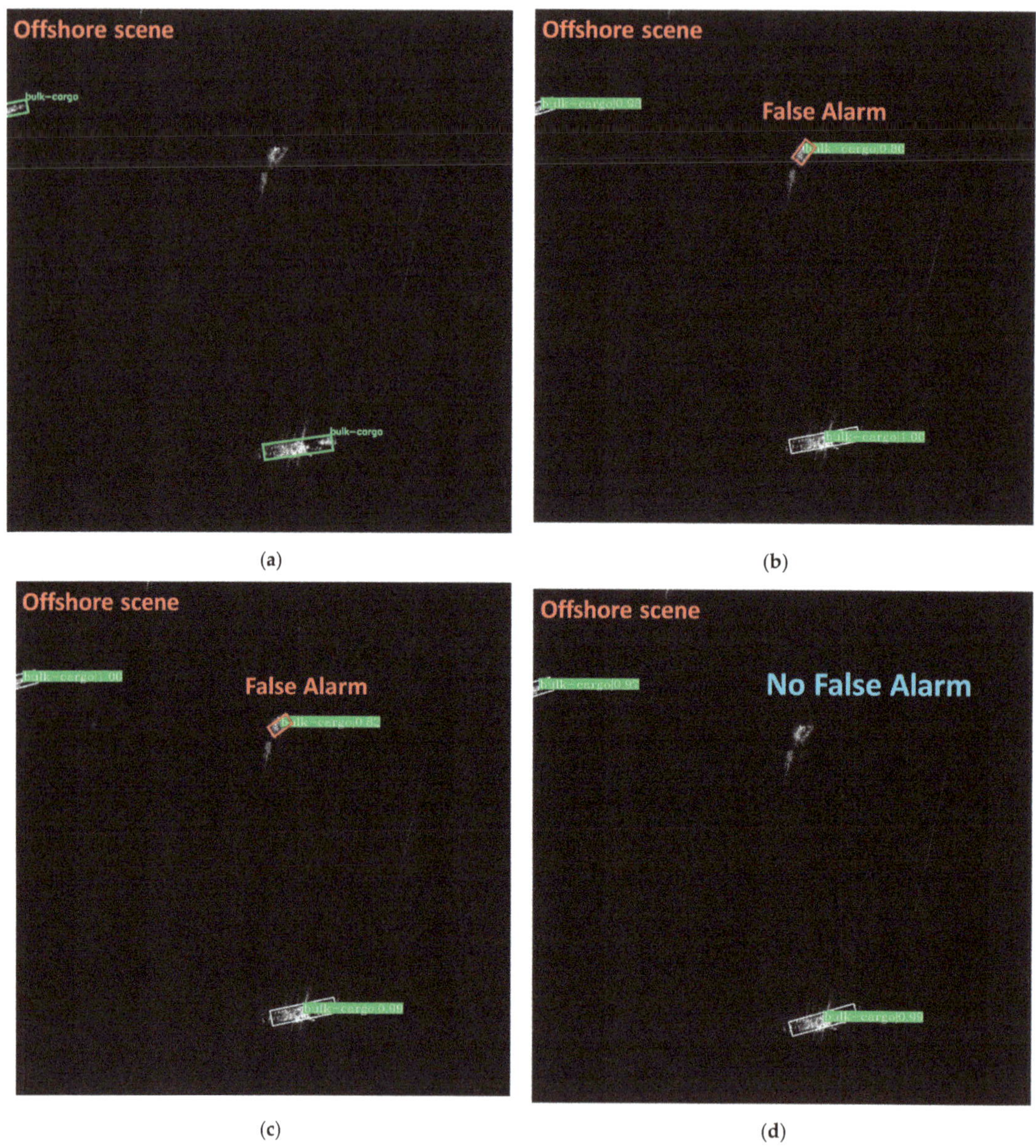

Figure 12. Detection results in offshore scenes. (**a**) Ground truth; (**b**) Result of ROI; (**c**) Result of O-RCNN; (**d**) Result of RBFA-Net.

Figure 13 shows the detection and classification results of the inshore ship. It can be seen from the picture that when the complex background occupies most part of the picture, our RBFA-Net successfully detects and classifies the bulk cargo target. O-RCNN and ROI fail to detect the ship targets. This is because our network adopts the alignment module to adjust the sampling position and improve the accuracy of ship detection.

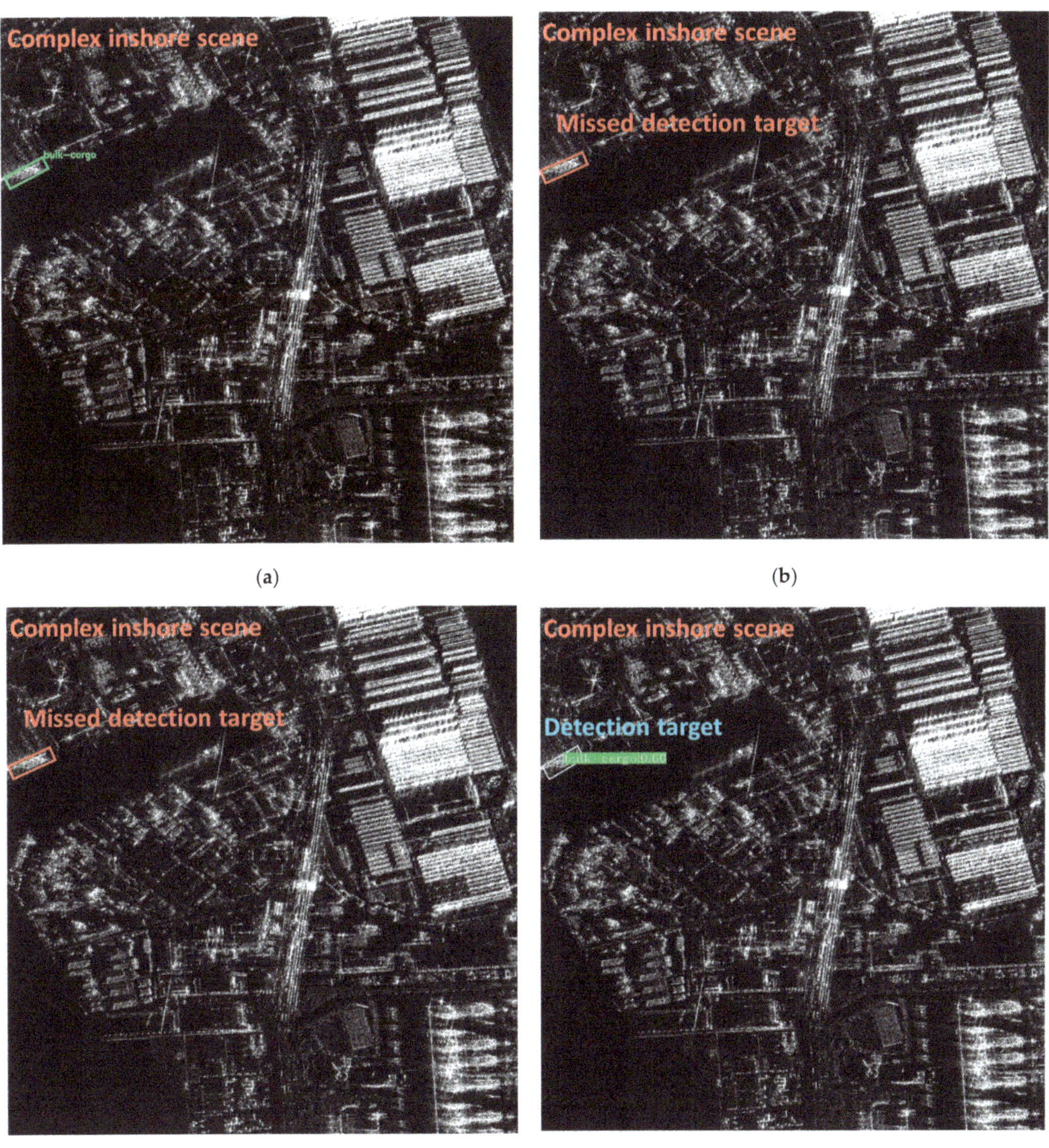

(a)

(b)

(c)

(d)

Figure 13. Detection results in inshore scenes. (**a**) Ground truth; (**b**) Result of ROI; (**c**) Result of O-RCNN; (**d**) Result of RBFA-Net.

Figure 14 shows the detection and classification results of densely arranged ship scenes. Generally, due to the complex background, false alarm often occurs in the detection results of inshore ships. In addition, the dense arrangement of inshore ships also brings challenges in detection and classification. From the detection results, we can see that our network not only correctly detects and classifies the densely arranged ships in the nearshore scene but also suppresses the false alarms in inshore scenes. Meanwhile, in the detection results of ROI and O-RCNN, these two networks mistakenly detect the coastal background as bulk cargo targets. This is because we introduce a task decoupling module, which can adjust the feature maps, respectively, for solving the conflict between the regression task and classification task. This adjustment also enhances RBFA-Net's ability to identify background and ship targets.

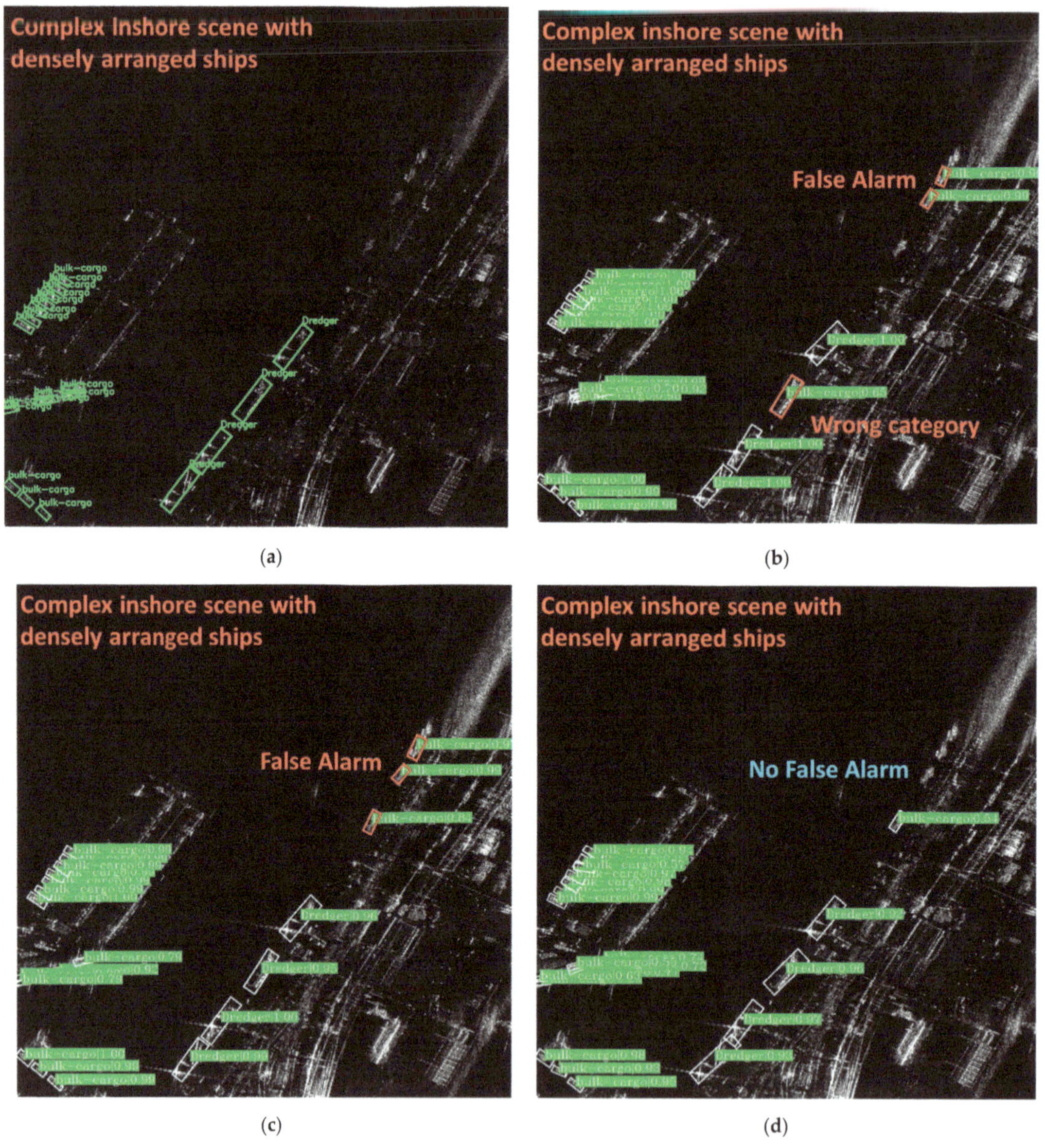

Figure 14. Detection results in densely arranged ship scenes. (**a**) Ground truth; (**b**) Result of ROI; (**c**) Result of O-RCNN; (**d**) Result of RBFA-Net.

4.2. Ablation Study

In this section, we conduct a series of experiments to verify the effectiveness of important improvements in RBFA-Net. They are a balanced-attention feature pyramid network (BAFPN), an anchor-guided feature alignment network (AFAN) and a task decoupling module (TDM). In addition, we also qualitatively explain the improvement brought about by each import module combined with the test results.

Table 4 quantitatively shows the impact of each important improvement on detection accuracy. The '✔' Table 4 means with the module, and the '--' means without the module. in Our RBFA-Net is designed on the basis of RetinaNet. It can be seen from the results that with the addition of these improvements, the detection accuracy of the network gradually improves.

Table 4. Effectiveness of each improvement in RBFA-Net.

BAFPN	AGFAN	TDM	mAP (%)
--	--	--	41.26
✔			45.74
✔	✔		59.08
✔	✔	✔	**63.42**

4.2.1. Effect of BAFPN

Table 5 shows the ablation study results on BAFPN. We compare the detection results using FPN. The two networks used in the experiment are completely consistent, except for the FPN part. One network uses FPN, the other uses BAFPN to extract feature maps. Both networks contain AGFAN and task decoupling module in RDN. The results show that using BAFPN can improve the detection and classification accuracy. By using BAFPN, we can fuse multi-scale ship features and pay attention to the global information, so as to improve our detection and classification accuracy.

Table 5. Effectiveness of Balanced-FPN.

Models	C1 (%)	C2 (%)	C3 (%)	C4 (%)	C5 (%)	C6 (%)	mAP (%)
FPN	60.01	60.85	35.07	43.80	73.43	76.50	58.28
Balanced-FPN (Ours)	59.39	57.36	41.51	73.48	77.17	71.62	**63.42**

4.2.2. Effect of AGFAN

Table 6 shows the ablation study results on AGFAN. In this experiment, we compare the detection results in the network with and without alignment module, respectively. The two networks used in the experiment are completely consistent, except for the alignment module. One network contains AGFAN, while the other does not. Both networks contain BAFPN and the task decoupling module in RDN. As can be seen from Table 6, the network with the alignment module has a higher detection accuracy. This is because the alignment module adjusts the sampling points of the network according to the rotated anchors. This enables the network to extract features according to the guidance of the anchor and reduces the dislocation caused by the use of rotated anchors. The '✔' Table 6 means with AGFAN, and the '--' means without AGFAN.

Table 6. Effectiveness of AGFAN.

Models	C1 (%)	C2 (%)	C3 (%)	C4 (%)	C5 (%)	C6 (%)	mAP (%)
×	35.77	43.98	33.03	27.27	75.87	49.67	44.27
✔	59.39	57.36	41.51	73.48	77.17	71.62	**63.42**

4.2.3. Effect of TDM

Table 7 shows the ablation study results on the task decoupling module (TDM). In this experiment, we compare the detection results in the network with and without the alignment module, respectively. The two networks used in the experiment are completely consistent, except for the task decoupling module. One network contains the task decoupling module, while the other does not. Both networks contain BAFPN and AGFAN. As can be seen from Table 7, the network with the alignment module has a higher detection accuracy. This is because the alignment module adjusts the sampling points of the network according to the rotated anchor. This enables the network to extract features according to the guidance of the anchor and reduces the dislocation caused by the use of rotating anchors. The '✔' Table 7 means with TDM, and the '--' means without TDM.

Table 7. Effectiveness of TDM.

Models	C1 (%)	C2 (%)	C3 (%)	C4 (%)	C5 (%)	C6 (%)	mAP (%)
×	62.20	57.15	36.87	48.25	73.68	82.91	60.17
✔	59.39	57.36	41.51	73.48	77.17	71.62	**63.42**

4.2.4. Effect of Balanced L1 Loss

Table 8 shows the ablation study results on balanced L1 loss. In this experiment, we test the smooth L1 used by most networks [25,28,29] as the control. In this ablation experiment, the networks used in the two experiments are exactly same, that is, BAFPN and decoupling module are used. The only difference is that one network uses balance L1 loss as the regression loss, and the other uses smooth L1 loss as the regression loss.

Table 8. Different types of regression loss.

Regression Loss Types	C1 (%)	C2 (%)	C3 (%)	C4 (%)	C5 (%)	C6 (%)	mAP (%)
Smooth L1	58.12	55.11	36.45	63.64	73.71	71.00	60.16
Balanced L1 loss (Ours)	59.39	57.36	41.51	73.48	77.17	71.62	**63.42**

5. Discussion

From the experimental results, it can be seen that compared with the other networks, our RBFA-Net achieves better performance indicators, in particular, the highest mAP. However, like other current networks, our RBFA-Net also has some similar problems. For example, the false alarm rate is still high. The classification accuracy of small sample ships, such as dredger ships and fishing ships, is still low. The rotation angle accuracy is not high enough, and alignment errors exist. These are the problems that we need to further study in the next stage. From the ablation study results, it can be seen that anchor-guided feature alignment network plays an important role in SAR ship detection with rotated bounding boxes. Therefore, it is very important to study how to better align the feature maps with the rotated anchors in the future.

6. Conclusions

In this paper, a balanced rotated feature-aligned network (RBFA-Net) is proposed for SAR ship recognition. Firstly, a balanced-attention FPN can better integrate multi-scale image information and reduce the impact of multi-scale ship feature differences. In the balanced-attention FPN, non-local module can effectively enhance the network's response to global information. Secondly, rotated anchors can effectively reduce the interference of complex background in SAR ship recognition. At the same time, they can also reduce the problem of true-value suppression caused by NMS for densely gathered ships. In addition, the feature alignment module aligns the feature map with the anchor boxes, which reduces the misalignment between the rotation anchor and the feature map. Furthermore, in the rotational detection network, we add a task decoupling module to adjust the feature maps and recalibrate them for the classification task and regression task. Lastly, we use balanced L1 loss to reduce the imbalance between regression loss and classification loss. We conduct extensive ablation experiments and confirm the effectiveness of each improvement. The experimental results show that RBFA-Net has the best performance compared with the other eight methods on the SRSDD dataset.

SAR ship detection methods based on deep learning have the advantages of full automation, high speed and strong model migration ability. On the basis of a large amount of data training, the deep-learning method can mine features that cannot be mined by traditional algorithms, so as to better realize SAR ship detection. The key advantage of SAR ship detection based on deep learning lies in the large amount of high-quality data and appropriate network models. In the future, there will be more high-quality datasets and more networks that can better mine SAR image features.

Our future works are as follows:

1. We will improve the detection accuracy of rotation angle of the rotated bounding boxes. The structure of the regression network is relatively simple, which may not meet the requirements of rotation detection.
2. We will improve the detection and classification accuracy of small sample ships, such as fishing boats and law enforcement ships. We suppose that BAFPN does not fully mine the unobvious features of small samples of ships. Ship segmentation can also be considered in the future.

Author Contributions: Conceptualization, Z.S.; methodology, Z.S.; software, Z.S.; validation, Z.S.; formal analysis, Z.S.; investigation, Z.S.; resources, Z.S.; data curation, Z.S.; writing—Original draft preparation, Z.S.; writing—Review and editing, X.Z. and T.Z. (Tianwen Zhang); visualization, Z.S. and X.X.; supervision, T.Z. (Tianjiao Zeng); project administration, X.Z.; funding acquisition, X.Z. All authors have read and agreed to the published version of the manuscript.

Funding: This work was supported by the National Natural Science Foundation of China (61571099).

Data Availability Statement: Not applicable.

Acknowledgments: We would like to thank all editors and reviewers for their valuable comments for improving this manuscript.

Conflicts of Interest: The authors declare no conflict of interest.

References

1. Moreira, A.; Prats-Iraola, P.; Younis, M.; Krieger, G.; Hajnsek, I.; Papathanassiou, K.P. A tutorial on synthetic aperture radar. *IEEE Geosci. Remote Sens. Mag.* **2013**, *1*, 6–43. [CrossRef]
2. Zhang, T.; Zhang, X. Injection of Traditional Hand-Crafted Features into Modern CNN-Based Models for SAR Ship Classification: What, Why, Where, and How. *Remote Sens.* **2021**, *13*, 2091. [CrossRef]
3. Liu, T.; Zhang, J.; Gao, G.; Yang, J.; Marino, A. CFAR Ship Detection in Polarimetric Synthetic Aperture Radar Images Based on Whitening Filter. *IEEE Trans. Geosci. Remote Sens.* **2020**, *58*, 58–81. [CrossRef]
4. Huang, X.; Yang, W.; Zhang, H.; Xia, G.-S. Automatic Ship Detection in SAR Images Using Multi-Scale Heterogeneities and an A Contrario Decision. *Remote Sens.* **2015**, *7*, 7695–7711. [CrossRef]
5. Schwegmann, C.P.; Kleynhans, W.; Salmon, B.P. Synthetic aperture radar ship detection using Haar-like features. *IEEE Geosci. Remote Sens. Lett.* **2016**, *14*, 154–158. [CrossRef]
6. Redmon, J.; Divvala, S.; Girshick, R.; Farhadi, A. You Only Look Once: Unified, real-time object detection. In Proceedings of the IEEE Conference on Computer Vision and Pattern Recognition, Las Vegas, NV, USA, 27–30 June 2016; pp. 779–788.
7. Zhang, T.; Zhang, X.; Shi, J.; Wei, S. Depthwise Separable Convolution Neural Network for High-Speed SAR Ship Detection. *Remote Sens.* **2019**, *11*, 2483. [CrossRef]
8. Xu, X.; Zhang, X.; Zhang, T. Lite-YOLOv5: A Lightweight Deep Learning Detector for On-Board Ship Detection in Large-Scene Sentinel-1 SAR Images. *Remote Sens.* **2022**, *14*, 1018. [CrossRef]
9. Liu, W.; Anguelov, D.; Erhan, D.; Szegedy, C.; Reed, S.; Fu, C.Y.; Berg, A.C. SSD: Single Shot MultiBox Detector. In Proceedings of the European Conference on Computer Vision, Amsterdam, The Netherlands, 8–16 October 2016; Springer: Cham, Switzerland, 2016; pp. 21–37.
10. Zhang, T.; Zhang, X. High-Speed Ship Detection in SAR Images Based on a Grid Convolutional Neural Network. *Remote Sens.* **2019**, *11*, 1206. [CrossRef]
11. Redmon, J.; Farhadi, A. YOLO9000: Better, faster, stronger. In Proceedings of the IEEE Conference on Computer Vision and Pattern Recognition, Honolulu, HI, USA, 21–26 July 2017; pp. 7263–7271.
12. Zhang, T.; Zhang, X. ShipDeNet-20: An Only 20 Convolution Layers and <1-MB Lightweight SAR Ship Detector. *IEEE Geosci. Remote Sens. Lett.* **2021**, *18*, 1234–1238. [CrossRef]

13. Lin, T.Y.; Goyal, P.; Girshick, R.; He, K.; Dollár, P. Focal loss for dense object detection. In Proceedings of the IEEE International Conference on Computer Vision, Venice, Italy, 22–29 October 2017; pp. 2980–2988.

14. Girshick, R.; Donahue, J.; Darrell, T.; Malik, J. Rich feature hierarchies for accurate object detection and semantic segmentation. In Proceedings of the IEEE Conference on Computer Vision and Pattern Recognition, Columbus, OH, USA, 24–27 June 2014; pp. 580–587.

15. Zhang, T.; Zhang, X.; Liu, C.; Shi, J.; Wei, S.; Ahmad, I.; Zhan, X.; Zhou, Y.; Pan, D.; Li, J.; et al. Balance learning for ship detection from synthetic aperture radar remote sensing imagery. *ISPRS J. Photogramm. Remote Sens.* **2021**, *182*, 190–207. [CrossRef]

16. Ren, S.; He, K.; Girshick, R.; Sun, J. Faster R-CNN: Towards Real-Time Object Detection with Region Proposal Networks. In *Advances in Neural Information Processing Systems*; MIT Press: Cambridge, MA, USA, 2015; pp. 91–99.

17. Cai, Z.; Vasconcelos, N. Cascade r-cnn: Delving into high quality object detection. In Proceedings of the IEEE Conference on Computer Vision and Pattern Recognition, Salt Lake City, UT, USA, 18–23 June 2018; pp. 6154–6162.

18. He, K.; Gkioxari, G.; Dollár, P. Mask r-cnn. In Proceedings of the IEEE International Conference on Computer Vision, Venice, Italy, 22–29 October 2017; pp. 2961–2969.

19. He, K.; Zhang, X.; Ren, S. Spatial pyramid pooling in deep convolutional networks for visual recognition. *IEEE Trans. Pattern Anal. Mach. Intell.* **2015**, *37*, 1904–1916. [CrossRef] [PubMed]

20. Shin, H.-C.; Roth, H.R.; Gao, M.; Lu, L.; Xu, Z.; Nogues, I.; Yao, J.; Mollura, D.; Summers, R.M. Deep convolutional neural networks for computer-aided detection: Cnn architectures, dataset characteristics and transfer learning. *IEEE Trans. Med. Imaging* **2016**, *35*, 1285–1298. [CrossRef] [PubMed]

21. Zhang, T.; Zhang, X.; Li, J.; Xu, X.; Wang, B.; Zhan, X.; Xu, Y.; Ke, X.; Zeng, T.; Su, H.; et al. SAR Ship Detection Dataset (SSDD): Official Release and Comprehensive Data Analysis. *Remote Sens.* **2021**, *13*, 3690. [CrossRef]

22. Lei, S.; Lu, D.; Qiu, X.; Ding, C. SRSDD-v1.0: A High-Resolution SAR Rotation Ship Detection Dataset. *Remote Sens.* **2021**, *13*, 5104. [CrossRef]

23. Zhang, T.; Zhang, X.; Ke, X. Quad-FPN: A Novel Quad Feature Pyramid Network for SAR Ship Detection. *Remote Sens.* **2021**, *13*, 2771. [CrossRef]

24. Sun, K.; Liang, Y.; Ma, X.; Huai, Y.; Xing, M. DSDet: A Lightweight Densely Connected Sparsely Activated Detector for Ship Target Detection in High-Resolution SAR Images. *Remote Sens.* **2021**, *13*, 2743. [CrossRef]

25. Wang, Y.; Wang, C.; Zhang, H.; Dong, Y.; Wei, S. Automatic Ship Detection Based on RetinaNet Using Multi-Resolution Gaofen-3 Imagery. *Remote Sens.* **2019**, *11*, 531. [CrossRef]

26. Jiao, J.; Zhang, Y.; Sun, H. A densely connected end-to-end neural network for multiscale and multiscene SAR ship detection. *IEEE Access* **2018**, *6*, 20881–20892. [CrossRef]

27. Liu, L.; Pan, Z.; Lei, B. Learning a rotation invariant detector with rotatable bounding box. *arXiv* **2017**, arXiv:1711.09405.

28. Yang, R.; Pan, Z.; Jia, X. A novel CNN-based detector for ship detection based on rotatable bounding box in SAR images. *IEEE J. Sel. Top. Appl. Earth Obs. Remote Sens.* **2021**, *14*, 1938–1958. [CrossRef]

29. Pan, Z.; Yang, R.; Zhang, Z. MSR2N: Multi-Stage Rotational Region Based Network for Arbitrary-Oriented Ship Detection in SAR Images. *Sensors* **2020**, *20*, 2340. [CrossRef] [PubMed]

30. Chen, S.; Zhang, J.; Zhan, R. R2FA-Det: Delving into High-Quality Rotatable Boxes for Ship Detection in SAR Images. *Remote Sens.* **2020**, *12*, 2031. [CrossRef]

31. Ding, J.; Xue, N.; Long, Y.; Xia, G.S.; Lu, Q. Learning RoI Transformer for Oriented Object Detection in Aerial Images. In Proceedings of the IEEE International Conference on Computer Vision and Pattern Recognition, Long Beach, CA, USA, 16–20 June 2019; pp. 2849–2858.

32. Wang, Y.; Wang, C.; Zhang, H.; Dong, Y.; Wei, S. A SAR Dataset of Ship Detection for Deep Learning under Complex Backgrounds. *Remote Sens.* **2019**, *11*, 765. [CrossRef]

33. Yang, R.; Wang, G.; Pan, Z.; Lu, H.; Zhang, H.; Jia, X. A Novel False Alarm Suppression Method for CNN-Based SAR Ship Detector. *IEEE Geosci. Remote Sens. Lett.* **2020**, *18*, 1401–1405. [CrossRef]

34. Jiang, B.; Luo, R.; Mao, J. Acquisition of localization confidence for accurate object detection. In Proceedings of the European Conference on Computer Vision, Munich, Germany, 8–14 September 2018; pp. 784–799.

35. Zhang, T.; Zhang, X. A Polarization Fusion Network with Geometric Feature Embedding for SAR Ship Classification. *Pattern Recognit.* **2021**, *123*, 108365. [CrossRef]

36. He, J.; Wang, Y.; Liu, H. Ship Classification in Medium-Resolution SAR Images via Densely Connected Triplet CNNs Integrating Fisher Discrimination Regularized Metric Learning. *IEEE Trans. Geosci. Remote Sens.* **2021**, *59*, 3022–3039. [CrossRef]

37. Zhang, T.; Zhang, X. Squeeze-and-Excitation Laplacian Pyramid Network with Dual-Polarization Feature Fusion for Ship Classification in SAR Images. *IEEE Geosci. Remote Sens. Lett.* **2021**, *19*, 4019905. [CrossRef]

38. Zeng, L.; Zhu, Q.; Lu, D.; Zhang, T.; Wang, H.; Yin, J.; Yang, J. Dual-Polarized SAR Ship Grained Classification Based on CNN With Hybrid Channel Feature Loss. *IEEE Geosci. Remote Sens. Lett.* **2021**, *19*, 4011905. [CrossRef]

39. Zhang, T.; Zhang, X.; Ke, X.; Liu, C.; Xu, X.; Zhan, X.; Wang, C.; Ahmad, I.; Zhou, Y.; Pan, D.; et al. HOG-ShipCLSNet: A Novel Deep Learning Network with HOG Feature Fusion for SAR Ship Classification. *IEEE Trans. Geosci. Remote Sens.* **2021**, *60*, 5210322. [CrossRef]

40. Law, H.; Deng, J. Cornernet: Detecting objects as paired keypoints. In Proceedings of the European Conference on Computer Vision (ECCV), Munich, Germany, 8–14 September 2018; pp. 765–781.

41. Zhou, K.; Zhang, M.; Wang, H.; Tan, J. Ship Detection in SAR Images Based on Multi-Scale Feature Extraction and Adaptive Feature Fusion. *Remote Sens.* **2022**, *14*, 755. [CrossRef]

42. Zhang, Y.; Sheng, W.; Jiang, J.; Jing, N.; Wang, Q.; Mao, Z. Priority Branches for Ship Detection in Optical Remote Sensing Images. *Remote Sens.* **2020**, *12*, 1196. [CrossRef]

43. Chen, P.; Li, Y.; Zhou, H.; Liu, B.; Liu, P. Detection of Small Ship Objects Using Anchor Boxes Cluster and Feature Pyramid Network Model for SAR Imagery. *J. Mar. Sci. Eng.* **2020**, *8*, 112. [CrossRef]

44. Zhang, T.; Zhang, X.; Shi, J.; Wei, S.; Wang, J.; Li, J.; Su, H.; Zhou, Y. Balance Scene Learning Mechanism for Offshore and Inshore Ship Detection in SAR Images. *IEEE Geosci. Remote Sens. Lett.* **2020**, *19*, 4004905. [CrossRef]

45. Lin, T.-Y.; Dollar, P.; Girshick, R.; He, K.; Hariharan, B.; Belongie, S. Feature Pyramid Networks for Object Detection. In Proceedings of the IEEE Conference on Computer Vision and Pattern Recognition (CVPR), Honolulu, HI, USA, 21–26 July 2017; pp. 936–944.

46. Li, J.; Qu, C.; Shao, J. Ship detection in SAR images based on an improved faster R-CNN. In Proceedings of the SAR in Big Data Era: Models, Methods and Applications, Beijing, China, 13–14 November 2017; pp. 1–6.

47. Zhang, T.; Zhang, X.; Shi, J.; Wei, S. HyperLi-Net: A hyper-light deep learning network for high-accurate and high-speed ship detection from synthetic aperture radar imagery. *ISPRS J. Photogramm. Remote Sens.* **2020**, *167*, 123–153. [CrossRef]

48. Pang, J.; Chen, K.; Shi, J.; Feng, H.; Ouyang, W.; Lin, D. Libra R-CNN: Towards Balanced Learning for Object Detection. *arXiv* **2019**, arXiv:1904.02701.

49. Wang, X.; Girshick, R.; Gupta, A.; He, K. Non-local Neural Networks. *arXiv* **2017**, arXiv:1711.07971.

50. Zhou, Y.; Yang, X.; Zhang, G. MMRotate: A Rotated Object Detection Benchmark using Pytorch. *arXiv* **2022**, arXiv:2204.13317.

51. Wang, J.; Chen, K.; Yang, S.; Loy, C.C.; Lin, D. Region proposal by guided anchoring. In Proceedings of the IEEE Conference on Computer Vision and Pattern Recognition, Long Beach, CA, USA, 16–20 June 2019; pp. 2965–2974.

52. Song, G.; Liu, Y.; Wang, X. Revisiting the sibling head in object detector. *arXiv* **2020**, arXiv:2003.07540.

53. Hu, J.; Shen, L.; Sun, G. Squeeze-and-excitation networks. In Proceedings of the IEEE Conference on Computer Vision and Pattern Recognition (CVPR), Salt Lake City, UT, USA, 18–23 June 2018; pp. 7132–7141.

54. Glorot, X.; Bordes, A.; Bengio, Y. Deep sparse rectifier neural networks. In Proceedings of the Fourteenth International Conference on Artificial Intelligence and Statistics, Ft. Lauderdale, FL, USA, 11–13 April 2011; pp. 315–323.

55. Yang, X.; Yan, J.; Feng, Z.; He, T. R3Det: Refined Single-Stage Detector with Feature Refinement for Rotating Object. *arXiv* **2019**, arXiv:1908.05612.

56. Yang, Z.; Liu, S.; Hu, H.; Wang, L.; Lin, S. RepPoints: Point Set Representation for Object Detection. In Proceedings of the 2019 IEEE/CVF International Conference on Computer Vision (ICCV), Seoul, Korea, 29 October–1 November 2019; pp. 9656–9665.

57. Chen, K.; Wang, J.; Pang, J. MMDetection: Open mmlab detection toolbox and benchmark. *arXiv* **2019**, arXiv:1906.07155.

58. He, K.; Zhang, X.; Ren, S.; Sun, J. Identity Mappings in Deep Residual Networks. In Proceedings of the 14th European Conference on Computer Vision, Amsterdam, The Netherlands, 8–16 October 2016; Part IV. pp. 630–645.

59. Zhang, T.; Zhang, X. A Full-Level Context Squeeze-and-Excitation ROI Extractor for SAR Ship Instance Segmentation. *IEEE Geosci. Remote Sens. Lett.* **2022**, *19*, 4506705. [CrossRef]

60. Goyal, P.; Dollár, P.; Girshick, R.; Noordhuis, P.; Wesolowski, L.; Kyrola, A.; Tulloch, A.; Jia, Y.; He, K. Accurate, Large Minibatch SGD: Training ImageNet in 1 Hour. *arXiv* **2017**, arXiv:1706.02677.

61. Yi, J.; Wu, P.; Liu, B.; Huang, Q.; Qu, H.; Metaxas, D. Oriented Object Detection in Aerial Images with Box Boundary-A ware V ectors. In Proceedings of the 2021 IEEE Winter Conference on Applications of Computer Vision (WACV), Virtual, 5–9 January 2021; pp. 2149–2158.

62. Xie, X.; Cheng, G.; Wang, J. Oriented R-CNN for Object Detection. *arXiv* **2021**, arXiv:2108.05699.

63. Xu, Y.; Fu, M.; Wang, Q.; Wang, Y.; Chen, K.; Xia, G.; Bai, X. Gliding Vertex on the Horizontal Bounding Box for Multi-Oriented Object Detection. *IEEE Trans. Pattern Anal. Mach. Intell.* **2021**, *43*, 1452–1459. [CrossRef] [PubMed]

Article

NeXtNow: A Convolutional Deep Learning Model for the Prediction of Weather Radar Data for Nowcasting Purposes

Alexandra-Ioana Albu [1,†], Gabriela Czibula [1,*,†], Andrei Mihai [1,†], Istvan Gergely Czibula [1,†] and Sorin Burcea [2] and Abdelkader Mezghani [3]

1   Department of Computer Science, Babeş-Bolyai University, 400084 Cluj-Napoca, Romania; alexandra.albu@ubbcluj.ro (A.-I.A.); andrei.mihai@ubbcluj.ro (A.M.); istvan.czibula@ubbcluj.ro (I.G.C.)
2   Romanian National Meteorological Administration, 013686 Bucharest, Romania; sorin.burcea@meteoromania.ro
3   Meteorologisk Institut, 0371 Oslo, Norway; abdelkader.mezghani@met.no
*   Correspondence: gabriela.czibula@ubbcluj.ro; Tel.: +40-264-405327
†   These authors contributed equally to this work.

Abstract: With the recent increase in the occurrence of severe weather phenomena, the development of accurate weather nowcasting is of paramount importance. Among the computational methods that are used to predict the evolution of weather, deep learning techniques offer a particularly appealing solution due to their capability for learning patterns from large amounts of data and their fast inference times. In this paper, we propose a convolutional network for weather forecasting that is based on radar product prediction. Our model (*NeXtNow*) adapts the ResNeXt architecture that has been proposed in the computer vision literature to solve the spatiotemporal prediction problem. *NeXtNow* consists of an encoder–decoder convolutional architecture, which maps radar measurements from the past onto radar measurements that are recorded in the future. The ResNeXt architecture was chosen as the basis for our network due to its flexibility, which allows for the design of models that can be customized for specific tasks by stacking multiple blocks of the same type. We validated our approach using radar data that were collected from the Romanian National Meteorological Administration (NMA) and the Norwegian Meteorological Institute (MET) and we empirically showed that the inclusion of multiple past radar measurements led to more accurate predictions further in the future. We also showed that *NeXtNow* could outperform *XNow*, which is a convolutional architecture that has previously been proposed for short-term radar data prediction and has a performance that is comparable to those of other similar approaches in the nowcasting literature. Compared to *XNow*, *NeXtNow* provided improvements to the critical success index that ranged from 1% to 17% and improvements to the root mean square error that ranged from 5% to 6%.

Keywords: weather nowcasting; deep learning; ResNeXt; radar data

Citation: Albu, A.-I.; Czibula, G.; Mihai, A.; Czibula, I.G.; Burcea, S.; Mezghani, A. *NeXtNow*: A Convolutional Deep Learning Model for the Prediction of Weather Radar Data for Nowcasting Purposes. *Remote Sens.* 2022, 14, 3890. https://doi.org/10.3390/rs14163890

Academic Editor: Gwanggil Jeon

Received: 4 July 2022
Accepted: 8 August 2022
Published: 11 August 2022

Publisher's Note: MDPI stays neutral with regard to jurisdictional claims in published maps and institutional affiliations.

1. Introduction

Short-term weather analysis and forecasting for the next 0 to 6 h, which is known as *weather nowcasting* [1,2], are of great interest in meteorology due to the increasing number of severe weather events that can severely affect the safety of the human population by causing damage and even mortality. For instance, precipitation nowcasting, which refers to the prediction of rainfall intensity in specific regions in the near future, plays an important role in our daily life [2] and represents a challenging topic in nowcasting. The problem is that short-term weather forecasting is complex and difficult, even for operational meteorologists, mainly due to the large volume of data that needs to be examined and interpreted in a short period of time. In addition, nowcasting [3] is highly dependent on various environmental conditions and requires a lot of human experience.

Significant progress has recently been made in the field of nowcasting, ranging from operational nowcasting systems to the numerous *computational intelligence* solutions that have been proposed in the literature for detecting the occurrence of severe weather events [4,5]. Radar data [6] are useful for nowcasting [7,8]. Most operational nowcasting systems use numerical weather prediction (NWP) models, which combine radar data and other meteorological observations to provide weather forecasts for up to 6 h in the future, e.g., the INCA system [9]. Still, there are a lot of challenges in issuing precise nowcasting warnings as most severe events (e.g., severe convective storms) occur within small spatial areas and have short overall life cycles. Despite the significant improvements that have been achieved by NWP models in precipitation nowcasting, there are still some limitations to their use, such as insufficient computational resources in operational centers and increased error rates at convection-permitting scales [10].

Most existing operational and semi-operational methods for nowcasting are based on cell-tracking algorithms that use radar data as inputs. A real-time cell-tracking algorithm named TITAN, which is useful for single cells, was introduced by Dixon and Wiener [7] and the SCIT system, which uses a more complex cell-tracking algorithm and reflectivity thresholds, was later proposed by Johnson et al. [8]. An operational nowcasting tool that uses a centroid cell-tracking method (named TRT) was also developed by Hering et al. [11] and used by Meteo Swiss, while Germany's National Meteorological Service uses a nowcasting system that is called NowCastMIX, which was introduced by James et al. [5] and employs fuzzy logic rules to analyze remote and ground observations. Fuzzy logic is also used in a cell-tracking algorithm that was proposed by Jung and Lee [4]. AROME [12] is a prediction model that is used to provide a forecast for up to 30 h in the future and has been used by Meteo France since 2008, while AROME-NWC [13] was later developed for nowcasting in the range of 0–6 h.

Machine learning (ML) techniques are useful computational intelligence tools that can assist operational meteorologists in decision-making as they are able to learn relevant patterns from weather-related data. *Deep learning* (DL) [14,15] models have become popular within the ML domain as they are able to express target functions that are more complex than those that are encoded by traditional ML models. Additionally, DL models can automatically extract useful features from raw data, thereby removing the difficult task of manual feature engineering that is required by classical ML models. DL architectures are characterized by a hierarchy of multiple levels of representations. Despite the complex aspects of this type of neural model, the key advantage of DL models is the fact that they are universal approximators, i.e., they have the ability to learn arbitrary functions as a composition of several operations.

Reflectivity (R) and Doppler radial velocity (V) are radar products that are used by operational meteorologists to monitor the spatiotemporal evolution of precipitating clouds and thus, they are useful in weather nowcasting.

These reflectivity and velocity values are also used by operational radar algorithms to estimate rainfall and to track and classify storms: R values that are higher than 35 dBZ [7,16] suggest the possible occurrence of convective storms, which are associated with heavy rainfall. The prediction of the values of these radar products based on their historical measurements is important for the early assessment of storm evolution, which can lead to improved nowcasting and timely severe weather warnings.

Time series data, which represent the values of the selected variables at specific time points, are generally used in forecasting because of the temporal characteristics of these data. This study used time series radar data to predict the values of the radar products in a specific geographical region, based on their historical values. The contribution of the paper is twofold. Firstly, the proposal of *NeXtNow*, which is a new convolutional weather forecasting model that was adapted from the ResNeXt [17] architecture that has been proposed in the computer vision literature for the task of spatiotemporal prediction. Our proposed model has an encoder–decoder architecture that consists of ResNeXt blocks and simple convolutions, which maps past radar measurements onto radar measurements that are recorded in the future. Therefore, *NeXtNow* is a customized version of ResNeXt for the short-term prediction of radar data. We opted for the convolutional architecture instead of a recurrent alternative due to its training and inference efficiency. The ResNeXt architecture was chosen due to the versatility of ResNet-type [18] architectures, which allows for the design of models that can be customized for specific tasks by stacking multiple blocks of the same type. Moreover, we empirically showed that the inclusion of multiple past radar measurements led to more accurate predictions further in the future. The performance of our proposed *NeXtNow* model was evaluated using two case studies, which consisted of real radar data that were collected from the Romanian National Meteorological Administration (NMA) and the Norwegian Meteorological Institute (MET). The obtained results were analyzed from a meteorological perspective to examine the ability of the *NeXtNow* model to capture relevant patterns in the evolution of radar echoes, i.e., patterns that could be relevant for nowcasting severe weather phenomena. Comparisons between *NeXtNow* and other models in the literature highlighted that *NeXtNow* outperformed a convolutional architecture [19] that was proposed for short-term radar data prediction and that its performance was also comparable to the performances of other similar approaches from the nowcasting literature. To the best of our knowledge, an approach that is similar to *NeXtNow* has not yet been proposed in the nowcasting literature.

To summarize, the following research questions guided this study:

RQ1 How can the ResNeXt deep learning architecture be adapted for the task of spatiotemporal prediction and customized for the short-term prediction of radar data? (This led to the development of the *NeXtNow* model.)

RQ2 How does the *NeXtNow* model perform using real radar data? Would adding multiple past radar measurements improve the future prediction performance?

RQ3 Does the ResNeXt model improve the performance of the short-term prediction of radar data? Is the performance improvement that is achieved by *NeXtNow* statistically significant from similar existing approaches?

The rest of the paper is organized as follows. Section 2 presents a literature review of recent machine learning and deep learning approaches for weather nowcasting that use radar data (Section 2.1) and an overview of the case studies that were used to evaluate our proposed model (Section 2.2). The methodology that was used in our work is detailed in Section 2.3. Section 3 presents the experimental evaluation of our proposed approach and the comparisons to similar models are presented in Section 4. The last section presents the conclusions of this study and potential directions for future research.

2. Materials and Methods

2.1. Recent Literature Advances in Nowcasting, Based on Radar Data Prediction

Various classical ML and DL models have been introduced in the literature for weather nowcasting. In the following section, we summarize the nowcasting techniques that are based on radar data that have been proposed recently.

Prudden et al. [20] reviewed the existing forecasting methods for precipitation prediction that are based on radar data and the machine learning techniques that are applicable for radar-based precipitation nowcasting. Four classes of methods for precipitation nowcasting were mentioned by the authors: persistence-based methods, probabilistic and stochastic methods, nowcasting convective development and ML-based approaches. The study emphasized the performance improvements that could be obtained by applying deep neural networks combined with domain knowledge about the physical system that was being modeled. The authors also highlighted the potential of generative adversarial networks, which are able to capture data uncertainty and generate new data that follow the same distribution patterns as the input data.

Han et al. [21] used support vector machines (SVMs) for radar data nowcasting, which was modeled as a binary classification task. The model was trained to identify whether the radar would detect a radar echo >35 dBZ in the following 30 min. The features that characterized the input data included temporal and spatial information. The experiments revealed a probability of detection (POD) of around 0.61, a critical success index (CSI) of 0.36 and a false alarm rate (FAR) of about 0.52.

Ji [22] employed artificial neural networks for short-term precipitation prediction using radar observations that were collected from China from 2010 to 2012. The reflectivity values were extracted from the raw data, then interpolated into 3D data and used to train the predictive model. The minimum and maximum values that were obtained for the root mean square error (RMSE) were 0.97 and 4.7, respectively [22].

A convolutional neural network (CNN) model was proposed by Han et al. [16] for predicting convective storms in the near future using radar data. The model was designed as a binary classification model to predict whether radar echo values would be higher than 35 dBZ in the next 30 min. The input radar data were represented by 3D images and the output was also a 3D image, in which each point of the image was "1" when the radar echo was predicted to be higher than 35 dBZ in the next 30 min and "0" when it was not. The experiments produced a CSI value of 0.44.

Socaci et al. [19] proposed an adaptation of the Xception deep learning model, which they named *XNow*, for the short-term prediction of radar data. Experiments were performed using radar data that were provided by the Romanian National Meteorological Administration and an average *normalized root mean square error* of less than 3% was obtained.

The U-Net convolutional architecture has been employed in multiple studies on weather nowcasting using radar data [23,24]. Agrawal et al. [23] proposed a U-Net model for precipitation nowcasting. Their proposed model surpassed several baselines of other methods in terms of short-term prediction (up to 1 h), namely the persistence model and an optical flow algorithm, as well as the high-resolution rapid refresh (HRRR) system, but was outperformed by the HRRR model in terms of forecasts for up to 5 h. The RainNet model, which was proposed by Ayzel et al. [25], is a U-Net model that was trained using a logcosh objective function. Trebing et al. [24] introduced a lightweight U-Net model that used depth-wise separable convolutions. Their model achieved a similar performance to that of the classical U-Net while only having a quarter of its parameters.

Ciurlionis and Lukosevicius [26] used a CNN model to forecast future precipitation using current precipitation data. They used precipitation data that were estimated using a radar and trained the model with four time steps as the inputs and the next step as the output (i.e., when t was the current step, the input data were $t-3, t-2, t-1$ and t and the model predicted data for $t+1$). To predict further in the future, they used consecutive predictions (using the predicted data from the previous step as the input for the next step). They compared their approach to four basic numerical algorithms: the persistence model, a basic translation algorithm, a step translation algorithm and a sequence translation algorithm. They measured whether the models correctly predicted zero or non-zero values (i.e., they transformed the task into a classification problem). When predicting one time step, both the CNN and the sequence translation algorithm had a CSI of 0.81, while the others had CSI values of under 0.8. For predictions further in the future, the CNN had a better performance than the sequence translation algorithm; for example, at 60 min, the CNN model had a CSI of 0.71 while the numerical algorithm had a CSI of 0.65.

Differentiating from the general trend of using deep learning for machine learning models, Mao and Sorteberg [27] proposed a model that was based on a random forest (RF) for precipitation nowcasting. The random forest was trained to predict precipitation data. The inputs for the model were multiple types of data, with the main ones being precipitation data that were estimated using a radar, AROME numerical model predictions and other various data from ground weather stations, such as air pressure, air temperature and/or wind speed. To evaluate the model, the predictions were transformed into two classes: below 0.1 and above or equal to 0.1. They obtained a CSI of 0.49 for the proposed model, while the automatic radar nowcasting had a CSI of 0.42 and a baseline numerical model had a CSI of 0.33.

Bonnet et al. [28] used a video prediction model named PredRNN++, which was based on ConvLSTM combined with gradient highway units (GHUs), to predict radar reflectivity and had radar reflectivity as the input. They only used the reflectivity from the lowest elevation angle, which was collected every 5 min. The input data consisted of 10 time steps and the model predicted 10 time steps into the future. In order to measure the performance of the model, they also transformed the predictions into classifications using the thresholds of 10 dBZ for predictions and 20 dBZ for observations. In terms of metrics, they used CSI and the equitable threat score (ETS), which is an improvement on CSI that also takes true negatives into consideration. Their model obtained a CSI of 0.52 and an ETS of 0.46 for prediction at 15 min and outperformed ENCAST, which is the model that is currently used in São Paulo, Brazil, based on the extrapolation of the data that were collected from the radar.

While the majority of nowcasting models that have been proposed so far have been based on a single machine learning model, Xiang et al. [29] proposed a model that combined two types of neural networks in order to improve the nowcasting results: decision trees and numerical methods. The goal of their model was to predict the amount of precipitation at a single point 1–2 h in the future (they targeted points where there were weather stations so they were able to compare the predictions to the ground truth values that were obtained by the stations). The dataset was processed so it only contained time steps with meteorological activity. Their model worked in three steps: first, they used a numerical model for trajectory tracking to compute the trajectory of the meteorological phenomenon (e.g., storm, clouds, etc.); then, there was a feature extraction phase, in which the best features were selected (some were just general features that were provided by the weather station and some were dependent on the previous phase, such as cropping images depending on the computed trajectory); the final phase consisted of using three models to separately predict the amount of precipitation. Each model used a different set of features from the features that were extracted in the second phase. For the final output, these three values were summed up with different weights. They tested the model using different features that were extracted in the second phase. The best results were 4.035 for the RMSE and 246.52 for the mean absolute percentage error (MAPE).

One of the main problems with using convolutional neural networks that were trained with conventional loss functions to predict images is that the predictions tend to be blurry or smoothed out. Hu et al. [30] proposed an improvement for nowcasting models by adding generative adversarial networks (GANs) as a second step after the usual predictive model. They proposed two types of GANs: a spatial GAN (acting on the actual image) and a spectral GAN (acting on the spectrum of the image following a fast Fourier transform). A masked-style loss function was introduced to improve the sharpness of the generated images. In addition, a new metric (the power spectral density score (PSDS)) was proposed, which was computed based on the spectrum of the images. In order to evaluate the quality of the predictions, another metric (the learned perceptual image patch similarity (LPIPS)) was used, which was measured according to the perceptual similarity between the observations and the predictions. The CSI metric was employed to measure the performance of the model using binarized values. In their experiments, U-Net and ConvLSTM were used as base models. The results that were obtained using both types of GANs were better than those that were obtained using only the spatial GAN, except when measuring CSI at the lowest threshold. Adding the mask-style loss yielded better results in most cases. As mentioned before, the original models yielded better results than the GANs for CSI at the lowest threshold, but this changed at higher thresholds. The GANs produced better LPIPS scores, which were even better when using the mask-style loss function (0.412 for the original ConvLSTM and 0.27 for the ConvLSM with both GANs and the loss function). The PSDS scores were significantly improved when using the GANs and the loss function (0.78 for the original ConvLSTM and 0.16 for the ConvLSTM with both GANs and the loss function).

Choi and Kim [31] also used GANs to improve the performance of U-Net models. Their goal was to predict radar reflectivity using radar reflectivity as the input data. The authors proposed a precipitation nowcasting model (Rad-cGAN) that was based on a conditional generative adversarial network (cGAN). To evaluate their model, they compared their estimated precipitation values using the ZR model to the observed ground truth precipitation values that were gathered at several dams. They obtained a Pearson correlation coefficient of 0.86, an RMSE of 0.42, a Nash–Sutcliffe efficiency (NSE) of 0.73 and a CSI of 0.81.

2.2. Case Studies

In the following section, we describe the case studies that were used to evaluate the proposed *NeXtNow* model. The two case studies were conducted using datasets from Romania (provided by the NMA) and Norway (provided by the MET), which were selected because they belonged to different geographical/climatic areas and contained different radar measurements (as further highlighted in Table 1), thus allowing us to test the performance of the *NeXtNow* model more thoroughly.

2.2.1. First Case Study (NMA Data)

The NMA dataset that was used in our first case study was collected by a Doppler single-polarization radar that is located in central Romania. During a full volume scan, which is completed every 6 min, the radar outputs many different products that are related to the location, intensity and movement of precipitating clouds and their associated meteorological phenomena. For the experiments, we used the base reflectivity (R) product and the base velocity (V) product. The radar collects these base products at nine elevation angles, effectively gathering nine sets of velocity and reflectivity data at each time step. For both products, we used the data from the lowest four elevation angles, which resulted in eight products in total: R01, R02, R03, R04, V01, V02, V03 and V04. The reflectivity and Doppler radial velocity were used for the NMA case study as these are the first products that are analyzed by forecasters to identify weather features. The use of velocity fields can be theoretically useful because they can introduce the effects of convergence zones into the model for the prediction of the initiation and evolution of convective storms. The experiments that are presented in Section 3 empirically sustained this hypothesis.

To train, validate and test the model, 20 summer days with heavy rain, wind and hail and without any meteorological events were extracted from the observations, which corresponded to events that were observed in June 2010 (2nd, 10th, 12th, 13th, 14th, 19th, 20th, 22nd, 23rd and 24th), June 2017 (from 3rd to 7th) and June 2018 (11th, 13th, 15th, 16th and 21st). The study area was the central Romania region (central Transylvania) as the radar is located near the village of Bobohalma. The month of June was selected for the NMA case study as, in Romania, it is the month that the most convective storms and convective systems develop in the Carpathian basin. The dataset included days both with and without severe meteorological events and thus, a diverse dataset was obtained. Out of the entire area that is scanned by the radar, we focused on a central square with a size of 256×256 cells (the radar is located in the middle of this square).

2.2.2. Second Case Study (MET Data)

The MET radar dataset that was used in our second case study consisted of composite reflectivity values that were obtained from the MET Norway Thredds Data Server [32]. The data, which are available at [33], were obtained by processing the raw reflectivity measurements that were retrieved from multiple radars. Thus, the reflectivity product that is stored at the MET Norway is a composite map that is obtained from all elevations and tilts by taking into account the radar scans that have the best quality and not the strongest reflectivity across the elevations. This composite reflectivity product is obtained by applying an interpolation procedure and using different weights for the various radars, depending on their quality and other meteorological or non-meteorological factors that can alter the radar measurements. The reflectivity values that were used in our experiments were collected at intervals of 5 min.

To train, validate and test the model, days with and without meteorological events were selected from December 2020 (23rd, 25th, 26th and 27th), January 2021 (17th and 18th), March 2021 (3rd and 4th), April 2021 (12th and 13th), June 2021 (the entire month) and January 2022 (1st–25th). The days were selected so as to obtain a diverse dataset that contained days both with and without severe meteorological events and included both summer and winter months. The analyzed geographical area was a region surrounding Oslo. From the entire map, a square of 256 × 256 pixels was selected.

Table 1 describes the datasets that were used in our case studies. The second column in the table indicates the radar products of interest and the last column shows the number of days on which the radar data that were used in each case study were collected.

Table 1. A description of the datasets that were used in our case studies.

Case Study	Radar Products	Number of Days
NMA	Reflectivity and velocity at the four lowest elevations (R01, R01, R03, R04, V01, V02, V03 and V04)	20
MET	Composite reflectivity (CR)	65

2.3. Methodology

With the goal of answering our first research question (RQ1), we developed and evaluated our *NeXtNow* deep learning model, which was customized for the short-term prediction of weather radar products. *NeXtNow* was adapted for radar data prediction from the ResNeXt [17] architecture, which is mainly used for image processing. To the best of our knowledge, no other architecture that is based on ResNeXt has been proposed for weather nowcasting or spatiotemporal prediction problems. While several works have proposed fully convolutional or convolutional–recurrent neural networks for weather forecasting, they have employed simple and causal 2D or 3D convolutional architectures [34] and architectures that were inspired by U-Net [24] or the Xception model [19]. The basic details of the ResNeXt deep learning model are presented in Section 2.3.1, then Section 2.3.2 introduces the model that was used in our approach. Our *NeXtNow* learning model is introduced in Section 2.3.3, while the testing stage of *NeXtNow* and the methodology that was employed for the performance evaluation is discussed in Section 2.3.4.

2.3.1. ResNeXt Architecture

The ResNeXt architecture was proposed by Xie et al. [17] as an improved version of the *ResNet* model [35]. The *ResNet* architecture [35] addresses the difficulty in training very deep neural networks by introducing shortcut connections, a technique in which the input of an architectural block is added to its output in order to obtain the final output. By passing information from earlier layers to deeper layers, the network can optimize residual mappings, thus making it possible to efficiently train very deep architectures. This process can be viewed as a form of feature fusion, in which features at different levels of depth are combined using an addition operation [36]. Multiple residual blocks are stacked to form deep networks [35].

ResNeXt further builds on this architectural blueprint by using *grouped convolutions* instead of plain convolutions inside the residual blocks. Grouped convolutions are a type of convolutional layer in which the input is split channel-wise into multiple groups, with each group being processed individually by convolutions and concatenated at the end to obtain the final result. This construction has been shown to be equivalent to applying a set of aggregated transformations, which can be formalized as follows.

Given an input x and a hyperparameter that is called *cardinality C*, an aggregated transformation can be obtained from a set of transformations $\{\tau_1, \ldots, \tau_C\}$ as:

$$\mathcal{F}(x) = \sum_{i=1}^{C} \tau_i(x)$$

Following the strategy in *ResNet*, the aggregated transformation is a residual connection, which leads to the following computation for the output:

$$y = x + \sum_{i=1}^{C} \tau_i(x)$$

Figure 1 shows a schematic representation of the two types of blocks that are used in the *ResNet* and ResNeXt architectures. It has been shown experimentally that tuning the hyperparameter C can lead to significant performance improvements in image classification tasks [17].

As in the case of *ResNet*, the ResNeXt architecture is composed of a succession of blocks [17].

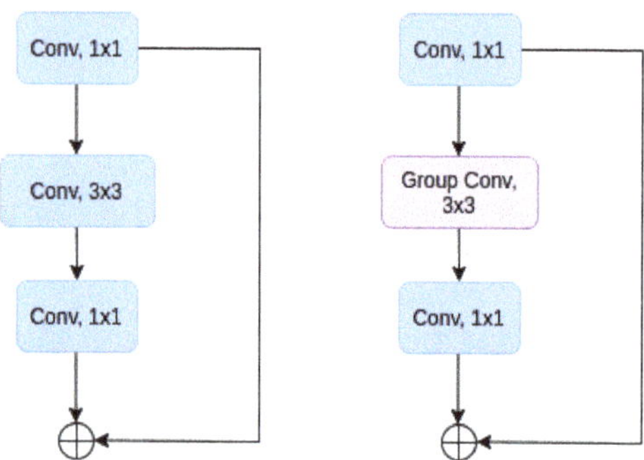

Figure 1. The *ResNet* (**left**) and *ResNeXt* (**right**) blocks. The kernel size in each convolution is shown in the figure.

2.3.2. Formalization, Data Modeling and Preprocessing

We denoted the radar products of interest by $P = \{r_1, r_2, \ldots, r_n\}$, where n is the dimensionality of P (the number of radar products that were used). For our case studies that were described in Section 2.2, we obtained the following values for P and n:

- For the first case study (NMA dataset), $n = 8$ and $P = \{R01, R02, R03, R04, V01, V02, V03, V04\}$;
- For the second case study (MET dataset), $n = 1$ and $P = \{CR\}$.

The radar data that were input at a certain time moment t were denoted by I_t and were modeled as 3D images with n channels (corresponding to the available radar products), with the i-th channel representing the value of the radar product r_i at time t. More specifically, the OX and OY axes represented the longitudinal and latitudinal values of the geographical area and the OZ axis represented the channels (i.e., the values of the radar products P at time moment t).

A sample 4-channel 3D image (with $n = 4$ products) is shown in Figure 2.

Given a certain step k, the goal of our learning problem was to predict the 3D image at time moment t from the 3D images that were collected at the time moments $t - k, t - k + 1, \ldots, t - 1$. In our model, the output was also an n-channel 3D image, in which the value of a point on the i-th channel of the image I_t was the value that was predicted for the radar product r_i at time t. We noted that one time step (i.e., the time period between two consecutive time moments $t - 1$ and t) represented the time resolution between two consecutive radar scans. More specifically, a time step was 6 min for the NMA case study and 5 min for the MET case study.

We denoted the sequence of 3D images that represented the radar data that were collected at time moments $t - k, t - k + 1, \ldots, t - 1$ by $Seq(t, k) = \; < I_{t-k}, I_{t-k+1}, \ldots, I_{t-1} >$. In this context, the target function of our learning problem was a function M that mapped the k-length sequence of n-channel 3D images ($Seq(t, k)$) onto another n-channel 3D image

(I_t), i.e., $I_t = M(Seq(t,k))$. The *NeXtNow* deep learning model learned hypothesis h, which was an approximation of M ($h \approx M$), i.e., $h(I_{t-k}, I_{t-k+1}, \ldots, I_{t-1}) \approx I_t \; \forall t$. Thus, for a sequence of images ($Seq(t,k)$), *NeXtNow* provided a multi-channel 3D image I_t that contained the estimation of the values of the radar products at time t.

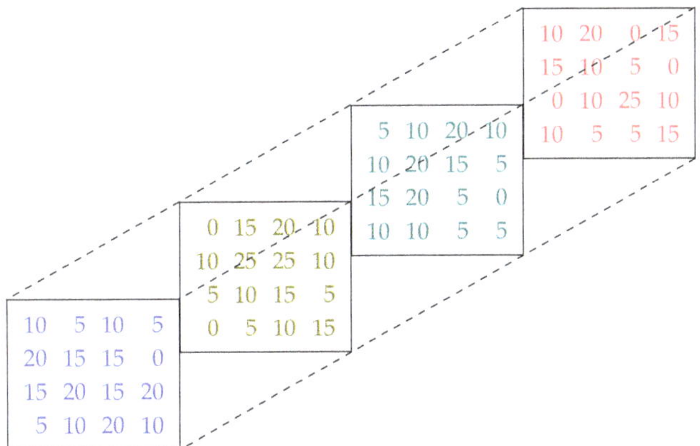

Figure 2. A sample 4-channel 3D image for I_t.

A sequence of 3D images that contained radar data that were collected at different time moments t was available. A dataset $\mathcal{D}$ was created from sequences in the form of $< I_{t-k}, I_{t-k+1}, \ldots, I_{t-1} >$, i.e., a sequence of n-channel 3D images that represented radar data that were collected at time moments $t - k, t - k + 1, \ldots, t - 1$. For each instance, the $Seq(t,k)$ from the ground truth $\mathcal{D}$ (i.e., the n-channel 3D image I_t that contained the values of the radar products at time t) was available and was used to train the model.

Before building the *NeXtNow* deep learning model, a preprocessing step was applied to the 3D images I_t to correct any possible errors that existed in the radar data. For the NMA dataset, two different preprocessing methods were used, depending on the product. For R, the only preprocessing that was carried out was to replace the "No Data" (NaN) values with "0". For the V product, there was a more complex preprocessing step. The issue with V was that it was a very noisy product because it represented the velocity *relative to the radar*, so there were some cases in which the radar could not properly estimate the direction or the speed, thus producing invalid values. These invalid values appeared often enough that they could interfere with the model learning [37]. We addressed this problem by introducing a cleaning step, which replaced the invalid values with valid values. The new values were computed as the weighted average of the values in the neighborhood surrounding the invalid value. The weight of a value in the neighborhood was inverse proportional to the difference between that value and the invalid value.

The raw MET data were preprocessed as follows. Since the raw data had negative reflectivity values, which were not important for nowcasting, these values were all replaced with a constant value of -1. Additionally, the NaN values that corresponded to missing radar measurements were replaced with values that were outside the domain of valid reflectivity values, i.e., -5, in order to be able to distinguish them from the negative values and the reflectivity values that were of interest (i.e., the positive values).

As well as the previous preprocessing steps, the data were normalized using the classical *min-max* normalization method. For the *min-max* normalization, we used the minimum and maximum values from the domain of the radar products instead of the minimum and maximum values from the training dataset. This way, we made sure that the same values in different datasets were assigned the same normalized values. In the case of the MET dataset, the minimum value that was used for normalization was -5, which corresponded to the missing radar measurements.

2.3.3. Building the *NeXtNow* Model

The predictive model *NeXtNow* was built using a training dataset that consisted of training samples in the form of $(Seq(T,k), I_t)$, where $I_t = M(Seq(T,k))$ represented the ground truth (the 3D image that consisted of the real values of the radar products at time t) that was used to train the instance $Seq(t,k) = < I_{t-k}, I_{t-k+1}, \dots, I_{t-1} >$.

The proposed model had a fully convolutional encoder–decoder architecture, which was formed of three main components. The first component was an encoder, which was inspired by the ResNeXt architecture.

The encoder consisted of two classical convolutions, which had the role of providing multiple feature maps for the inputs, followed by three ResNeXt blocks. The blocks were constructed according to the original ResNeXt paper [17], as presented in Section 2.3.1. The final convolution in the block multiplied the filter size by four, while the group convolution downsampled the input image by a factor of two. Each convolution in the block was followed by a batch normalization layer and the ReLU activation function. The convolutions that were used in the encoder had a kernel size of 3×3.

The second component was a series of eight identical ResNeXt blocks, with 1024 filters each. In contrast to the blocks that were used in the encoder, the blocks that were included in this component did not change the resolution or number of filters of their inputs, but they did have the aim of obtaining refined representations for the feature maps that were retrieved from the encoder. Empirically, we found that the addition of these additional blocks was beneficial to the model's overall performance.

While the first two components benefited from the use of ResNeXt blocks, we opted for a succession of simple convolutional layers for the decoder as experimenting with more complex architectural components did not lead to a better performance for the forecasting model. Therefore, in our proposed model, the decoder consisted of a series of upsampling layers, followed by convolutions. Following a standard approach to designing architectures for image-to-image tasks, the number of filters was progressively increased in the encoder and decreased in the decoder.

A schematic representation of the *NeXtNow* architecture is shown in Figure 3.

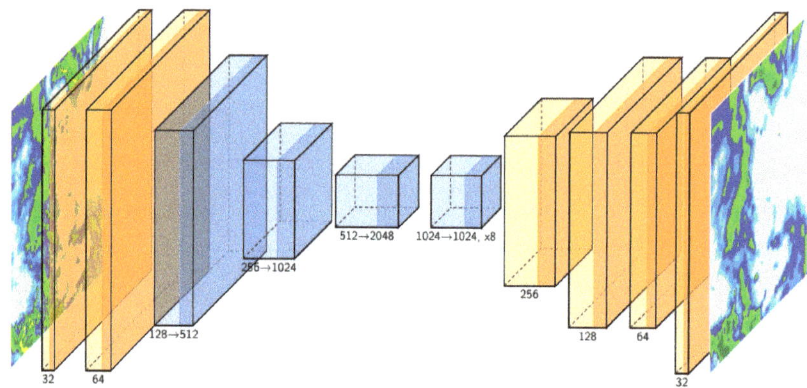

Figure 3. The proposed *NeXtNow* architecture. The ResNeXt blocks are depicted in blue, while classical convolutional layers are shown in orange. In the case of the ResNeXt blocks, the filters that correspond to the first and last convolutions in the block are shown, while only only the number of filters is shown for the plain convolutions. This figure was created using the PlotNeuralNet package [38].

The proposed architecture represented a new purely convolutional approach to weather nowcasting. The main advantage of our model was the simplicity and flexibility of the architecture, which allowed it to be easily adapted for other spatiotemporal prediction tasks with few hyperparameters that needed to be tuned. A limitation of our approach was that it did not incorporate a recurrent component for modeling the time dimension, relying instead on a simple concatenation operation for the time steps. Our model could be extended, however, by including modules from our architecture in recurrent architectures as feature extractors.

The datasets for both case studies (NMA and MET) were split into train, validation and testing subsets from the total number of days that were available (i.e., 20 days for the NMA dataset and 65 days for the MET dataset): 80% for training, 10% for model validation and the remaining of 10% for testing. From each subset (training/validation/testing), the complete days (with no missing time steps) were used.

2.3.4. Performance Evaluation and Testing Methodology

As shown in Section 2.3.3, after the *NeXtNow* model was trained, it was evaluated using 10% of the instances from the datasets $\mathcal{D}$, which were unseen during the training stage.

Various performance metrics were computed to assess the performance of *NeXtNow* using a testing subset. The experiments were repeated three times using three different training–validation–testing splits and the values for each of the performance metrics were averaged over the three runs.

Depending on the type of the input data that was used in the forecasting problem, there were three types of verification methods that were used for the performance evaluation: categorical, continuous (real values) or probabilistic approaches. Our experiments used the continuous approach since we modeled the problem as a regression task and used continuous input data that were mapped onto a continuous output.

The first set of evaluation metrics that we considered used the continuous ground truth data and the continuous forecasts that were made by the *NeXtNow* model. Given a testing dataset with n ground truth data samples in which each sample was an image containing m points, we denoted the ground truth (observation) value for the i-th point in the t-th testing instance by $O_{t,i}$ and the prediction (forecast) value for the i-th point in the t-th testing instance by $F_{t,i}$. The following evaluation metrics that have been used in the regression literature were computed for each testing sample [39]:

- *Root mean square error (RMSE)*, which was computed as the square root of the mean square errors that were obtained using the tth testing data sample:

$$RMSE(t) = \sqrt{\frac{\sum_{i=1}^{m}(O_{t,i} - F_{t,i})^2}{m}}.$$

  Lower *RMSE* values indicated better predictions.
- *Correlation coefficient (CC)*, which expressed a linear relationship between the forecast and the actual observation (ground truth) and was computed as

$$CC(t) = \frac{\sum_{i=1}^{m}(F_{t,i} - \overline{F}_t)(O_{t,i} - \overline{O}_t)}{\sqrt{\sum_{i=1}^{m}(F_{t,i} - \overline{F}_t)^2}\sqrt{\sum_{i=1}^{m}(O_{t,i} - \overline{O}_t)^2}},$$

  where $\overline{O}_t$ represents the average of the actual observations $(\overline{O}_t = \frac{1}{m} \cdot \sum_{i=1}^{m} O_{t,i})$ and $\overline{F}_t$ is the average of the forecasts $(\overline{F}_t = \frac{1}{m} \cdot \sum_{i=1}^{m} F_{t,i})$. CC produced values between $[-1, 1]$, where $CC = 1$ represented a perfect fit between the forecast and the true observation that was obtained. Higher values of CC indicated better predictions.

- The radar reflectivity data included numerous missing points, which corresponded to the regions for which the radar did not detect any signals. In the NMA case study, these points were associated with 0 values, while for the MET dataset, which contained negative values, we encoded missing radar reflectivity data using a value of -5, as presented in Section 2.3.2. In order to present common terminology and notations, we referred to these points as zero-labeled points. Since we were not interested in the prediction performance at these points, we only computed the values for the $RMSE$ and CC performance metrics for the non-zero labeled instances , i.e.,:

$$RMSE_{nz}(t) = \sqrt{\frac{\sum_{i,O_{t,i}\neq 0} (O_{t,i} - F_{t,i})^2}{n_z(t)}}$$

where $n_z(t) = |\{i \in \{1,\dots,m\}|O_{t,i} \neq 0\}|$ is the number of non-zero points in testing sample t and

$$CC_{nz}(t) = \frac{\sum_{i,O_{t,i}\neq 0} (F_{t,i} - \overline{F}_t)(O_{t,i} - \overline{O}_t)}{\sqrt{\sum_{i,O_{t,i}\neq 0} (F_{t,i} - \overline{F}_t)^2}\sqrt{\sum_{i,O_{t,i}\neq 0} (O_{t,i} - \overline{O}_t)^2}}$$

where $\overline{O}_t$ and $\overline{F}_t$ represent the mean observations and forecasts that were computed across the non-zero points.

The values that were obtained for all of the testing samples were averaged in order to obtain the final evaluation metrics for the testing subset: $RMSE = \frac{\sum_{t=1}^{n} RMSE(t)}{n}$, $RMSE_{nz} = \frac{\sum_{t=1}^{n} RMSE_{nz}(t)}{n}$, $CC = \frac{\sum_{t=1}^{n} CC(t)}{n}$ and $CC_{nz} = \frac{\sum_{t=1}^{n} CC_{nz}(t)}{n}$.

For a thorough assessment of *NeXtNow*'s performance, we discretized its continuous output by applying a threshold in order to evaluate the performance of our model using additional evaluation metrics. For meteorologists, the classes of the values of the radar products are particularly relevant, for example, for stratiform and convective rainfall classification. By applying a threshold τ to the continuous output values that were provided by *NeXtNow*, the set of evaluation metrics was enlarged with the performance metrics that were used for binary classification: values that were higher than τ could be considered as belonging to the *positive* class, while values that were lower than τ belonged to the *negative* class.

For the testing dataset, after computing the confusion matrix that corresponded to the binary classification task (TP, number of true positives; FP, number of false positives; TN, number of true negatives; FN, number of false negatives), the evaluation metrics that are described below were calculated:

- *Critical success index* (CSI), which was obtained as $CSI = \frac{TP}{TP + FN + FP}$;
- *False alarm rate* (FAR), which was computed as $FAR = \frac{FP}{FP + TP}$;
- *Probability of detection* (POD), which represented the *recall* of the classifier and was computed as $POD = \frac{TP}{TP + FN}$;
- *Bias* ($BIAS$), which was used for categorical forecasts and was equal to the total number of events that were positively predicted divided by the total number of actual positive events, i.e., $BIAS = (TP + FP)/(TP + FN)$.

We note that the CSI, FAR, POD and $BIAS$ metrics have been widely used for performance assessment in the forecasting literature. CSI, FAR and POD ranged between $[0, 1]$, while the domain of $BIAS$ was $[0, \infty)$. Higher values of CSI and POD and lower

FAR values were expected for better predictions, while *BIAS* values of closer to 1 were expected for better forecasting models.

3. Results

This section aims to answer the second research question (RQ2) and present the experimental results that were obtained when evaluating the performance of the *NeXtNow* forecasting model, which was built to predict the values of radar products using the methodology that was introduced in Section 2.3. The model was implemented using the TensorFlow library [40]. The experiments were run on two laptops, which had the following configurations: an Intel i9-10980HK CPU, 32 GB of RAM and an Nvidia RTX 2080 Super for GPU acceleration; an Intel i7-9750H CPU and an Nvidia GeForce GTX 1660 GPU.

3.1. Datasets

The datasets that were used in our experiments were collected during the two case studies that were described in Section 2.2 and are publicly available at [41] (NMA dataset) and [33] (MET dataset).

3.1.1. NMA Dataset

The meteorological radar at NMA collects data at every 6 min (240 acquisitions per day) at nine elevation angles. Every collection at every elevation is a matrix of floating point numbers and contains $460 \times 460 = 211{,}600$ data points, ranging from 0 to 70. A custom cleaning process was applied to the raw radar from NMA and the cleaned data were used in the experiments. The data were stored in .NetCDF (network common data form) files [42] and each collection (for each elevation and each radar product) was stored in a separate file ($460 \times 460 \times 240 \times 9 = 457{,}056{,}000$ data points for a single product for an entire day).

Figure 4 summarizes the reflectivity radar product for all elevation levels and all acquisitions during one full day and highlights the area that is covered by the radar, both in terms of raw data and geographical location (projected onto the map).

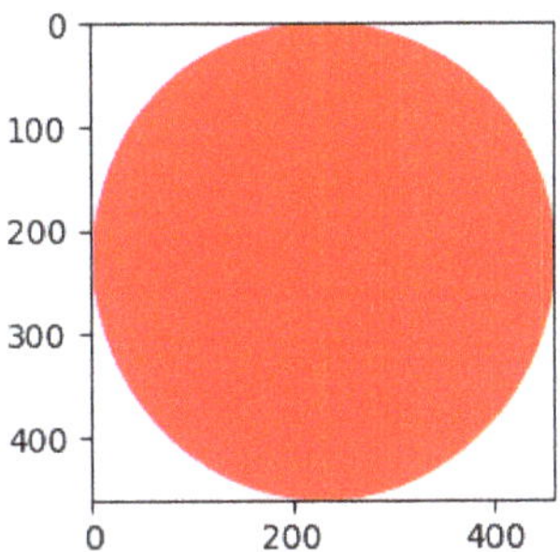
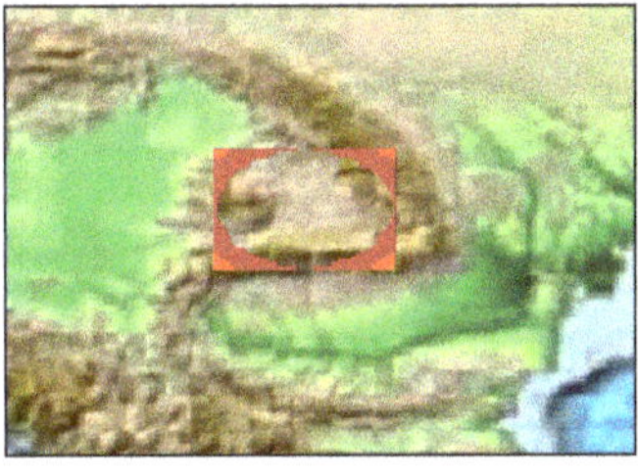

Figure 4. A visualization of the data points that are missing for all acquisitions at all elevation levels during an entire day in the NMA dataset for the reflectivity product. On the left, there is a visualization of the data matrix, in which each red pixel indicates that the value is present all day for all elevations. On the right, there is a visualization of the missing data projected onto the map, in which the red color indicates that the values are missing all day for all acquisitions at all elevation levels.

The histogram in Figure 5 reveals the imbalance between the larger values and smaller values and also highlights the fact that during the cleaning process, the data points were grouped into categories ([0–5), [5–10), [10–15)...). There were significantly fewer high values (> 50 usually indicated severe meteorological phenomena) than small values (the x axis used a logarithmic scale) and this imbalance in the data made the prediction problem more difficult.

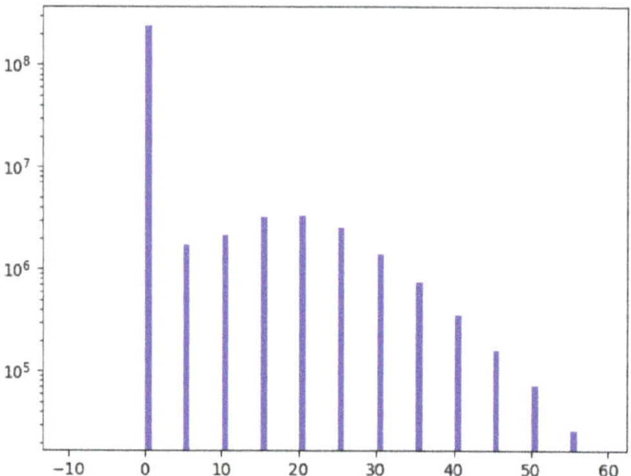

Figure 5. A histogram of the non-missing values in the NMA dataset for the entire region for a whole day, including all acquisitions at all elevations. A logarithmic scale was used on the OY axis.

3.1.2. MET Dataset

The meteorological radars at MET collect data at every 5 min (288 acquisitions per day). The data that were used in the experiments were measurements of *composite reflectivity*, which is a derived (computed) radar product that aggregates the actual radar data for all elevations. Every collection is a matrix of floating point numbers and contains $2134 \times 1694 = 3{,}614{,}996$ data points, ranging from -33 to 80. All data that correspond to a single day are stored inside a single .NetCDF file ($2134 \times 1694 \times 288 = 1{,}041{,}118{,}848$ data points).

Figure 6 graphically presents the data points that are missing for all acquisitions during a day in the MET dataset. Each pixel in the picture represents 288 data points (all acquisitions during the day). A pixel is red when the data were missing for the entire day in that region. The left-hand picture illustrates the data matrix, while the right-hand figure shows the data when projected onto a map.

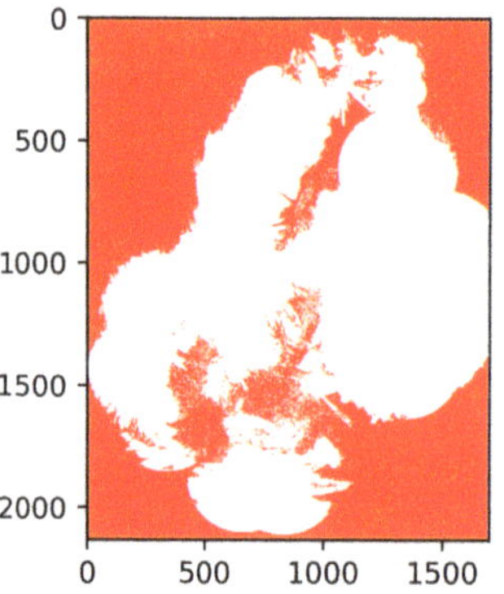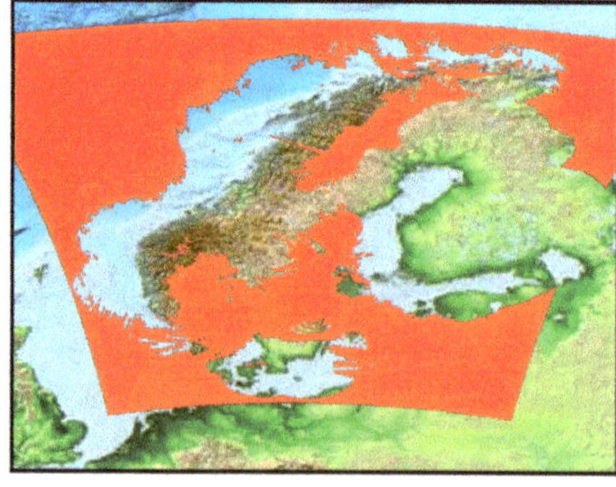

Figure 6. A visualization of data points that are missing for all acquisitions during a day from the MET dataset. Each pixel in the picture represents 288 data points (all acquisitions during a day). A pixel is colored in red when the data were missing for the entire day at that pixel. On the left, there is a visualization of the data matrix; on the right, there is a visualization of the data projected onto a map.

The available MET data presented some challenges in terms of applying deep learning methods. Approximately 50% of the data were missing (zero-value) due to various factors. Figure 7 depicts the number of missing data points during a day. Each pixel in the picture represents the number of times the data were present during the day. A pixel is colored in dark red when the data were present for all acquisitions during a day and white/transparent when the data were missing for the entire day. On the left, there is a visualization of the data matrix; on the right, there is a visualization of the data projected onto a map. As shown in Figure 7, the data are never collected in some regions because those regions are not covered by the radars or the geographic topology prevents data collection. For other areas, as shown in Figure 7, data are sometimes present and sometimes not (data are temporarily unavailable at a given point because measurements are eliminated from the composite product, etc.).

A histogram of the non-missing values in the MET dataset for the entire region and for all acquisitions during a day is presented in Figure 8. As shown in the figure, the distribution of the actual values in the dataset was highly imbalanced, as for the NMA dataset (Section 3.1.1). Larger values were of more interest from a meteorological viewpoint as these indicated severe weather phenomena, but those were relatively rare. This severe imbalance was a challenge from a supervised learning viewpoint.

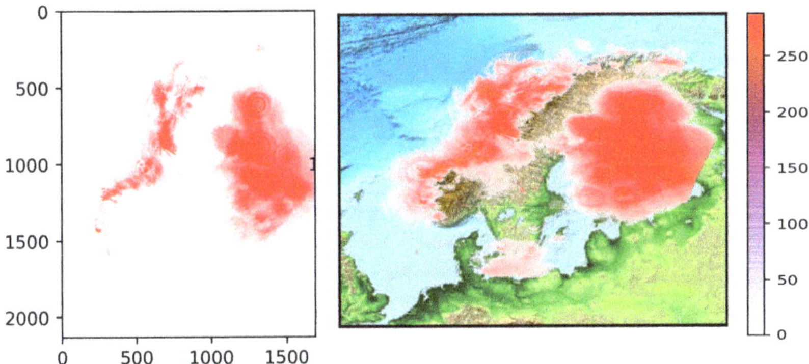

Figure 7. A visualization of the number of missing data points in the MET dataset during a day. Each pixel in the picture represents the number of times the data were present during the day. A pixel is colored in dark red when the data were present for all acquisitions during the day and white/transparent when data were missing for the entire day. On the left, there is a visualization of the data matrix; on the right, there is a visualization of the data projected onto a map.

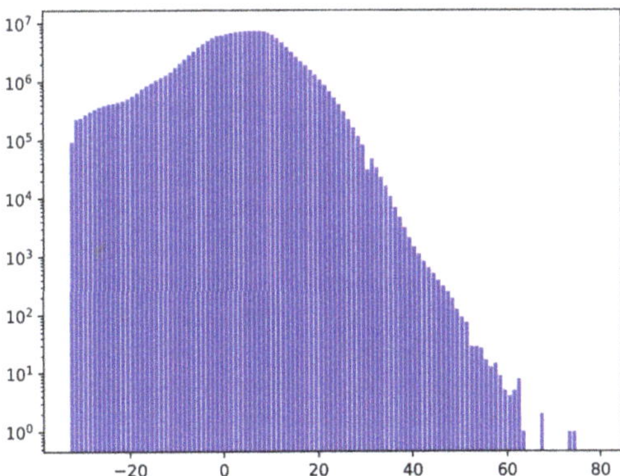

Figure 8. A histogram of the non-missing values in the MET dataset for the entire region and for all acquisitions during a day. A logarithmic scale was used on the OY axis.

3.2. Parameter Setting

Our model was trained to minimize the errors between the ground truth radar data values and the predicted outputs using the root mean square error loss. The Adam optimizer was also used with an initial learning rate of 0.0005 for the MET experiments and an initial learning rate of 0.001 for the NMA experiments. A learning rate scheduler was applied to reduce the learning rate by a factor of two after every five epochs with no improvements in the validation loss. For the MET case study, the 32 grouped convolutions were used in the ResNeXt blocks while for the NMA case study, 64 groups were used.

As described in Section 2.3.4, for performance evaluation purposes, a threshold τ was used to transform the continuous output of the *NeXtNow* model into a discrete output. The values that we considered for the threshold τ were 5, 10, 15, 20 and 30, which corresponded to light to moderate rainfall.

3.3. Experimental Results

This section presents the results that were obtained for our proposed model using the two case studies that were presented in Section 2.2. We first evaluated our model in terms of predicting the evolution of the radar data products using a lead time of one time step (5 or 6 min) in the future. Table 2 shows the results that were obtained for the regression metrics using the two case studies for predicting one time step in the future, where k denotes the number of previous time steps that were used in the prediction (as presented in Section 2.3). The results for the classification evaluation metrics that were obtained for the considered thresholds are presented in Table 3. The best values are highlighted.

By analyzing the performance metrics that are shown in Table 3, we observed that the performance decreased with the increases in the threshold value, which was to be expected since high reflectivity values were scarce in the dataset, thus were challenging to predict accurately. For both the NMA and MET datasets, the results in Tables 2 and 3 revealed that the average performance for predicting one time step in the future using four previous time steps ($k = 4$) was better than that when only using one previous time step ($k = 1$) in terms of the *FAR*, *CC* and CC_{nz} performance metrics. We noted that for the NMA dataset, the *FAR* value was better for $k = 1$ with high values for the threshold τ (i.e., 20 and 30). For the other performance metrics (*RMSE*, $RMSE_{nz}$, *CSI*, *POD* and *BIAS*) the performance was better for $k = 1$. This suggested that when the *NeXtNow* model used four previous time steps, it reduced the number of forecasts that were false alarms; however, it forecasted a smaller number of events than when it only used one previous time step.

The values for the performance metrics (Tables 2 and 3) that were obtained by our *NeXtNow* model using one previous time step ($k = 1$) for both the NMA and MET datasets were tested against those that were obtained using four previous time steps ($k = 4$) using a two-tailed paired Wilcoxon signed-rank test [43,44]. A p-value of less than 0.00001 was obtained, which highlighted that the differences between the *NeXtNow* performances when using one and four previous time steps were statistically significant, with a significance level of $\alpha = 0.01$. Thus, the results revealed that using multiple time steps did not improve the performance of predictions for one time step in the future, which corresponded to a lead time of 6 min for the NMA case study and 5 min for the MET case study.

While this result might seem counter-intuitive, it was not completely unexpected for the NMA experiments. In our previous work on radar data from NMA [45], we used unsupervised neural network techniques (self-organizing maps) to mine relevant patterns from the data and empirically showed that when predicting the value of the next step at a location, there were no significant differences between the patterns that were mined from one previous time step and those that were mined from five previous time steps. In other words, when the patterns were similar, adding more time steps did not add much more information. We hypothesized that this occurred because we were using both reflectivity and velocity, thus the network had the possibility to find the trajectory of the meteorological event from a single time step because of the velocity product, so multiple time steps did not add much information.

Table 2. The results for the regression evaluation metrics for predicting one time step in the future, where k denotes the number of previous time steps that were used in the predictions. The means and standard deviations that were computed across the three experimental runs are shown. The best values for the performance metrics are marked with bold and colored with yellow (for the NMA case study) and with blue (for the MET case study).

Case Study	k	$RMSE$ ($\downarrow$)		$RMSE_{nz}$ ($\downarrow$)		CC ($\uparrow$)		CC_{nz} ($\uparrow$)	
		Mean	Stdev	Mean	Stdev	Mean	Stdev	Mean	Stdev
NMA	1	**2.442**	0.091	**7.988**	0.371	0.550	0.005	0.666	0.020
	4	2.539	0.079	8.038	0.184	**0.557**	0.025	**0.674**	0.033
MET	1	**1.582**	0.584	**5.213**	1.817	0.747	0.060	0.603	0.042
	4	1.606	0.542	5.295	1.584	**0.823**	0.057	**0.693**	0.059

Table 3. The results for the classification evaluation metrics for predicting one time step in the future. The means and standard deviations that were computed across the three experimental runs are shown. The best values for the performance metrics are marked with bold and colored with yellow (for the NMA case study) and with blue (for the MET case study).

Case Study	k	Threshold τ	CSI ($\uparrow$)		FAR ($\downarrow$)		POD ($\uparrow$)		$BIAS$ ($\uparrow$)	
			Mean	Stdev	Mean	Stdev	Mean	Stdev	Mean	Stdev
NMA	1	5	**0.683**	0.009	0.134	0.043	**0.767**	0.046	**0.888**	0.100
		10	**0.595**	0.068	0.074	0.035	**0.629**	0.094	**0.682**	0.130
		15	**0.459**	0.126	0.060	0.034	**0.477**	0.145	**0.511**	0.174
		20	**0.311**	0.170	**0.076**	0.043	**0.325**	0.187	**0.359**	0.220
		30	**0.135**	0.144	**0.204**	0.115	**0.151**	0.167	**0.210**	0.245
	4	5	0.674	0.025	**0.116**	0.033	0.741	0.052	0.841	0.089
		10	0.578	0.072	**0.066**	0.027	0.605	0.089	0.650	0.112
		15	0.434	0.101	**0.056**	0.029	0.448	0.111	0.477	0.130
		20	0.266	0.099	0.081	0.057	0.275	0.107	0.305	0.131
		30	0.094	0.089	0.309	0.266	0.106	0.103	0.192	0.179
MET	1	5	**0.735**	0.060	0.114	0.043	**0.809**	0.038	**0.913**	0.009
		10	**0.660**	0.048	0.136	0.027	**0.737**	0.044	**0.853**	0.036
		15	**0.584**	0.061	0.160	0.017	**0.657**	0.067	**0.781**	0.068
		20	**0.517**	0.061	0.177	0.013	**0.581**	0.071	**0.705**	0.078
		30	**0.212**	0.195	0.204	0.018	**0.232**	0.216	**0.293**	0.270
	4	5	0.663	0.094	**0.020**	0.011	0.673	0.099	0.687	0.105
		10	0.467	0.175	**0.011**	0.005	0.470	0.177	0.476	0.180
		15	0.293	0.176	**0.005**	0.002	0.293	0.176	0.295	0.178
		20	0.139	0.126	**0.008**	0.009	0.139	0.126	0.140	0.127
		30	0.003	0.005	**0.001**	0.002	0.003	0.005	0.003	0.005

In light of these new results, it might just be that because meteorological data change very slowly from one time step to another [46], the trajectory of the meteorological event was not so relevant when only predicting one time step (5 to 6 min) in the future. While the velocity product might still be enough to make multiple past time steps redundant, this could explain why using multiple previous time steps did not improve the results for the MET experiments, in which the velocity product was not used. This meant that using multiple previous time steps, which would allow the network to learn to compute the trajectory of an event, could be more useful when predicting further than 5 min in the future.

In other words, in the absence of radar products that relate to motion, we hypothesized that including multiple radar measurements could encapsulate information regarding the direction of movements and hence, improve the predictive performance for *larger lead times* than just 5 min. In order to validate this hypothesis for the MET case study, we further evaluated the predictive performance of forecasts that were performed by our model for 15 min in the future. The model was trained in a similar manner to before, with the difference that it was optimized to predict the radar reflectivity values at a time point 15 min in the future using a series of k consecutive time steps. As previously, we performed the experiments for $k \in \{1,4\}$. The results for the regression and classification metrics are presented in Tables 4 and 5, respectively. The best obtained results are highlighted for each regression metric (Table 4) and each classification metric for each value of the threshold τ (Table 5).

By analyzing the results for both the regression and classification metrics in Tables 4 and 5, we observed that the results that were obtained for $k = 4$ time steps were better than the results that were obtained using only one past time step, which confirmed our hypothesis. Only for the threshold value of 30 were the values for FAR and $BIAS$ slightly better for $k = 1$ than for $k = 4$, but this might be due to the data imbalance (reflectivity values that were higher than $\tau = 30$ were scarce in the dataset). By comparing the regression metrics in Tables 2 and 4 and the classification metrics in Tables 3 and 5, we also observed that our model obtained better results for a 5-min lead time than for a 15-min lead time, which could be explained by the fact that the forecasts were more challenging for lead times that were further in the future.

Figure 9 illustrates some sample predictions from our *NeXtNow* model, which was trained using the NMA dataset. The first column depicts the inputs, the second column presents the predictions and the last column shows the actual radar observations. Each row in the figure shows a different product (R01 to R04).

Table 4. The results for the regression evaluation metrics for predicting three time steps in the future using the MET case study, where k denotes the number of previous time steps that were used in the predictions. The means and standard deviations that were computed across the three experimental runs are shown. The best values for the performance metrics are marked with bold.

k	$RMSE$ ($\downarrow$)		$RMSE_{nz}$ ($\downarrow$)		CC ($\uparrow$)		CC_{nz} ($\uparrow$)	
	Mean	**Stdev**	**Mean**	**Stdev**	**Mean**	**Stdev**	**Mean**	**Stdev**
1	2.423	0.834	7.184	1.829	0.623	0.055	0.450	0.066
4	**1.810**	0.612	**5.805**	1.722	**0.747**	0.059	**0.597**	0.064

Table 5. The results for the classification evaluation metrics for predicting three time steps in the future using the MET case study. The means and standard deviations that were computed across the three experimental runs are shown. The best values for the performance metrics are marked with bold.

k	Threshold τ	CSI ($\uparrow$)		FAR ($\downarrow$)		POD ($\uparrow$)		$BIAS$ ($\uparrow$)	
		Mean	**Stdev**	**Mean**	**Stdev**	**Mean**	**Stdev**	**Mean**	**Stdev**
1	5	0.467	0.136	0.155	0.105	0.522	0.167	0.633	0.232
	10	0.342	0.208	0.155	0.116	0.384	0.240	0.478	0.313
	15	0.230	0.179	0.143	0.129	0.252	0.200	0.320	0.265
	20	0.133	0.139	0.107	0.132	0.143	0.154	0.181	0.210
	30	0.006	0.010	**0.208**	0.294	0.006	0.010	**0.010**	0.017
4	5	**0.660**	0.026	**0.096**	0.041	**0.710**	0.008	**0.786**	0.032
	10	**0.511**	0.081	**0.082**	0.035	**0.537**	0.094	**0.587**	0.115
	15	**0.359**	0.118	**0.064**	0.027	**0.370**	0.128	**0.397**	0.146
	20	**0.195**	0.154	**0.060**	0.029	**0.199**	0.159	**0.211**	0.171
	30	**0.007**	0.011	0.431	0.497	**0.007**	0.011	0.007	0.012

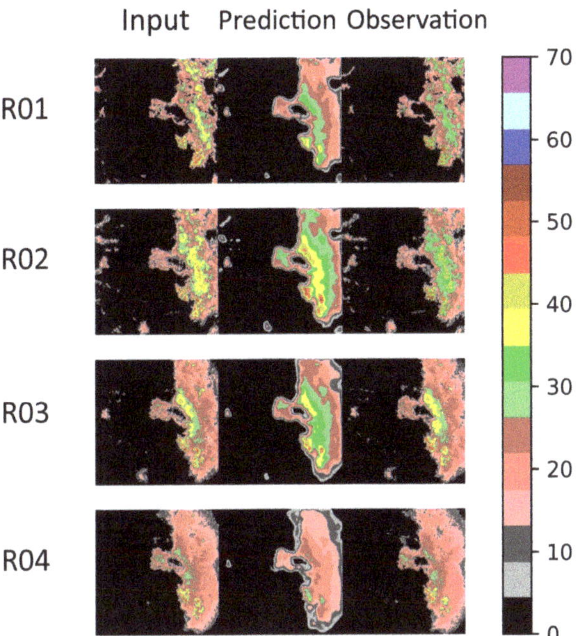

Figure 9. Sample predictions from our model that was trained using the NMA dataset. The first column depicts the inputs, the second column presents the predictions and the last column shows the ground truth observations. Each row shows a different product (R01 to R04). The illustrated observations and predictions correspond to radar measurements that were gathered in an area of approximately 250 × 250 km.

Figures 10 and 11 show sample predictions that were obtained using our model that was trained using the MET case study for lead times of 5 min and 15 min, respectively. In both figures, the first four columns show the inputs, the fifth column depicts the predictions and the last column shows the actual radar observations. As can be observed from the figures, the predictions that were produced by the model were smoother than the actual observations.

The experimental results that were previously presented revealed a decrease in *NeXtNow*'s performance at higher reflectivity values, which was likely due to the imbalance between the amounts of smaller and higher reflectivity values in the datasets. Extending the dataset by including a much larger number of convective events could improve the prediction at higher reflectivity values. Nevertheless, when dealing with storm-based nowcasting, as per the NMA case study, the prediction of reflectivity spatial patterns is equally important to assess the evolution of convective storms. Our future work is envisaged to address this challenge.

By comparing the predictions of *NeXtNow* for one time step in the future using the NMA (Figure 9) and MET (Figure 10) datasets, we could observe better predictions at high reflectivity values using the NMA dataset. This improvement in *NeXtNow*'s prediction performance at higher reflectivity values could be due to the velocity field introducing supplementary information about convections.

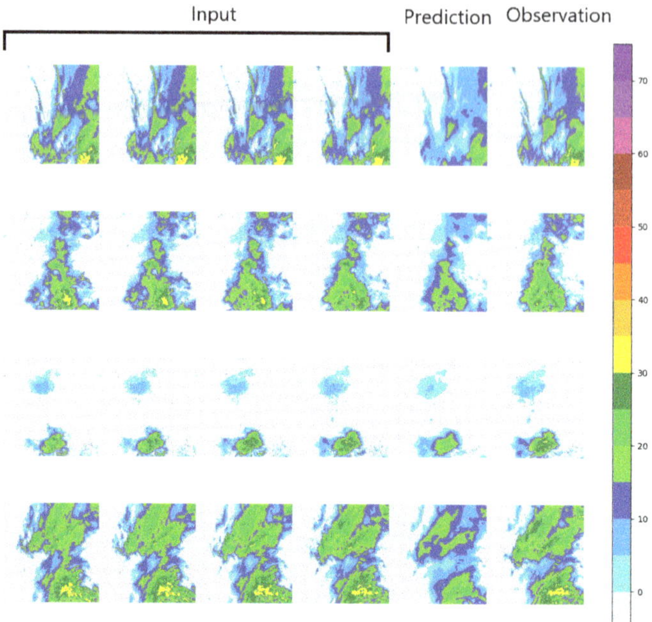

Figure 10. Sample predictions from our model that was trained using the MET dataset for a 5-min lead time. The first four columns show the inputs, the fifth column depicts the predictions and the last column shows the observations. The illustrated observations and predictions correspond to radar measurements that were gathered in an area of approximately 250 × 250 km.

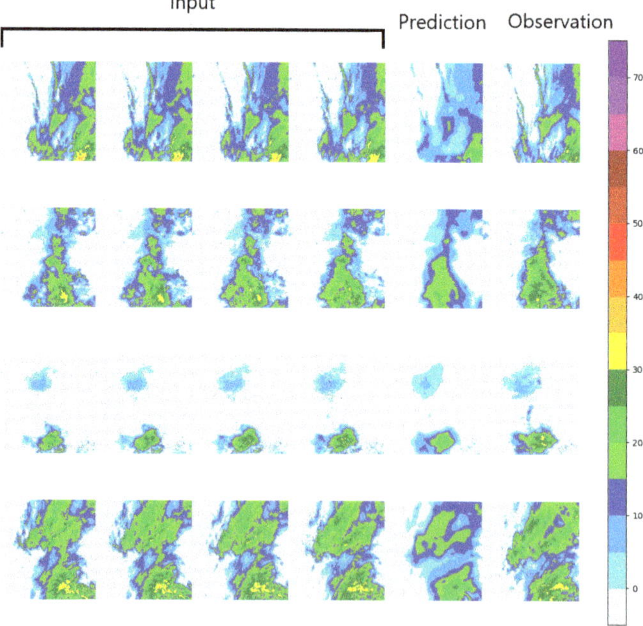

Figure 11. Sample predictions from out model that was trained using the MET dataset for a 15-min lead time. The first four columns show the inputs, the fifth column depicts the predictions and the

last column shows the observations. The illustrated observations and predictions correspond to radar measurements that were gathered in an area of approximately 250×250 km.

4. Discussion

The literature review that was presented in Section 2.1 discussed the recent advancements within the field of deep learning nowcasting. Still, there has been limited research on short-term prediction of radar products' values. Most of the related work has focused on the precipitation nowcasting problem.

To answer our third research question (RQ3), the proposed *NeXtNow* model was further compared to a convolutional architecture that has been previously proposed in the nowcasting literature for the short-term prediction of radar data and has a goal similar to ours (*XNow* [19]). *XNow* is an Xception-based deep learning model that was trained using radar data that were collected at time $t - 1$ for a specific geographic area for predicting one time step in the future (i.e., predicting the radar data at time t).

For an exact comparison between the *NeXtNow* and *XNow* models, *XNow* was evaluated using the methodology that was employed for our evaluation of *NeXtNow*. The experiments were repeated three times using three different training–validation–testing splits and the values for each of the performance metrics that were described in Section 2.3.4 were averaged over the three runs.

Tables 6 and 7 illustrate the results for the regression and classification metrics for the *NeXtNow* and *XNow* models, which were trained using one previous time step ($k = 1$) for predicting one time step in the future using both the NMA and MET datasets. The results for the classification metrics that are shown in Table 7 were evaluated for various values of the threshold τ. The means and standard deviations that were computed across the three runs are also shown in the tables. The best values are highlighted.

The comparative results in Tables 6 and 7 highlighted that for both the NMA and MET datasets, *NeXtNow* outperformed *XNow* in most of the evaluation metrics and at most of the considered thresholds. In all of these cases, the standard deviation was lower for *NeXtNow*, which showed that the *NeXtNow* model was more stable than *XNow*. For the NMA dataset, we noted that *NeXtNow* was outperformed by *XNow*, but only in terms of *FAR* at all thresholds. This suggested that *NeXtNow* forecasted a higher number of events than *XNow*, but it erroneously forecasted a slightly higher number of normal weather conditions. For the MET dataset, on the other hand, there were only four cases when *NeXtNow* was only slightly outperformed by *XNow*.

The improvement in the performance of *NeXtNow* with respect to *XNow* was statistically significant, with a significance level of $\alpha = 0.01$, as shown by a one-tailed paired Wilcoxon signed-rank test [43,44]. A *p*-value of less than 0.00001 was obtained, which highlighted the statistical significance of the differences that were observed between the performances of *NeXtNow* and *XNow*, as shown in Tables 6 and 7.

The performance of *NeXtNow* could not be precisely compared to that of other approaches in the literature that focused on the prediction of the values of radar products as the datasets that were used for their evaluation differed from ours (considering the radar products that were employed, i.e., reflectivity, velocity and composite reflectivity in our case) and the learning tasks were not formulated exactly as in this paper. When we were only looking at the magnitude of the performance metrics that have been provided by the literature and we disregarded the datasets that were used, we noted the following: *RMSE* values ranging from 0.97 to 4.7 [22], *CSI* values ranging from 0.36 [21] to 0.81 [31], a *POD* value of 0.61 [21] and a *FAR* value of 0.52. The experimental results that were presented in Section 3.3 revealed that the performance of the *NeXtNow* model for the classification task for predicting one step in the future (for $\tau = 5$) compared favorably to the performances of the models in the related work: a maximum *RMSE* of 2.442 for the regression task; *CSI* values of 0.683 and 0.735 (for the two case studies); *POD* values ranging from 0.673 to 0.809; and a maximum *FAR* of 0.134.

Table 6. The results for the regression metrics for the *NeXtNow* and *XNow* models, which were trained using one previous time step ($k = 1$) for predicting one time step in the future using both the NMA and MET datasets. The means and standard deviations that were computed across the three experimental runs are shown. The best values for the performance metrics are marked with bold and colored with yellow (for the NMA case study) and with blue (for the MET case study).

Case Study	Model	$RMSE$ ($\downarrow$)		$RMSE_{nz}$ ($\downarrow$)		CC ($\uparrow$)		CC_{nz} ($\uparrow$)	
		Mean	Stdev	Mean	Stdev	Mean	Stdev	Mean	Stdev
NMA	*NeXtNow*	**2.442**	0.091	**7.988**	0.371	0.550	0.005	0.666	0.020
	XNow	2.507	0.137	8.174	0.388	**0.595**	0.016	**0.692**	0.003
MET	*NeXtNow*	**1.582**	0.584	**5.213**	1.817	**0.747**	0.060	**0.603**	0.042
	XNow	1.630	0.615	5.410	1.857	0.737	0.068	0.580	0.053

Table 7. The results for the classification metrics for the *NeXtNow* and *XNow* models, which were trained using one previous time step ($k = 1$) for predicting one time step in the future using both the NMA and MET datasets. The means and standard deviations that were computed across the three experimental runs are shown. The best values for the performance metrics are marked with bold and colored with yellow (for the NMA case study) and with blue (for the MET case study).

Case Study	Model	Threshold τ	CSI ($\uparrow$)		FAR ($\downarrow$)		POD ($\uparrow$)		$BIAS$ ($\uparrow$)	
			Mean	Stdev	Mean	Stdev	Mean	Stdev	Mean	Stdev
NMA	*NeXtNow*	5	**0.683**	0.009	0.134	0.043	**0.767**	0.046	**0.888**	0.100
		10	**0.595**	0.068	0.074	0.035	**0.629**	0.094	**0.682**	0.130
		15	**0.459**	0.126	0.060	0.034	**0.477**	0.145	**0.511**	0.174
		20	**0.311**	0.170	0.076	0.043	**0.325**	0.187	**0.359**	0.220
		30	**0.135**	0.144	0.204	0.115	**0.151**	0.167	**0.210**	0.245
	XNow	5	0.637	0.017	**0.102**	0.009	0.687	0.024	0.766	0.033
		10	0.480	0.066	**0.053**	0.013	0.495	0.073	0.523	0.084
		15	0.293	0.121	**0.038**	0.014	0.297	0.125	0.311	0.134
		20	0.148	0.112	**0.038**	0.013	0.149	0.114	0.156	0.121
		30	0.029	0.036	**0.065**	0.018	0.029	0.036	0.032	0.040
MET	*NeXtNow*	5	**0.735**	0.060	**0.114**	0.043	0.809	0.038	0.913	0.009
		10	**0.660**	0.048	**0.136**	0.027	**0.737**	0.044	0.853	0.036
		15	**0.584**	0.061	**0.160**	0.017	**0.657**	0.067	**0.781**	0.068
		20	**0.517**	0.061	0.177	0.013	**0.581**	0.071	**0.705**	0.078
		30	**0.212**	0.195	**0.204**	0.018	**0.232**	0.216	**0.293**	0.270
	XNow	5	0.723	0.064	0.131	0.045	**0.810**	0.044	**0.932**	0.015
		10	0.641	0.068	0.153	0.041	0.725	0.065	**0.856**	0.067
		15	0.558	0.099	0.167	0.032	0.629	0.121	0.755	0.145
		20	0.468	0.139	**0.165**	0.021	0.521	0.172	0.626	0.217
		30	0.182	0.189	0.223	0.025	0.200	0.212	0.253	0.270

5. Conclusions

In this paper, we proposed a convolutional deep learning model called *NeXtNow*, which was inspired by the ResNeXt architecture for weather radar data forecasting. *NeXtNow* adapted the ResNeXt [17] architecture that has been proposed in the computer vision literature for the task of spatiotemporal prediction. Our proposed model has an encoder–decoder architecture, which maps past radar measurements onto radar measurements that are recorded in the future. We noted the generality of the *NeXtNow* model, which was proposed for short-term radar data prediction. The model could be applied not only to nowcasting but also to predicting other meteorological phenomena, such as heatwaves or droughts.

To evaluate the performance of *NeXtNow* using radar data that were obtained from different geographical/climatic areas and contained different radar measurements, two case studies were considered: one using data that were collected from Romania and the other employing data that were collected from Norway.

The research questions that were formulated in Section 1 were answered. The *NeXtNow* model that was designed for short-term radar data prediction answered RQ1. To the best of our knowledge, the ResNeXt architecture has yet not been adapted for the task of spatiotemporal prediction or, more specifically, radar data prediction. RQ2 was answered by the experiments that were performed using the time series of radar data that were provided by the Romanian National Meteorological Administration and the Norwegian Meteorological Institute. We empirically showed through these case studies that including multiple past radar measurements did not improve predictions for one time step in the future, but they did provide more accurate predictions for multiple time steps further in the future. Our experimental evaluation of *NeXtNow* also highlighted an improvement in performance when predicting the values of the radar products based on their historical values compared to the performance of a convolutional architecture that has been previously proposed in the nowcasting literature for short-term prediction of radar data (*XNow* [19]).

To answer to RQ3, the improvement in the performance of *NeXtNow* with respect to that of *XNow* was proven to be statistically significant, with a significance level of $\alpha = 0.01$, as shown by a one-tailed paired Wilcoxon signed-rank test. Additionally, through our experiments, we empirically showed that including multiple past radar measurements led to more accurate predictions at time steps that were further in the future.

The *NeXtNow* architecture that was proposed in this paper offers one step toward the broader goal of our research, which is to develop accurate ML-based prediction models that can be integrated into both Romanian and Norwegian weather nowcasting systems.

Future work will be carried out with the aim of improving the predictive performance of our model for extreme weather phenomena by using weighted loss functions, which could emphasize the errors that are obtained for high reflectivity values. Comparisons between our *NeXtNow* model and other classic methods for radar echo extrapolation are also envisaged for a more thorough experimental validation. For instance, advection methods (e.g., Lagrangian advection), which use reflectivity values from multiple past time steps as inputs, are known for their very good performances and may offer comparable or even better performances than *NeXtNow* by providing sharper predictions. Nevertheless, the spatial resolution of our model's predictions could be enhanced by using perceptual losses or adversarial training techniques, which will also be investigated in future work. Data that are collected from geographical areas other than Romania and Norway will be used to further validate the *NeXtNow* model. Future performance improvements are also envisaged for predicting multiple times steps in the future by extending our approach to include a recurrent architecture.

Author Contributions: Conceptualization, A.-I.A., G.C., A.M. (Andrei Mihai) and I.G.C.; methodology, A.-I.A., G.C., A.M. (Andrei Mihai) and I.G.C.; software, A.-I.A. and A.M. (Andrei Mihai); validation, A.-I.A., G.C., A.M. (Andrei Mihai) and I.G.C.; formal analysis, A.-I.A., G.C., A.M. (Andrei Mihai) and I.G.C.; investigation, A.-I.A., G.C., A.M. (Andrei Mihai) and I.G.C.; resources, A.-I.A., G.C., A.M. (Andrei Mihai), I.G.C., S.B. and A.M. (Abdelkader Mezghani); data curation, S.B.; writing—original draft preparation, G.C.; writing—review and editing, A.-I.A., G.C., A.M. (Andrei Mihai), S.B. and A.M. (Abdelkader Mezghani); visualization, A.-I.A., G.C., A.M. (Andrei Mihai) and I.G.C.; funding acquisition, G.C., S.B. and A.M. (Abdelkader Mezghani). All authors have read and agreed to the published version of the manuscript.

Funding: The research leading to these results has received funding from the NO Grants 2014–2021 under project contract number 26/2020.

Data Availability Statement: The data that were used in this study are available at [41] (NMA data) and [33] (MET data).

Acknowledgments: The authors would like to thank the editors and anonymous reviewers for their useful suggestions and comments that helped us to improve this paper and presentation. The research leading to these results has received funding from the NO Grants 2014–2021 under project contract number 26/2020.

Conflicts of Interest: The authors declare no conflict of interest. The funders had no role in the design of the study, the collection, analyses or interpretation of data, the writing of the manuscript or the decision to publish the results.

References

1. Franch, G.; Nerini, D.; Pendesini, M.; Coviello, L.; Jurman, G.; Furlanello, C. Precipitation Nowcasting with Orographic Enhanced Stacked Generalization: Improving Deep Learning Predictions on Extreme Events. *Atmosphere* **2020**, *11*, 267. [CrossRef]
2. Chen, L.; Cao, Y.; Ma, L.; Zhang, J. A Deep Learning-Based Methodology for Precipitation Nowcasting With Radar. *Earth Space Sci.* **2020**, *7*, e2019EA000812. [CrossRef]
3. Alonso-Montesinos, J.; Monterreal, R.; Fernandez-Reche, J.; Ballestrín, J.; López, G.; Polo, J.; Barbero, F.J.; Marzo, A.; Portillo, C.; Batlles, F.J. Nowcasting System Based on Sky Camera Images to Predict the Solar Flux on the Receiver of a Concentrated Solar Plant. *Remote Sens.* **2022**, *14*, 1602. [CrossRef]
4. Jung, S.H.; Lee, G. Radar-based cell tracking with fuzzy logic approach. *Meteorol. Appl.* **2015**, *22*, 716–730. [CrossRef]
5. James, P.; Reichert, B.; Heizenreder, D. NowCastMIX–optimized automatic warnings from continuously monitored nowcasting systems based on fuzzy-logic evaluations of storm attributes. In Proceedings of the 8th European Conference on Severe Storms, Wiener Neustadt, Austria, 14–18 September 2015.
6. Li, K.; Zhang, M.; Xu, M.; Tang, R.; Wang, L.; Wang, H. Ship Detection in SAR Images Based on Feature Enhancement Swin Transformer and Adjacent Feature Fusion. *Remote Sens.* **2022**, *14*, 3186. [CrossRef]
7. Dixon, M.; Wiener, G. TITAN: Thunderstorm Identification, Tracking, Analysis, and Nowcasting—A radar-based methodology. *J. Atmos. Ocean. Technol.* **1993**, *10*, 785–797. [CrossRef]
8. Johnson, J.T.; MacKeen, P.L.; Witt, A.; Mitchell, E.D.W.; Stumpf, G.J.; Eilts, M.D.; Thomas, K.W. The Storm Cell Identification and Tracking Algorithm: An Enhanced WSR-88D Algorithm. *Weather Forecast.* **1998**, *13*, 263–276. [CrossRef]
9. Haiden, T.; Kann, A.; Wittmann, C.; Pistotnik, G.; Bica, B.; Gruber, C. The Integrated Nowcasting through Comprehensive Analysis (INCA) System and Its Validation over the Eastern Alpine Region. *Weather Forecast.* **2011**, *26*, 166–183. [CrossRef]
10. Sun, J.; Xue, M.; Wilson, J.W.; Zawadzki, I.; Ballard, S.; onvlee hooiMeyer, J.; Joe, P.; Barker, D.; Li, P.W.; Golding, B.; et al. Use of NWP for Nowcasting Convective Precipitation: Recent Progress and Challenges. *Bull. Am. Meteorol. Soc.* **2014**, *95*, 409–426. [CrossRef]
11. Hering, A.M.; Morel, C.; Galli, G.; Sénési, S.; Ambrosetti, P.; Boscacci, M. Nowcasting thunderstorms in the Alpine region using a radar based adaptive thresholding scheme. In Proceedings of the Third European Conference on Radar Meteorology (ERAD), Island of Gotland, Sweden, 6–10 September 2004; pp. 206–211.
12. Auger, L.; Dupont, O.; Hagelin, S.; Brousseau, P.; Brovelli, P. AROME–NWC: A new nowcasting tool based on an operational mesoscale forecasting system. *Q. J. R. Meteorol. Soc.* **2015**, *141*, 1603–1611. [CrossRef]
13. Merlet, N.; Cau, P.; Jauffret, C.; Maynard, K.; Sanchez, I.; Brousseau, P. AROME-NWC Overview, Results, Evolution and Perspectives. *Eur. Forecast.* **2017**, *22*, 40–43.
14. Goodfellow, I.; Bengio, Y.; Courville, A. *Deep Learning*; MIT Press: Cambridge, MA, USA, 2016.
15. Su, H.; Jiang, J.; Wang, A.; Zhuang, W.; Yan, X.H. Subsurface Temperature Reconstruction for the Global Ocean from 1993 to 2020 Using Satellite Observations and Deep Learning. *Remote Sens.* **2022**, *14*, 3198. [CrossRef]
16. Han, L.; Sun, J.; Zhang, W. Convolutional Neural Network for Convective Storm Nowcasting Using 3-D Doppler Weather Radar Data. *IEEE Trans. Geosci. Remote Sens.* **2020**, *58*, 1487–1495. [CrossRef]
17. Xie, S.; Girshick, R.B.; Dollár, P.; Tu, Z.; He, K. Aggregated Residual Transformations for Deep Neural Networks. In Proceedings of the 2017 IEEE Conference on Computer Vision and Pattern Recognition (CVPR), Honolulu, HI, USA, 21–26 July 2017; pp. 5987–5995.
18. Mu, K.; Zhang, Z.; Qian, Y.; Liu, S.; Sun, M.; Qi, R. SRT: A Spectral Reconstruction Network for GF-1 PMS Data Based on Transformer and ResNet. *Remote Sens.* **2022**, *14*, 3163. [CrossRef]
19. Socaci, I.A.; Czibula, G.; Ionescu, V.S.; Mihai, A. *XNow*: A deep learning technique for nowcasting based on radar products' values prediction. In Proceedings of the IEEE 14th International Symposium on Applied Computational Intelligence and Informatics (SACI 2020), Timisoara, Romania, 21–23 May 2020; pp. 117–122.
20. Prudden, R.; Adams, S.V.; Kangin, D.; Robinson, N.H.; Ravuri, S.V.; Mohamed, S.; Arribas, A. A review of radar-based nowcasting of precipitation and applicable machine learning techniques. *arXiv* **2020**, arXiv:2005.04988.
21. Han, L.; Sun, J.; Zhang, W.; Xiu, Y.; Feng, H.; Lin, Y. A machine learning nowcasting method based on real-time reanalysis data. *J. Geophys. Res. Atmos.* **2017**, *122*, 4038–4051. [CrossRef]
22. Yan, J.I. Short-term Precipitation Prediction Based on a Neural Network Method from Radar Observations. In Proceedings of the 3rd International Conference on Artificial Intelligence and Industrial Engineering (AIIE 2017), Shanghai, China, 26–27 November 2017; pp. 246–251.
23. Agrawal, S.; Barrington, L.; Bromberg, C.; Burge, J.; Gazen, C.; Hickey, J. Machine learning for precipitation nowcasting from radar images. In Proceedings of the 33rd Conference on Neural Information Processing Systems (NeurIPS 2019), Vancouver, BC, Canada, 8–14 December 2019; pp. 1–6.
24. Trebing, K.; Stanczyk, T.; Mehrkanoon, S. SmaAt-UNet: Precipitation nowcasting using a small attention-UNet architecture. *Pattern Recognit. Lett.* **2021**, *145*, 178–186. [CrossRef]

25. Ayzel, G.; Scheffer, T.; Heistermann, M. RainNet v1. 0: A convolutional neural network for radar-based precipitation nowcasting. *Geosci. Model Dev.* **2020**, *13*, 2631–2644. [CrossRef]
26. Aivaras, C.; Mantas, L. Nowcasting precipitation using weather radar data for Lithuania: The first results. In Proceedings of the CEUR Workshop Proceedings: System 2018, Gliwice, Poland, 28 May 2018; Volume 2147, pp. 55–60.
27. Mao, Y.; Sorteberg, A. Improving Radar-Based Precipitation Nowcasts with Machine Learning Using an Approach Based on Random Forest. *Weather Forecast.* **2020**, *35*, 2461–2478. [CrossRef]
28. Bonnet, S.M.; Evsukoff, A.; Morales Rodriguez, C.A. Precipitation Nowcasting with Weather Radar Images and Deep Learning in São Paulo, Brasil. *Atmosphere* **2020**, *11*, 1157. [CrossRef]
29. Xiang, Y.; Ma, J.; Wu, X. A Precipitation Nowcasting Mechanism for Real-World Data Based on Machine Learning. *Math. Probl. Eng.* **2020**, *2020*, 8408931. [CrossRef]
30. Hu, Y.; Chen, L.; Wang, Z.; Pan, X.; Li, H. Towards a More Realistic and Detailed Deep-Learning-Based Radar Echo Extrapolation Method. *Remote Sens.* **2022**, *14*, 24. [CrossRef]
31. Choi, S.; Kim, Y. Rad-cGAN v1.0: Radar-based precipitation nowcasting model with conditional Generative Adversarial Networks for multiple domains. *Geosci. Model Dev.* **2022**, *15*, 5967–5985. [CrossRef]
32. MET Norway Thredds Data Server. Available online: https://thredds.met.no/thredds/catalog.html (accessed on 7 May 2021).
33. Composite Reflectivity Product–MET Norway Thredds Data Server. Available online: https://thredds.met.no/thredds/catalog/remotesensing/reflectivity-nordic/catalog.html (accessed on 15 January 2022).
34. Castro, R.; Souto, Y.M.; Ogasawara, E.; Porto, F.; Bezerra, E. Stconvs2s: Spatiotemporal convolutional sequence to sequence network for weather forecasting. *Neurocomputing* **2021**, *426*, 285–298. [CrossRef]
35. He, K.; Zhang, X.; Ren, S.; Sun, J. Deep residual learning for image recognition. In Proceedings of the IEEE Conference on Computer Vision and Pattern Recognition, Las Vegas, NV, USA, 27–30 June 2016; pp. 770–778.
36. Dai, Y.; Gieseke, F.; Oehmcke, S.; Wu, Y.; Barnard, K. Attentional feature fusion. In Proceedings of the IEEE/CVF Winter Conference on Applications of Computer Vision, Waikoloa, HI, USA, 3–8 January 2021; pp. 3560–3569.
37. Czibula, G.; Mihai, A.; Mihuleţ, E. *NowDeepN*: An Ensemble of Deep Learning Models for Weather Nowcasting Based on Radar Products' Values Prediction. *Appl. Sci.* **2021**, *11*, 125. [CrossRef]
38. PlotNeuralNet. Available online: https://github.com/HarisIqbal88/PlotNeuralNet (accessed on 16 June 2022).
39. WCRP. Methods for Probabilistic Forecasts. Available online: https://www.cawcr.gov.au/projects/verification/#Methods_for_probabilistic_forecasts (accessed on 20 December 2021).
40. Abadi, M. TensorFlow: Learning functions at scale. In Proceedings of the 21st ACM SIGPLAN International Conference on Functional Programming, Nara, Japan, 18–22 September 2016.
41. Mihai, A. NMA Data Set. Available online: http://www.cs.ubbcluj.ro/~mihai.andrei/datasets/nextnow/ (accessed on 31 July 2022).
42. Network Common Data Form. Available online: https://www.unidata.ucar.edu/software/netcdf/ (accessed on 15 January 2022).
43. Siegel, S.; Castellan, N. *Nonparametric Statistics for the Behavioral Sciences*, 2nd ed.; McGraw–Hill, Inc.: NewYork, NY, USA, 1988.
44. Social Science Statistics. Available online: http://www.socscistatistics.com/tests/ (accessed on 15 April 2022).
45. Mihai, A. Using self-organizing maps as unsupervised learning models for meteorological data mining. In Proceedings of the 2020 IEEE 14th International Symposium on Applied Computational Intelligence and Informatics (SACI), Timisoara, Romania, 21–23 May 2020; pp. 000023–000028.
46. Czibula, G.; Mihai, A.; Mihulet, E.; Teodorovici, D. Using self-organizing maps for unsupervised analysis of radar data for nowcasting purposes. *Procedia Comput. Sci.* **2019**, *159*, 48–57. [CrossRef]

Article

Spectral-Spatial Interaction Network for Multispectral Image and Panchromatic Image Fusion

Zihao Nie [1], Lihui Chen [1], Seunggil Jeon [2] and Xiaomin Yang [1,*]

1   College of Electronics and Information Engineering, Sichuan University, Chengdu 610064, China
2   Samsung Electronics Co., Ltd., 129, Samseong-ro, Yeongtong-gu, Suwon-si 16677, Gyeonggi-do, Korea
*   Correspondence: arielyang@scu.edu.cn

Abstract: Recently, with the rapid development of deep learning (DL), an increasing number of DL-based methods are applied in pansharpening. Benefiting from the powerful feature extraction capability of deep learning, DL-based methods have achieved state-of-the-art performance in pansharpening. However, most DL-based methods simply fuse multi-spectral (MS) images and panchromatic (PAN) images by concatenating, which can not make full use of the spectral information and spatial information of MS and PAN images, respectively. To address this issue, we propose a spectral-spatial interaction Network (SSIN) for pansharpening. Different from previous works, we extract the features of PAN and MS, respectively, and then interact them repetitively to incorporate spectral and spatial information progressively. In order to enhance the spectral-spatial information fusion, we further propose spectral-spatial attention (SSA) module to yield a more effective spatial-spectral information transfer in the network. Extensive experiments on QuickBird, WorldView-4, and WorldView-2 images demonstrate that our SSIN significantly outperforms other methods in terms of both objective assessment and visual quality.

Keywords: deep learning; spectral-spatial interaction network; spectral-spatial attention; pansharpening

Citation: Nie, Z.; Chen, L.; Jeon, S.; Yang, X. Spectral-Spatial Interaction Network for Multispectral Image and Panchromatic Image Fusion. *Remote Sens.* **2022**, *14*, 4100. https://doi.org/10.3390/rs14164100

Academic Editor: Adrian Stern

Received: 28 June 2022
Accepted: 19 August 2022
Published: 21 August 2022

Publisher's Note: MDPI stays neutral with regard to jurisdictional claims in published maps and institutional affiliations.

1. Introduction

Due to the limitations of remote sensing satellite-imaging system, remote sensing satellite images with both high spatial and high spectral resolution are difficult to obtain. This problem can be mitigated by improving the hardware. However, it turns out to be an arduous task due to the strict limit of the signal-to-noise ratio of satellite products [1]. To alleviate this problem, the pansharpening technique was proposed. The main purpose of pansharpening is to generate a high-spatial-resolution (HR) multispectral (MS) image, which contains spatial information of the panchromatic (PAN) image and spectral information of the corresponding MS images, by fusing a low-spatial-resolution (LR) MS image with a HR PAN image. As one of the most basic and dynamic research topics in remote sensing, pansharpening has a significant impact on many remote sensing applications, such as crop mapping [2], land cover classification [3], and target detection [4]. In recent years, with the wide application of deep learning in various computer vision tasks, DL-based pansharpening methods have developed rapidly. Inspired by SRCNN [5], Masi et al. [6] first attempted to use convolutional neural networks (CNN) for pansharpening and stacks of three convolutional layers (PNN) for pansharpening, achieving state-of-the-art results. Motivated by the results of the PNN, many pansharpening methods based on deep learning have emerged in recent years [7–13].

Although they have different structures and achieve the desired effect, they usually underutilize the advantage of spectral information that exists in the MS images and spatial information that exists in the PAN images. Most of them tend to concatenate PAN and MS images at the beginning of the network and extract the input feature maps with a single network, which is simple and easy to implement but is not conducive to spectral-spatial information fusion. Even though some DL-based methods [14–18] are designed to be multi-branch structures, and the features of the input PAN and MS images are extracted respectively, they fail to consider the interaction and the impact of spectral-spatial information, which is adverse to information transmission and conversion in networks.

In view of the above issues, we propose a spectral-spatial interaction network for pansharpening. Specifically, to fully extract the features of MS and PAN images, SSIN is designed as dual-branch structures, where the spatial branch is used to extract spatial information from PAN images and the spectral branch extracts spectral information from MS images. Inspired by [19], we take into account information guidance between different branches and design a spectral-spatial attention (SSA) module to fully extract the advantageous information from the two branches. Moreover, we introduce an information interaction block (IIB) into our network for information interaction of the spectral branch and spatial branch. Furthermore, we assemble IIB and SSA into an information interaction group (IIG) as the basic structure of our network. It is worth noting that we use a long skip connection to pass the upsampled MS image to the end of the network directly; many DL-based methods [8,20–22] have demonstrated the effectiveness of this approach.

In summary, the main contributions of this article are as follows:

(1) We propose a spectral-spatial interaction network (SSIN) based on information interaction group for pansharpening. The network is designed as a dual-branch architecture. It extracts spectral and spatial information independently from the two branches and are interacted repetitively to incorporate spectral and spatial information progressively.

(2) We propose the information interaction block (IIB) to enhance the conversion and transmission of spectral and spatial information between the two branches. Because it is a dual-input and dual-output structure, it can be embedded in dual-branch networks efficiently.

(3) We design a lightweight and effective spectral-spatial attention module, which is able to calculate spatial attention from the PAN branch to guide the MS branch. Similarly, it calculates spectral attention from the MS branch to guide the PAN branch. In this way, the advantages of information of MS and PAN images can be fully utilized, which facilitates the fusion of different information.

The remainder of this paper is organized as follows. Section 2 introduces the related work, while Section 3 introduces the proposed SSIN and each part of the network in detail. Section 4 presents the data sets, evaluation index, ablation study, parameters analysis and comparison with SOTA methods on three data sets. Section 5 presents the efficiency study. Finally, Section 6 draws conclusions.

2. Related Work

Over the past decades, many pansharpening methods have been put forward. These methods fall into four categories [23]: component substitution (CS)-based methods, multi-resolution analysis (MRA)-based methods, variational optimization (VO)-based methods, and deep learning (DL)-based methods. In CS-based methods, the LR MS image is decomposed into spectral and spatial components, then the decomposed spatial components of the LR MS image are replaced by the histogram-matched PAN image. Finally, the HR MS image is obtained by inverse transformation. Several widely known CS-based methods include intensity-hue-saturation (IHS) [24,25], principal component analysis (PCA) [26], the Gram–Schmidt (GS) conversion [27], the adaptive GS (GSA) [28], the partial replacement adaptive CS (PRACS) [29], and the band-dependent spatial details (BDSD) [30]. These CS-based methods are simple and efficient and can directly extract spatial information from PAN images. However, spectral distortion is prone to occurs during pansharpening when the correlation between the PAN and MS images is low [31]. As for MRA-based methods, MRA-based methods typically consist of three steps: (1) the upsampled MS image and PAN image are decomposed into multiple scales, (2) fusion at every scale, (3) utilizes an inverse transform to get the reconstructed image. MRA-based methods use multi-scale decomposition to decompose the source image into multiple scales, for instance, discrete wavelet transform (DWT) [32], Laplacian pyramid (LP) [33], generalized Laplacian pyramid (GLP) [34], and contourlet transformation [35]. These methods are generally superior to CS-based methods in spectral fidelity. However, multiscale transformation brings a large amount of computation and may lead to spatial distortion [36].

In order to balance spectral and spatial distortions, some methods based on variational optimization are proposed.

VO-based pansharpening methods transformed the pansharpening process into an optimization problem. The key to solve the problem is the establishment of energy function and the selection of optimization algorithm [37].

This category was developed in the 1990s [38]. Since Ballester et al. [39] proposed the pioneer variational method for pansharpening, VO-based pansharpening methods attracted more and more attention and developed rapidly. Model-based methods [40,41] and sparse-based methods [42,43] are two representative VO-based methods. Even though VO-based methods can produce high-quality fusion results, the optimization is time-consuming [44].

In recent years, with the rapid development of DL-based image super-resolution, many DL-based pansharpening methods have been put forward which greatly improve the performance and efficiency of pansharpening. Masi et al. [6] first introduced a simple three-layer CNN into pansharpening. Inspired by VDSR [20], Wei et al. [8] proposed a deep residual neural network with 11 convolutional layers for pansharpening. Yang et al. [7] proposed a deep network architecture for the pansharpening called PanNet, which trained the network in the high-pass domain to preserve the spatial structure. To capture multiscale detailed information, Yuan et al. [12] introduce multiscale feature extraction and residual learning into CNN for pansharpening. These pansharpening methods process at the pixel level. Different from the previous method, Liu et al. [13] presented a two-stream fusion network (TFNet) to fuse PAN and MS images in the feature level and reconstructed the pan-sharpened image from the fused features. To take full advantage of gradient characteristics in pansharpening. Lai et al. [45] utilized gradient information to guide the pansharpening process. The above method simply concatenated the upsampled MS image and PAN image, input it into the network, and then learned the mapping relationship between the input and HRMS image directly, which resulted in spatial distortion. To solve this problem. Wang et al. [15] integrated a multiscale U-shaped CNN into pansharpening for make full use of multispectral information. To explore the intra-image characteristics and the inter-image correlation concurrently. Guan et al. [46] proposed a three-stream structure network to fully extract the valuable information that encoded in the HR PAN images and LR hyperspectral images.

As mentioned above, although most of the current DL-based methods have significantly improved the fusion performance, they do not explore the advantage information of MS and PAN images. Unlike the above methods, we consider the interaction and impact of spectral-spatial information on the basis of the dual-branch structure. We use both spatial attention and spectral attention mechanisms in the SSA module to take full advantage of the advantageous information of the MS and PAN images.

3. Proposed Method

In this section, the implementation details of the SSIN will be described in detail, including the overall architecture, information interaction block (IIB), spectral-spatial attention (SSA) module, and information fusion (IF) module. We first introduce the overall structure of the network, and then the other network components in turn.

3.1. Network Architecture

Figure 1 shows the network architecture of SSIN, which consists of three main parts: spectral-spatial information extraction, spectral-spatial information interaction, and spatial-spectral information fusion. Specifically, we use a single convolutional layer to extract information and then stack the information interaction group (IIG) to achieve the information interaction. Finally, we use the IF module to fuse spectral and spatial information. The input LRMS image and PAN image are denoted as $I_{MS} \in \mathbb{R}^{h \times w \times c}$ and $I_{PAN} \in \mathbb{R}^{H \times W \times 1}$. The spectral and spatial information is first extracted by a single convolution layer, respectively.

$$I_{spe}^0 = f_{spe}\left(f_{up}(I_{MS})\right), I_{spa}^0 = f_{spa}(I_{PAN}), \tag{1}$$

where $I_{spe}^0 \in \mathbb{R}^{H \times W \times B}$ and $I_{spa}^0 \in \mathbb{R}^{H \times W \times B}$ represent the extracted spectral and spatial information, respectively. $f_{spe}(\cdot)$ and $f_{spa}(\cdot)$ are 3×3 convolutional layers used to extract the spectral-spatial information from the input image I_{MS} and I_{PAN}. $f_{up}(\cdot)$ represents the bicubic interpolation.

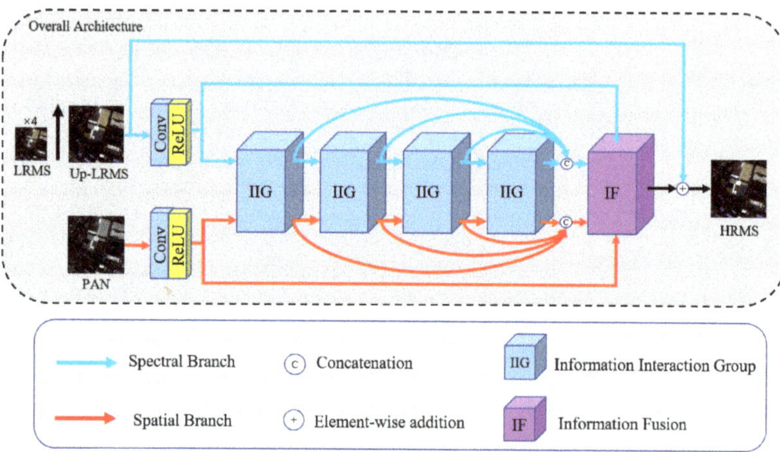

Figure 1. The architecture of the proposed SSIN.

Then, the extracted spectral and spatial information are fed into a series of IIG to realize spectral-spatial information interaction, which can be formulated as follows:

$$\left(I_{spe}^n, I_{spa}^n\right) = f_{IIG}^n\left(I_{spe}^{n-1}, I_{spa}^{n-1}\right), n = (1, 2, ..., N) \tag{2}$$

where $f_{IIG}^n(\cdot)$ stands for the the nth IIG, and I_{spe}^n, I_{spa}^n represent the output spectral and spatial information of the nth IIG, respectively. N denotes the total numbers of IIG.

Inspired by [47], we cascade all these IIGs to fully use the information interacted at different stages. Then, the spectral and spatial information extracted by each IIG are concatenated and fed into an information fusion module to fuse the spectral-spatial information. Simultaneously, we feed the spectral information I_{spe}^0 and spatial information I_{spa}^0 into the IF module to maintain the original information concentration. Finally, to maintain the integrity of the spectral information, we add the upsampled LRMS image to the fused information forming global residual learning. This process can be expressed as:

$$I_{fuse} = f_{IF}\left(\left[I_{spe}^1, \cdots, I_{spe}^N\right], \left[I_{spa}^1, \cdots, I_{spa}^N\right], I_{spe}^0, I_{spa}^0\right) \tag{3}$$

$$I_{SRMS} = I_{fuse} + f_{up}(I_{MS}) \tag{4}$$

where $f_{IF}(\cdot)$ and $[\cdot]$ denote the IF module and concatenation operation. I_{fuse} represents the fused information by the IF module. I_{SRMS} indicates the pansharpened MS image.

3.2. Information Interaction Group

Figure 2 shows the network architecture of the IIG, which is constructed by IIB, SSA, and residual channel attention blocks (RCABs) [48]. In IIG, the SSA is in the middle of two IIBs and connected by two cascade RCABs in each branch, which can be formulated as follows:

$$\left(I_{spe}^{n'}, I_{spa}^{n'}\right) = f_{IIB,1}^{n}\left(I_{spe}^{n-1}, I_{spa}^{n-1}\right) \tag{5}$$

$$\left(I_{spe}^{n''}, I_{spa}^{n''}\right) = f_{SSA}^{n}\left(f_{RCAB}\left(I_{spe}^{'}\right), f_{RCAB}\left(I_{spa}^{'}\right)\right) \tag{6}$$

$$\left(I_{spe}^{n}, I_{spa}^{n}\right) = f_{IIB,2}^{n}\left(f_{RCAB}\left(I_{spe}^{n''}\right), f_{RCAB}\left(I_{spa}^{n''}\right)\right) \tag{7}$$

where $f_{IIB,1}^{n}(\cdot)$, $f_{IIB,2}^{n}(\cdot)$ represent the first and second IIB in IIG, respectively. $I_{spe}^{n'}\left(I_{spa}^{n'}\right)$ and $I_{spe}^{n''}\left(I_{spa}^{n''}\right)$ represent the intermediate process of IIG. f_{RCAB} indicates two cascaded RCAB. The symmetrical structure of IIG is helpful to maintain the synchronization of spectral and spatial information.

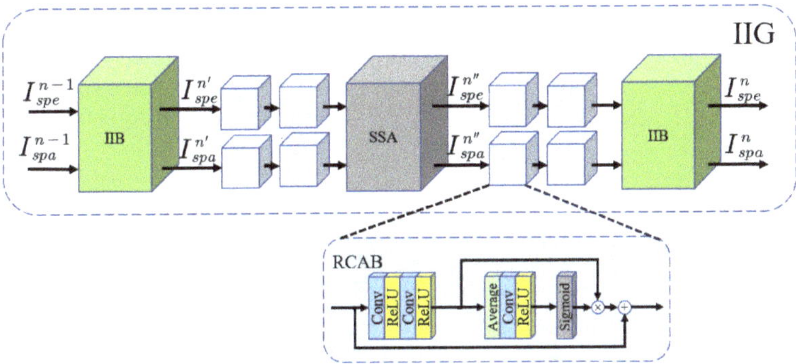

Figure 2. Schematic diagram of the information interaction group, "⊕" denotes elementwise addition, and "⊗" denotes matrix multiplication.

3.3. Information Interaction Block

The main function of the IIB is to realize information interaction in SSIN. As shown in Figure 3, we first extract spectral information of the input spectral-branch feature by a convolutional layer with 3×3 kernel size, then the extracted spectral information is concatenated with the input spatial-branch feature and further use a 1×1 convolutional layer to update the spatial information. Note that we add the input spatial-branch feature to the updated spatial information to achieve local residual learning.

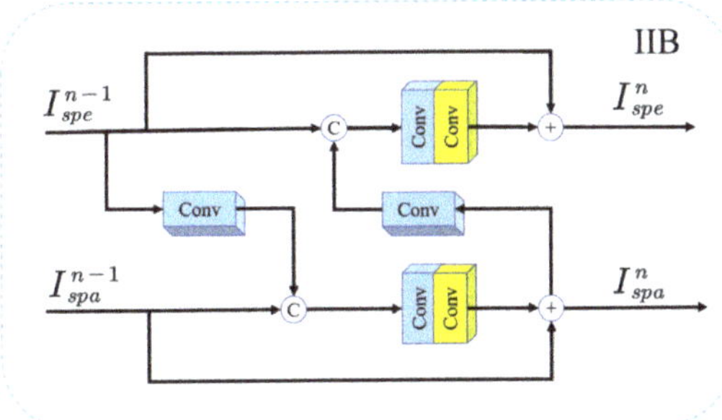

Figure 3. Schematic diagram of the information interaction block, "⊕"denotes elementwise addition, and "⊗" denotes matrix multiplication.

Simultaneously, we extract the spatial information of the spatial branch using a convolutional layer and then concatenated it with the input of the spectral-branch feature, but the difference is that we extract the spatial information of the spatial-branch output port. In this way, the updated spatial information can be used to guide spectral information updata. In summary, the information interaction block can be formulated as:

$$I_{spa}^{n} = \text{ReLu}\left(H_{1\times1}\left(\left[H_{3\times3}\left(I_{spe}^{n-1}\right), I_{spa}^{n-1}\right]\right)\right) + I_{spa}^{n-1} \tag{8}$$

$$I_{spe}^{n} = \text{ReLu}\left(H_{1\times1}\left(\left[H_{3\times3}\left(I_{spa}^{n}\right), I_{spe}^{n-1}\right]\right)\right) + I_{spe}^{n-1} \tag{9}$$

where $H_{1\times1}$ and $H_{3\times3}$ represent convolution operation with 1×1 kernel size and 3×3 kernel size, respectively. $ReLU(\cdot)$ represents the ReLU activation function [49].

3.4. Spectral-Spatial Attention Module

In order to take full advantage of the advantageous information of the spectral branch and spatial branch, we design a lightweight and effective spectral-spatial attention (SSA) module to guide spectral-spatial information integration. We compute the spatial and spectral attention from the branch of spatial and spectral, respectively. Then, we multiply the original features with the attention maps from another branch to transfer the corresponding information. Finally, we add the original features with the above weighted features in each branch to maintain the original information concentration. The schematic of SSA is shown in Figure 4.

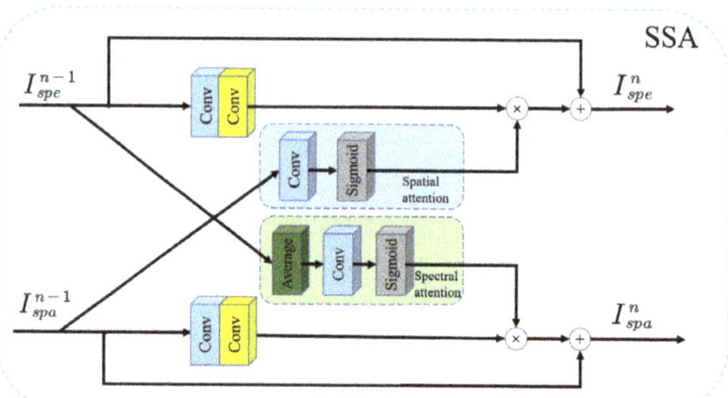

Figure 4. Schematic diagram of the spectral-spatial attention module, "$\oplus$" denotes elementwise addition, and "$\otimes$" denotes matrix multiplication.

Similar to [50], we use global average pooling and 1D convolution to achieve spectral attention which is lightweight and effective. The weights of spectral attention $w_{spe} \in \mathbb{R}^{1\times1\times B}$ can be computed as:

$$w_{spe} = \delta\left(H_{1D}\left(g\left(I_{spe}^{n-1}\right)\right)\right) \tag{10}$$

where $g(x) = \frac{1}{WH}\sum_{i=1}^{W}\sum_{j=1}^{H} I_{spe(i,j)}^{n-1}$ is channel-wise global average pooling (GAP) and σ is the Sigmoid function. $H_{1D}(\cdot)$ indicates 1D convolution.

For spatial attention, we use 1×1 convolution instead of the max pooling to generate the spatial attention map $w_{spa} \in \mathbb{R}^{H\times W\times1}$. It can be formulated as:

$$w_{spa} = \delta\left(H_{1\times1}\left(I_{spa}^{n-1}\right)\right) \tag{11}$$

To sum up, the spectral-spatial attention module can be formulated as:

$$I_{spe}^{n} = \text{ReLU}\left(H_{3\times3}\left(I_{spe}^{n-1}\right)\right) \otimes w_{spa} + I_{spe}^{n-1} \tag{12}$$

$$I_{spa}^{n} = \text{ReLU}\left(H_{3\times3}\left(I_{spa}^{n-1}\right)\right) \otimes w_{spe} + I_{spa}^{n-1} \tag{13}$$

where $\otimes$ denotes the element-wise multiplication with the auto broadcast mechanism of Pytorch.

3.5. Information Fusion Module

The main purpose of the information fusion (IF) module is to fuse the interacted spectral-spatial information to reconstruct an HR-MS image I_{SRMS}. The schematic of IF is shown in Figure 5.

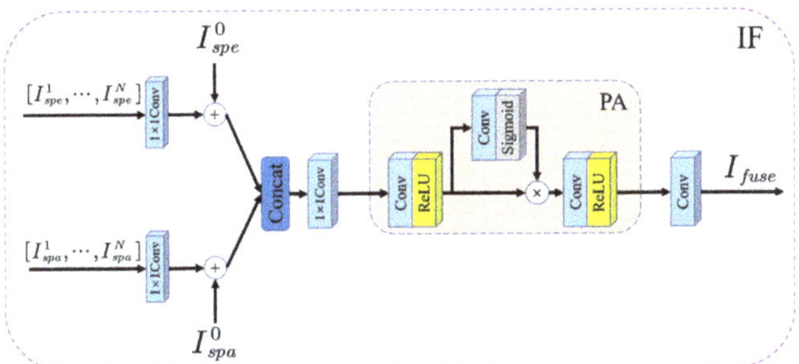

Figure 5. Schematic diagram of the information fusion module, "$\oplus$" denotes elementwise addition, and "$\otimes$" denotes matrix multiplication.

To fully use the information interacted at different stages, we take as input the concatenation of the spectral and spatial information extracted by each IIG in each branch. First, the concatenated inputs $\left[I_{spe}^1, \cdots, I_{spe}^N\right] \in \mathbb{R}^{H \times W \times NB}$ and $\left[I_{spa}^1, \cdots, I_{spa}^N\right] \in \mathbb{R}^{H \times W \times NB}$ are fed to a 1×1 convolution to squeeze the number of channels, and then the squeezed information is added with its initial extracted spectral I_{spe}^0 and spatial I_{spa}^0 information to generate feature maps $F_{spe} \in R^{H \times W \times B}$ and $F_{spa} \in R^{H \times W \times B}$, respectively. Next, we concatenate the feature maps in each branch to generate $F_{fuse} \in \mathbb{R}^{H \times W \times B}$ and the 1×1 convolution is used for squeeze channels again. This process can be described as:

$$F_{spe} = H_{1 \times 1}\left(\left[I_{spe}^1, \cdots, I_{spe}^N\right]\right) + I_{spe}^0 \tag{14}$$

$$F_{spa} = H_{1 \times 1}\left(\left[I_{spa}^1, \cdots, I_{spa}^N\right]\right) + I_{spa}^0 \tag{15}$$

$$F_{fuse} = H_{1 \times 1}\left(\left[F_{spe}, F_{spa}\right]\right) \tag{16}$$

Inspired by [51], we adopt the pixel attention (PA) block at the end of the IF to incorporate the spectral and spatial information, which consists of two convolution layers and a PA layer between them. PA obtains the attention maps, and only goes through a 1×1 convolution and a Sigmoid function, which will then be used to weight the input features. It could effectively improve the final performance at lower parameter cost [51], which is validated by the ablation study in Section 4.4. We denote the proposed PA as $f_{PA}(\cdot)$, the F_{fuse} is further fed into the PA block:

$$I_{fuse}' = f_{PA}\left(F_{fuse}\right) \tag{17}$$

Finally, to match the channel number of the input MS image, a convolutional layer is used to generate the final fusion result:

$$I_{fuse} = H_{3 \times 3}\left(I_{fuse}'\right). \tag{18}$$

4. Experiments

4.1. Datasets

We conduct experiments using partial datasets provided in [37]. To verify the performance of SSIN, we chose the dataset from three different satellites for our experiments including QuickBird (QB), WorldView-4 (WV4), and WorldView-2 (WV2). The detailed information of the datasets is shown in Table 1. For each dataset, we randomly divide the data into the training set and test set by a ratio of 8:2, and twenty percent of the test set is chosen as the validation set. Due to the lack of data volume, the data augmentation is utilized to generate training samples, including random cropping, random horizontal flips, and rotating. After the data augmentation, 400×2, 380×2 and 400×2 image pairs are used as training samples for QB, WV4 and WV2, respectively. The sizes of PAN and LRMS images are 64×64 and 16×16. As for the test data, 80 reduced-scale and 80 full-scale image pairs are utilized. The sizes of reduced-scale PAN and LRMS images are 256×256 and 64×64, respectively. The sizes of full-scale PAN and MS images are 1024×1024 and 256×256, respectively. All the deep learning methods use the same dataset for training and testing.

Table 1. The detailed information of the datasets.

Satellite Sensors	Image Type	Spatial Dimension	Spectral Dimension	Dimension Size	Bits
QuickBird	MS	2.44 m	Four band	$256 \times 256 \times 4$	11 bit
	PAN	0.61 m	one band	1024×1024	
WorldView4	MS	1.24 m	Four band	$256 \times 256 \times 4$	11 bit
	PAN	0.31 m	one band	1024×1024	
WorldView2	MS	2 m	Eight band	$256 \times 256 \times 8$	11 bit
	PAN	0.5 m	one band	1024×1024	

Following the Wald protocol [52], we downsample the PAN and MS images at a four-fold scale using spatial degradation based on a modulation transfer function (MTF), then we can use the degraded images as the network input and the HR MS images as ground truth image to train the network.

4.2. Train Details

In the training stage, we use the Adam optimization algorithm [53] to optimize our network, and the initial learning rate is set to 0.0005. Every 200 epochs, the learning rate drops by a factor of two. We set the batch size to 10 and the patch size of LRMS images to 16 for training with 1200 epochs. Our network uses the $\ell 1$ norm as the loss function.

In our experiments, all the DL-based approaches are implemented in Pytorch framework and are trained on a GTX-1080Ti GPU, while traditional methods are conducted by MATLAB. As for the test phase, all the objective evaluation indexes are calculated by MATLAB, and the results were averaged on the corresponding test set.

4.3. Evaluation Index

In the reduced-resolution experiment, the HR MS image is usually used as the reference image for evaluation, therefore, commonly used objective evaluation indicators, including spectral angle mapper (SAM) [54], the erreur relative global adimensionnelle de synthese (ERGAS) [55], the correlation coefficient(CC) [56] and the Q2n index [57], is used to evaluate all the approaches at reduced resolution. The optimal values of CC and Q2n are 1, and the best values of for ERGAS and SAM are 0.

Since there is no reference image in the full-resolution experiment, we adopt the popular non-reference quality evaluation index, i.e., the quality with no reference (QNR) [58] to evaluate the pansharpening performance. QNR consists of a spectral distortion index, D_λ

and a spatial distortion index, D_s, which can be computed as: $QNR = (1 - D_\lambda)^\alpha (1 - D_S)^\beta$. In the following experiments, α and β are both set as to 1. The ideal value is 1 for QNR, and 0 for D_λ and D_S

4.4. Ablation Study

To verify the validity of IIB, SSA, and PA, we conduct an ablation study on the WV2 validation dataset. In this subsection, the number of IIG and the number of RCAB in each IIG are set to four and two, respectively. The results of ablation experiments are given in Table 2. The meanings of each method are as follows:

- Base: Baseline, ablate the PA block, disconnect the interaction cables in each IIB, and removes both spatial and spectral attention from each SSA.
- Base+PA: Add the PA block in IF on the baseline
- Base+PA+SSA: Add both spatial attention and spectral attention on the Base+PA
- Base+PA+SSA+IIB: The method proposed in this paper (SSIN) that turn on the interaction connection in each IIB on the basis of Base+PA+SSA.

For the baseline, we remove PA from IF and turn off the interaction connection in SSA and IIB. By doing this, spatial and spectral information can only be processed separately, and the spatial and spectral branches are independent of each other. From Table 2, it is not hard to see that baseline achieved the worst performance compared to other methods, which illustrates the importance of interaction between the two branches.

Table 2. The quantiative evaluation result of ablation study, the best value is in bold.

Methods	SAM↓	ERGAS↓	Q2n↑	CC↑
Base	3.7354	2.8492	0.7783	0.9634
Base+PA	3.4086	2.605	0.7839	0.968
Base+PA+SSA	3.3495	2.563	0.7854	0.969
Base+PA+SSA+IIB	**3.2598**	**2.4239**	**0.7882**	**0.9711**

4.4.1. Effect of the PA

Inspired of [50], we add pixel attention (PA) block in IF to improve information fusion performance. To verify the significance of the PA in the IF. We add the PA in IF based on the baseline, which is called "Base+PA".

As can be seen in Table 2, compared with baseline, "Base+PA" has significantly improved in all evaluation indexes. The reason is that PA can improve the expression ability of convolutions [50]. In particular, PA can automatically calculate the importance of each neuron in the feature maps for reconstruction according to the input features, and then rescale these neurons with the importance.

4.4.2. Effect of the SSA

To take full advantage of the advantageous information of the spectral branch and spatial branch, we use both spatial attention and spectral attention mechanisms in the SSA module. It can be seen in Table 2 that "Base+PA+SSA" has achieved better fusion performance than "Base+PA". This is because SSA can make use of the cross-attention mechanism. That is to enhance the spectral characteristics of the spatial branch by spectral attention in the spectral branch, while enhancing spatial characteristics of the spectral branch by using spatial attention of the spatial branch. In this way, the two branches can make up the weakness of each other by using their respective advantages, which are conducive to the final information fusion.

4.4.3. Effect of the IIB

In this section, we assess the effectiveness of the IIB. Comparing "Base+PA+SSA" with "Base+PA+SSA+IIB" in Table 2, we can observe that "Base+PA+SSA", without any information interaction, has worse results in all objective indicators. This is because SSA can take into account the information from another branch before generating attention. Through the information interaction, the network with SSA can obtain more appropriate and accurate attention weights. Moreover, the spectral-spatial information cannot be effectively incorporated by the IF module without information interactions. With the increase of information interaction, the fusion performance of the network is steadily improved as shown in Figure 6.

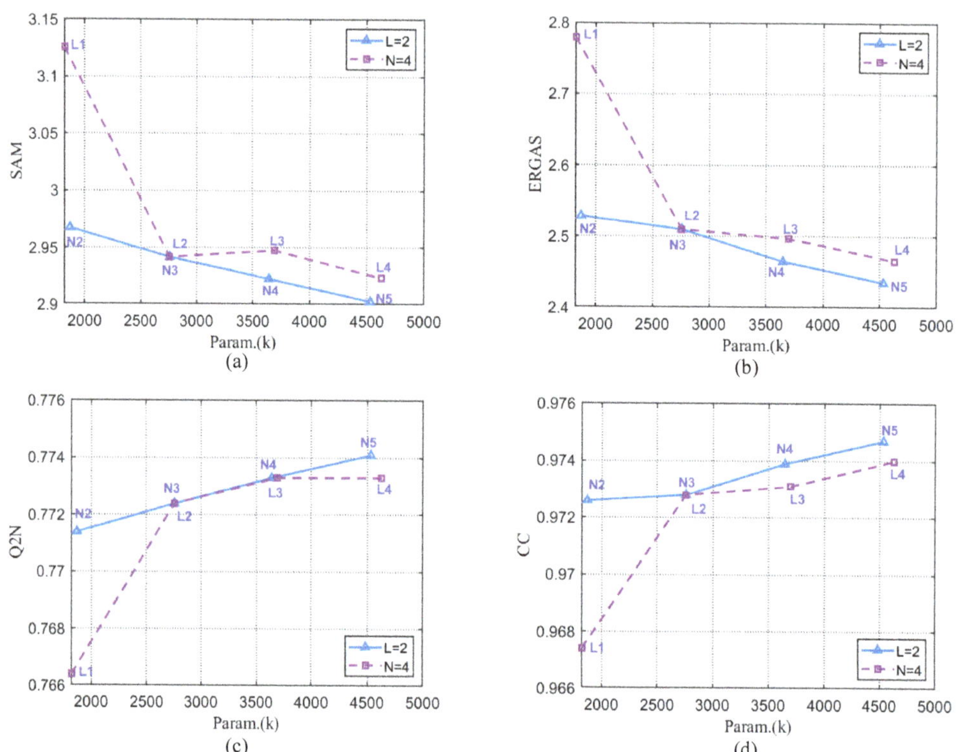

Figure 6. Compare network performance and parameters configured with different parameters. (**a**) the result of SAM. (**b**) the result of ERGAS. (**c**) the result of Q2n. (**d**) the result of CC.

4.5. Parameters Analysis

SSIN is constructed mainly by IIG. The number of IIG, N, and the number of RCAB in each IIG, L, are two key factors affecting network performance.

To explore the impacts of N and L on performance, we conduct eight groups of contrast experiments. In order to keep the number of parameters within a reasonable range, we adjust L from 1 to 4 with a step of 1 by keeping $N = 3$, then we adjust N from 2 to 5 with a step of 1 by keeping $L = 2$, while the other settings remain unchanged. The experiment is conducted on the WV2 test dataset. The objective evaluation indices results are given in Table 3. To obtain the influence of N and L on network performance more intuitively, we show the experimental results in Figure 6.

Table 3. The quantiative evaluation results with different number of IIG and RCAB on the wv2 test dataset. The best value is in bold.

Methods	SAM↓	ERGAS↓	Q2n↑	CC↑	#Params
SSIN (N = 2, L = 2)	2.9677	2.5289	0.7714	0.9726	1.87 M
SSIN (N = 3, L = 2)	2.9418	2.5101	0.7724	0.9728	2.76 M
SSIN (N = 4, L = 2)	2.9227	2.4653	0.7733	0.9739	3.64 M
SSIN (N = 5, L = 2)	**2.9031**	**2.4354**	**0.7741**	**0.9747**	**4.53 M**
SSIN (N = 3, L = 1)	3.1256	2.7793	0.7664	0.9674	1.82 M
SSIN (N = 3, L = 2)	2.9418	2.5101	0.7724	0.9728	2.75 M
SSIN (N = 3, L = 3)	2.9476	2.4972	0.7733	0.9731	3.70 M
SSIN (N = 3, L = 4)	**2.9234**	**2.4658**	**0.7733**	**0.974**	**4.63 M**

We first discuss the effect of L with N fixed to 4. The purple curve in Figure 6 shows the experimental results, from which we can find that the SSIN performance and the number of parameters grow with the increase of L, especially when L increase from 1 to 2. However, the increase of L has little impact on network performance when L reaches 2, but brings more parameters. So for the rest of the experiment, the L is set to 2.

Then, we fixed L to 2 to study the effect of N, the results of this part are shown by the blue curve in Figure 6. As we can see, with the increase of N, the performance and the number of parameters of SSIN increase steadily, especially in Figure 6a,c. This is because the number of IIB grows as N increases, and information interaction can enhance the performance of the SSA module and significantly improve the final information fusion performance, as we discussed in Section 4.4.

In addition, it is worth noting that increasing N can bring greater performance improvement than increasing L under the condition of a similar number of parameters. The main reason is that increase of L can enhance the ability of information extraction but hinder the dissemination of information, while the increase of N can significantly increase the number of information interactions. The interaction between spectral information and spatial information has a more important impact on the performance of SSIN.

To sum up, considering the performance and the number of parameters, we selected $N = 4$ and $L = 2$ as our final setting for our proposed SSIN.

4.6. Comparison with SOTA Methods

In this section, to verify the effectiveness of SSIN, several SOTA pansharpening methods are used to conduct comparative experiments at both reduced resolution and full resolution. For a fair comparison, all parameters of the DL-based approach are kept consistent with the original papers for best performance. In this article we use an implementation of traditional methods that can be downloaded for free [23,59].

The SOTA pansharpening methods used for comparison includes the MS image interpolation (EXP) [34], the robust band-dependent spatial detail (BDSD-PC) [60], GS adaptive (GSA) [28], partial replacement adaptive CS (PRACS) [29], MTF-GLP [61], and seven DL-based methods, such as PNN [6], PanNet [7], MSDCNN [12], TFNET [13], GGPCRN [45], MUCNN [15], MDA-Net [46].

4.6.1. Reduced-Resolution Experiments

The reduced-resolution experimental results of different methods on the QB datasets sets are shown in Table 4. It can be clearly observed from Table 4 that DL-based methods outperform the traditional methods on the evaluation indexes, which reflect the powerful fusion performance of deep learning. Among them, our proposed SSIN performs the best for all the reduced-resolution indexes, followed by MDA-Net [46], which demonstrates the effectiveness of SSIN. Similar results also appear on WV4 and WV2 datasets, as shown in Tables 5 and 6. SSIN achieves the best result except the Q2n on WV2.

Table 4. Quantiative results of different methods on the QB dataset. The best results are in bold and the second-best results are underlined.

Methods	SAM↓	ERGAS↓	Q2N↑	CC↑
EXP	1.8768	1.8316	0.6913	0.8776
GSA	1.4291	1.1440	0.8329	0.9447
PRACS	1.4681	1.1957	0.8210	0.9406
BDSD-PC	1.4268	1.1099	0.8388	0.9470
MTF-GLP	1.4162	1.1602	0.8282	0.9438
PNN	1.1802	0.9451	0.8660	0.9601
PanNet	1.0350	0.8068	0.8825	0.9686
MSDCNN	1.0011	0.7787	0.8830	0.9696
TFNET	0.8475	0.6494	0.8991	0.9778
GGPCRN	0.7714	0.5951	0.9091	0.9809
MUCNN	0.9134	0.7121	0.8950	0.9737
MDA-Net	<u>0.7740</u>	<u>0.5908</u>	<u>0.9104</u>	<u>0.9815</u>
SSIN	**0.7292**	**0.5427**	**0.914**	**0.9833**

Table 5. Quantiative evaluation comparison of different methods on the WV4 dataset. The best results are in bold and the second-best results are underlined.

Methods	SAM↓	ERGAS↓	Q2N↑	CC↑
EXP	2.5639	3.2561	0.6957	0.9026
GSA	2.6454	2.5375	0.7735	0.9372
PRACS	2.5875	2.4452	0.7753	0.9412
BDSD-PC	2.6018	2.4424	0.7876	0.9435
MTF-GLP	2.5785	2.5715	0.7783	0.9409
PNN	1.9677	1.9709	0.8371	0.9583
PanNet	1.9605	1.9341	0.8361	0.9594
MSDCNN	1.8648	1.8522	0.8492	0.9618
TFNET	1.4589	1.3066	0.8909	0.9788
GGPCRN	1.3167	1.1445	0.9025	0.9826
MUCNN	1.7308	1.7138	0.8676	0.9675
MDA-Net	<u>1.3068</u>	<u>1.1487</u>	<u>0.9040</u>	<u>0.9828</u>
SSIN	**1.1994**	**1.0399**	**0.9095**	**0.9853**

Table 6. Quantiative evaluation comparison of different methods on the WV2 dataset. The best results are in bold and the second-best results are underlined.

Methods	SAM↓	ERGAS↓	Q2N↑	CC↑
EXP	5.3078	7.3730	0.4887	0.8017
GSA	4.9165	5.0016	0.6764	0.9089
PRACS	5.1933	5.8450	0.6111	0.8844
BDSD-PC	4.8530	4.7933	0.6853	0.9169
MTF-GLP	4.7501	4.8380	0.6899	0.9150
PNN	3.7671	3.4542	0.7422	0.9535
PanNet	3.6355	3.2757	0.7545	0.9574
MSDCNN	3.4789	3.0815	0.7528	0.9608
TFNET	3.1212	2.6381	0.7642	0.9702
GGPCRN	3.0029	2.5669	0.7699	0.9715
MUCNN	3.3052	2.9356	0.7576	0.9642
MDA-Net	<u>2.9554</u>	<u>2.5030</u>	**0.7737**	<u>0.9729</u>
SSIN	**2.9227**	**2.4653**	<u>0.7733</u>	**0.9739**

Although the quantitative evaluation shows the excellent performance of SSIN, in order to demonstrate the effect of SSIN in subjective evaluation, we make a subjective visual comparison of some samples in the above datasets. To prove the universality of SSIN, we selected images of three different scenes from the above datasets for comparison.

Specifically, the harbor image is from the QB dataset, the forest image is from the WV4 dataset, and the city image is from the WV2 dataset.

Figures 7–9 present the visual comparison of different methods on the three satellite datasets. For intuitive comparison, residual results between fusion image and the ground truth are also presented. The concrete method is to take the average of the absolute values of each band residuals.

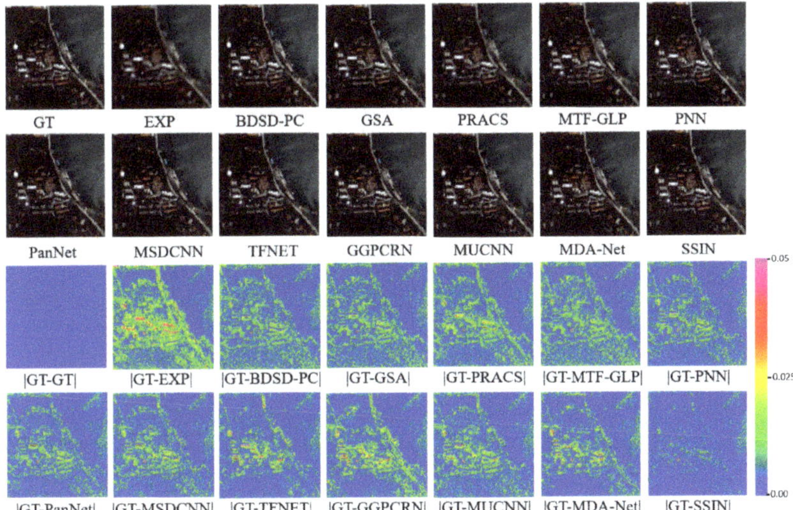

Figure 7. The reduced-resolution experiments results of different methods on QB dataset.

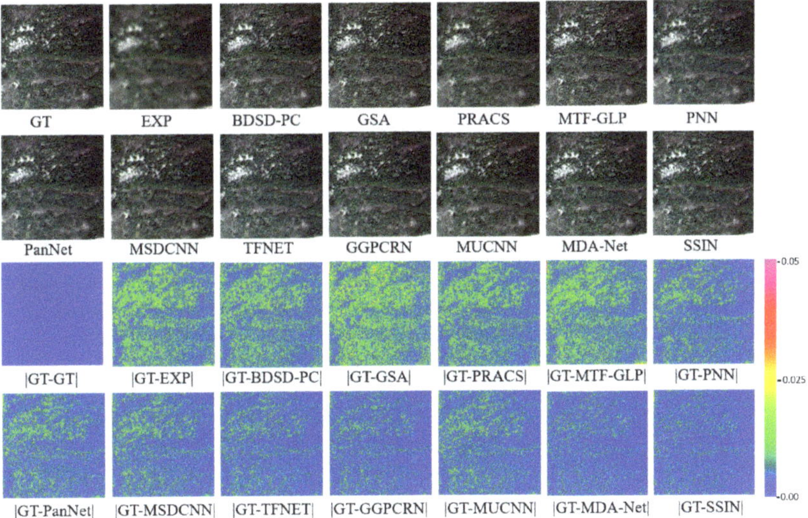

Figure 8. The reduced-resolution experiments results of different methods on WV4 dataset.

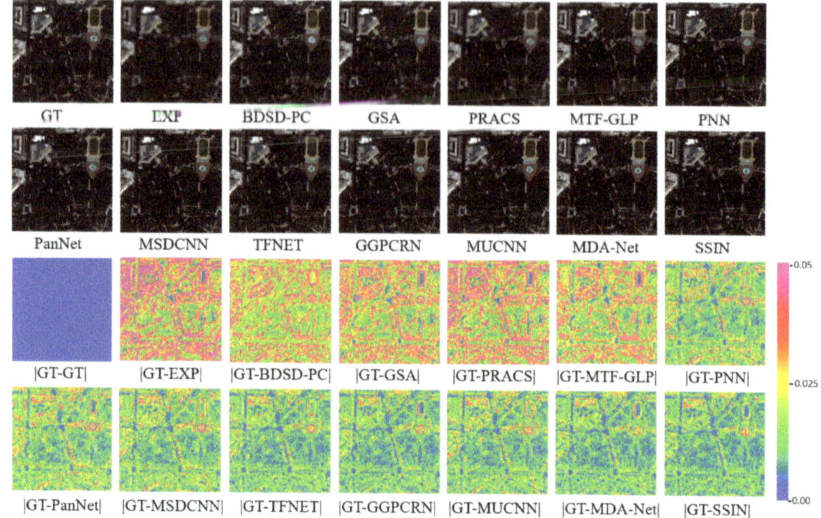

Figure 9. The reduced-resolution experiments results of different methods on WV2 dataset.

Figure 7 displays the fusion results of an image from the QB dataset. We can clearly see from Figure 7 that the results of EXP and PRACS methods are blurred and contain serious spatial distortion compared with GT. The results of the BDSD-PC, PNN, and MUCNN are slightly darker than the ground truth (GT). According to the residual result from Figure 7, it can be found that the error between the traditional method and the GT is greater, while the DL-based methods have less error. Moreover, comparing various DL-based methods, we can see that our proposed SSIN is the closest to the GT, which indicates that our model has better performance in spatial recovery and spectral preservation.

A visual comparison of WV4 dataset is shown in Figure 8. As can be seen, EXP, PRACS, PNN, and MSDCNN produce very blurry images with serious spatial distortion, while the results of BDSD-PC, GSA, and MTF-GLP generate significant spectral distortion in the forest area. The results of the MDA-Net, GGPCRN and our proposed methods are difficult to discern visually. We can further see the residual result from Figure 8, similarly to the experimental results of the QB dataset case, DL-based methods are closer to GT than traditional methods. Among DL-based methods, the residual image of MDA-Net and SSIN is closer to GT than others, which is consistent with the results of the quantitative evaluation shown in Table V. Although the residual image of MDA-Net is very close to that of SSIN, further observation reveals that MDA-Net is slightly brighter than SSIN.

The visual comparison of the WV2 dataset is depicted in Figure 9. It can be observed that, compared with the first two datasets, the test results of each method have larger errors in the WV2 dataset. The reason is that the number of bands in the WV2 dataset is twice that of the first two datasets, making reconstruction more difficult. This can also be seen by comparing objective indicators of the three datasets. As we can see from the residual result in Figure 8, the results of the proposed SSIN are closer to the GT. In particular, it can be clearly seen from the circle in the upper right corner of the residual image that SSIN has the smallest error.

The above comparison at reduced resolution demonstrates the superior performance of SSIN.

4.6.2. Full-Resolution Experiments

In order to evaluate the generality of the above method, we also conducted a full-resolution experiment on the QB and WV4 datasets. Table 7 shows the average quantitative results of the full-resolution experiments from the QB and WV4 datasets. Since the EXP does not inject the details of the PAN image into the LRMS image, the result of the EXP shows spectral features similar to those of the LRMS image, which can be regarded as a reference for evaluating spectral preservation [9,59] and excluded from the comparison.

As shown in Table 7, our method is mediocre in D_λ on both QB and WV4 datasets but achieves the best results in D_s and the QNR of our method is high, which demonstrated our method still achieves satisfactory fusion results. One potential reason is that the spatial attention of SSA module changes the spectral features excessively during the interaction process, resulting in the reduction of the fidelity of the original spectral information.

Table 7. Quantitative evaluation comparison of different methods on the QB and WV4 dataset at the full-resolution experiments. The best results are in bold and the second-best results are underlined.

Methods	QB			WV4		
	$D_\lambda\downarrow$	$D_s\downarrow$	QNR↑	$D_\lambda\downarrow$	$D_s\downarrow$	QNR↑
EXP	**0**	**0.1016**	**0.8984**	**0**	**0.0819**	**0.9181**
GSA	0.0875	0.1743	0.7584	0.0766	0.1576	0.7803
PRACS	**0.0465**	0.1096	0.8510	**0.0305**	0.0975	0.8758
BDSD-PC	0.0622	0.1515	0.7998	0.0478	0.1258	0.8350
MTF-GLP	0.1261	0.2004	0.7056	0.0914	0.1332	0.7907
PNN	0.0622	0.1115	0.8374	0.0473	0.0612	0.8944
PanNet	0.0604	0.0990	0.8502	<u>0.0326</u>	0.0620	**0.9076**
MSDCNN	0.0572	0.1025	0.8493	0.0449	0.0665	0.8927
TFNET	0.0492	0.0728	0.8840	0.0569	0.0562	0.8905
GGPCRN	0.0509	0.0688	0.8858	0.0555	0.0581	0.8902
MUCNN	0.0488	0.0886	0.86	0.0611	0.0591	0.8847
MDA-Net	<u>0.0473</u>	<u>0.0656</u>	**0.8921**	0.0560	0.0607	0.8873
SSIN	0.0532	**0.0609**	<u>0.8910</u>	0.0483	**0.0534**	<u>0.9012</u>

Figures 10 and 11 show the visualized results of different methods on QB and WV4 datasets in the full-resolution experiment, respectively. In addition, we enlarge the region marked in the red box in the fused images for better subjective evaluation. To intuitively observe the differences between different methods, the residual image between the results of EXP and the results of other methods in Figures 10 and 11 are shown in Figures 12 and 13, respectively.

As shown in Figure 10, The results of PNN, PanNet, and MSDCNN suffer from obvious spectral distortion. The results of BDSD-PC, GSA, MTF-GLP, and PRACS yield different levels of spatial distortion compared with the PAN. Specifically, they produce thicker strip structures. The other methods produce better visual results. From the residual images in Figure 12, we can see that the traditional methods inject fewer details than the compared DL-based methods but have better spectral preservation. On the DL based category, we can find that the residual image of PNN, PanNet, and MSDCNN has obvious spatial distortion. Because there are noise pixels in the whole residual image, however, the residual images of the rest of the methods are difficult to discriminate.

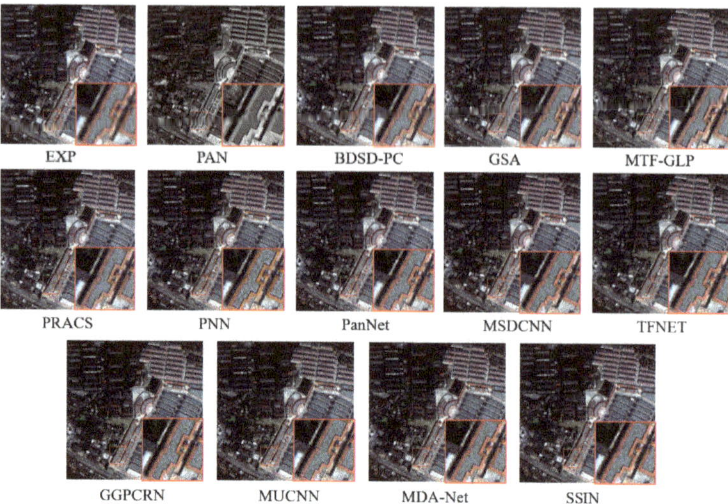

Figure 10. The full-resolution experiments results of different methods on QB dataset.

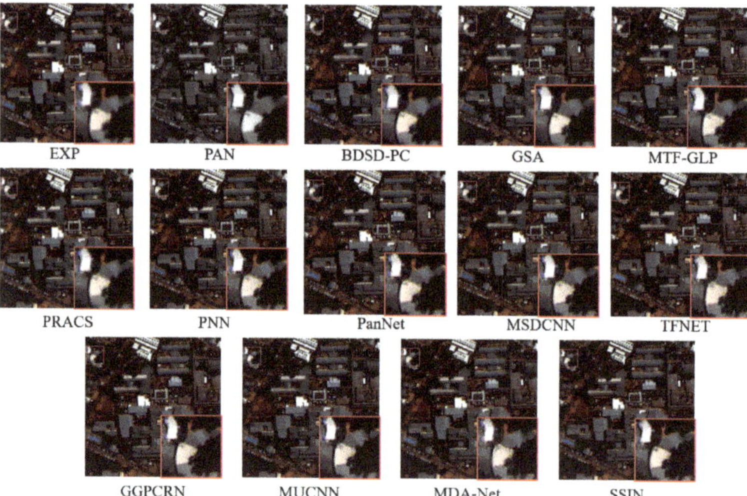

Figure 11. The full-resolution experiments results of different methods on WV4 dataset.

Figure 11 shows results of a full-resolution sample from WV4. As it shows, the result of the GSA and PRACS exhibits serious spectral distortion. Specifically, the grass color in the lower-left corner of the result images is lighter than the result of EXP. The results of the BDSD-PC, MTF-GLP, and PRACS are blurred and present serious spatial distortion. As for DL-based methods, MSDCNN, MUCNN, PanNet, PNN, TFNET, and MDA-Net produce obvious artifacts and spatial distortion with varying degrees. GGPCRN and SSIN generate relatively clearer images. As shown in the enlarged area, we can observe that the results of PNN, PanNet, TFNET, MUCNN, and MDA-Net contain obvious spectral distortions evidenced by distinct color pixels from the result of EXP. As we can see in Figure 13, MTF-GLP and BDSD-PC inject fewer details than DL-based methods. However, GSA injects more details than PNN, PanNet, MSDCNN, and MDA-Net. PRACS observably injects the spatial details, but there is significant spectral distortion, as the residual image of PRACS is obvious color deviation. Furthermore, the residual images of PNN, PanNet, MSDCNN,

TFNET, MUCNN, MDA-Net, and SSIN exhibit different levels of spectral distortion as the residual images have many distinct colors of pixels. Although our SSIN suffers from some spectral distortion and is slightly worse than BDSD-PC, GSA, and MTF-GLP in spectral preservation, it injects more edges into the fusion result.

Figure 12. Residual results of the enlarged region of Figure 10.

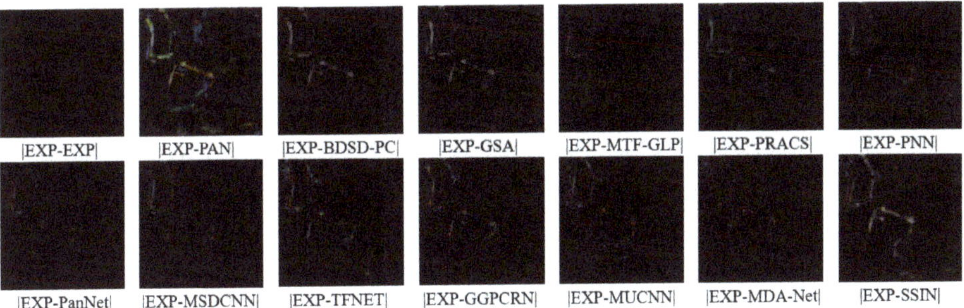

Figure 13. Residual results of the enlarged region of Figure 11.

Overall, from the above results of both reduced- and full-resolution experiments, our SSIN achieves favorable and promising performance in spatial detail injection and spectral preservation.

5. Efficiency Study

In order to compare the computational time and learnable parameters of different methods, we record the results of different methods on the full-resolution experiments from WV4 datasets as shown in Table 8. All traditional methods are implemented using MATLAB and tested on an Intel Core i7-12700K CPU. DL-based methods are tested on a desktop with an Nvidia RTX 3090 GPU. As shown in Table 8, the traditional methods take more computational time than the DL-based methods. PRACS is the slowest method, whose computational time is about 1s per image. As for DL-based methods, although MDA-Net has achieved good results in previous experiments, it needs more time and computational cost to generate the fused images. Although SSIN has a slightly larger number of parameters than other DL-based methods except for MDA-Net, it has better performance.

Table 8. Performance comparison of different methods.

	Methods	Time(s)	#Parameters
Traditional methods	EXP	0.1136	-
	BDSD-PC	0.199	-
	GSA	0.5937	-
	PRACS	0.97	-
	MTF-GLP	0.4555	-
DL-based methods	PNN	0.008	80 K
	PanNet	0.009	77 k
	MSDCNN	0.0089	190 K
	TFNET	0.0095	2.36 M
	GGPCRN	0.014	1.77 M
	MUCNN	0.0081	1.36 M
	MDA-Net	0.0172	12 M
	SSIN	0.0157	3.63 M

6. Conclusions

In this paper, we proposed a dual-branch network named SSIN with spectral-spatial interaction for pansharpening. In the proposed SSIN, the PAN and the LRMS images are processed separately to fully extract their features. SSIN extracted spatial information from the PAN image and spectral information from the MS image. To make the most of the spatial-spectral information, we propose an information interaction block based on a dual-branch network to promote the interaction between spectral and spatial information. Furthermore, the spectral-spatial attention module is used to guide information integration and enhance the characteristics of the another branch. The performance improvement of the two modules for the dual-branch network was proved in the ablation study. Moreover, we used pixel attention in the information fusion module to adjust the importance of each pixel in the feature maps, thereby further improving the network performance. Extensive experiments have demonstrated the effectiveness of our proposed method on pansharpening.

Author Contributions: Conceptualization, Z.N.; methodology, Z.N.; software, Z.N. and L.C.; validation, Z.N. and L.C.; formal analysis, L.C.; investigation, S.J., X.Y.; resources, S.J., X.Y.; data curation, X.Y.; writing—original draft preparation, Z.N.; writing—review and editing, Z.N.; visualization, Z.N. and L.C.; supervision, S.J., X.Y.; project administration, X.Y.; funding acquisition, X.Y. All authors have read and agreed to the published version of the manuscript.

Funding: The research in our paper is sponsored by the funding from Science Foundation of Sichuan Science and Technology Department (2021YFH0119) and Sichuan University under grant 2020SCUNG205.

Data Availability Statement: Not applicable.

Acknowledgments: We thank all the editors and reviewers in advance for their valuable comments that will improve the presentation of this paper.

Conflicts of Interest: The authors declare no conflict of interest.

References

1. Meng, X.; Shen, H.; Li, H.; Zhang, L.; Fu, R. Review of the pansharpening methods for remote sensing images based on the idea of meta-analysis: Practical discussion and challenges. *Inf. Fusion* **2019**, *46*, 102–113. [CrossRef]
2. Gilbertson, J.K.; Kemp, J.; van Niekerk, A. Effect of pan-sharpening multi-temporal Landsat 8 imagery for crop type differentiation using different classification techniques. *Comput. Electron. Agric.* **2017**, *134*, 151–159. [CrossRef]
3. Du, P.; Liu, S.; Xia, J.; Zhao, Y. Information fusion techniques for change detection from multi-temporal remote sensing images. *Inf. Fusion* **2013**, *14*, 19–27. [CrossRef]
4. Qu, Y.; Qi, H.; Ayhan, B.; Kwan, C.; Kidd, R. DOES multispectral / hyperspectral pansharpening improve the performance of anomaly detection? In Proceedings of the 2017 IEEE International Geoscience and Remote Sensing Symposium (IGARSS), Fort Worth, TX, USA, 23–28 July 2017; pp. 6130–6133. [CrossRef]

5. Dong, C.; Loy, C.C.; He, K.; Tang, X. Learning a Deep Convolutional Network for Image Super-Resolution. In *Proceedings of the Computer Vision—ECCV 2014*; Fleet, D., Pajdla, T., Schiele, B., Tuytelaars, T., Eds.; Springer International Publishing: Cham, Switzerland, 2014; pp. 184–199.

6. Masi, G.; Cozzolino, D.; Verdoliva, L.; Scarpa, G. Pansharpening by Convolutional Neural Networks. *Remote Sens.* **2016**, *8*, 594. [CrossRef]

7. Yang, J.; Fu, X.; Hu, Y.; Huang, Y.; Ding, X.; Paisley, J. PanNet: A Deep Network Architecture for Pan-Sharpening. In Proceedings of the 2017 IEEE International Conference on Computer Vision (ICCV), Venice, Italy, 22–29 October 2017; pp. 1753–1761. [CrossRef]

8. Wei, Y.; Yuan, Q. Deep residual learning for remote sensed imagery pansharpening. In Proceedings of the 2017 International Workshop on Remote Sensing with Intelligent Processing (RSIP), Shanghai, China, 18–21 May 2017; pp. 1–4. [CrossRef]

9. Chen, L.; Lai, Z.; Vivone, G.; Jeon, G.; Chanussot, J.; Yang, X. ArbRPN: A Bidirectional Recurrent Pansharpening Network for Multispectral Images With Arbitrary Numbers of Bands. *IEEE Trans. Geosci. Remote Sens.* **2022**, *60*, 1–18. [CrossRef]

10. Lei, D.; Chen, H.; Zhang, L.; Li, W. NLRNet: An Efficient Nonlocal Attention ResNet for Pansharpening. *IEEE Trans. Geosci. Remote. Sens.* **2022**, *60*, 1–13. [CrossRef]

11. Li, C.; Zheng, Y.; Jeon, B. Pansharpening via Subpixel Convolutional Residual Network. *IEEE J. Sel. Top. Appl. Earth Obs. Remote. Sens.* **2021**, *14*, 10303–10313. [CrossRef]

12. Yuan, Q.; Wei, Y.; Meng, X.; Shen, H.; Zhang, L. A multiscale and multidepth convolutional neural network for remote sensing imagery pan-sharpening. *IEEE J. Sel. Top. Appl. Earth Obs. Remote Sens.* **2018**, *11*, 978–989. [CrossRef]

13. Liu, X.; Liu, Q.; Wang, Y. Remote sensing image fusion based on two-stream fusion network. *Inf. Fusion* **2020**, *55*, 1–15. [CrossRef]

14. Yang, Y.; Tu, W.; Huang, S.; Lu, H.; Wan, W.; Gan, L. Dual-Stream Convolutional Neural Network With Residual Information Enhancement for Pansharpening. *IEEE Trans. Geosci. Remote Sens.* **2022**, *60*, 1–16. [CrossRef]

15. Wang, Y.; Deng, L.J.; Zhang, T.J.; Wu, X. SSconv: Explicit Spectral-to-Spatial Convolution for Pansharpening. In Proceedings of the 29th ACM International Conference on Multimedia, Virtual Event, 20–24 October 2021; pp. 4472–4480.

16. Fu, S.; Meng, W.; Jeon, G.; Chehri, A.; Zhang, R.; Yang, X. Two-Path Network with Feedback Connections for Pan-Sharpening in Remote Sensing. *Remote Sens.* **2020**, *12*, 1674. [CrossRef]

17. Zhong, X.; Qian, Y.; Liu, H.; Chen, L.; Wan, Y.; Gao, L.; Qian, J.; Liu, J. Attention FPNet: Two-Branch Remote Sensing Image Pansharpening Network Based on Attention Feature Fusion. *IEEE J. Sel. Top. Appl. Earth Obs. Remote Sens.* **2021**, *14*, 11879–11891. [CrossRef]

18. Wu, X.; Huang, T.Z.; Deng, L.J.; Zhang, T.J. Dynamic Cross Feature Fusion for Remote Sensing Pansharpening. In Proceedings of the IEEE/CVF International Conference on Computer Vision (ICCV), Montreal, BC, Canada, 11–17 October 2021; pp. 14687–14696.

19. Yao, J.; Hong, D.; Chanussot, J.; Meng, D.; Zhu, X.; Xu, Z. Cross-Attention in Coupled Unmixing Nets for Unsupervised Hyperspectral Super-Resolution. In *Proceedings of the Computer Vision—ECCV 2020*; Vedaldi, A., Bischof, H., Brox, T., Frahm, J.M., Eds.; Springer International Publishing: Cham, Switzerland, 2020; pp. 208–224.

20. Kim, J.; Lee, J.K.; Lee, K.M. Accurate Image Super-Resolution Using Very Deep Convolutional Networks. In Proceedings of the IEEE Conference on Computer Vision and Pattern Recognition (CVPR), Las Vegas, NV, USA, 27–30 June 2016.

21. Lim, B.; Son, S.; Kim, H.; Nah, S.; Mu Lee, K. Enhanced Deep Residual Networks for Single Image Super-Resolution. In Proceedings of the IEEE Conference on Computer Vision and Pattern Recognition (CVPR) Workshops, Honolulu, HI, USA, 21–26 July 2017.

22. Ledig, C.; Theis, L.; Huszar, F.; Caballero, J.; Cunningham, A.; Acosta, A.; Aitken, A.; Tejani, A.; Totz, J.; Wang, Z.; et al. Photo-Realistic Single Image Super-Resolution Using a Generative Adversarial Network. In Proceedings of the IEEE Conference on Computer Vision and Pattern Recognition (CVPR), Honolulu, HI, USA, 21–26 July 2017.

23. Vivone, G.; Dalla Mura, M.; Garzelli, A.; Restaino, R.; Scarpa, G.; Ulfarsson, M.O.; Alparone, L.; Chanussot, J. A new benchmark based on recent advances in multispectral pansharpening: Revisiting pansharpening with classical and emerging pansharpening methods. *IEEE Geosci. Remote Sens. Mag.* **2020**, *9*, 53–81. [CrossRef]

24. Haydn, R. Application of the IHS color transform to the processing of multisensor data and image enhancement. In Proceedings of the International Symposium on Remote Sensing of Arid and Semi-Arid Lands, Cairo, Egypt, 19–25 January 1982.

25. Carper, W.; Lillesand, T.; Kiefer, R. The use of intensity-hue-saturation transformations for merging SPOT panchromatic and multispectral image data. *Photogramm. Eng. Remote Sens.* **1990**, *56*, 459–467.

26. Kwarteng, P.; Chavez, A. Extracting spectral contrast in Landsat Thematic Mapper image data using selective principal component analysis. *Photogramm. Eng. Remote Sens.* **1989**, *55*, 339–348.

27. Laben, C.A.; Brower, B.V. Process for Enhancing the Spatial Resolution of Multispectral Imagery Using Pan-Sharpening. U.S. Patent 6,011,875, 4 January 2000.

28. Aiazzi, B.; Baronti, S.; Selva, M. Improving component substitution pansharpening through multivariate regression of MS + Pan data. *IEEE Trans. Geosci. Remote Sens.* **2007**, *45*, 3230–3239. [CrossRef]

29. Choi, J.; Yu, K.; Kim, Y. A new adaptive component-substitution-based satellite image fusion by using partial replacement. *IEEE Trans. Geosci. Remote Sens.* **2010**, *49*, 295–309. [CrossRef]

30. Garzelli, A.; Nencini, F.; Capobianco, L. Optimal MMSE Pan Sharpening of Very High Resolution Multispectral Images. *IEEE Trans. Geosci. Remote Sens.* **2008**, *46*, 228–236. [CrossRef]

31. Ghassemian, H. A review of remote sensing image fusion methods. *Inf. Fusion* **2016**, *32*, 75–89. [CrossRef]

 *remote sensing*

Article

M-O SiamRPN with Weight Adaptive Joint MIoU for UAV Visual Localization

Kailin Wen [1,2], Jie Chu [1,*], Jiayan Chen [1], Yu Chen [1,3] and Jueping Cai [1]

[1] School of Microelectronics, Xidian University, Xi'an 710071, China
[2] Suzhou Honghu Qiji Electronic Technology Co., Ltd., Suzhou 215008, China
[3] The 54th Research Institute of China Electronics Technology Group Corporation, Shijiazhuang 050081, China
* Correspondence: chujie@xidian.edu.cn

Abstract: Vision-based unmanned aerial vehicle (UAV) localization is capable of providing real-time coordinates independently during GNSS interruption, which is important in security, agriculture, industrial mapping, and other fields. owever, there are problems with shadows, the tiny size of targets, interfering objects, and motion blurred edges in aerial images captured by UAVs. Therefore, a multi-order Siamese region proposal network (M-O SiamRPN) with weight adaptive joint multiple intersection over union (MIoU) loss function is proposed to overcome the above limitations. The normalized covariance of 2-O information based on1-O features is introduced in the Siamese convolutional neural network to improve the representation and sensitivity of the network to edges. We innovatively propose a spatial continuity criterion to select 1-O features with richer local details for the calculation of 2-O information, to ensure the effectiveness of M-O features. To reduce the effect of unavoidable positive and negative sample imbalance in target detection, weight adaptive coefficients were designed to automatically modify the penalty factor of cross-entropy loss. Moreover, the MIoU was constructed to constrain the anchor box regression from multiple perspectives. In addition, we proposed an improved Wallis shadow automatic compensation method to pre-process aerial images, providing the basis for subsequent image matching procedures. We also built a consumer-grade UAV acquisition platform to construct an aerial image dataset for experimental validation. The results show that our framework achieved excellent performance for each quantitative and qualitative metric, with the highest precision being 0.979 and a success rate of 0.732.

Keywords: UAV visual navigation; Siamese network; multi-order feature; MIoU

Citation: Wen, K.; Chu, J.; Chen, J.; Chen, Y.; Cai, J. M-O SiamRPN with Weight Adaptive Joint MIoU for UAV Visual Localization. *Remote Sens.* **2022**, *14*, 4467. https://doi.org/10.3390/rs14184467

Academic Editor: Gwanggil Jeon

Received: 30 July 2022
Accepted: 3 September 2022
Published: 7 September 2022

Publisher's Note: MDPI stays neutral with regard to jurisdictional claims in published maps and institutional affiliations.

1. Introduction

UAVs are widely used in national defense, agriculture, mapping, and other industries, with the advantages of autonomous flight, high flexibility, extensible function, low costs, and so on [1,2]. Precise overhead localization is the foundation for UAV navigation and other extended functions [3]. Currently, the dominant UAV localization and navigation technologies in use are inertial navigation, Global Navigation Satellite Systems (GNSS), and combinations of these technologies [4]. Inertial navigation is commonly considered an auxiliary navigation technology attributed to high short-term accuracy but with large long-term errors [5]. GNSS is susceptible to an electromagnetic environment and interference attacks, with unstable signals and poor autonomy [6,7]. Without relying on external information, scene matching-based UAV visual localization has the advantage of strong independence and anti-interference ability, which has become a research hotspot [8,9].

The visual localization is realized by matching the aerial image collected by a UAV in real time with the pre-stored reference image (usually a satellite image), where the coordinates of the matched block in the reference image are the specific location of the UAV [10]. The key to visual localization is feature extraction and matching of images, whose performance directly determines the performance of the localization and navigation system. The combination of local features and classifiers or clusters enables visual localization. Liu et al. [11] designed a visual compass based on point and line features for UAV high-altitude orientation estimation, using the appearance and geometry structure of the point and line features in the remote sensing images. Majdik et al. [12] proposed a textured three-dimensional model and similar discrete places in the topological map and to address the air-ground matching problem. Obviously, it is necessary to design the local features for specific tasks considering specific invariants, so experienced researchers and time consumption are indispensable. Therefore, deep learning-based visual localization is proposed to automate feature extraction and detection end-to-end [13,14]. In particular, the convolutional neural network-based architecture approach with strong feature extraction capability has shown excellent performance in tasks such as target detection and image retrieval, and has been applied as a general-purpose feature extractor for visual localization tasks [15,16]. Wu et al. [17] introduced information theoretic regularization into reinforcement learning for visual navigation and improved the success rate by 10% over some state-of-the-art models. Bertinetto et al. [18]. first proposed a fully convolutional Siamese network (SiamFC) by converting target matching into similarity learning. Subsequently, a series of CNN-based Siamese networks has been proposed because of their simple structure, end-to-end training, and high matching efficiency and speed. The improved DSiam [19], SA-Siam [20], etc., have achieved excellent performance, where the backbone networks are usually AlexNet [21], ResNet [22], or DenseNet [23] for feature extraction and correlation. Among them, Li et al. [24] constructed SiamRPN by combining the region proposal network (RPN) [25] in the Siamese network, and the matching accuracy and speed were improved at the same time, which outputs the location and prediction score of the target by box regression.

Although local feature and semantic-driven approaches are capable of visual localization, the challenge of tiny targets in a large overall image has not been addressed. There are still obvious drawbacks as follows:

- Pseudo-features caused by shadows: The height variation in the terrain and tall buildings lead to shadows in the collected aerial images [26], and the flight time of the UAV is uncertain, so the changing light conditions during a day cause different shadow appearance. These non-ideal effects result in the original information being masked and generating pseudo-edges, which make it difficult to extract the desired features.

- Blurred edge of tiny target: Because of the large size of aerial images, the target in the hole image occupies a small number of pixels with tiny sizes, which are easily disturbed by the background [27]. Since aerial images are usually acquired in motion, there exist blurred edges of the target image due to shaking. The high-resolution features and high-level semantic features of the deep network cannot be obtained at the same time [28]. After multi-layer feature extraction, the features of small targets may be lost. Consequently, it is necessary to explore feature extraction for blurred edges of tiny targets.

- Information imbalance: The regions whose overlapping rate with the target region are less than the threshold value are defined as negative samples, resulting in an unavoidable class imbalance in the target detection task [29]. The loss in backpropagation is obtained by accumulating all samples in a batch, so the training process is dominated by negative samples and the positive samples are not sufficiently trained.

To overcome these insufficiencies, we propose a stretched Wallis shadow compensation method and a multi-order Siamese region proposal network (M-O SiamRPN) with weight adaptive joint multiple intersection over union loss function. The former is used for aerial image preprocessing, and the latter improves the edge detection of tiny targets and the robustness of information imbalance. The contributions are summarized as follows:

- An improved Wallis shadow automatic compensation method is proposed. A pixel contrast-based stretching factor is constructed to increase the effectiveness of shadow compensation of the Wallis filter. The recovered images are used for searching and matching to reduce the effect of shadows on localization results.

- Multi-order features based on spatial continuity are generated to extract the blurred edges of tiny targets. Here, a feature map selection criterion based on spatial continuity is proposed to construct the second-order feature via a first-order feature map with richer spatial information Normalized covariance is introduced in the Siamese network structure to increase the second-order information and enhance the sensitivity of features to edges.
- To improve the classification and location regression performance degradation caused by positive and negative sample imbalance, we propose a weight adaptive joint multiple intersection over union loss function. Depending on the number of positive and negative samples, the weight adaptive scale automatically changes the loss penalty factor for classification. In addition, multiple intersection over union (MIoU) based on generalized intersection over union (*GIoU*) and distance intersection over union (*DIoU*) are constructed to constrain the regression anchor box from different perspectives.

The rest of this paper is organized as follows. Section 2 is dedicated to describing the proposed pre-processing shadow compensation and M-O SiamRPN with weight adaptive joint multiple intersection over union loss function framework. In Section 3, the effectiveness of the proposed framework is verified using aerial images acquired by a self-built UAV platform with satellite images to construct the dataset. The discussion and conclusions are in Sections 4 and 5, respectively.

2. Materials and Methods

2.1. Pre-Processing: Stretched Wallis Shadow Compensation Method

Due to the light angle, terrain undulation, and building blockage, light is blocked in some regions of the UAV aerial image. The resulting shadows cover parts of the image, so the associated pseudo-information is unavoidable [30]. Recovery of aerial images by shadow recovery methods to remove shadow information is capable to improve the subsequent matching accuracy.

Here, a stretched Wallis shadow compensation method is proposed. The pixel contrast-based stretching factor is introduced in the Wallis filtering method, aiming for adaptive compensation for different degrees of shadow occlusions, to improve the contrast within the shadow region. Wallis is based on the property that the mean and variance of different regions in the whole image (without shadows) are essentially constant. Using the mean and variance of the image, the filter coefficients are constructed to achieve the shaded areas being restored to the normal lighting situation [31]. With the shaded regions being restored, the difference between these two parts is reduced and the brightness of the shaded regions is increased. The Wallis is described as:

$$\begin{cases} I^t(x,y) = \alpha I(x,y) + \beta \\ \alpha = \frac{\sigma^t}{\sigma^t + \frac{\sigma}{a^2}} \\ \beta = bm^t + (1 - b - \alpha)m \end{cases} \tag{1}$$

where I denotes the pixel value of the original image at (x, y), I^t is the target pixel value, a, b are hyperparameters, and the σ, m, σ^t, m^t are variance in the neighborhood, pixel mean in the neighborhood, target variance, and target pixel mean, respectively. Different α are obtained via adjusting a, which is used to modify the variance of the shaded areas relative to the whole image, contributing to the increase in contrast in the shaded areas. Similarly, by changing the parameter β through b, the mean value of the shaded regions is increased and the brightness is improved. The result of shading recovery is shown in Figure 1. It can be observed that the recovery method based on the overall image does not address the effects of different lighting in the shadow regions, suffering from inadequate compensation.

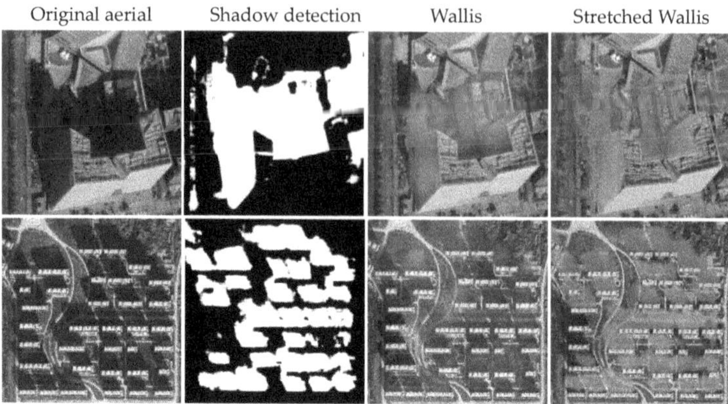

Figure 1. Shadow compensation results using Stretched Wallis.

Considering the diversity of shadow intensities in different regions, a pixel contrast-based stretching factor is proposed and integrated to α. The pixel value I is calculated as:

$$I(x,y) = R \cdot V(x,y) \tag{2}$$

where R is the reflectance and V is the light intensity V at (x, y), which is determined by the direct light intensity V^d and environmental light intensity V^e:

$$V = V^d + V^e \tag{3}$$

Given the part of the obscured image, the shadow regions can be approximated as if only the environmental light intensity is available. Thus, the mean pixel value I can be written separately according to the shaded and unshaded areas as:

$$I^{shadow-free} = R \cdot (V^d + V^e) = \frac{\sum\limits_{i=1}^{n} I^{s-f}{}_i(x,y)}{n} \tag{4}$$

$$I^{shadow} = R \cdot V^e = \frac{\sum\limits_{i=1}^{n} I_i(x,y)}{n} \tag{5}$$

Therefore, the proportional relationship between environmental light intensity and direct light intensity in an aerial image can therefore be defined as:

$$r = \frac{I^{shadow-free} - I^{shadow}}{I^{shadow}} = \frac{1}{3}\left(\sum_{R,G,B} \frac{\sum\limits_{j=1}^{m} I^{s-f} - \sum\limits_{i=1}^{n} I}{\sum\limits_{i=1}^{n} I} \right) \tag{6}$$

It can be seen that a larger r means that the contrast between the shadow-free region and shaded region is larger and more information needs to be recovered. Most of the r values in the shaded regions are in the range of 2 to 6. To ensure adequate compensation for strong shading and smoothness of stretching, the stretching factor S is defined as:

$$S = \log(1 + e^r) \tag{7}$$

Then, with the addition of the stretching factor S, α can be expressed as:

$$\alpha = \frac{\sigma^t}{\sigma^t + \frac{\sigma}{(Sa)^2}} = \frac{\sigma^t}{\sigma^t + \frac{\sigma}{(\log(1+e^r)a)^2}} \tag{8}$$

The region of width K around the shadow detected by reference [32] is represented as the non-shaded region associated with this shaded region, which is obtained by morphological expansion. The mean and variance of the shaded and non-shaded regions are calculated according to the above regions to solve for the parameters α and β of the stretched Wallis shading compensation. The recovery of the shaded area is shown in Figure 1. Obviously, the overall brightness and contrast of the shadows are significantly enhanced by the stretched Wallis compensation method. In addition, the pixel-based contrast stretching factor compensation is more adaptive and the contrast enhancement effect is more targeted.

2.2. M-O SiamRPN with Weight Adaptive Joint Multiple Intersection over Union for Visual Localization

Visual localization is accomplished by comparing real-time aerial imagery with pre-stored satellite images, matching the most similar region in the satellite image to the target, and marking the coordinate position. Considering the non-ideal effects of small target size, blurred edges, and non-balanced information in aerial images, we propose an M-O SiamRPN and a weight adaptive joint multiple intersection over union loss function. The former is designed to address the localization task as a classification and detection problem in satellite maps using real-time aerial images as a template through a SiamRPN backbone, where multi-order with multi-resolution features is used to improve the expression ability and sensitivity to edge textures. The latter balances the effectiveness of a large number of negative samples and a small number of positive samples by simultaneously constraining the anchor box with weight adaptive scale and multiple intersection over union.

2.2.1. M-O SiamRPN Framework

As shown in Figure 2, the proposed M-O SiamRPN consists of three critical components: (1) Siamese ResNet backbone for feature extraction applied to the classification branch and regression branch; (2) Multiple-order information generation module based on selected first-order features; (3) Region proposal network (RPN) under multiple constraints:

- Siamese ResNet backbone: Inspired by SiamRPN, the first-order features used for classification and regression to perform correlation are extracted by the Siamese structure of shared weights. Here, to enhance the feature representation, AlexNet is replaced with Resnet50 [22] in the original SiamRPN structure to increase the network depth.
- Multiple-order information generation module is constructed to sufficiently exploit the complementary characteristics of texture sensitivity of second-order information with the semantic information of first-order features and the different resolution features. For further increasing the representation of second-order features for texture, as well as removing redundant information to avoid overly complex networks, a convolution kernel selection criterion based on spatial continuity is proposed. Using this criterion, first-order features with richer texture information are selected for the generation of second-order information. The second-order information is introduced by adding the covariance matrix to the residual module.
- RPN with multiple constraints: Corresponding to the Siamese structure used for feature extraction, RPN also contains two branches: the classification branch and the regression branch, where both branches are performed with correlation operations as:

$$Cor^{\mathrm{cls}} = T^{\mathrm{cls}} \otimes S^{\mathrm{cls}} \tag{9}$$

$$Cor^{reg} = T^{reg} \otimes S^{reg} \tag{10}$$

where Cor^{cls} and Cor^{reg} denote the correlation calculation of the classification branch and regression branch, T^{cls} and T^{reg} represent the features that the template branch fed into RPN for classification and regression, S^{cls} and S^{reg} are search branch features fed into RPN for classification and regression, and $\otimes$ is correlation calculation. The RPN outputs the location and category scores of the targets through different anchor boxes sliding the features input into the two branches. In the above regression procedure, a weight adaptive joint multiple intersection over union loss function is proposed to optimize the framework, where the weight adaptive scale aims to balance the information between positive and negative samples, and multiple intersection over union is used to improve the accuracy of anchor boxes regression.

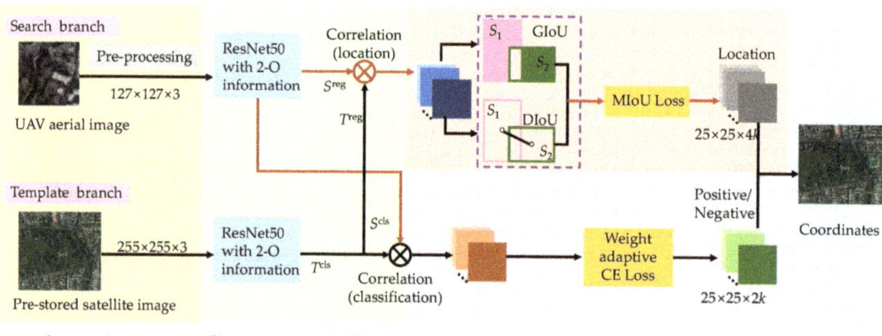

Figure 2. M-O SiamRPN with weight adaptive joint multiple intersection over union framework.

2.2.2. First-Order Feature Selection Criteria Based on Spatially Continuity

The essence of convolution is to extract the efficient information of samples with kernels, so the corresponding important feature can be selected through the reasonable judgment of convolution kernels. In practical image processing, the large contiguous region extracted by the convolution kernel represents the corresponding feature maps with large blocks of contiguous pixels, which means that the feature map retains more essential information. Attributed to the different emphases of the features extracted by the kernels with different parameters, the spatial information contained in the convolutional kernel weights is the key to texture and edge extraction.

Here, we propose the concept of spatial continuity of convolutional kernels to integrate the originally scattered spatial information in the kernels, using the spatial information extraction ability as measurement criteria. The spatial continuity coefficient is calculated quantitatively and used to rank the convolution kernels, selecting the feature map with richer spatial information, as shown in Figure 3. For convolutional layers close to the input, this means that the top-ranked feature maps extract more sufficient local information, as shown in Figure 4.

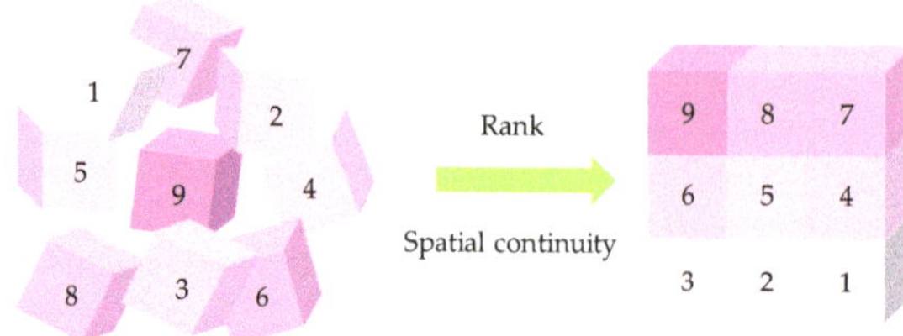

Figure 3. Schematic of convolutional kernel spatial information ranking.

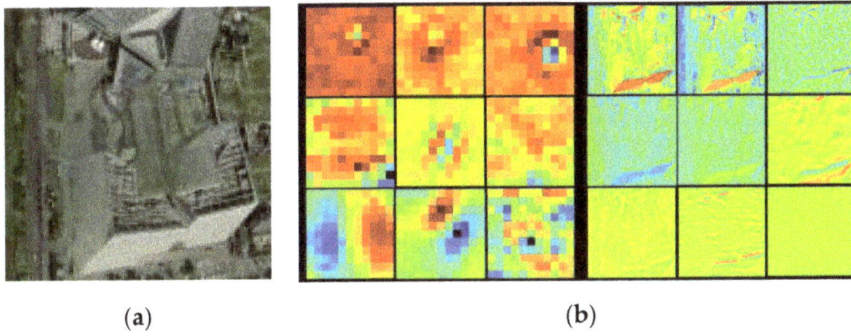

(a)   (b)

Figure 4. Convolutional kernels and feature maps arranged by spatial continuity from largest to smallest (9 randomly selected): (**a**) Input sample; (**b**) convolutional kernels and the corresponding feature maps.

The quantitative calculation of spatial continuity includes connected component analysis and weight information content. It is mathematically described as the sum of the products of the connected areas of all connected domains and the Frobenius norm [33] of the corresponding connected blocks in the matrix, and can be written as:

$$SC = \sum_{k=1}^{m} C_k \|W\|_F, \ (x,y) \in A_k \tag{11}$$

$$\|W\|_F = \sqrt{Tr(W^\mathrm{T} W)} = \sqrt{\sum W_{x,y}^2} \tag{12}$$

where W is the weight matrix corresponding to the convolution kernel, m is the total number of connected blocks in W, A_k is the kth connected block of W, and C_k is the area of the A_k.

The connected domain is an effective analysis tool in the fields of pattern recognition and image segmentation, which mainly consists of four-neighborhood and eight-neighborhood domains, as shown in Figure 5. Here we choose the eight-neighborhood domain. To obtain the connected domain area C_k, the binarized weight matrix M firstly needs to be calculated:

$$M_{x,y} = \begin{cases} 1, & |W_{x,y}| > \text{threshold value} \\ 0, & |W_{x,y}| < \text{threshold value} \end{cases} \tag{13}$$

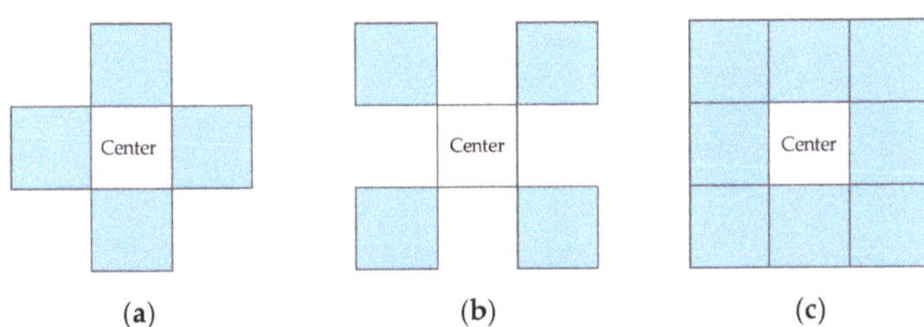

(a)   (b)   (c)

Figure 5. Diagram of 4-neighborhood and 8-neighborhood: (**a**) 4-neighborhood; (**b**) 4-diagonal neighborhood; (**c**) 8-neighborhood.

Here, the threshold value is the hyperparameter. After calculating the M matrix, all the eight-connected domains in the matrix are traversed, and the connected areas of all the eight-connected domains are calculated at the same time. The exploration directions and location during the search process are defined as:

$$directions = [[1,1], [-1,0], [-1,1], [0,-1], [0,1], [1,-1], [1,0], [1,1]] \tag{14}$$

$$\begin{cases} x' = x + directions[k][0] \\ y' = y + directions[k][0] \end{cases} 0 \leq k \leq 8 \tag{15}$$

where *directions* denotes the array of searching directions, x and y are the horizontal and vertical coordinates of the current position, and x', y' are next position coordinates to be explored. The visited elements larger than the threshold are added to the connected block, and then the elements linked to the connected block continue to be explored. The total number of elements larger than the threshold that has been explored is the area of the concatenated domain C_k.

Finally, each convolutional kernel is sorted according to the value of spatial continuity, and the corresponding features are selected according to the selection proportion p. In the M-O SiamRPN framework, the first block in ResNet50 is chosen as a candidate to ensure that the features retain sufficient local information, which is conducive to the representation of texture by second-order features. Meanwhile, to combine the simplicity of the framework, p is set to 30% after several experiments, i.e., the features corresponding to the top 30% of the SC ranked convolutional kernels are selected.

2.2.3. Multiple-Order Feature for Blurred Edges of Tiny Target

The backbone of SiamRPN is replaced with a ResNet50, which increases the network depth and improves the semantic representation of the features, but it is still essentially a first-order feature. For the description of blurred edges of tiny targets, the assistance of second-order features is required [34]. The second-order information based on normalized covariance is added to the residual block as an embedded injection layer, shown in Figure 6b. The first-order features applied to produce the second-order information are selected by the *SC* proposed in Section 2.2.2. This subsection focuses on the generation of second-order information with normalized covariance and the fusion of multi-order features.

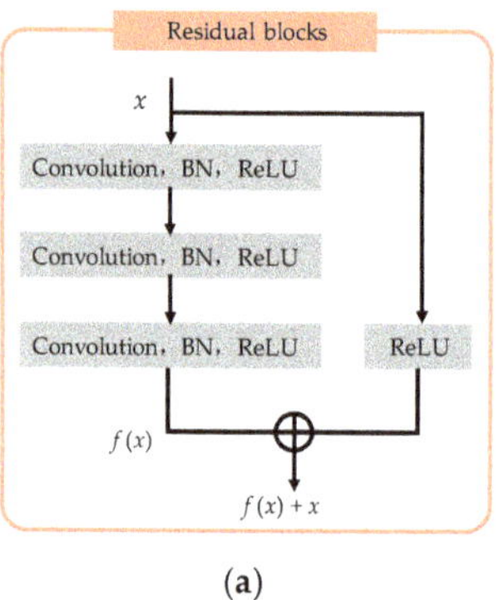

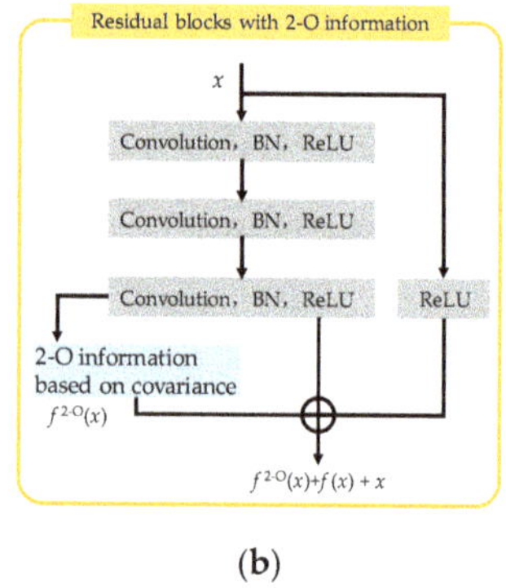

Figure 6. Generation of multi-order features in residual blocks: (**a**) Traditional residual block; (**b**) proposed multi-information residual block.

After a pair of samples is fed into the Siamese frame, a set of first-order features, i.e., $f^{1\text{-}O} \in h \times w \times c$, is obtained after extraction and *SC* selection. $f^{1\text{-}O}$ can be rewritten in vector form according to the same position of different channels as $f^{1\text{-}O} = X = \{X_1, X_1, \ldots, X_n\}$, then its corresponding covariance matrix $f^{2\text{-}O}$ is calculated as:

$$f^{2\text{-}O} = \frac{1}{n}\left(\sum_{i=1}^{n} XX^{\mathrm{T}}\right) + \varepsilon I \tag{16}$$

$$XX^{\mathrm{T}} = \begin{pmatrix} \mathrm{cov}(X_1, X_1) & \mathrm{cov}(X_1, X_2) & \cdots & \mathrm{cov}(X_1, X_n) \\ \mathrm{cov}(X_1, X_1) & \mathrm{cov}(X_2, X_2) & \cdots & \mathrm{cov}(X_2, X_n) \\ \vdots & \vdots & \ddots & \vdots \\ \mathrm{cov}(X_n, X_1) & \mathrm{cov}(X_n, X_2) & \cdots & \mathrm{cov}(X_n, X_n) \end{pmatrix} \tag{17}$$

$$\mathrm{cov}(X, Y) = E[(X - \overline{X})(Y - \overline{Y})] \tag{18}$$

where I is the unit matrix, ε denotes a small positive number, cov represents the covariance calculation, and $E(\bullet)$ is the mean value. The second-order covariance with channel interactions is obtained by transposed outer product calculations, whose powerful mathematical statistical guarantees are verified in various classification applications. The vector used to calculate the covariance contains descriptions of different channels for the same location, as shown in Figure 7. Therefore, the obtained second-order information simultaneously yields the higher-order characteristics and the channel complementary [35]. Different from other second-order methods, feature maps with high spatial continuity are selected for conducting second-order calculations, which do not involve dimensional surges and do not require an additional dimensionality reduction step.

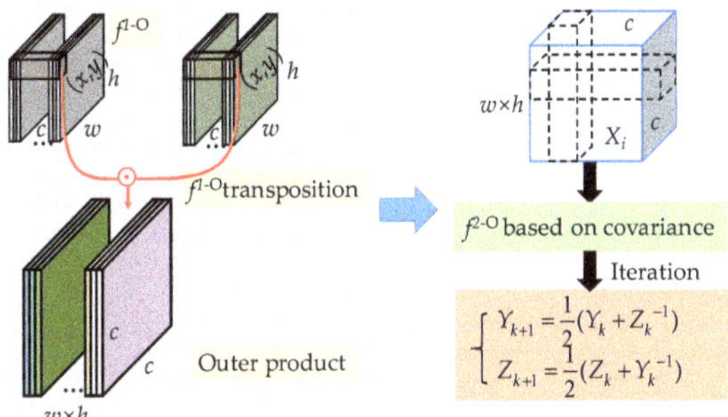

Figure 7. The generation of 2-O information based on covariance.

The resulting second-order $f^{2\text{-}O}$ belongs to Riemann flow and cannot be directly fused, which needs to be decomposed into Euclidean space [36]. So, the $f^{2\text{-}O}$ is normalized based on eigenvalue decomposition, as shown in Figure 7. Since the covariance matrix $f^{2\text{-}O}$ is the symmetric and positive definite, it can be decomposed by the singular value decomposition (SVD) as:

$$f^{2\text{-}O} = U\Lambda U^{\mathrm{T}} \tag{19}$$

where $\Lambda = (\lambda_1, \lambda_2, \ldots \lambda_n)$ is the diagonal matrix consisting of the eigenvalues λ_n and $U = [u_1, u_1, \ldots, u_n]$ denotes the corresponding eigenvector. It can be seen from Equation (19) that the power normalization processing of covariance matrix $f^{2\text{-}O}$ can be expressed as the power operation of eigenvalue:

$$(f^{2\text{-}O})^{\omega} = U\Lambda^{\omega} U^{\mathrm{T}} \tag{20}$$

When $\omega = 0.5$, it is the square-root normalization of the matrix. Due to the slow computational speed of GPU for matrix decomposition, the iterative method to calculate the square root is employed to meet the demand for high real-time performance of localization. Let the $f^{2\text{-}O} = Z^2$, for the equation:

$$F(Z) = Z^2 - f^{2\text{-}O} = 0 \tag{21}$$

Denman-Beavers [37] formula is applied to the iterative solution, avoiding the drawbacks of the pathological case of SVD and poor support on the GPU. The mathematical definition of Denman-Beavers is:

$$K_{k+1} = \frac{1}{2}(K_k + O_k^{-1}) \tag{22}$$

$$O_{k+1} = \frac{1}{2}(O_k + K_k^{-1}) \tag{23}$$

where the initial K_0 is set to $f^{2\text{-O}}$, O_0 is the unit array I, and K_k and O_k are converged globally at $(f^{2\text{-O}})^{1/2}$ and $(f^{2\text{-O}})^{-1/2}$, respectively. Fifteen iterations have been experimentally demonstrated to be sufficient.

As an injection layer in the residual block, the second-order information based on normalized covariance is segmentally differentiable, which is able to be trained end-to-end by back-propagation. From the chain rule, the partial derivative of the loss function L with respect to $f^{2\text{-O}}$ can be described as:

$$\frac{\partial L}{\partial f^{2\text{-O}}} = U\{(Q^\mathrm{T} \odot (U^\mathrm{T}\frac{\partial L}{\partial U})) + (\frac{\partial L}{\partial \Lambda})_{diag}\}U^\mathrm{T} \tag{24}$$

where the Q is interleaved matrix consisting of elements q as:

$$q_{ij} = \left\{ \begin{array}{l} 1/(\lambda_i - \lambda_j), i \neq j \\ 0, i = j \end{array} \right. \tag{25}$$

From the combination of the above equations, the different partial derivatives are obtained as:

$$\frac{\partial L}{\partial U} = \{\frac{\partial L}{\partial Z} + (\frac{\partial L}{\partial Z})^\mathrm{T}\}U\Lambda \tag{26}$$

$$\frac{\partial L}{\partial \Lambda} = (\Lambda)'U^\mathrm{T}\frac{\partial L}{\partial Z}U \tag{27}$$

$$(\Lambda)' = diag(\frac{1}{2\sqrt{\lambda_1}}, \frac{1}{2\sqrt{\lambda_2}}, \cdots, \frac{1}{2\sqrt{\lambda_n}}) \tag{28}$$

In a similar manner to the summation of features in the residual module, second-order features are fused by summing matrix elements after channel adjustment:

$$f^{\text{M-O}}(x) = f^{1\text{-O}}(x) \oplus x \oplus f^{2\text{-O}} \tag{29}$$

where x denotes the input of the current module, $\oplus$ is the sum of matrix elements, and element alignment by convolution is omitted here. With the nonlinearity appearing throughout the R^2 space, the output of the overall network is further increased by adding second-order information to enhance the expression of textures and edges.

2.2.4. Weight Adaptive Joint Multiple Intersection over Union Loss Function

During the training process, only anchor frame coverage exceeding the threshold is determined as a positive sample, while all other cases are negative samples. This setting leads to an imbalanced proportion of positive and negative samples [38]. In an attempt to improve the overall classification performance, the classifier tends to focus less on the minority class and favor the majority class, resulting in the minority samples being difficult to be recognized [39]. In addition, the fast and accurate location regression of the anchor box is also crucial to improving network optimization.

The loss function L of the framework based on SiamRPN consists of two parts, one of which is classification loss L_{cls} represented by cross-entropy loss and the other is loss function L_{reg} used to optimize the target position. Therefore, we redesign the loss function for each of the two branches of classification and regression, and the weight adaptive joint multiple intersection over union loss function is proposed to cope with the issue of imbalanced sample information.

For classification loss L_{cls}, the classical binary cross-entropy loss function is expressed as:

$$L_{\text{CE}}(y_i, p_i) = -(y_i \cdot \log(p_i) + (1 - y_i) \cdot \log(1 - p_i)) \tag{30}$$

where y_i is the label of the sample x_i and p_i is the output probability. The errors are accumulated equally for each sample, without considering the effect of the majority of negative sample information on the training process. According to the cost-sensitive principle, the penalty weight of minority category samples is additionally increased to enhance the focus on positive samples, improving the identification efficiency of positive samples. Here, a sample mean distribution coefficient AG is defined as:

$$AG = S_{\text{all}}/n \tag{31}$$

where S_{all} and n are the numbers of sample categories. The implication of AG is the number of samples contained in each category in the ideal case of a uniform sample. Then, a weighting scale r is defined:

$$r = AG/n_c \tag{32}$$

where the n_c denotes the number of samples actually classified into category c. For negative samples, the number of samples within the category is larger than AG, so the scale r is less than 1, which serves to reduce the weights. On the contrary, in the case of positive samples, whose number is less than AG, the scale r is greater than 1 and the weight is increased. The scaling of the weights is automatically adapted to the imbalance of the sample categories. By adding the weight adaptive scale r to L_{ce}, the new loss function of the classification branch is:

$$L_{\text{cls}} = -\frac{1}{N}\sum_{i=1}^{N} r_i[y_i \log(p_i) + (1 - y_i) \log(1 - p_i)] \tag{33}$$

Deriving the above Equation (34), the gradient is obtained as:

$$g = r_i[y_i \cdot (p_i - 1) + (1 - y_i) \cdot p_i] \tag{34}$$

It can be seen that the gradient values are scaled up when the samples are positive, so that the total gradient contribution of both categories to the model optimization process is balanced.

The second part of the loss function is L_{reg} for predicting the position of the target. Smooth-l_1 is used in SiamRPN for regression optimization, but the four important vertices of the anchor boxes are not included in the optimization metric, which is less robust to rotations, scale changes, and non-overlapping boundaries. The intersection over union (IoU) [25] focuses on the overlap between the prediction anchor box and the groundtruth, considering the whole process as a regression calculation, which is essentially the cross-entropy of the overlap rate:

$$IoU = \frac{S_1 \cap S_2}{S_1 \cup S_2} \tag{35}$$

$$L_{IoU} = -p\ln(IoU) - (1 - p)\ln(1 - IoU) \tag{36}$$

where S_1 and S_2 are prediction box and groundtruth, respectively, p is the corresponding probability. IoU indicates the overlapping relationship of two borders, which obviously ranges from 0 to 1. However, the case of non-overlapping is not captured by the IoU. In this work, multiple intersection over union (MIoU) is proposed to optimize the regression of the prediction anchor box, designing a more complete loss function, which contains generalized intersection over union ($GIoU$) and distance intersection over union ($DIoU$) [40,41].

The non-overlapping part of the two boxes is the concern of the $GIoU$, aiming to make the two boxes infinitely close from the perspective of minimizing the non-overlapping area, as shown in Figure 8b. Firstly, it is necessary to find the smallest box S_3 that completely wraps S_1 and S_2, and then the area not occupied by the two boxes in S_3 is calculated as:

$$GIoU = IoU - \frac{|S_3 - (S_1 \cup S_2)|}{S_3} \tag{37}$$

$$L_{GIoU} = 1 - IoU + \frac{|S_3 - (S_1 \cup S_2)|}{S_3} \tag{38}$$

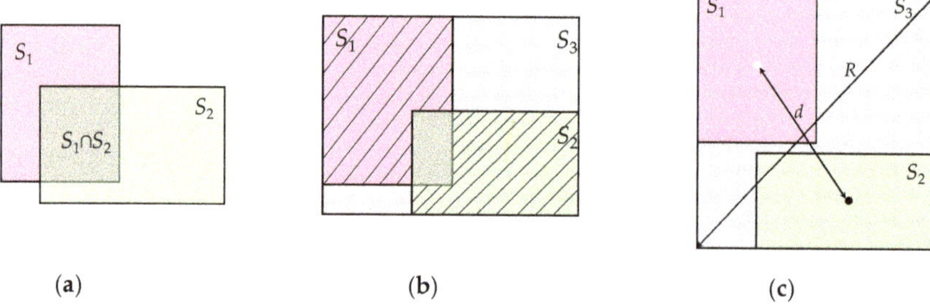

Figure 8. Schematic diagram of different intersection over union: (**a**) *IoU*; (**b**) *GIoU*; (**c**) *DIoU*.

Obviously, when the prediction box S_1 and the groundtruth S_2 do not overlap, i.e., $IoU = 0$, the value of $GIoU$ is not zero, ensuring that the gradient of the loss function can be propagated backwards. However, when S_1 contains S_2, $GIoU$ degenerates to IoU and the exact regression calculation cannot be implemented through both $GIoU$ and IoU. Therefore, $DIoU$ is introduced and its visualization is shown in Figure 8c. Constraints on the distance between the centers of the two boxes S_1 and S_2 are incorporated in the regression process through $DIoU$ to better characterize the overlap information, which is calculated as:

$$DIoU = IoU - \frac{d(S_1, S_2)}{R(S3)} \tag{39}$$

$$L_{DIoU} = 1 - IoU + \frac{d(S_1, S_2)}{R(S_3)} \tag{40}$$

where $d(S_1, S_2)$ denotes the distance between the centroids of the two boxes and R is the diagonal S_3. Thus, the regression loss function L_{reg} based on MIoU is a combination of $DIoU$ of $GIoU$:

$$L_{reg} = \alpha(1 - IoU + \frac{|S_3 - (S_1 \cup S_2)|}{S_3}) + \beta(1 - IoU + \frac{d(S_1, S_2)}{R(S_3)}) \tag{41}$$

where α and β are hyperparameters. The loss function of the overall framework L can be expressed as:

$$Loss = L_{cls} + L_{reg} \tag{42}$$

$$Loss = \frac{1}{n}\sum_{i=1}^{n} L(y_i, p_i) + \alpha(1 - IoU + \frac{|S_3 - (S_1 \cup S_2)|}{S_3}) + \beta(1 - IoU + \frac{d(S_1, S_2)}{R(S_3)}) \tag{43}$$

2.3. Experiment Setup

2.3.1. Details about UAV Platform

The experimental platform is based on a DJI M600 UAV and a homemade camera based on an Imx327 sensor with a resolution of 1080P and a frame rate of 60. The UAV platform is shown in Figure 9.

Figure 9. Homemade UAV platform based on DJI M600 and 1080P camera.

2.3.2. Dataset

The dataset is divided into two parts: aerial images and satellite maps. The aerial images are collected by the self-built UAV platform mentioned above, with a total of 3960 images. The number of aerial images is then expanded to 10,000 by data enhancement processes including panning, zooming, blurring, and flipping, which cover categories such as mountains, forests, rivers, lakes, coastlines, urban intersections, playgrounds, factory buildings, single-family buildings, neighborhood buildings, viaducts, parks, roads without green belts, roads with green belts, etc. Satellite images corresponding to the aerial images are taken from publicly available satellite images.

2.3.3. Network Parameters

Inspired by SiamRPN++ [42], the stride of Stage3 and Stage4 of the backbone ResNet50 are halved and combined with null convolution. The dimensions of the features input to the RPN head for interaction are 25×25 and 25×25, as illustrated in Figure 2. The threshold of *IoU* is set to 0.75, which is attributed to the fact that the final output is the location rather than the proposed region.

The proposed framework is constructed on PyTorch with an epoch of 50 and batch size of 128, and the framework is optimized by SGD with a momentum factor of 0.9 and weight decay of 0.0001. The learning rate is increased from 0.005 to 0.01 using Warmup in the first 5 epochs, after which the learning rate decayed from 0.01 to 0.005 in exponential form. The backbone is not trained for the first 10 epochs and the whole framework is trained after the 10th epoch. The hardware platforms used for the experiments are Intel-Core i7-8700K CPU@3.70GHz and NVIDIA GeForce RTX3090 GPU.

2.3.4. Evaluation Metrics

Reasonable evaluation metrics are an important task for UAV localization. Precision, success rate, frames per second (FPS), and center location error (CLE) are chosen to demonstrate the performance of M-O SiamRPN with multi-optimization loss function.

3. Results

3.1. Training Process Results

During the experiment, we recorded the performance of the M-O SiamRPN with a multi-optimization loss function. The losses in the training process are illustrated in Figure 10. Obviously, the loss decreased rapidly within 10 epochs and leveled off after the 15th epoch. It can be observed that the proposed framework did not suffer from underfitting and overfitting problems, which indicates that our dataset is reasonable and the proposed model is matched to the dataset.

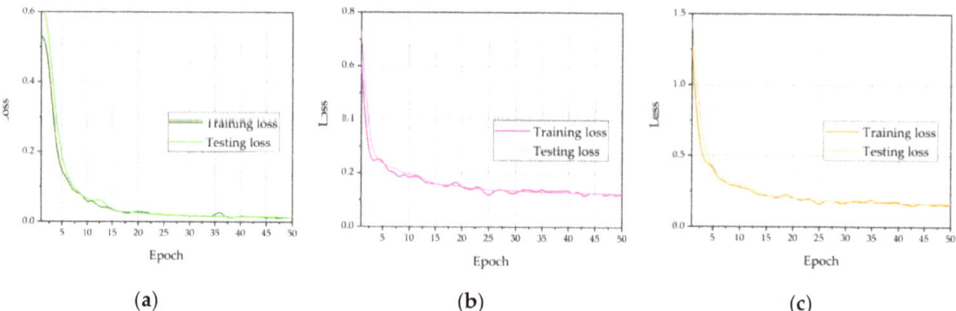

Figure 10. Losses of the M-O SiamRPN with multi-optimization: (**a**) The loss of classification branch; (**b**) The loss of location regression branch; (**c**) The total loss.

3.2. Performance Evaluation

To fairly validate the proposed framework, comprehensive comparison experiments are conducted. Siamese structure networks similar to our approach i.e., SiamFC [18], SiamRPN [24], SiamRPN++ [42], and other current best performing localization networks such as CFNet [43] and Global Tracker [44]. Here, we record the precision and success rate of all the matching networks, and the details of the different comparison results in different networks are given below. From Figure 11, it is obvious that the M-O SiamRPN with multi-optimization achieved significantly higher precision and success rate than those of other frameworks improving 0.016 and 0.019 over the previous best result of SiamRPN++. The results indicate that the multi-order features proposed in this work enhance the texture representation as well as the weight adaptive multiple optimization being able to reduce the influence of non-equilibrium information.

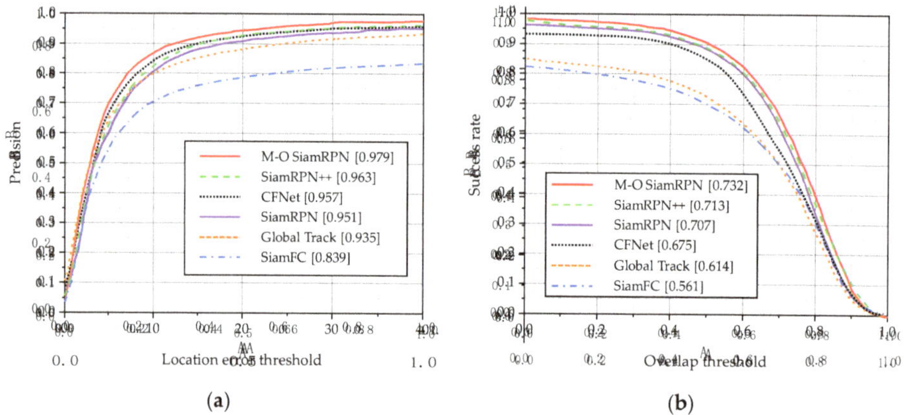

Figure 11. (**a**) Precision and (**b**) success plots of M-O SiamRPN with multi-optimization and state-of-the-art frameworks. The mean precision and AUC scores are reported for each framework.

The performance improvement of the network is usually accompanied by an increase in time complexity. However, for a UAV visual localization task with a high real-time requirement, the framework must complete inference quickly to guarantee localization timeliness. Here, we compare the FPS and CLE of different localization frameworks to verify the processing speed and robustness, and the results are shown in Table 1. Since the proposed spatial continuity criterion is effective to select a small amount of significant first-order features, the processing time of M-O SiamRPN with multi-optimization is not remarkably increased even with the additional injection of second-order information. Compared to the structurally similar SiamRPN, our method is slightly slower, which is attributed to the replacement of the first-order feature extraction network and the generation of second-order information. In the proposed framework, 35 frames are processed per second, which is faster than the SiamRPN++ with the same backbone ResNet50.

Table 1. The FPS and CLE of different localization frameworks.

Method	FPS	CLE
This work	35	5.47
SiamRPN++ [42]	30	5.98
SiamRPN [24]	37	6.26
CFNet [43]	47	6.11
SiamFC [18]	41	6.56
Global Tacker [44]	40	5.83

3.3. Contribution of Multiple-Order Feature

To illustrate the contribution of multiple-order features to promote performance, we compare the AlexNet and ResNet, which are the backbone in SiamRPN and SiamRPN++, respectively, with the backbone proposed in this paper. The results are listed in Table 2, embedding the above-mentioned backbones in our framework and discussing the percentage of the added second-order information. By comparing the first-order features, it can be seen that as the depth increases, the network feature description capability is improved and ResNet50 achieved better results than AlexNet with a 0.08 and 0.04 improvement in precision and success rate, respectively. With the injection of second-order information, the performance of the network increases significantly. The framework with 30% and 50% second-order features obtained the highest precision of 0.979 and 0.980, an improvement of 0.020 and 0.021 compared to the first-order Resnet50. The multi-order features benefit the localization success rate, which was improved by 0.11 ($p = 10\%$), 0.21 ($p = 30\%$), 0.19 ($p = 50\%$), and 0.19 ($p = 100\%$), respectively, compared with the first-order backbone ResNet50. Considering the processing speed and centering error, $p = 30\%$ is used in practical applications.

Table 2. Ablation studies on the backbone of M-O SiamRPN with weight adaptive joint multiple intersection over union loss function.

Backbone	Precision	Success Rate	FPS	CLE
AlexNet [21]	0.951	0.707	38	6.26
ResNet18 [22]	0.954	0.710	37	6.11
ResNet50 [22]	0.959	0.711	37	6.07
ResNet50 with M-O feature ($p = 10\%$)	0.968	0.722	33	5.54
ResNet50 with M-O feature ($p = 30\%$)	0.979	0.732	35	5.47
ResNet50 with M-O feature ($p = 50\%$)	0.980	0.730	32	5.43
ResNet50 with M-O feature ($p = 100\%$)	0.978	0.730	30	5.50

3.4. Contribution of Weight Adaptive Joint MIoU Loss Function

The uniform metric Average Precision (AP) is chosen for performance measurement in order to demonstrate in detail the role of weight adaption and multiple intersection over union. AP is the area under the curve of precision versus recall, which is a widely accepted criterion for target detection tasks. We set different *IoU* thresholds, i.e., *IoU* = {0.5, 0.6, 0.7, 0.8, 0.9} and used *IoU*, *DIoU*, and *GIoU* as comparisons. The results are reported in Table 3. L_{MIoU} with weight adaptive gained the highest AP and L_{MIoU} achieved the second-best results for all different thresholds of IoU. In addition, L_{MIoU} with weight adaptive showed the most significant improvement under the harsher conditions with higher thresholds.

Table 3. Quantitative comparison of M-O SiamRPN using L_{IoU}, L_{CIoU}, L_{DIoU}, L_{GIoU}, L_{MIoU}.

Loss Function	AP50	AP60	AP70	AP80	AP90
L_{IoU}	82.46%	72.75%	60.26%	39.57%	8.75%
Relative improve	3.96%	4.70%	3.87%	3.09%	5.12%
L_{CIoU}	83.31%	74.19%	60.71%	39.15%	8.62%
Relative improve	3.11%	3.26%	3.42%	3.51%	5.25%
L_{DIoU}	83.57%	75.14%	61.43%	40.60%	8.81%
Relative improve	2.85%	2.31%	2.70%	2.06%	5.06%
L_{GIoU}	84.18%	76.33%	61.87%	39.64%	9.22%
Relative improve	2.24%	1.12%	2.26%	3.02%	4.65%
L_{MIoU}	85.36%	76.57%	62.61%	40.15%	9.81%
Relative improve.	1.06	0.88%	1.52 %	2.51%	4.06%
L_{MIoU} with weight adaptive	86.42%	77.45%	64.13%	42.66%	13.87%

3.5. Qualitative Evaluation

The qualitative results of the M-O SiamRPN with multi-optimization and state-of-the-art methods are compared in Figure 12. The left column shows pairs that were correctly localized by all frameworks, and the right column shows pairs that were successfully localized by M-O SiamRPN with multi-optimization but failed to be localized by other frameworks. It can be seen that all methods perform well when obvious features (e.g., bright colors, special edges) appear in the image. In contrast, when the target in the search image is blurred or has similar contours to the surroundings, other localization methods are prone to mislocalization or missed detection, while our framework can still localize correctly. The improvement in localization performance can be attributed to the following aspects. Firstly, the fusion of the second-order covariance information enables the backbone CNN to extract the target features more effectively and characterize the blurred edges more adequately. Secondly, the negative impact of information imbalance on the classifier is reduced due to the integration of weight adaptive scale into the classification branch loss function. Finally, the intersection over union constrains the anchor boxes from different aspects, resulting in more accurate localization of the location regression branches.

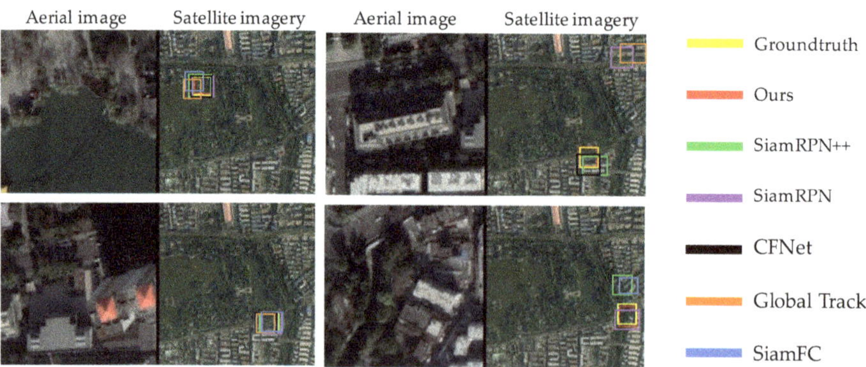

Figure 12. Qualitative comparison of M-O SiamRPN with multi-optimization and state-of-the-art frameworks.

4. Discussion

Autonomous high-precision real-time localization in the case of GNSS and other navigation module failures is extremely important for the safe and full-scene application of UAVs. By matching aerial images with satellite maps, vision-based UAV localization can acquire precise coordinates without additional auxiliary information. However, the real-time collected aerial images are limited by non-ideal effects such as shadow occlusion, few pixels occupied by the target to be matched, and blurred edges caused by flight jitter, resulting in vision-based localization that is still challenging. The experimental results show that the M-O SiamRPN with weight adaptive joint multiple intersection over union loss can achieve accurate pure visual localization with an accuracy of 0.974 and a success rate of 0.732.

ResNet50 embedded with second-order information as the backbone of feature extraction can significantly improve the feature representation of the framework. The increase in precision and success rate can be attributed to two aspects: (1) edges are essentially mutations of pixels, and second-order information is more sensitive to mutations. This is consistent with the effectiveness of second-order pooling [45] and second-order features in bilinear CNN [46] in fine-grained classification. In addition, the proposed spatial continuity as an evaluation criterion for first-order features can select feature maps that retain more adequate local information. (2) The fusion of multi-order features enhances the overall nonlinearity of the neural network at the feature map level, enabling better performance of network fitting. In the optimization phase of the framework, we designed a new loss function to address the sample imbalance problem. For the classification branch of the Siamese frame, the penalty coefficient of cross-loss is automatically adjusted to increase the contribution of a few samples to the loss by comparing the distribution of positive samples within the batch to the ideal equilibrium distribution. In the regression branch, the performance of the prediction box and groundtruth box inclusion, as well as non-overlapping cases, is improved by constraining from both the non-overlapping area and the diagonal of the anchor boxes.

Although the proposed model has proven to be accurate and efficient, it does have limitations. By analyzing the failure samples, the precision and success rate of localization are lower in densely vegetated mountainous areas. This is due to the small spacing and dense canopy in mountainous and forested areas, where even different tree species have similar canopy shapes and colors. In addition, the different slopes of the terrain are not clearly reflected in the aerial images due to the absence of other significant references, further increasing the difficulty of localization. Currently, UAV vision provides mainly visible images. In some special applications, it can be supplemented with information from other wavelength bands, such as adding multispectral through hyperspectral cameras. With other spectral information, the UAV obtains both flight altitude and terrain slope features in visual localization, enriching the information contained in aerial images in complex environments. Among them, the reconstruction of targets by multispectral information and the effective fusion of multi-source features will be worth further investigation.

5. Conclusions

In this paper, we focus on the challenge of a tiny target and blured edge detection problem in UAV visual localization. M-O SiamRPN with weight adaptive joint multiple intersection over union loss and a stretched Wallis shadow compensation method are proposed. The pre-processed aerial images significantly reduce the effect of shadows and are the basis for subsequent image matching procedures. M-O SiamRPN with weight adaptive joint multiple intersection over union loss consists of three parts: a dual-stream Siamese first-order framework based on ResNet50; a covariance-based second-order information generation module; and an RPN module under weight adaptive multiple constraints. To exploit the first-order features adequately, we used Resnet50 as the backbone and proposed the concept of spatial continuity to rank the convolution kernels and select the features with richer local information for 2-O feature generation. For the blurred edges of tiny targets, we presented M-O features, which incorporate second-order information obtained by normalized covariance calculation of selected 1-O features to enhance the network description capability. To address the problem of positive and negative sample imbalance in the detection task, a weight adaptive cross-entropy loss and MIoU were designed to improve the regression precision. The former automatically adjusts the penalty coefficient of the loss function according to the number of positive and negative samples, while the latter constrains the interest anchor box from multiple perspectives. We built a consumer-grade UAV acquisition platform to construct an aerial image dataset for experimental validation. The results show that our framework obtained excellent performance for each quantitative and qualitative metric.

The proposed visual framework is only based on a UAV platform, which is capable of autonomously achieving high-precision real-time localization in the event of GNSS and other navigation module failure. This provides the basis for higher-level extended functionality, which is important for safe, full-scene applications of UAVs.

Author Contributions: Conceptualization, K.W. and J.C. (Jie Chu); methodology, K.W., J.C. (Jie Chu) and J.C. (Jueping Cai); software, J.C. (Jiayan Chen) and Y.C.; experiment validation: K.W., Y.C. and J.C. (Jiayan Chen); writing—review and editing, K.W. and J.C. (Jie Chu). All authors have read and agreed to the published version of the manuscript.

Funding: This research was funded by Shaanxi Province Key Research and Development Program (grant number 2021ZDLGY02-01), Wuhu-Xidian University Industry-University-Research Cooperation Special Fund (XWYCXY-012021003) and Supported by the National 111 Center (B12026).

Data Availability Statement: Not applicable.

Conflicts of Interest: The authors declare no conflict of interest.

References

1. Li, W.; Li, H.; Wu, Q.; Chen, X.; Ngan, K.N. Simultaneously detecting and counting dense vehicles from drone images. *IEEE Trans. Ind. Electron.* **2019**, *66*, 9651–9662. [CrossRef]
2. Ye, Z.; Wei, J.; Lin, Y.; Guo, Q.; Zhang, J.; Zhang, H.; Deng, H.; Yang, K. Extraction of Olive Crown Based on UAV Visible Images and the U2-Net Deep Learning Model. *Remote Sens.* **2022**, *14*, 1523. [CrossRef]
3. Workman, S.; Souvenir, R.; Jacobs, N. Wide-Area Image Geolocalization with Aerial Reference Imagery. In Proceedings of the IEEE International Conference on Computer Vision (ICCV), Santiago, Chile, 7–13 December 2015; pp. 3961–3969.
4. Morales, J.J.; Kassas, Z.M. Tightly Coupled Inertial Navigation System with Signals of Opportunity Aiding. *IEEE Trans. Aerosp. Electron. Syst.* **2021**, *57*, 1930–1948. [CrossRef]
5. Zhang, F.; Shan, B.; Wang, Y.; Hu, Y.; Teng, H. MIMU/GPS Integrated Navigation Filtering Algorithm under the Condition of Satellite Missing. In Proceedings of the IEEE CSAA Guidance, Navigation and Control Conference (GNCC), Xiamen, China, 10–12 August 2018.
6. Liu, Y.; Luo, Q.; Zhou, Y. Deep Learning-Enabled Fusion to Bridge GPS Outages for INS/GPS Integrated Navigation. *IEEE Sens. J.* **2022**, *22*, 8974–8985. [CrossRef]
7. Guo, Y.; Wu, M.; Tang, K.; Tie, J.; Li, X. Covert spoofing algorithm of UAV based on GPS/INS-integrated navigation. *IEEE Trans. Veh. Technol.* **2019**, *68*, 6557–6564. [CrossRef]
8. Wortsman, M.; Ehsani, K.; Rastegari, M.; Farhadi, A.; Mottaghi, R. Learning to Learn How to Learn: Self-Adaptive Visual Navigation Using Meta-Learning. In Proceedings of the IEEE/CVF Conference on Computer Vision and Pattern Recognition (CVPR), Long Beach, CA, USA, 15–20 June 2019; pp. 6743–6752.
9. Qian, J.; Chen, K.; Chen, Q.; Yang, Y.; Zhang, J.; Chen, S. Robust Visual-Lidar Simultaneous Localization and Mapping System for UAV. *IEEE Geosci. Remote Sens. Lett.* **2022**, *19*, 1–5. [CrossRef]
10. Zheng, Z.; Wei, Y.; Yang, Y. University-1652: A multi-view multi-source benchmark for drone-based geo-localization. *arXiv* **2020**, arXiv:2002.12186.
11. Liu, Y.; Tao, J.; Kong, D.; Zhang, Y.; Li, P. A Visual Compass Based on Point and Line Features for UAV High-Altitude Orientation Estimation. *Remote Sens.* **2022**, *14*, 1430. [CrossRef]
12. Majdik, A.L.; Verda, D.; Albers-Schoenberg, Y.; Scaramuzza, D. Air-ground matching: Appearance-based gps-denied urban localization of micro aerial vehicles. *J. Field Robot.* **2015**, *32*, 1015–1039. [CrossRef]
13. Chang, K.; Yan, L. LLNet: A Fusion Classification Network for Land Localization in Real-World Scenarios. *Remote Sens.* **2022**, *14*, 1876. [CrossRef]
14. Zhai, R.; Yuan, Y. A Method of Vision Aided GNSS Positioning Using Semantic Information in Complex Urban Environment. *Remote Sens.* **2022**, *14*, 869. [CrossRef]
15. Nassar, A.; Amer, K.; ElHakim, R.; ElHelw, M. A Deep CNN-Based Framework for Enhanced Aerial Imagery Registration with Applications to UAV Geolocalization. In Proceedings of the IEEE/CVF Conference on Computer Vision and Pattern Recognition Workshops (CVPRW), Salt Lake City, UT, USA, 18–22 June 2018; pp. 1513–1523.

16. Ahn, S.; Kang, H.; Lee, J. Aerial-Satellite Image Matching Framework for UAV Absolute Visual Localization using Contrastive Learning. In Proceedings of the International Conference on Control, Automation and Systems (ICCAS), Jeju, Korea, 12–15 October 2021.
17. Wu, Q.; Xu, K.; Wang, J.; Xu, M.; Manocha, D. Reinforcement learning based visual navigation with information-theoretic regularization. *IEEE Robot. Autom. Lett.* **2021**, *6*, 731–738. [CrossRef]
18. Cen, M.; Jung, C. Fully Convolutional Siamese Fusion Networks for Object Tracking. In Proceedings of the IEEE International Conference on Image Processing (ICIP), Athens, Greece, 7–10 October 2018; pp. 3718–3722.
19. Zhu, Z.; Wang, Q.; Li, B.; Wu, W.; Yan, J.; Hu, W. Distractor-aware Siamese Networks for Visual Object Tracking. In Proceedings of the European Conference on Computer Vision (ECCV), Munich, Germany, 8–14 September 2018.
20. He, A.; Luo, C.; Tian, X.; Zeng, W. A Twofold Siamese Network for Real-Time Object Tracking. In Proceedings of the IEEE/CVF Conference on Computer Vision and Pattern Recognition (CVPR), Salt Lake City, UT, USA, 18–23 June 2018; pp. 4834–4843.
21. Russakovsky, O.; Deng, J.; Su, H.; Krause, J.; Satheesh, S.; Ma, S. Imagenet large scale visual recognition challenge. *Int. J. Comput. Vis.* **2015**, *115*, 211–252. [CrossRef]
22. He, K.; Zhang, X.; Ren, S.; Sun, J. Deep residual learning for image recognition. In Proceedings of the IEEE Conference on Computer Vision and Pattern Recognition (CVPR), Las Vegas, NV, USA, 27–30 June 2016; pp. 770–778.
23. Huang, G.; Liu, Z.; Laurens, V.; Weinberger, K.Q. Densely Connected Convolutional Networks. In Proceedings of the IEEE Conference on Computer Vision and Pattern Recognition (CVPR), Honolulu, HI, USA, 21–26 July 2017; pp. 2261–2269.
24. Li, B.; Yan, J.; Wu, W.; Zhu, Z.; Hu, X. High Performance Visual Tracking with Siamese Region Proposal Network. In Proceedings of the IEEE/CVF Conference on Computer Vision and Pattern Recognition (CVPR), Salt Lake City, UT, USA, 18–23 June 2018; pp. 8971–8980.
25. Girshick, R. Fast R-CNN. In Proceedings of the IEEE International Conference on Computer Vision (ICCV), Santiago, Chile, 7–13 December 2015; pp. 1440–1448.
26. Jiang, H.; Chen, A.; Wu, Y.; Zhang, C.; Chi, Z.; Li, M.; Wang, X. Vegetation Monitoring for Mountainous Regions Using a New Integrated Topographic Correction (ITC) of the SCS + C Correction and the Shadow-Eliminated Vegetation Index. *Remote Sens.* **2022**, *14*, 3073. [CrossRef]
27. Chen, J.; Huang, B.; Li, J.; Wang, Y.; Ren, M.; Xu, T. Learning Spatio-Temporal Attention Based Siamese Network for Tracking UAVs in theWild. *Remote Sens.* **2022**, *14*, 1797. [CrossRef]
28. Yu, W.; Yang, K.; Yao, H.; Sun, X.; Xu, P. Exploiting the complementary strengths of multi-layer CNN features for image retrieval. *Neurocomputing* **2017**, *237*, 235–241. [CrossRef]
29. Feng, J.; Xu, P.; Pu, S.; Zhao, K.; Zhang, H. Robust Visual Tracking by Embedding Combination and Weighted-Gradient Optimization. *Pattern Recognit.* **2020**, *104*, 107339. [CrossRef]
30. Zhu, X.X.; Tuia, D.; Mou, L.; Xia, G.S.; Zhang, L.; Xu, F. Deep learning in remote sensing: A comprehensive review and list of resources. *IEEE Trans. Geosci. Remote Sens.* **2018**, *5*, 8–36. [CrossRef]
31. Gao, X.; Wan, Y.; Zheng, S. Automatic Shadow Detection and Compensation of Aerial Remote Sensing Images. *Geomat. Inf. Sci. Wuhan Univ.* **2012**, *37*, 1299–1302.
32. Zhou, T.; Fu, H.; Sun, C.; Wang, S. Shadow Detection and Compensation from Remote Sensing Images under Complex Urban Conditions. *Remote Sens.* **2021**, *13*, 699. [CrossRef]
33. Powell, M. Least frobenius norm updating of quadratic models that satisfy interpolation conditions. *Math Program.* **2004**, *100*, 183–215. [CrossRef]
34. Gatys, L.A.; Ecker, A.S.; Bethge, M. Image Style Transfer Using Convolutional Neural Networks. In Proceedings of the IEEE Conference on Computer Vision and Pattern Recognition (CVPR), Las Vegas, NV, USA, 27–30 June 2016.
35. Wu, Y.; Sun, Q.; Hou, Y.; Zhang, J.; Wei, X. Deep covariance estimation hashing. *IEEE Access* **2019**, *7*, 113225. [CrossRef]
36. Li, P.; Chen, S. Gaussian process approach for metric learning. *Pattern Recognit.* **2019**, *87*, 17–28. [CrossRef]
37. Denman, E.D.; Beavers, A.N. The matrix Sign function and computations in systems. *Appl. Math. Comput.* **1976**, *2*, 63–94. [CrossRef]
38. Fan, Q.; Zhuo, W.; Tang, C.K.; Tai, Y.W. Few-Shot Object Detection with Attention-RPN and Multi-Relation Detector. In Proceedings of the IEEE/CVF Conference on Computer Vision and Pattern Recognition (CVPR), Seattle, WA, USA, 13–19 June 2020; pp. 4012–4021.
39. Saqlain, M.; Abbas, Q.; Lee, J.Y. A Deep Convolutional Neural Network for Wafer Defect Identification on an Imbalanced Dataset in Semiconductor Manufacturing Processes. *IEEE Trans. Semicond. Manuf.* **2020**, *33*, 436–444. [CrossRef]
40. Rezatofighi, H.; Tsoi, N.; Gwak, J.Y.; Sadeghian, A.; Reid, I.; Savarese, S. Generalized intersection over union: A metric and a loss for bounding box regression. In Proceedings of the IEEE/CVF Conference on Computer Vision and Pattern Recognition (ICCV), Long Beach, CA, USA, 15–20 June 2019; pp. 658–666.
41. Zheng, Z.; Wang, P.; Liu, W.; Li, J.; Ye, R.; Ren, D. Distance-iou loss: Faster and better learning for bounding box regression. *arXiv* **2019**, arXiv:1911.08287. [CrossRef]
42. Li, B.; Wu, W.; Wang, Q.; Zhang, F.; Xing, J.; Yan, J. SiamRPN++: Evolution of Siamese Visual Tracking with Very Deep Networks. In Proceedings of the IEEE/CVF Conference on Computer Vision and Pattern Recognition (CVPR), Long Beach, CA, USA, 15–20 June 2019; pp. 4277–4286.

43. Shen, Z.; Dai, Y.; Rao, Z. CFNet: Cascade and Fused Cost Volume for Robust Stereo Matching. In Proceedings of the IEEE/CVF Conference on Computer Vision and Pattern Recognition (CVPR), Nashville, TN, USA, 20–25 June 2021; pp. 13901–13910.
44. Huang, L.; Zhao, X.; Huang, K. GlobalTrack: A Simple and Strong Baseline for Long-term Tracking. *arXiv* **2019**, arXiv:1912.08531. [CrossRef]
45. Li, P.; Xie, J.; Wang, Q.; Zuo, W. Is Second-Order Information Helpful for Large-Scale Visual Recognition? In Proceedings of the IEEE International Conference on Computer Vision (ICCV), Venice, Italy, 22–29 October 2017; pp. 2089–2097.
46. Lin, T.-Y.; RoyChowdhury, A.; Maji, S. Bilinear Convolutional Neural Networks for Fine-Grained Visual Recognition. *IEEE Trans. Pattern Anal. Mach. Intell.* **2018**, *40*, 1309–1322. [CrossRef]

Article

SORAG: Synthetic Data Over-Sampling Strategy on Multi-Label Graphs

Yijun Duan [1,*], Xin Liu [1], Adam Jatowt [2], Hai-tao Yu [3], Steven Lynden [1], Kyoung-Sook Kim [1] and Akiyoshi Matono [1]

1   National Institute of Advanced Industrial Science and Technology Tokyo Waterfront, 2 Chome-3-26 Aomi, Tokyo 135-0064, Japan
2   Department of Computer Science, University of Innsbruck, Innrain 52, 6020 Innsbruck, Austria
3   Faculty of Library, Information and Media Science, University of Tsukuba, 1 Chome-1-1 Tennodai, Tsukuba 305-8577, Japan
*   Correspondence: yijun.duan@aist.go.jp

Abstract: In many real-world networks of interest in the field of remote sensing (e.g., public transport networks), nodes are associated with multiple labels, and node classes are imbalanced; that is, some classes have significantly fewer samples than others. However, the research problem of imbalanced multi-label graph node classification remains unexplored. This non-trivial task challenges the existing graph neural networks (GNNs) because the majority class can dominate the loss functions of GNNs and result in the overfitting of the majority class features and label correlations. On non-graph data, minority over-sampling methods (such as the synthetic minority over-sampling technique and its variants) have been demonstrated to be effective for the imbalanced data classification problem. This study proposes and validates a new hypothesis with unlabeled data over-sampling, which is meaningless for imbalanced non-graph data; however, feature propagation and topological interplay mechanisms between graph nodes can facilitate the representation learning of imbalanced graphs. Furthermore, we determine empirically that ensemble data synthesis through the creation of virtual minority samples in the central region of a minority and generation of virtual unlabeled samples in the boundary region between a minority and majority is the best practice for the imbalanced multi-label graph node classification task. Our proposed novel data over-sampling framework is evaluated using multiple real-world network datasets, and it outperforms diverse, strong benchmark models by a large margin.

Keywords: imbalanced data classification; data over-sampling; generative adversarial network; graph convolutional network; semi-supervised learning; remote sensing

Citation: Duan, Y.; Liu, X.; Jatowt, A.; Yu, H.-t.; Lynden, S.; Kim, K.-S.; Matono, A. SORAG: Synthetic Data Over-Sampling Strategy on Multi-Label Graphs. *Remote Sens.* **2022**, *14*, 4479. https://doi.org/10.3390/rs14184479

Academic Editors: Jungho Im and Gwanggil Jeon

Received: 26 June 2022
Accepted: 26 August 2022
Published: 8 September 2022

Publisher's Note: MDPI stays neutral with regard to jurisdictional claims in published maps and institutional affiliations.

1. Introduction

Graphs are becoming ubiquitous across a large spectrum of real-world applications in the form of social networks, citation networks, telecommunication networks, biological networks, etc. [1]. In addition, numerous applications involving multimedia, such as video surveillance, video streaming, healthcare systems, and intelligent indoor security systems depend on using graphs as research objects [2–4]. For a considerable number of real-world graph node classification tasks, the training data follow a *long-tail* distribution, and node classes are *imbalanced*. In other words, each of a few "majority" classes has a large number of samples, while most classes only contain a handful of instances. Taking the NCI chemical compound graph as an example, only approximately 5% of molecules are labeled as active in the anticancer bioassay test [5]. Graph node classification tasks are often further complicated by the fact that graph nodes can be associated with multiple labels in many real-world network data. Many social media sites, such as BlogCatalog, Flickr, and YouTube, allow users to use a diverse set of labels representing their various interests. A person can join several interest groups on Flickr, such as *Landscape* and *Travel*, and different video genres on YouTube, such as *Cooking* and *Wrestling*. Furthermore, many networks are characterized by imbalanced label distribution and multi-label nodes at the same time, as shown in Figure 1.

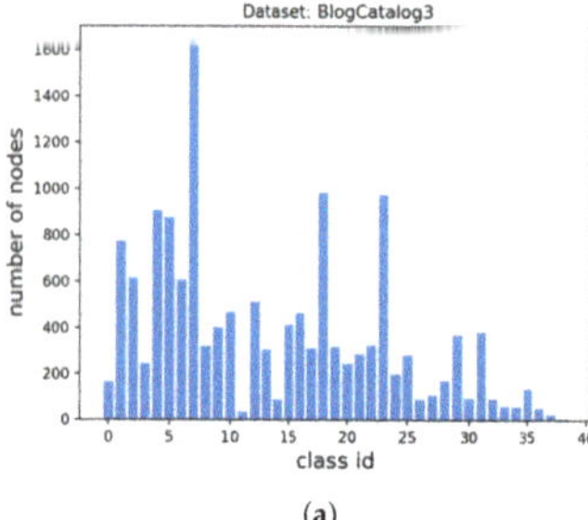

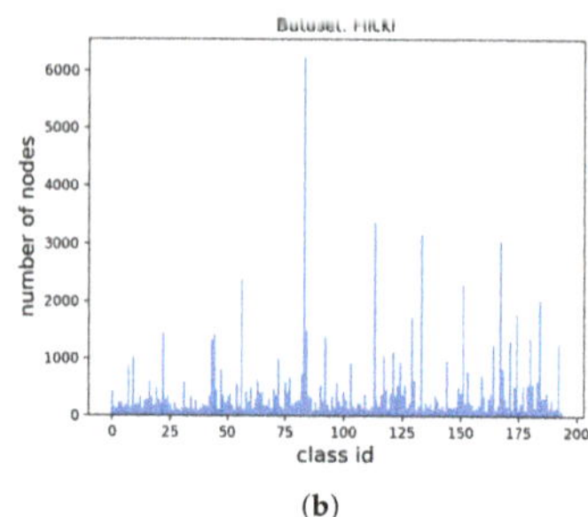

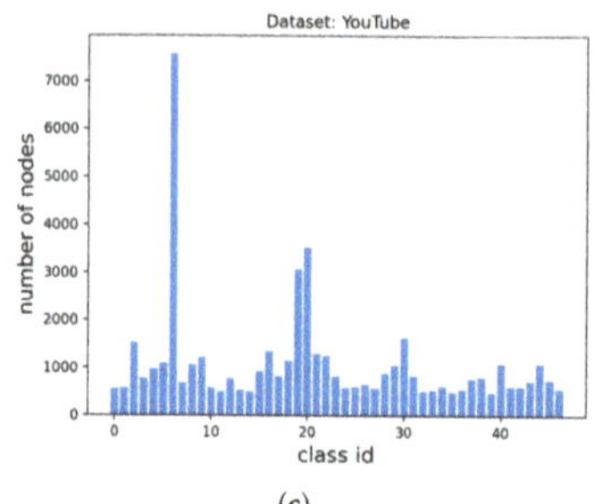

Figure 1. Label distribution of three real-world multi-label network datasets: (**a**) BlogCatalog3 [6], (**b**) Flickr [6], and (**c**) YouTube [7]. The horizontal axis represents the label id, and the vertical axis represents the number of nodes containing each label. It can be clearly observed that these label distributions are all highly imbalanced, where a few classes contain many more nodes than the rest of the classes.

To date, a large body of work has focused on graph representation learning (GRL) with balanced node classes and simplex labels [8–12]. However, these models do not perform well when graphs exhibit the aforementioned characteristics of being imbalanced and multi-label for the following reasons. (1) *The problem caused by the imbalanced setting*: The imbalanced data make the classifier overfit the majority class, and the features of the minority class cannot be sufficiently learned [13]. Furthermore, the above problem is aggravated by the presence of the *topological interplay* effect [5] between graph nodes, making the feature propagation dominated by the majority classes. (2) *The problem caused by the multi-label setting*: Multi-label graph architectures typically encode very complex interactions between nodes with shared labels [5], which is challenging to capture. Therefore, it is essential to develop a specific graph learning method for class imbalanced multi-label graph data. However, research in this direction is still in its infancy. Thus, in this study, we propose *imbalanced multi-label GRL* to address this challenge while also contributing to graph learning theory.

For imbalanced data, *minority over-sampling* is an effective measure to improve the classification accuracy [14–16]. This strategy has recently been confirmed to be effective for graph data as well [17]. Traditional over-sampling techniques consist of a two-step process: (1) the selection of some minority instances as "seed examples"; (2) the generation of synthetic data with features and labels similar to the seed examples, which are then added to the training data. For example, the most popular over-sampling technique— the synthetic minority over-sampling technique (SMOTE) [14]—addresses the problem of minority generation by performing interpolation between randomly selected minority instances and their nearest neighbors. Cost-sensitive learning is another type of effective approach for alleviating the problem of imbalanced data applied to a classification [16], where the basic assumption is that the cost resulting from different types of misclassification varies significantly (e.g., the cost of treating an intruder as a non-intruder is much greater than treating a non-intruder as an intruder). The principle of applying cost-sensitive learning methods to imbalanced learning problems is to assign a larger penalty cost to misclassified minority class samples [18]. Existing cost-sensitive classification algorithms can generally be grouped into three categories [19]: algorithms that (1) pre-process the training data, (2) post-process the output, and (3) apply direct cost-sensitive learning methods. Data pre-processing aims to make the classification results on the new training set equivalent to cost-sensitive classification decisions on the original training set, typically along the lines of sampling [18] and weighting [16]. Post-processing the output makes the classifier biased toward minority classes by adjusting the classifier decision threshold, as represented by MetaCost [20] and ETA [21]. Direct cost-sensitive learning methods embed the cost information into the objective function of the learning algorithm to obtain the minimal expected cost, such as cost-sensitive decision trees [22] and cost-sensitive SVM [23].

However, mainstream over-sampling techniques have significant shortcomings when applied to graph data, as the selection of seed examples prioritizes global minority nodes while ignoring local minority nodes, and each synthetic instance is always assigned a label based on some specific strategy, which may be incorrect. This is because, in contrast to non-graph data, the relationships between graph nodes are explicitly expressed by the edges connecting them, meaning that the representation learning of a node can be heavily dependent on its neighboring *unlabeled* nodes through the feature propagation mechanism inherent to graphs.

Motivated by the above observations, we propose and test the following hypothesis. In addition to synthetic minority samples, synthetic *unlabeled* samples can also facilitate the debiasing of graph neural networks (GNNs) on an imbalanced training set. In particular, for nearby global minority samples that are a local majority, we can "safely" produce virtual samples of the same class and add them into the training sets to balance the class distribution. Global minority samples that are also a local minority are more likely to be local outliers, and thus, they are risky for selection as seed examples for further over-sampling. For nearby global minority samples whose neighbors are class-balanced, it is difficult to determine the labels of virtual samples. Thus, the production of *unlabeled* virtual nodes should be encouraged, which can help minorities by "blocking" the over-aggregation of the majority features delivered through edges. This idea is illustrated in Figure 2. **We argue that the key to over-sampling on an imbalanced multi-label graph is to flexibly combine the synthesis of both labeled and unlabeled instances enriched by label correlations.**

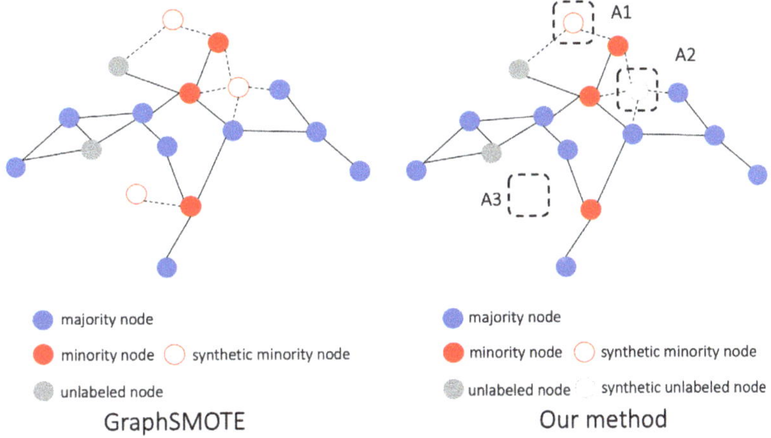

Figure 2. A comparison between our method and the current state-of-the-art graph over-sampling method **GraphSMOTE** [17]. In the current method, the idea is to generate new minority instances near randomly selected minority nodes and create virtual edges (dotted lines in the figure) between those synthetic nodes and real nodes. Instead, we synthesize minority instances in safe areas (i.e., A1), generate *unlabeled* instances in locally balanced areas (i.e., A2), and do not conduct data over-sampling near minority nodes that are outliers (i.e., A3). For the simplicity of illustration, only a single-label scenario is shown.

We extend the existing over-sampling algorithms to a novel framework for the imbalanced multi-label graph node classification task based on the above considerations. We extend the classic global minority-based seed example selection to the local minority perspective (see Section 3.2). Distinct from interpolation, which is commonly used in mainstream over-sampling techniques [24], we use a generative adversarial network (GAN) [25] to generate new instances. As a representative deep generative model, a GAN can capture label correlation information by estimating the probability distribution of seed examples [26]. We propose an ensemble architecture of a GAN and conditional GAN (CGAN) [27] for the flexible generation of both unlabeled and labeled synthetics (see Section 3.3). To make use of the graph topology information, we propose a method to obtain new edges between the generated samples and existing data with an edge predictor (see Section 3.4). The augmented graph is finally sent to a graph convolutional network (GCN) [12] for representation learning, together with the learned label correlations (see Section 3.5). We name our proposed framework **SORAG**, abbreviated from **S**ynthetic data **O**versampling St**RA**tegy on **G**raph.

In summary, our contribution is three-fold:

- We study an *unexplored* and novel research problem. We advance the traditional simplex-label graph learning to an imbalanced multi-label graph learning setting, which has diverse real-world applications (e.g., long-tailed graph node classification, link prediction, community detection, and ranking). To the best of our knowledge, this study is the first to focus on this task.
- We propose a new, general, and efficient GNN that addresses the deficiencies of previous graph over-sampling methods. Our framework flexibly ensembles the synthesis of labeled and *unlabeled* nodes to support the minority classes and leverage label correlations to generate more natural nodes.
- Extensive experiments on multiple *standard* real-world datasets demonstrate the high effectiveness of our approach. Compared with the current state-of-the-art model **GraphSMOTE** [17], our method has an improvement of 1.5% in terms of Micro-F1 and 3.3% in terms of Macro-F1 on average. The experimental results demonstrate the high effectiveness of the proposed approach. Detailed analyses of **SORAG** under different experimental environments and parameters are also presented.

2. Related Works

2.1. Graph Representation Learning

A growing number of applications use non-Euclidean methods to generate data, which are then represented as graphs with complex relationships and inter-object dependencies. Past feature representation and extraction algorithms face substantial difficulties in handling the complexity of graph data. Over the past decade, many studies on extending traditional feature extraction approaches for graph data have emerged. Among them, GRL has evolved considerably and can be roughly divided into three generations, including traditional graph embedding, modern graph embedding, and deep learning on graphs. The first generation of methods are classic dimension reduction techniques, such as IsoMap [28] and LLE [29]. The second generation of feature extraction methods on graphs are modern graph embedding methods, such as DeepWalk [30] and LINE [31]. GNNs can be broadly regarded as the third (and latest) generation of GRL after the traditional and modern graph embedding, and they are reported to achieve the most promising performance in a wide range of computational tasks on graphs [32].

A growing body of research has shown that GNNs are extremely effective for both traditional GRL tasks (e.g., recommender systems and social network analysis) and new research areas (e.g., healthcare, physics, and combinatorial optimization) [32]. A typical GNN consists of graph filters and/or graph pooling layers. The former take the node features and graph structure as inputs and output new node features. The latter take the graph as an input and output a coarsened graph with a few nodes.GNNs can be broadly classified into spatial and spectral approaches based on their graph filters. The former explicitly leverages the graph structure, for example, spatially close neighbors, whereas the latter analyzes the graph using a graph Fourier transform and an inverse graph Fourier transform [33].

Classical spatial-based GNNs include [9–11,34–37]. Ref. [9] is a very early GNN that uses the local transition function as a graph filter. A GraphSAGE filter [10] uses different aggregators (mean/LSTM/pooling) to aggregate information regarding the one-hop neighbors of the nodes. In addition, a GAT-filter [11] relies on a self-attention mechanism to distinguish the importance of neighboring nodes during the aggregation process. An ECC-filter [34] was also proposed to handle graphs with different types of edges. Similarly, a GGNN-filter [36] was designed for graphs with different types of directed edges. By contrast, a Mo-filter [35] is based on a Gaussian kernel. Finally, an MPNN [37] is a more general framework, with the GraphSAGE-filter and GAT-filter mentioned above being special cases. In general, spatial-based GNNs are more generalized and flexible.

Spectral-based graph filters use the graph spectral theory in the design of the filtering operations within the spectral domain. Early studies [33] dealt with the eigen decomposition of the Laplacian matrix and the matrix multiplication between dense matrices; thus, they are computationally expensive. To overcome this problem, a Poly-filter [38] based on a K-order truncated polynomial was proposed. To solve the problem of a Poly-filter in which the basis of the polynomial is not orthogonal, a Cheby-filter [38] based on the Chebyshev polynomial was introduced. A GCN-filter [12] is a simplified version of a Cheby-filter. The latter involves a K-hop neighborhood of a node during the filtering process, whereas in the former, K = 1. A GCN-filter can also be regarded as a spatial-based filter. Currently, GCNs are among the most widely used types of GNN. In our model, we used a GCN as the key component.

Feature extraction methods such as graph embedding and graph kernel techniques are strongly related to the study on GNNs. Compared to GNNs, the former only focus on representing network nodes as low-dimensional vector representations without targeting subsequent tasks, such as graph node classification and link prediction. Many graph embedding techniques are linear and not in an end-to-end manner, such as random walk [39] and matrix factorization [40]. Graph kernel techniques employ a kernel function to obtain the vector representations of graphs. They are also an important type of approaches to solve the graph classification problem. However, compared to GNNs, they are not learnable and far less efficient.

GNNs are designed for different graph-based tasks, such as node classification, link prediction, graph classification, and community detection. In particular, our task is semi-supervised, which means that we need to learn the representation of all nodes from a few labeled nodes and the remaining unlabeled nodes. GNNs are a rapidly growing field at the present time. For a more comprehensive and detailed introduction to this field, we refer the reader to [32].

2.2. Imbalanced Learning

If the classification classes are not approximately equal, and a few classes contain many more samples than the others, the dataset is called imbalanced. Representative techniques to handle imbalanced datasets include sampling methods, ensemble algorithms, and cost-sensitive approaches.

Re-sampling the original dataset is a strategy for balancing the majority and minority classes at the data level. This type of method constructs a well-balanced training set by over-sampling the minority class or under-sampling the majority class during the pre-processing step. Following that, any learning algorithm might be trained on the new dataset to reduce the system's bias towards the majority classes. A typical example of sampling models is SMOTE [14]. To shift the learning bias toward the minority class, it generates synthetic samples depending on their nearest neighbors. Numerous extensions using various distance measures or selection criteria of seed samples are proposed based on the regular SMOTE algorithm. Among them, representative methods include the borderline SMOTE [41], the safe-level SMOTE [42], and the density-based SMOTE [43].

The methods of over-sampling and under-sampling are not limited to any modeling algorithm, which could be essentially regarded as data pre-processing processes. Ensemble classifiers and cost-sensitive approaches, which are algorithm-specific, could also be used to address data imbalance concerns as algorithm-level enhancement. The former type of method includes diverse hybrid sampling/boosting algorithms, such as SMOTEBoost [44], random under-sampling boost (RUSB) [45], and the balance cascade approach [46]. Besides the boosting algorithms, other ensemble methods such as balanced random forest [47] can also be applied for imbalanced datasets. Cost-sensitive approaches, which penalize the misclassification of the minority more severely, have also been reported to be effective in addressing the problem of class imbalance. Two popular examples of them are AdaCost [48] and weighted random forest [47].

2.3. Connections to Our Work

The focus of this study is to investigate solutions for semi-supervised GRL and graph node classification on imbalanced multi-label graphs. The most closely related works can be found in the emerging area of imbalanced graph node classification [5,17,49]. Among these works, the dual-regularized graph convolutional network (DR-GCN) [5] model relies on a class-conditioned adversarial training process to facilitate the separation of labeled nodes and the identification of minority class nodes. GraphSMOTE [17] attempts to transfer the classical SMOTE method [14], which deals with imbalanced data, to graph data. In addition, the relaxed GCN network (RECT) [49] has reported the best performance on imbalanced graph node classification tasks, and its core idea is based on the design and optimization of a class-semantic-related objective function. Unlike GraphSMOTE, which is based on labeled minority generation, we present the first graph node over-sampling model that utilizes synthetic *unlabeled* nodes to inhibit the tendency of GNNs to overfit to majority under the topological effect. The new supervision information resulting from labeled synthetics and the blocking of over-propagated majority features by unlabeled synthetics facilitates balanced learning between different classes, taking advantage of the strong topological interdependence between nodes on a graph. We specify the details of the proposed model in the next section.

3. Methodology

3.1. System Overview

An illustration of the proposed framework **SORAG** is shown in Figure 3. **SORAG** is composed of four components: (1) the first part is in charge of determining the global minority degree (GMD) and local minority degree (LMD) of each node and constructing training data (i.e., seed examples) for the virtual node generator; (2) the second part, the node generator, is an ensemble of a GAN [25] network and a CGAN [27] network, where the GAN is responsible for creating unlabeled nodes and the CGAN is used for generating labeled nodes; (3) the third component is an edge generator. Its job is to create virtual edges between the synthetic and real nodes so that the generated nodes can participate in the message passing on the graph more effectively; (4) finally, a GCN-based node classifier is designed for learning the node representations of the augmented graph as well as the inter-label dependencies for multi-label node classification. We elaborate on each component as follows.

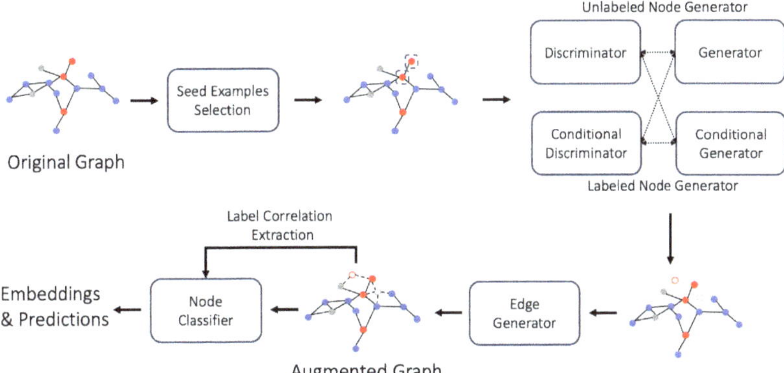

Figure 3. Overview of the **SORAG** framework. First, based on the feature matrix and the adjacency matrix of the input graph, we calculate the local minority degree (LMD) value of each node in the training set and classify it into one of the four types: safe (SF), borderline (BD), rare (RR), and outlier (OT). After that, we calculate the seed probability (SP) values of nodes in the SF and BD classes and select the seed nodes based on such values (see Section 3.2). Then, we use the seed nodes to train the node generator (i.e., the ensemble of GAN and CGAN) to generate high-quality unlabeled synthetic nodes and labeled synthetic nodes. Notice that by adjusting the objective function, we can flexibly manipulate the data distribution simulated by the node generator (see Section 3.3). With virtual nodes, an edge generator, which is essentially a feed-forward neural network, is used to generate virtual edges connecting the virtual nodes and the real nodes. The role of the generated edges is to facilitate feature propagation between two types of nodes (see Section 3.4). Finally, the new graph containing virtual nodes and virtual edges is fed into a GCN that learns the discriminative graph embeddings and performs effective node classification. During this process, the label correlation matrix provides helpful label correlation and interaction information (see Section 3.5).

3.2. Imbalance Measurement

In multi-label learning, a commonly used measure that evaluates the global imbalance of a particular label is *IRLbl*. Let $|C_i|$ be the number of instances whose i-th label value is 1; *IRLbl* is then defined as follows:

$$IRLbl_j = \frac{max\{|c_1|, |c_2|, \ldots, |c_m|\}}{c_i}. \tag{1}$$

Therefore, the larger the value of *IRLbl* for a label, the more minority class it is. For a node v_i, its GMD is defined as follows:

$$GMD_i = \frac{IRLbl_j \cdot [B_{ij} = 1]}{\sum_{j=1}^{m} [B_{ij} = 1]}, \tag{2}$$

where $[B_{ij} = 1]$ means v_i has the j-th label, and $\sum_{j=1}^{m} [B_{ij} = 1]$ counts the number of labels that v_i has.

The LMD of a node can be measured by the proportion of opposite class values in its local neighborhood. For v_i, let N_i^k denote its k-hop neighbor nodes. Then, for label c_j, the proportion of neighbors having an opposite class to the class of v_i is computed as

$$S_{ij} = \frac{\sum_{v_m \in N_i^k} [B_{ij} \neq B_{mj}]}{|N_i^k|}, \tag{3}$$

where $S \in \mathbb{R}^{n \times m}$ is a matrix defined to store the local imbalance of all nodes for each label. Given S, a straightforward way to compute the LMD for v_i is to average its S_{ij} for all labels as follows:

$$LMD_i = \frac{\sum_{j=1}^{m} S_{ij}[B_{ij} = g_j]}{m}, \tag{4}$$

where $g_j \subset \{0, 1\}$ denotes the minority class of the j-th label. Namely, if $|c_j| \geq 0.5 \cdot n$, $g_j = 1$; otherwise, $g_j = 0$. Here, n is the total number of vertices. Further, we group the global minority nodes into different types based on the LMD, and each type is identified correctly by the classifier with different difficulties. Following [50,51], we discretize the range $[0, 1]$ of LMD_i to define four types of nodes, namely safe (SF), borderline (BD), rare (RR), and outlier (OT), according to their local imbalance:

- SF: $0 \leq LMD_i < 0.3$. Safe nodes are basically surrounded by nodes containing similar labels.
- BD: $0.3 \leq LMD_i < 0.7$. Borderline nodes are located in the decision boundary between different classes.
- RR: $0.7 \leq LMD_i < 1.0$. Rare nodes are located in a region overwhelmed by different nodes and are distant from the decision boundary.
- OT: $LMD_i = 1.0$. Outliers are totally connected to different nodes.

Based on the above categories, we are confident about generating new virtual samples for the global minority samples belonging to SF by imitating their features and labels to balance the distorted class distribution. The global minority samples belonging to BD are located in the decision boundary; hence, it is challenging to determine the label for virtual samples similar to them. Therefore, we keep the new samples unlabeled and use them to weaken the over-propagation of majority class features by taking advantage of the feature smoothing mechanism on the graph. The global minority samples belonging to RR/OT are more likely to be outliers and should not be selected as seeds to generate new samples.

Furthermore, for v_i, we define two metrics: labeled seed probability (LSP) and unlabeled seed probability (USP) to describe the probability of being selected as a seed example to generate labeled synthetic nodes and unlabeled synthetic nodes, respectively. The seed probabilities (LSP/USP) are calculated as follows:

$$SP_i = GMD_i \cdot LMD_i = \begin{cases} LSP_i, & \text{if } v_i \in \text{SF} \\ USP_i, & \text{if } v_i \in \text{BD} \end{cases} \tag{5}$$

We compute the LSP and USP scores for all nodes and sort them in descending order. The top-ranked nodes (controlled by the hyper-parameter seed example rate ρ) will be selected as seed examples. A min-max normalization processes all the GMD and LMD scores to improve the computation stability.

3.3. Node Generator

We denote the joint distribution of node feature x and label y in the SF region as $P_{SF}(x, y)$, the marginal distribution of y as $P_{SF}(y)$, and the marginal distribution of x in the BD region as $P_{BD}(x)$. Generator G_l is expected to generate labeled instances in the SF region, while generator G_u should output unlabeled synthetics in the BD region. Let the data distribution produced by G_l and G_u be denoted as $P_l(x, y)$ and $P_u(x)$, respectively; then, we expect $P_{BD}(x) \approx P_u(x)$ and $P_{SF}(x, y) \approx P_l(x, y)$. Furthermore, a more flexible goal is to have $P_{BD}(x) \approx \alpha \cdot P_u(x) + (1 - \alpha) \cdot P_l(x)$, $P_{SF}(x, y) \approx \beta \cdot P_l(x, y) + (1 - \beta) \cdot P_u(x, y)$, $\alpha \approx 1, \beta \approx 1$. α and β are parameters used to control G_l and G_u to produce various data distributions to fit the original data. Here, $P_u(x, y)$ is the joint distribution of $P_u(x)$ and $P_{SF}(y)$, and $P_l(x)$ is the marginal distribution of $P_l(x, y)$.

To achieve the above goal, we propose a node generator, which is essentially an ensemble of a GAN [25] and a CGAN [27]. The GAN is responsible for generating unlabeled synthetic nodes, whose generator and discriminator are, respectively, denoted as G_u and D_u. The CGAN is used for generating labeled synthetic instances, where its generator and discriminator are denoted as G_l and D_l, respectively. Our loss function for training the GAN is

$$\min_{G_u} \max_{D_u} \mathcal{L}_{GAN} = \mathbb{E}_{x \sim P_{BD}(x)} log D_u(x) + \alpha \cdot \mathbb{E}_{x \sim P_u(x)} log(1 - D_u(x)) \tag{6}$$

For the CGAN, our objective is given as

$$\min_{G_l} \max_{D_l} \mathcal{L}_{cGAN} = \mathbb{E}_{(x,y) \sim P_{SF}(x,y)} log D_l(x,y) + \beta \cdot \mathbb{E}_{(x,y) \sim P_l(x,y)} log(1 - D_l(x,y)) \tag{7}$$

To achieve flexible control over G_l and G_u, we design the following loss function based on the interaction between the GAN and CGAN:

$$\begin{aligned}\min_{G_u,G_l} \max_{D_u,D_l} \mathcal{L}_{GAN-cGAN} &= (1 - \alpha) \cdot \mathbb{E}_{x \sim P_l(x)} log(1 - D_u(x)) \\ &+ (1 - \beta) \cdot \mathbb{E}_{(x,y) \sim P_u(x,y)} log(1 - D_l(x,y))\end{aligned} \tag{8}$$

Combining these equations, our final loss for node generation $\mathcal{L}_{node}$ is

$$\mathcal{L}_{node} = \min_{G_u,G_l} \max_{D_u,D_l} \mathcal{L}_{GAN} + \mathcal{L}_{cGAN} + \mathcal{L}_{GAN-cGAN} \tag{9}$$

For our proposed generator, the following theoretical analysis is performed.

Proposition 1. *For any fixed G_u and G_l, the optimal discriminator D_u and D_l of the game defined by $\mathcal{L}_{node}$ is*

$$D_u^*(x) = \frac{P_{BD}(x)}{P_{BD}(x) + P_\alpha(x)}, D_l^*(x,y) = \frac{P_{SF}(x,y)}{P_{SF}(x,y) + P_\beta(x,y)} \tag{10}$$

where $P_\alpha(x) = \alpha \cdot P_u(x) + (1 - \alpha) \cdot P_l(x)$, and $P_\beta(x,y) = \beta \cdot P_l(x,y) + (1 - \beta) \cdot P_u(x,y)$.

Proof. We have

$$\begin{aligned}\mathcal{L}_{node} &= \int_x P_{BD}(x) log D_u(x) dx + \int_{x,y} P_{SF}(x,y) log D_l(x,y) dx dy \\ &+ \alpha \cdot \int_x P_u(x) log(1 - D_u(x)) dx + \beta \cdot \int_{x,y} P_l(x,y) log(1 - D_l(x,y)) dx dy \\ &+ (1 - \alpha) \cdot \int_x P_l(x) log(1 - D_u(x)) dx + (1 - \beta) \cdot \int_{x,y} P_u(x,y) log(1 - D_l(x,y)) dx dy \\ &= \int_x P_{BD}(x) log D_u(x) + P_\alpha(x) \cdot log(1 - D_u(x)) dx \\ &+ \int_{x,y} P_{SF}(x,y) log D_l(x,y) + P_\beta(x,y) \cdot log(1 - D_l(x,y)) dx dy\end{aligned} \tag{11}$$

For any $(a,b) \in \mathbb{R}^2 \backslash \{0,0\}$, the function $f(y) = a log y + b log(1 - y)$ achieves its maximum in $[0,1]$ at $\frac{a}{a+b}$. This concludes the proof. $\square$

Proposition 2. *The equilibrium of $\mathcal{L}_{node}$ is achieved if and only if $P_{BD}(x) = P_\alpha(x)$ and $P_{SF}(x,y) = P_\beta(x,y)$ with $D_u^*(x) = D_l^*(x,y) = \frac{1}{2}$, and the optimal value of $\mathcal{L}_{node}$ is $-4log2$.*

Proof. When $D_u(x) = D_u^*(x), D_l(x,y) = D_l^*(x,y)$, we have

$$\mathcal{L}_{node} = \int_x P_{BD}(x) log \frac{P_{BD}(x)}{P_{BD}(x) + P_\alpha(x)} dx + \int_{x,y} P_{SF}(x,y) log \frac{P_{SF}(x,y)}{P_{SF}(x,y) + P_\beta(x,y)} dxdy$$

$$+ \int_x P_\alpha(x) log \frac{P_\alpha(x)}{P_{BD}(x) + P_\alpha(x)} dx + \int_{x,y} P_\beta(x,y) log \frac{P_\beta(x,y)}{P_{SF}(x,y) + P_\beta(x,y)} dxdy \qquad (12)$$

$$= -4log2 + 2 \cdot JSD(P_{BD}(x)||P_\alpha(x)) + 2 \cdot JSD(P_{SF}(x,y)||P_\beta(x,y))$$

$$\geq -4log2$$

where the optimal value is achieved when the two Jensen–Shannon divergences are equal to 0, namely, $P_{BD}(x) = P_\alpha(x)$, and $P_{SF}(x,y) = P_\beta(x,y)$. When $\alpha = \beta = 1$, we have $P_{BD}(x) = P_u(x), P_{SF}(x,y) = P_l(x,y)$. $\square$

In the implementation, both G_u and G_l are designed as a three-layer feed-forward neural network. In contrast, D_u and D_l are designed with a relatively weaker structure: a one-layer feed-forward neural network for facilitating the training.

3.4. Edge Generator

The edge generator described in this section is responsible for estimating the relation between virtual nodes and real nodes, which facilitates feature propagation, feature extraction, and node classification. Such edge generators will be trained on real nodes and existing edges. Following a previous work [17], the inter-node relation is embodied in the weighted inner product of node features. Specifically, for two nodes v_i and v_j, let E_{ij} denote the probability of the existence of an edge between them, which is computed as

$$E_{ij} = \sigma(x_i \cdot W^{edge} \cdot x_j^T) \qquad (13)$$

where x_i and x_j are the feature vectors of v_i and v_j, respectively. $W^{edge} \in \mathbb{R}^{k \times k}$ is the weight parameter matrix to be learned, and $\sigma = Sigmoid()$. Then, the extended adjacency matrix A' is defined as follows:

$$A'_{ij} = \begin{cases} A_{ij}, & \text{if } v_i \text{ and } v_j \text{ are real nodes} \\ E_{ij}, & \text{if } v_i \text{ or } v_j \text{ is synthetic node} \end{cases} \qquad (14)$$

Compared to A, A' contains new information about virtual nodes and edges, which will be sent to the node classifier in Section 3.5. As the edge generator is expected to be partially trained based on the final node classifier (see Section 3.6), the predicted edges should be set as continuous so that the gradient can be calculated and propagated from the node classifier. Thus, E_{ij} is not discretized to some value in {0,1}. The edge generator should be capable of accurately predicting real edges to generate realistic virtual nodes. Then, the pre-trained loss function for training the edge generator is

$$\mathcal{L}_{edge} = ||E - A||^2 \qquad (15)$$

where E refers to predicted edges between real nodes.

3.5. Node Classifier

We now obtain an augmented balanced graph $G' = \{V', A', X', B'\}$, where V' consists of both real nodes and synthetic labeled and unlabeled nodes; further, A', X', and B' denote the edge, feature, and label information of the enlarged vertex set, respectively. A classic two-layer GCN structure [12] is adopted for node classification, given its high accuracy and efficiency. Its first and second layers are denoted as L^1 and L^2, respectively, and their corresponding outputs $\{O^1, O^2\}$ are

$$O^1 = ReLU(\tilde{D}^{-\frac{1}{2}} \tilde{A}' \tilde{D}^{-\frac{1}{2}} X' W^1) \qquad (16)$$

$$O^2 = \sigma(F\tilde{D}^{-\frac{1}{2}}\tilde{A}'\tilde{D}^{-\frac{1}{2}}O^1 W^2) \tag{17}$$

where $\tilde{A}' = A' + I$, I is an identity matrix of the same size as A'. $\tilde{D}$ is a diagonal matrix, and $\tilde{D}_{ii} = \sum_j \tilde{A}'_{ij}$. $\tilde{D}^{-\frac{1}{2}}\tilde{A}'\tilde{D}^{-\frac{1}{2}}$ is the normalized adjacency matrix. Further, W^1 and W^2 are the learnable parameters in the first and second layers, respectively. *ReLU* and σ are the respective activation functions of the first and the second layer, where $ReLU(Z)_i = max(0, Z_i)$, $\sigma(Z)_i = Sigmoid(Z)_i = \frac{1}{1+exp(-Z_i)}$. O^2 is the posterior probability of the class to which the node belongs. F is the label correlation matrix that is computed in the same way as in [5], which provides helpful label correlation and interaction information.

In Equations (16) and (17), the role of $\tilde{D}^{-\frac{1}{2}}\tilde{A}'\tilde{D}^{-\frac{1}{2}}$, or the normalized adjacency matrix, is to enrich the feature vector of a node by linearly adding all feature vectors of its neighbors. This is because the basic assumption of a GCN is that neighboring nodes (and thus those having similar neighbors) are more likely to belong to the same class. The role of W^1, W^2 is to transform the feature dimension of the nodes, making sparse high-dimensional node features dense at low dimensions. In addition, Equations (16) and (17) can also be equivalently described as the process by which the input signal (i.e., node feature X') is filtered through a graph Fourier transform in the graph spectral domain [32]; however, in this study, we consider the spatial domain.

Eventually, given the training labels B^{train}, we minimize the following cross-entropy error to learn the classifier, where p is the number of training samples, m is the size of the label set, and nc stands for node classifier. By minimizing $\mathcal{L}_{nc}$, we can learn the parameters of the GCN such that it predicts the posterior probability of the class to which the unlabeled node belongs.

$$\mathcal{L}_{nc} = -\sum_{i=1}^{p}\sum_{j=1}^{m} B_{ij}^{train} ln O_{ij}^2 \tag{18}$$

3.6. Optimization Objective

Based on the above content, the final objective function of our framework is given as

$$\min_{\Theta,\Phi,\Psi} \mathcal{L}_{nc} + \lambda \cdot \mathcal{L}_{node} + \mu \cdot \mathcal{L}_{edge} \tag{19}$$

where Θ, Φ, and Ψ are the sets of parameters for the synthetic node generator (Section 3.3), edge generator (Section 3.4), and node classifier (Section 3.5), respectively. λ and μ in Equation (19) are weight parameters. The best training strategy in our experiments is to first pre-train the node generator and the edge generator and then minimize Equation (19) to train the node classifier and fine-tune the node generator and edge generator at the same time. Our entire framework is easy to implement, general, and flexible. Different structural choices can be adopted for each component, and different regularization terms can be enforced to provide prior knowledge.

3.7. Training Algorithm

Algorithm 1 illustrates the proposed framework. **SORAG** is trained through the following components: (1) the selection of seed examples based on node LSP and USP scores; (2) the pre-training of the node generator (i.e., the ensemble of GAN and CGAN) for synthetic data generation; (3) the pre-training of the edge generator to produce new relation information; and finally, (4) the training of the node classifier on top of the over-sampled graph and the fine-tuning of the node generator and edge generator. The computational complexity of our model is approximately the sum of the computational complexity of the contained GAN, CGAN, and GCN.

Algorithm 1 Full Training Algorithm

Inputs: Graph data: $G = \{V, A, X, L, B\}$
Outputs: Network parameters, node representations, and predicted node class
1: Initialize the node generator, edge generator, and node classifier
2: Compute the node LSP and USP scores based on Equation (5)
3: Select the fraction of nodes with the highest LSP and USP scores as seed examples for D_l and D_u, respectively
4: **while** Not Converged **do**   ▷ Pre-train the node generator
5:      Update D_l by ascending along its gradient based on $\mathcal{L}_{node}$ (Equation (9))
6:      Update G_l by descending along its gradient based on $\mathcal{L}_{node}$
7:      Update D_u by ascending along its gradient based on $\mathcal{L}_{node}$
8:      Update G_u by descending along its gradient based on $\mathcal{L}_{node}$
9: **end while**
10: **while** Not Converged **do**   ▷ Pre-train the edge generator
11:     Update the edge generator by descending along its gradient based on $\mathcal{L}_{edge}$ (Equation (15))
12: **end while**
13: Construct the label–occurrence network and extract label correlations [5]
14: **while** Not Converged **do**▷ Train the node classifier and pre-train the other components
15:     Generate new unlabeled nodes using G_u
16:     Generate new labeled nodes using G_l
17:     Generate the new adjacency matrix A' using the edge generator
18:     Update the full model based on $\mathcal{L}_{nc} + \lambda \cdot \mathcal{L}_{node} + \mu \cdot \mathcal{L}_{edge}$ (Equation (19))
19: **end while**
20: Predict the test set labels with the trained model

---

4. Experimental Settings

4.1. Datasets

We use three popular multi-label networks: BLOGCATALOG3, FLICKR, and YOUTUBE as benchmark datasets. In Table 1, we list the statistical information of all datasets used, including the number of nodes, number of edges, number of node classes, and the tuned optimal value of key parameters of **SORAG**$_F$: {learning rate, weight decay, dropout rate, k (Section 3.2), ρ (Section 3.2), α (Section 3.3), β (Section 3.3), λ (Section 3.6), μ (Section 3.6)} (see the detailed parameter tuning discussion in Section 5.5). For each dataset, we assume that a majority class is one with more samples than the average class size, while a minority class is one with less samples. Below is a brief description of each dataset used.

- BLOGCATALOG3 [6] is the dataset crawled in July 2009 from BlogCatalog, which is a social blog directory website. This contains the friendship network crawled. The labels represent the topic categories provided by the authors, such as *Education*, *Food*, and *Health*. This network contains 10,312 nodes, 333,983 edges, and 39 labels. The edge type is undirected.

- FLICKR [6] is a crawl of the Flickr photo-sharing social network. Nodes are users, and edges represent that a user added another user to their list of contacts. The labels represent the interest groups of the users, such as *black and white photos*. This network contains 80,513 nodes, 5,899,882 edges, and 195 labels. The edge type is undirected.

- YOUTUBE [7] is a video-sharing website that includes a social network. The dataset contains a list of all the user-to-user links. The labels represent groups of viewers that enjoy common video genres such as *anime* and *wrestling*. This network contains 1,138,499 nodes, 2,990,443 edges, and 47 labels. The edge type is undirected.

For all datasets, we attribute each node with a 64-dim embedding vector obtained by performing dimensionality reduction on the adjacency matrix using PCA [52], similar to [5,17]. All of the above datasets are available at http://zhang18f.myweb.cs.uwindsor.ca/datasets/ (accessed on 25 June 2022).

Table 1. Dataset statistics.

Dataset	BLOGCATALOG3	FLICKR	YOUTUBE
# Nodes	10,312	333,983	39
# Edges	80,513	5,899,882	195
# Classes	1,138,499	2,990,443	47
learning rate	0.05	0.01	0.1
weight decay	5×10^{-4}	10^{-4}	10^{-3}
dropout rate	0.5	0.5	0.9
k	2	2	2
ρ	0.5	0.5	0.5
α	0.9	0.5	0.8
β	0.8	0.9	0.8
λ	0.1	1	1
μ	1	1	1

4.2. Analyzed Methods

To validate the performance of our approach, we compared it with several state-of-the-art and representative methods for multilabel graph learning and imbalanced graph learning, including **GCN** [12], **ML-GCN** [5], **SMOTE** [14], **GraphSMOTE** [17], and **RECT** [49]. The analyzed baseline methods are briefly introduced as follows:

- **GCN** [12] is a representative GNN structure that can naturally learn node representations from node features and network structures, where each node forms its representation by adopting a spectral-based convolutional filter to aggregate features recursively from all its neighborhood nodes.
- **GCN + under-sampling** [15]. For the **GCN**, we tested its combination with one conventional imbalanced learning technique: under-sampling, which reduces the degree of imbalance in the training set by under-sampling the majority class samples. This method is denoted as **GCN**$_{US}$.
- **GCN + threshold-moving** [15]. We also tested the combination of **GCN** with another imbalanced learning technique: threshold-moving, which moves the classification decision boundary to increase the classifier's preference for the minority class. This method is denoted as **GCN**$_{TM}$.
- **ML-GCN** [5] is a state-of-the-art multi-label graph learning approach. It considers a two-layer graph structure that allows the preservation of label correlations and enables the label-correlation enhanced node representation learning.
- **SMOTE** [14] is the most representative over-sampling technique, which generates synthetic minority samples by interpolating a minority sample and its nearest neighbors of the same class. Synthetic nodes are set to have the same edges as their seed node when applying it on the graph. Node representations are then learned by **GCN** [12].
- **GraphSMOTE** [17] is the state-of-the-art approach for the imbalanced node classification task, which is an adaption of **SMOTE** on graph data.
- **RECT** [49] is also an imbalanced graph learning method. It merges a GNN and proximity-based embeddings for the imbalanced setting and utilizes imbalanced labels by deducing the class-semantic descriptions for minority classes.

In addition, three variants of the proposed method were implemented:

- **SORAG**$_F$: The full model. The synthetic nodes include both unlabeled and labeled types;
- **SORAG**$_L$: Only labeled nodes are generated;
- **SORAG**$_U$: Only unlabeled nodes are generated.

It is necessary to mention that all the baselines above, except **ML-GCN** (which is intrinsically designed as a multi-label classifier), are manually set to conduct multi-label node classification by modifying the last layer of their network structure. The implementation of the baseline approach relies on publicly released code from relevant sources (**SMOTE**: https://github.com/analyticalmindsltd/smote_variants (accessed on 25 June 2022), **GraphSmote**: https://github.com/TianxiangZhao/GraphSmote (accessed on 25 June 2022), **RECT**: https://github.com/zhengwang100/RECT (accessed on 25 June 2022), **GCN**: https://github.com/tkipf/pygcn) (accessed on 25 June 2022).

4.3. Evaluation Metrics

Following many previous studies in the field of multi-label GRL [5,30,31], we adopt Micro-F1 and Macro-F1 to evaluate the node classification performance, which are defined as follows:

$$Micro - F1 = \frac{\sum_{i=1}^{m} 2 \cdot TP^i}{\sum_{i=1}^{m} (2 \cdot TP^i + FP^i + FN^i)} \tag{20}$$

$$Macro - F1 = \frac{1}{m} \sum_{i=1}^{m} \frac{2 \cdot TP^i}{2 \cdot TP^i + FP^i + FN^i} \tag{21}$$

where TP^i, FP^i, and FN^i represent the number of true positives, false positives, and false negatives for the i-th label, respectively. m denotes the total number of labels. Micro-F1 measures the F1-score of the aggregated contributions of all the classes. Macro-F1 was defined as the arithmetic mean of the label-wise F1-scores. Compared to Micro-F1, Macro-F1 does not consider the class size. Micro-F1 and Macro-F1 combine the precision and recall of the model, the range of which is [0, 1]. The larger the value, the stronger is the model performance.

4.4. Training Configurations

Following the semi-supervised learning setting, we randomly sampled a portion of the labeled nodes (i.e., sampling ratio) of each dataset and used them for evaluation. Then, we randomly split the sampled nodes into 60%/20%/20% for training, validation, and testing. As in [30], the sampling ratios for the BLOGCATALOG3, FLICKR, and YOUTUBE networks were set to 10%, 1%, and 1%, respectively. To balance the class size, we experimented with different amounts of synthetic unlabeled nodes and synthetic labeled nodes (i.e., different oversampling rates); finally, they were set as those in Section 5.3. All the analyzed models were trained using the Adam optimizer [53] in PyTorch (2020.2.1, community edition) [54]. Each result is presented as the mean of ten replicated experiments. All models were trained until they converged with a typical number of training epochs of 200.

5. Experimental Results

5.1. Imbalanced Multi-Label Classification Performance

Tables 2 and 3 show the performance of all the methods in terms of Micro-F1 and Macro-F1. The results are presented as the mean of ten repeated experiments. Based on these results, we reached the following conclusions:

- When compared with the GCN and ML-GCN methods, which do not consider class distribution, the three variants of **SORAG** show significant improvements. For example, compared with ML-GCN, the improvement brought by $\textbf{SORAG}_F$ is 7.4%, 4.2%, and 5.2% in terms of Micro-F1 and 9.6%, 5.3%, and 9.1% in terms of Macro-F1, respectively. This demonstrates that our proposed data over-sampling strategy effectively enhances the classification performance of GNNs on imbalanced multi-label graph data.
- **SORAG** provides many more benefits than when applying the previous imbalanced graph node classifier (SMOTE, GraphSMOTE, RECT). On average, it outperforms earlier methods by 3.3%, 3.0%, and 1.1% in terms of Micro-F1 and 2.5%, 2.9%, and 4.5% in terms of Macro-F1, respectively. This result validates the advantage of **SORAG** over previous over-sampling techniques in combining the generation of minority and unlabeled samples.
- Both minority over-sampling and unlabeled data over-sampling can improve the classification performance. In particular, the former is more effective. A combination

of the two strategies works the best. As supporting evidence, $\textbf{SORAG}_F$ is the best performer in 5/6 tasks and the second-best performer in the remaining task.

Table 2. Imbalanced multi-label classification comparison in terms of Micro-F1. The first and second best results are boldfaced and underscored, respectively.

Metrics	Micro-F1 (%)		
Methods\Datasets	BLOGCATALOG3	FLICKR	YOUTUBE
GCN	37.36	34.03	36.19
GCN_{US}	39.42	36.53	37.85
GCN_{TM}	38.50	34.13	38.25
ML-GCN	37.51	38.91	37.64
SMOTE	40.24	39.30	39.01
GraphSMOTE	42.82	40.01	**43.70**
RECT	41.72	41.23	42.66
$\textbf{SORAG}_L$	<u>44.58</u>	<u>41.61</u>	41.98
$\textbf{SORAG}_U$	43.21	37.92	40.83
$\textbf{SORAG}_F$	**44.89**	**43.13**	<u>42.86</u>

Table 3. Imbalanced multi-label classification comparison in terms of Macro-F1. The first and second best results are boldfaced and underscored, respectively.

Metrics	Macro-F1 (%)		
Methods\Datasets	BLOGCATALOG3	FLICKR	YOUTUBE
GCN	30.27	21.17	26.53
GCN_{US}	30.28	23.05	27.69
GCN_{TM}	30.32	21.72	26.59
ML-GCN	30.39	21.56	27.52
SMOTE	30.65	23.08	28.53
GraphSMOTE	35.58	24.25	33.81
RECT	<u>38.66</u>	24.47	33.94
$\textbf{SORAG}_L$	38.45	<u>26.48</u>	<u>35.01</u>
$\textbf{SORAG}_U$	37.28	26.15	32.53
$\textbf{SORAG}_F$	**40.01**	**26.85**	**36.57**

5.2. Influence of Training Data

Similar to [30], we increased the sampling ratio of the BLOGCATALOG3 network from 10% to 90% to observe the performance of **SORAG** on larger training sets. Because the FLICKR and YOUTUBE networks are considerably larger, we varied the sampling ratio from 1% to 10%, which is also consistent with [30]. We also tested the performance of the state-of-the-art method, **GraphSMOTE**, for comparison.

Figure 4 shows the Micro-F1 and Macro-F1 of each analyzed model with respect to the sampling ratio on each dataset. It can be observed that with an increase in the training data, $\textbf{SORAG}_F$ exhibits the most stable and promising performance, whereas the performances of $\textbf{SORAG}_L$ and $\textbf{SORAG}_U$ fluctuate considerably under different test conditions. This finding supports our argument that the most effective oversampling strategy for multi-label graphs is to conjoin the generation of both unlabeled data and labeled data flexibly. It is also worth mentioning that **GraphSMOTE** shows competitive performance, especially on the YOUTUBE dataset. On average, compared with **GraphSMOTE**, the improvements brought by **SORAG**$_F$ are 3.9% (BLOGCATALOG3), 6.4% (FLICKR), and 0.4% (YOUTUBE) in terms of Micro-F1 and 6.2% (BLOGCATALOG3), −0.1% (FLICKR), and 1.4% (YOUTUBE) in terms of Macro-F1, respectively.

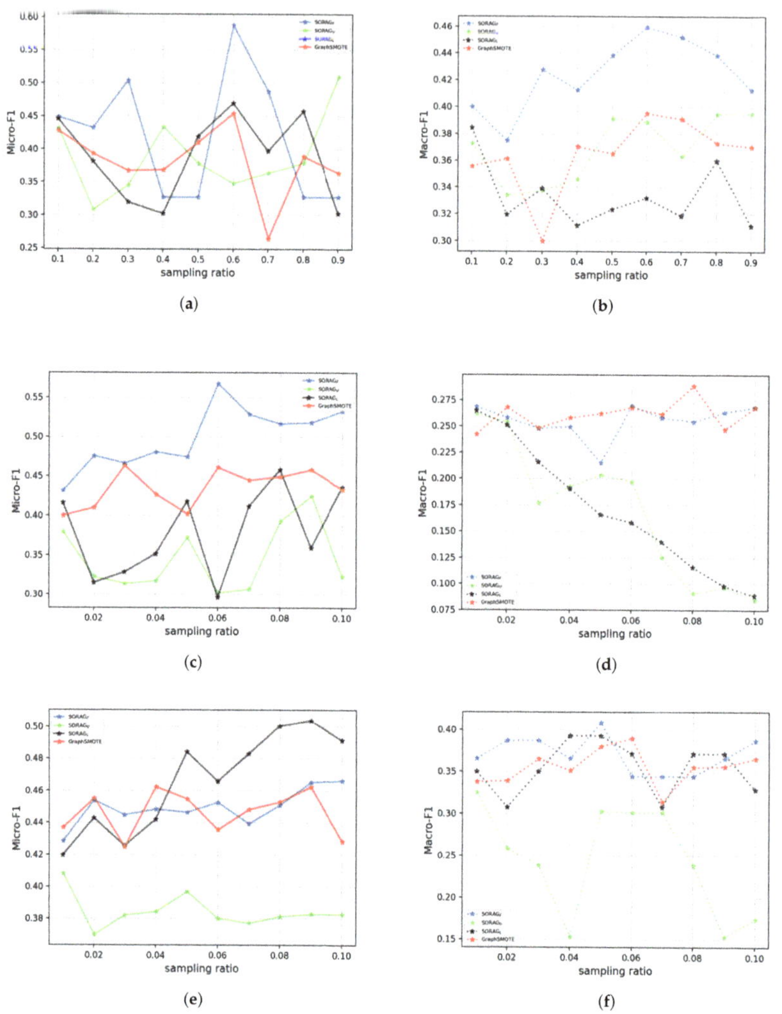

Figure 4. Performance of selected methods with reference to sampling ratio: (**a**) BLOGCATALOG3, MICRO-F1; (**b**) BLOGCATALOG3, MACRO-F1; (**c**) FLICKR, MICRO-F1; (**d**) FLICKR, MACRO-F1; (**e**) YOUTUBE, MICRO-F1; (**f**) YOUTUBE, MACRO-F1.

5.3. Influence of Over-Sampling Rate

In this section, we explore how the performance of **SORAG** varies with the oversampling rate. We varied the number of synthetic unlabeled nodes and the number of synthetic labeled nodes in [10%, 20%, ..., 90%, 100%] of the size of the training set on each dataset and recorded the performance change in **SORAG** as follows (see Figure 5). The sampling ratios for the BLOGCATALOG3, FLICKR, and YOUTUBE networks were set to 10%, 1%, and 1%, respectively, and all the parameters were the same as those in Section 5.2.

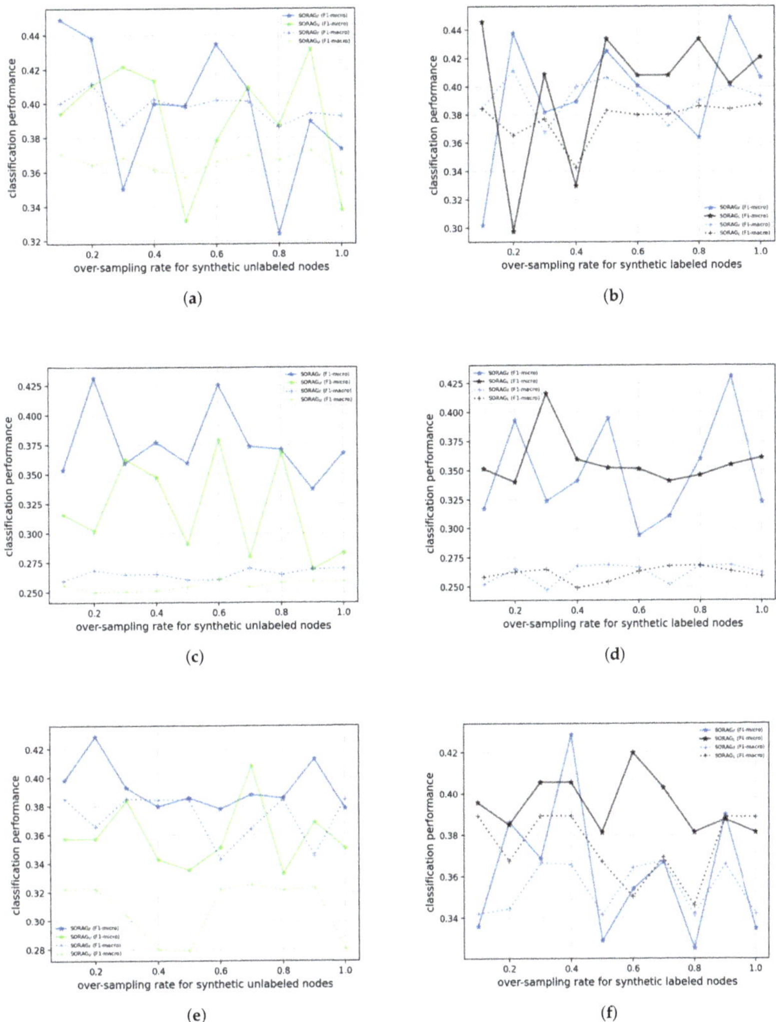

Figure 5. Influence of over-sampling rate on the model performance for each dataset: (**a**) BLOG-CATALOG3, UNLABELED NODE GENERATION; (**b**) BLOGCATALOG3, LABELED NODE GENERATION; (**c**) FLICKR, UNLABELED NODE GENERATION; (**d**) FLICKR, LABELED NODE GENERATION; (**e**) YOUTUBE, UNLABELED NODE GENERATION; (**f**) YOUTUBE, LABELED NODE GENERATION.

One clear observation is that the performance of **SORAG** fluctuates wildly with changes in the oversampling rate on all datasets. Unlike non-graph data, **SORAG** generates new edges for new samples during which more random noise is introduced. Therefore, as the over-sampling rate varies, the features of the virtual nodes affect the formation of the virtual edges. As a result, there is substantial variation in the feature propagation process on the new graph, which highly influences the classification results. This may explain the high sensitivity of **SORAG** to the oversampling rate. Below, in Table 4, we show the selected optimal over-sampling rates for all datasets.

Table 4. Optimal over-sampling rates for the synthetic unlabeled nodes (denoted as Rate$_U$) and synthetic labeled nodes (denoted as Rate$_L$) on each dataset. N/A abbreviates for "not applicable".

	BLOGCATALOG3		FLICKR		YOUTUBE	
	Rate$_U$	Rate$_L$	Rate$_U$	Rate$_L$	Rate$_U$	Rate$_L$
SORAG$_U$	0.9	N/A	0.6	N/A	0.7	N/A
SORAG$_L$	N/A	0.1	N/A	0.3	N/A	0.6
SORAG$_F$	0.1	0.9	0.2	0.9	0.2	0.4

5.4. Influence of Imbalance Ratio

This section discusses how the performance of the analyzed models varies as the imbalance of the training set changes. Similar to [5,17,49], we varied the percentage of global minority samples removed in [10%, 20, ..., 80%, 90%] on each dataset, according to the GMD ranking (see Section 3.2). The more samples were removed, the more imbalanced the training set became. The sampling ratios for the BLOGCATALOG3, FLICKR, and YOUTUBE networks were set to 10%, 1%, and 1%, respectively. The oversampling rates were set as presented in Table 4. All the parameters were the same as those in Section 5.2. The performance variation of each method is as follows (Figure 6). For comparison, we also report the performance of the state-of-the-art approach **GraphSMOTE**.

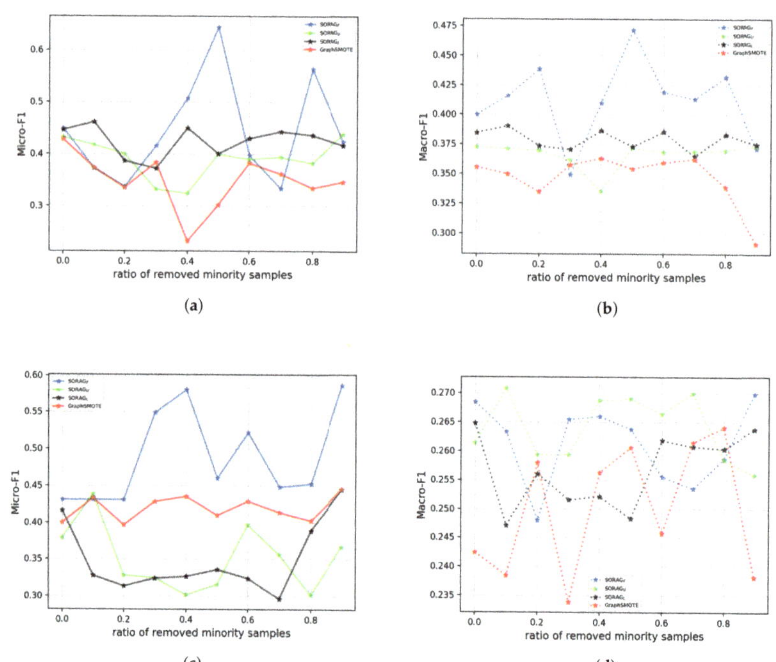

Figure 6. *Cont.*

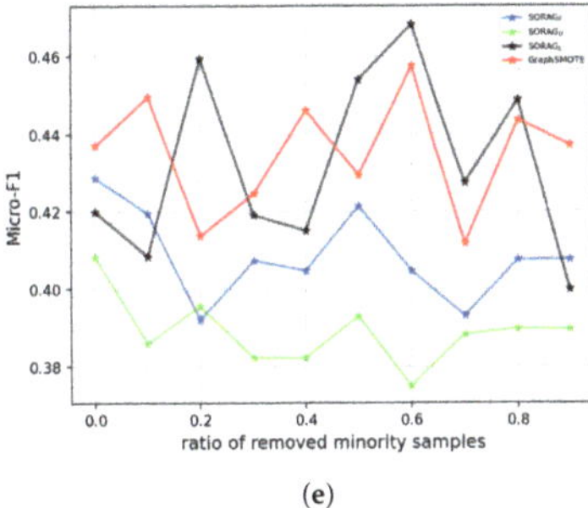
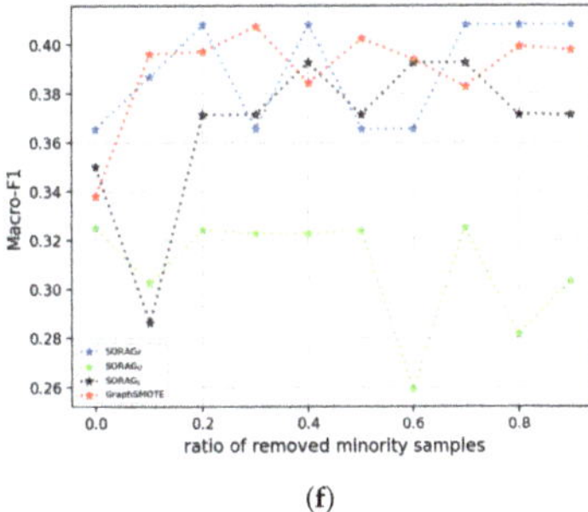

(e)   (f)

Figure 6. Influence of imbalance ratio: (**a**) BLOGCATALOG3, MICRO-F1; (**b**) BLOGCATALOG3, MACRO-F1; (**c**) FLICKR, MICRO-F1; (**d**) FLICKR, MACRO-F1; (**e**) YOUTUBE, MICRO-F1; (**f**) YOUTUBE, MACRO-F1.

Figure 6 demonstrates the strong robustness of **SORAG** in a variety of scenarios. As the imbalance ratio increases, the performance of **SORAG** is maintained at a high level. It can be observed that **SORAG** outperformed **GraphSMOTE** in nearly all comparisons. In particular, **SORAG**$_F$ was the best method. It achieved the best performance in 15 out of 30 test scenarios in terms of Micro-F1 and in 17 out of 30 test scenarios in terms of Macro-F1. In contrast, **GraphSMOTE** performed best only on the YOUTUBE network.

5.5. Parameter Tuning

For the validation set for each dataset, we used a grid search to tune the parameters in the following order: learning rate (range of $\{0.001, 0.005, 0.01, 0.05, 0.1\}$) $\rightarrow$ weight decay (range of $\{10^{-5}, 5 \times 10^{-5}, 10^{-4}, 5 \times 10^{-4}, 10^{-3}\}$) $\rightarrow$ dropout rate (range of $\{0.1–0.9\}$, step size 0.1) $\rightarrow$ k (range of $\{1, 2, 3\}$) $\rightarrow$ ρ (range of $\{0.1–0.9\}$, step size 0.1) $\rightarrow$ α (range of $\{0.1–0.9\}$, step size 0.1) $\rightarrow$ β (range of $\{0.1–0.9\}$, step size 0.1) $\rightarrow$ λ (range of $\{0.1, 0.5, 1, 5, 10\}$) $\rightarrow$ μ (range of $\{0.1, 0.5, 1, 5, 10\}$). Among these, the last six parameters are specific to the **SORAG** family. When tuning the parameters, the sampling ratios for the BLOGCATALOG3, FLICKR, and YOUTUBE networks were set to 10%, 1%, and 1%, respectively.

Figure 7 shows the variation in the performance of **SORAG**$_F$ on the validation set of each dataset as the values of key parameters change. We first note that for the three generic parameters—learning rate, weight decay, and dropout rate—their values have a significant effect on classification performance and thus should be determined carefully. For k, ρ, and μ, we observed that their optimal values were consistent for all three datasets. The experimental results suggest that the best local minority of the nodes should be computed based on their two-hop neighbors (i.e., $k = 2$). We assume that this is because the one-hop information will lead to the omission of valid neighbors, while the three-hop information will introduce the noise of irrelevant nodes. Moreover, for the objective function, $\mathcal{L}_{edge}$ has the same weight as $\mathcal{L}_{nc}$, which indicates that the generation of virtual edges is of considerable importance. On the two larger datasets (FLICKR and YOUTUBE), the synthesis of virtual nodes and virtual edges are demonstrated to have equal weights, whereas the construction of virtual nodes has a smaller weight on the BLOGCATALOG3 network.

By contrast, the optimal values of α and β are close to 1 under almost all conditions (the only exception is that $\alpha = 0.5$ for FLICKR). This observation verifies our expectation that $P_{BD}(x) \approx P_u(x)$ and $P_{SF}(x, y) \approx P_l(x, y)$. By introducing these two parameters, we can regulate the distribution of the synthetic nodes in a more flexible manner.

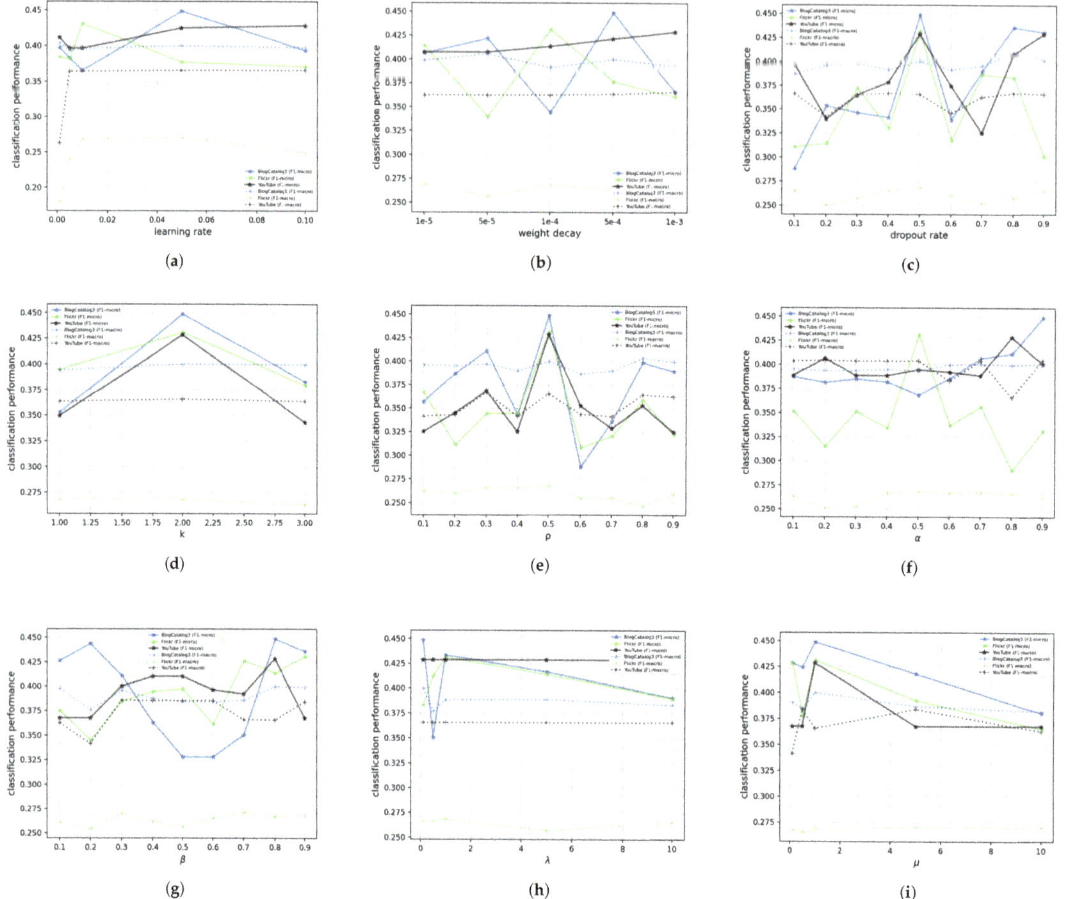

Figure 7. The effect of key parameters on the performance of **SORAG**$_F$ for each dataset: (**a**) learning rate; (**b**) weight decay; (**c**) dropout rate; (**d**) k; (**e**) ρ; (**f**) α; (**g**) β; (**h**) λ; (**i**) μ.

5.6. Validation of Key Procedures in Training SORAG

In this section, we answer the following research question: Do pre-training and fine-tuning the network components (i.e., two node generators and one edge generator) improve the model performance (see Algorithm 1)? For all datasets, we tested the following variants: {A, only pre-training the unlabeled node generator without fine-tuning; B, only pre-training the labeled node generator without fine-tuning; C, only pre-training the edge generator without fine-tuning; D, training the unlabeled node generator jointly with other components without pre-training; E, only training the labeled node generator jointly with other components without pre-training; F, only training the edge generator jointly with other components without pre-training; and G, the full model}. Naturally, {G-A, G-B, G-C} describe the effects of fine-tuning each focused component on performance, whereas {G-D, G-E, G-F} describe the performance difference between pre-training each component and non-pre-training. The results are presented in Table 5. The test method used was **SORAG**$_F$, and all the experimental settings and parameters were the same as those in Section 5.2.

Table 5. Performance comparisons of **SORAG**$_F$ with different training procedures.

	BLOGCATALOG3		FLICKR		YOUTUBE	
	Micro-F1	Macro-F1	Micro-F1	Macro-F1	Micro-F1	Macro-F1
G-A	11.8%	3.8%	3.7%	0.4%	4.3%	2.1%
G-B	3.5%	−1.8%	10.4%	1.5%	1.2%	6.4%
G-C	1.4%	0.0%	11.6%	1.5%	5.5%	4.3%
G-D	2.3%	2.9%	11.5%	0.2%	6.1%	4.3%
G-E	0.0%	0.0%	8.2%	0.2%	4.6%	4.3%
G-F	2.3%	2.9%	9.3%	1.0%	2.4%	2.1%

As presented in Table 5, we found that for all datasets, pre-training **SORAG**$_F$ improved Micro-F1. Macro-F1 was also improved in most cases, with the exception of pre-training the labeled generator on the BLOGCATALOG3 dataset, which slightly reduced Macro-F1. The average Micro-F1 and Macro-F1 improvements were 5.9.

Therefore, we concluded that training strategy G, which combined pre-training and fine-tuning of key components, was the best practice.

5.7. Case Study: Performance of SORAG on a Geographic Knowledge Graph

In this section, we exhibit the performance of **SORAG** on a standard geographic knowledge graph named US-AIRPORT as an example of the application of our approach in the field of remote sensing, as geographic knowledge graphs are reported to play an increasing role in the computation, analysis, and visualization of large-scale remote sensing data [55]. US-AIRPORT is the dataset used in struc2vec [8] (https://pytorch-geometric.readthedocs.io/en/latest/modules/datasets.html (accessed on 25 June 2022)), where nodes denote airports and labels correspond to activity levels. One-hot encodings were used as features. We randomly selected 20% of all nodes as the test set and 80% as the training set. Additionally, the average results from 10 separate experiments were used. Table 6 shows the performance of the analyzed methods in terms of Micro-F1 and Macro-F1. As shown, the performance gain from the state-of-the-art method GraphSMOTE towards **SORAG**$_F$ is considerable, which demonstrates our data over-sampling strategy is able to weaken the drawbacks of the imbalanced node distribution.

Table 6. Comparison in terms of Micro-F1 and Macro-F1 on US-AIRPORT. The best results are boldfaced.

	GCN	SMOTE	GraphSMOTE	SORAG$_L$	SORAG$_U$	SORAG$_F$
Micro-F1 (%)	22.69	56.44	59.46	60.57	60.21	**62.63**
Macro-F1 (%)	9.25	52.60	56.41	59.52	58.83	**60.10**

5.8. Synthetic Node Visualization

To analyze the synthetic nodes generated by **SORAG**$_F$ and how they differ from the real nodes more intuitively, we projected the feature vectors of both the real and virtual nodes generated by **SORAG**$_F$ for all datasets into two dimensions, as illustrated in Figures 8–10. For the dimensional-reduction method, we used t-SNE [56].

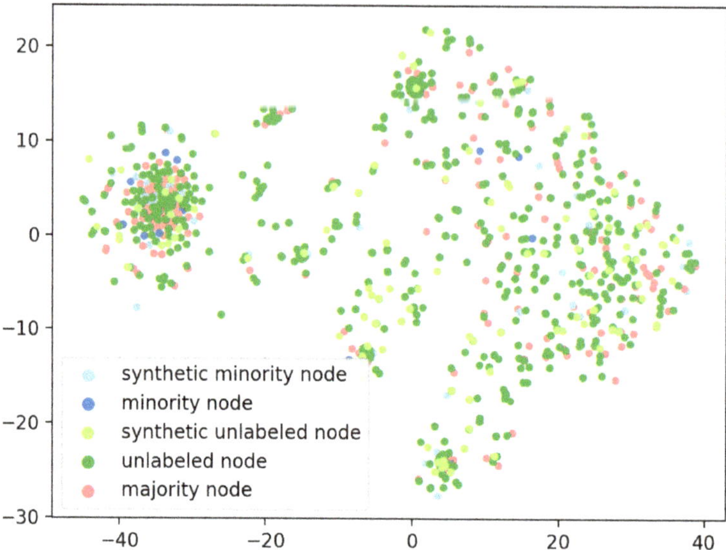

Figure 8. A t-SNE visualization of synthetic and real node features of the BLOGCATALOG3 dataset obtained from **SORAG**$_F$. The node colors denote the labels.

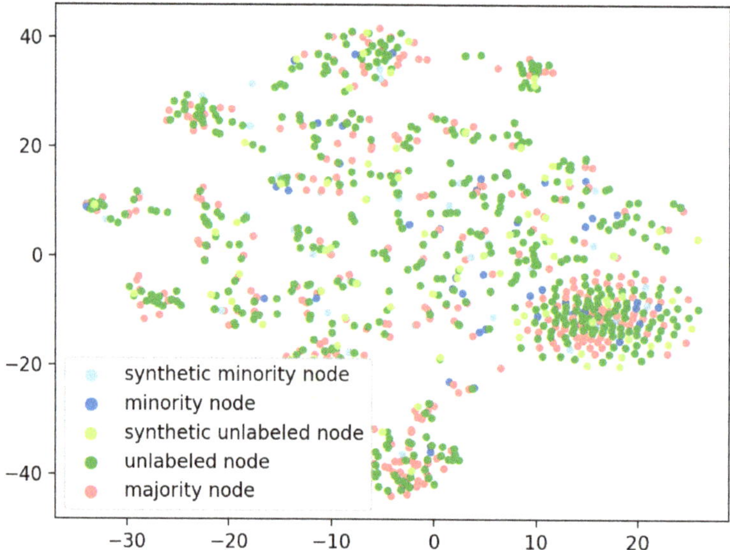

Figure 9. A t-SNE visualization of synthetic and real node features of the FLICKR dataset obtained from **SORAG**$_F$. The node colors denote the labels.

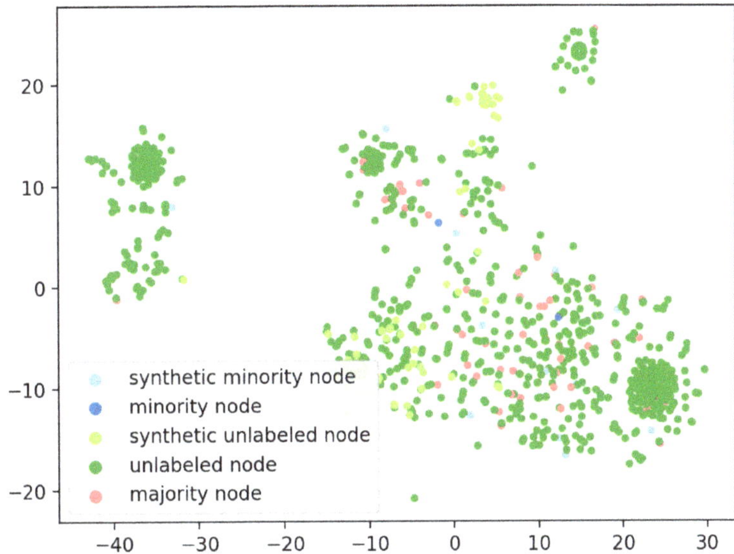

Figure 10. A t-SNE visualization of synthetic and real node features of the YOUTUBE dataset obtained from **SORAG**$_F$. The node colors denote the labels.

6. Conclusions

This study investigates a new research problem: imbalanced multilabel graph node classification. In contrast to existing over-sampling algorithms, which generate only new minority instances to balance the class distribution, we propose a novel data generation strategy called **SORAG**, which conjoins the synthesis of labeled instances in minority class centers and unlabeled instances in minority class borders. The new supervision information brought about by the labeled synthetics and the blocking of overpropagated majority features by the unlabeled synthetics facilitates balanced learning between different classes, taking advantage of the strong topological interdependence between nodes on a graph.

We conducted extensive comparative studies to evaluate the proposed framework on diverse, naturally imbalanced multilabel networks. The experimental results demonstrated the high effectiveness and robustness of **SORAG** in handling imbalanced data. In the future, we will develop GNN models that are more adapted to the nature of real-world networks (e.g., scale-free and small-world features).

Author Contributions: Methodology/system implementation/experiments/original draft preparation: Y.D.; Supervision: X.L., A.J., H.-t.Y., S.L., K.-S.K. and A.M. All authors have read and agreed to the published version of the manuscript.

Funding: This paper is based on results obtained from a project, JPNP20006, commissioned by the New Energy and Industrial Technology Development Organization (NEDO). We would also like to acknowledge partial support from JSPS Grant-in-Aid for Scientific Research (Grant Number 21K12042).

Data Availability Statement: The datasets used in this study are available at http://zhang18f.myweb.cs.uwindsor.ca/datasets/ (accessed on 25 June 2022).

Acknowledgments: We would like to thank the reviewers for the time and effort necessary to review the manuscript. We sincerely appreciate all the valuable comments and suggestions, which helped us improve the quality of the manuscript. This manuscript is an extension of the authors' earlier work, which is to be presented at the 2022 European Conference on Machine Learning and Principles and Practice of Knowledge Discovery in Databases (ECML-PKDD 2022).

Conflicts of Interest: The authors declare no conflict of interest.

Abbreviations

The following abbreviations are used in this manuscript:

GRL	Graph representation learning
GNN	Graph neural network
GCN	Graph convolutional network
GAN	Generative adversarial network
CGAN	Conditional generative adversarial network
SMOTE	Synthetic minority over-sampling technique
SORAG	Synthetic data over-sampling strategy on graph

References

1. Zhang, D.; Yin, J.; Zhu, X.; Zhang, C. Network representation learning: A survey. *IEEE Trans. Big Data* **2018**, *6*, 3–28.
2. Jalal, A.; Kamal, S.; Kim, D. A depth video sensor-based life-logging human activity recognition system for elderly care in smart indoor environments. *Sensors* **2014**, *14*, 11735–11759. [PubMed]
3. Ren, H.; Xu, G. Human action recognition in smart classroom. In Proceedings of the Fifth IEEE International Conference on Automatic Face Gesture Recognition, Washinton, DC, USA, 21 May 2002; pp. 417–422.
4. Puwein, J.; Ballan, L.; Ziegler, R.; Pollefeys, M. Joint camera pose estimation and 3d human pose estimation in a multi-camera setup. In Proceedings of the Asian Conference on Computer Vision, Singapore, 1–5 November 2014; pp. 473–487.
5. Shi, M.; Tang, Y.; Zhu, X.; Liu, J. Multi-label graph convolutional network representation learning. *IEEE Trans. Big Data* **2020**, 1169–1181. [CrossRef]
6. Tang, L.; Liu, H. Relational learning via latent social dimensions. In Proceedings of the 15th ACM SIGKDD International Conference on Knowledge Discovery and Data Mining, Paris, France, 28 June–1 July 2009; pp. 817–826.
7. Tang, L.; Liu, H. Scalable learning of collective behavior based on sparse social dimensions. In Proceedings of the 18th ACM Conference on Information and Knowledge Management, Hong Kong, China, 2–6 November 2009; pp. 1107–1116.
8. Ribeiro, L.F.; Saverese, P.H.; Figueiredo, D.R. struc2vec: Learning node representations from structural identity. In Proceedings of the 23rd ACM SIGKDD International Conference on Knowledge Discovery and Data Mining, Halifax, NS, Canada, 13–17 August 2017; pp. 385–394.
9. Scarselli, F.; Gori, M.; Tsoi, A.C.; Hagenbuchner, M.; Monfardini, G. The graph neural network model. *IEEE Trans. Neural Netw.* **2008**, *20*, 61–80.
10. Hamilton, W.L.; Ying, R.; Leskovec, J. Inductive representation learning on large graphs. In Proceedings of the 31st International Conference on Neural Information Processing Systems, Long Beach, CA, USA, 4–9 December 2017; pp. 1025–1035.
11. Veličković, P.; Cucurull, G.; Casanova, A.; Romero, A.; Lio, P.; Bengio, Y. Graph attention networks. *arXiv* **2017**, arXiv:1710.10903.
12. Kipf, T.N.; Welling, M. Semi-supervised classification with graph convolutional networks. *arXiv* **2016**, arXiv:1609.02907.
13. He, H.; Garcia, E.A. Learning from imbalanced data. *IEEE Trans. Knowl. Data Eng.* **2009**, *21*, 1263–1284.
14. Chawla, N.V.; Bowyer, K.W.; Hall, L.O.; Kegelmeyer, W.P. SMOTE: Synthetic minority over-sampling technique. *J. Artif. Intell. Res.* **2002**, *16*, 321–357.
15. Zhou, Z.H.; Liu, X.Y. Training cost-sensitive neural networks with methods addressing the class imbalance problem. *IEEE Trans. Knowl. Data Eng.* **2005**, *18*, 63–77.
16. Zhou, Z.H.; Liu, X.Y. On multi-class cost-sensitive learning. *Comput. Intell.* **2010**, *26*, 232–257.
17. Zhao, T.; Zhang, X.; Wang, S. Graphsmote: Imbalanced node classification on graphs with graph neural networks. In Proceedings of the 14th ACM International Conference on Web Search and Data Mining, Virtual, 8–12 March 2021; pp. 833–841.
18. Elkan, C. The foundations of cost-sensitive learning. In *International Joint Conference on Artificial Intelligence*; Lawrence Erlbaum Associates Ltd.: Seattle, WA, USA, 2001; Volume 17, pp. 973–978.
19. Fernández, A.; García, S.; Galar, M.; Prati, R.C.; Krawczyk, B.; Herrera, F. Cost-sensitive learning. In *Learning from Imbalanced Data Sets*; Springer: Berlin/Heidelberg, Germany, 2018; pp. 63–78.
20. Domingos, P. Metacost: A general method for making classifiers cost-sensitive. In Proceedings of the Fifth ACM SIGKDD International Conference on Knowledge Discovery and Data Mining, San Diego, CA, USA, 15–18 August 1999; pp. 155–164.
21. Sheng, V.S.; Ling, C.X. Thresholding for making classifiers cost-sensitive. In *AAAI*; AAAI Press: Boston, MA, USA, 2006; Volume 6, pp. 476–481.
22. Lomax, S.; Vadera, S. A survey of cost-sensitive decision tree induction algorithms. *ACM Comput. Surv. CSUR* **2013**, *45*, 1–35. [CrossRef]
23. Morik, K.; Brockhausen, P.; Joachims, T. Combining Statistical Learning with a Knowledge-Based Approach: A Case Study in Intensive Care Monitoring. In *ICML*; ACM Press: Bled, Slovenia, 1999; pp. 268–277. [CrossRef]
24. More, A. Survey of resampling techniques for improving classification performance in unbalanced datasets. *arXiv* **2016**, arXiv:1608.06048.

25. Goodfellow, I.; Pouget-Abadie, J.; Mirza, M.; Xu, B.; Warde-Farley, D.; Ozair, S.; Courville, A.; Bengio, Y. Generative adversarial nets. *Adv. Neural Inf. Process. Syst.* **2014**, *27*, 2672–2680. [CrossRef]

26. Xu, L.; Skoularidou, M.; Cuesta-Infante, A.; Veeramachaneni, K. Modeling Tabular data using Conditional GAN. *Adv. NIPS* **2019**, *659*, 7335–7345.

27. Mirza, M.; Osindero, S. Conditional generative adversarial nets. *arXiv* **2014**, arXiv:1411.1784.

28. Balasubramanian, M.; Schwartz, E.L. The isomap algorithm and topological stability. *Science* **2002**, *295*, 7. [CrossRef]

29. Roweis, S.T.; Saul, L.K. Nonlinear dimensionality reduction by locally linear embedding. *Science* **2000**, *290*, 2323–2326. [CrossRef]

30. Perozzi, B.; Al-Rfou, R.; Skiena, S. Deepwalk: Online learning of social representations. In Proceedings of the 20th ACM SIGKDD International Conference on Knowledge Discovery and Data Mining, New York, NY, USA, 24–27 August 2014; pp. 701–710.

31. Tang, J.; Qu, M.; Wang, M.; Zhang, M.; Yan, J.; Mei, Q. Line: Large-scale information network embedding. In Proceedings of the 24th International Conference on World Wide Web, Florence, Italy, 18–22 May 2015; pp. 1067–1077.

32. Ma, Y.T. *Deep Learning on Graphs*; Cambridge University Press: Cambridge, UK, 2021.

33. Bruna, J.; Zaremba, W.; Szlam, A.; LeCun, Y. Spectral networks and locally connected networks on graphs. *arXiv* **2013**, arXiv:1312.6203.

34. Simonovsky, M.; Komodakis, N. Dynamic edge-conditioned filters in convolutional neural networks on graphs. In Proceedings of the IEEE Conference on Computer Vision and Pattern Recognition, Honolulu, HI, USA, 21–26 July 2017; pp. 3693–3702.

35. Monti, F.; Bronstein, M.M.; Bresson, X. Geometric matrix completion with recurrent multi-graph neural networks. *arXiv* **2017**, arXiv:1704.06803.

36. Li, Y.; Tarlow, D.; Brockschmidt, M.; Zemel, R. Gated graph sequence neural networks. *arXiv* **2015**, arXiv:1511.05493.

37. Gilmer, J.; Schoenholz, S.S.; Riley, P.F.; Vinyals, O.; Dahl, G.E. Neural message passing for quantum chemistry. In Proceedings of the International Conference on Machine Learning, PMLR, Sydney, Australia, 6–11 August 2017; pp. 1263–1272.

38. Defferrard, M.; Bresson, X.; Vandergheynst, P. Convolutional neural networks on graphs with fast localized spectral filtering. *Adv. Neural Inf. Process. Syst.* **2016**, *29*, 3844–3852.

39. Spitzer, F. *Principles of Random Walk*; Springer Science & Business Media: New York, NY, USA, 2013; Volume 34.

40. Shen, X.; Pan, S.; Liu, W.; Ong, Y.S.; Sun, Q.S. Discrete network embedding. In Proceedings of the 27th International Joint Conference on Artificial Intelligence, Stockholm, Sweden, 13–19 July 2018; pp. 3549–3555.

41. Han, H.; Wang, W.Y.; Mao, B.H. Borderline-SMOTE: A new over-sampling method in imbalanced data sets learning. In Proceedings of the International Conference on Intelligent Computing, Hefei, China, 23–26 August 2005; pp. 878–887.

42. Bunkhumpornpat, C.; Sinapiromsaran, K.; Lursinsap, C. Safe-level-smote: Safe-level-synthetic minority over-sampling technique for handling the class imbalanced problem. In Proceedings of the Pacific-Asia Conference on Knowledge Discovery and Data Mining, Bangkok, Thailand, 27–30 April 2009; pp. 475–482.

43. Bunkhumpornpat, C.; Sinapiromsaran, K.; Lursinsap, C. DBSMOTE: Density-based synthetic minority over-sampling technique. *Appl. Intell.* **2012**, *36*, 664–684. [CrossRef]

44. Chawla, N.V.; Lazarevic, A.; Hall, L.O.; Bowyer, K.W. SMOTEBoost: Improving prediction of the minority class in boosting. In Proceedings of the European Conference on Principles of Data Mining and Knowledge Discovery, Cavtat-Dubrovnik, Croatia, 22–26 September 2003; pp. 107–119.

45. Seiffert, C.; Khoshgoftaar, T.M.; Van Hulse, J.; Napolitano, A. RUSBoost: A hybrid approach to alleviating class imbalance. *IEEE Trans. Syst. Man Cybern. Part A Syst. Hum.* **2009**, *40*, 185–197. [CrossRef]

46. Liu, X.Y.; Wu, J.; Zhou, Z.H. Exploratory undersampling for class-imbalance learning. *IEEE Trans. Syst. Man Cybern. Part B Cybern.* **2008**, *39*, 539–550.

47. Chen, C.; Liaw, A.; Breiman, L. *Using Random Forest to Learn Imbalanced Data*; University of California: Berkeley, CA, USA, 2004; Volume 110, p. 24.

48. Fan, W.; Stolfo, S.J.; Zhang, J.; Chan, P.K. AdaCost: Misclassification cost-sensitive boosting. In *ICML*; Citeseer: Bled, Slovenia, 1999; Volume 99, pp. 97–105.

49. Wang, Z.; Ye, X.; Wang, C.; Cui, J.; Yu, P. Network embedding with completely-imbalanced labels. *IEEE Trans. Knowl. Data Eng.* **2020**, *33*, 3634–3647. [CrossRef]

50. Napierala, K.; Stefanowski, J. Types of minority class examples and their influence on learning classifiers from imbalanced data. *J. Intell. Inf. Syst.* **2016**, *46*, 563–597. [CrossRef]

51. Liu, B.; Blekas, K.; Tsoumakas, G. Multi-label sampling based on local label imbalance. *Pattern Recognit.* **2022**, *122*, 108294. [CrossRef]

52. Pearson, K. LIII. On lines and planes of closest fit to systems of points in space. *Lond. Edinb. Dublin Philos. Mag. J. Sci.* **1901**, *2*, 559–572. [CrossRef]

53. Kingma, D.P.; Ba, J. Adam: A method for stochastic optimization. *arXiv* **2014**, arXiv:1412.6980.

54. Paszke, A.; Gross, S.; Massa, F.; Lerer, A.; Bradbury, J.; Chanan, G.; Killeen, T.; Lin, Z.; Gimelshein, N.; Antiga, L.; et al. Pytorch: An imperative style, high-performance deep learning library. *Adv. Neural Inf. Process. Syst.* **2019**, *32*, 8026–8037.

55. Wang, Z.; Yang, X.; Zhou, C. Geographic knowledge graph for remote sensing big data. *J. Geo-Inf. Sci.* **2021**, *23*, 13. [CrossRef]

56. Van der Maaten, L.; Hinton, G. Visualizing data using t-SNE. *J. Mach. Learn. Res.* **2008**, *9*, 2579–2605.

Article

Using Multi-Source Real Landform Data to Predict and Analyze Intercity Remote Interference of 5G Communication with Ducting and Troposcatter Effects

Kai Yang [1], Xing Guo [2,*], Zhensen Wu [1], Jiaji Wu [2], Tao Wu [3], Kun Zhao [2], Tan Qu [2] and Longxiang Linghu [2]

1 School of Physics, Xidian University, Xi'an 710071, China
2 School of Electronic Engineering, Xidian University, Xi'an 710071, China
3 Xi'an Institute of Space Radio Technology, Xi'an 710199, China
* Correspondence: guox@xidian.edu.cn

Abstract: At present, 5G base stations are densely distributed in major cities based on improving user concentrations and the large demand for services in urban hotspot areas. Moreover, 5G communication requires more accurate communication propagation loss (PL) monitoring. Low-build areas, such as suburbs and rural areas, are prone to forming relatively stable tropospheric ducts, which can bend the signal to the surface in the duct-trapping layer for multiple reflections. Due to the random flow of the atmospheric air mass, each reflection of the communication signal is re-scattered in the troposphere through the top of the duct layer, thereby expanding the propagation range of the signal and changing the expected effect of radio wave propagation. If ducting and troposcatter effects happen in the 5G base station antenna layer, co-channel interference (CCI) could occur, affecting the quality of electromagnetic propagation. Urban links in the plain area have no major terrain obstacles, but ground fluctuations and land cover scattering have a greater impact on the signal scattering at the bottom of the duct. On the basis of the forward-propagation theory, this paper adds factors, such as ducting the forecast value using real weather parameters, terrain, and land cover-type distributions to evaluate the CCIs of over-the-horizon communications on the intercity link. Based on 1300 sets of randomly generated terrains and landforms, two deep learning (DL) models were used to predict the PL of over-the-horizon communications between cities in a land-based ducting environment. The accuracy of LSTM prediction could reach 98.4%. The verification of PL prediction using DL in this paper allows for quick and efficient prediction of PL in the land-based ducting of intercity links using land cover characteristics.

Keywords: atmospheric duct; troposcatter; tropospheric turbulence; intercity co-channel interference; deep learning

Citation: Yang, K.; Guo, X.; Wu, Z.; Wu, J.; Wu, T.; Zhao, K.; Qu, T.; Linghu, L. Using Multi-Source Real Landform Data to Predict and Analyze Intercity Remote Interference of 5G Communication with Ducting and Troposcatter Effects. *Remote Sens.* **2022**, *14*, 4515. https://doi.org/10.3390/rs14184515

Academic Editor: Gang Zheng

Received: 20 July 2022
Accepted: 1 September 2022
Published: 9 September 2022

Publisher's Note: MDPI stays neutral with regard to jurisdictional claims in published maps and institutional affiliations.

1. Introduction

The fifth-generation (5G) wireless system is in the rapid development stage of China's mobile telecommunication standards. Moreover, 5G relies on unique advantages, such as high speed, low latency, large capacity, and wide connectivity to rapidly integrate applications in various fields of production and life, and its application scope will eventually range from mobile broadband services to next-generation car and device networking services [1]. Meanwhile, in the sixth-generation mobile communication technology planning, satellite internet communication will play an important role in realizing the transition from the interconnection of everything to the intelligent connection of everything. With the rapid development of coastal cities, the surrounding sea–land junction meteorological environment and the distribution of ground objects will change and become more complex. The modeling requirements for the forward-propagation of communication signals when mixed ducts and non-uniform turbulence occur are more accurate.

The 5G network must coordinate high, medium, and low frequencies according to user service coverage and speed requirements to form an efficient and coordinated layered coverage. According to public data, by the end of 2020, operators have opened a total of 718,000 5G base stations, and the number of 5G mobile terminal connections has exceeded 200 million. The dense distribution of base stations has higher requirements for accurate reception and anti-interference of communication links. We selected three sub-6 GHz in the 5G plan for the simulation, and evaluated the weather-induced intercity CCI that may have occurred in different urban links when ducts and tropospheric turbulence existed.

An atmospheric duct is mainly caused by the increase of atmospheric temperature with the increase of altitude or the rapid decrease of water vapor density with the increase of altitude. Turton [2] proposed five synoptic processes that can form atmospheric ducts—radiative temperature inversion, subsidence temperature inversion on land surface, temperature advection inversion at the sea–land interface, topographic inversion in local valleys, and frontal temperature inversion are prone to occur. According to data statistics, communication disconnections and other phenomena often occur at night after rain in the plains of China. The monitoring found that it was related to the interference of ducts. This paper selected urban links in the East China Plain (where ducts are prone to occur) to analyze the CCI of over-the-horizon communications.

In 1946, Leontovich and Fock [3] proposed a forward full-wave analytical method—the parabolic equation (PE) method—to solve the problem of forward-propagation in vertically layered inhomogeneous tropospheric ducts. In 1973, Hardin and Tappert [4] developed a split-step Fourier transform (SSFT) algorithm suitable for underwater acoustic transmission to solve parabolic equations. Subsequently, the PE model has been theoretically deduced and optimized by a series of scholars. Based on its framework, the Feit–Fleck wide-angle parabolic equation (WAPE) [5] method that can calculate a larger range of grazing angles was proposed. The grazing angle can reach 30°. In 1983, Thomson and Chapman [6] formally proposed a new WAPE form based on the Feit–Fleck method, and Kuttler [7] verified the good performance of WAPE at wide elevation angles.

The PE method takes into account the refraction and diffraction effects of radio wave propagation. Moreover, it can simply and accurately describe the dielectric parameters of the irregular surface and the refractive index distribution of non-uniform atmospheric structures to simulate the propagation characteristics of radio waves in a complex environment. The step-by-step iterative solution method can predict the propagation loss of point-to-point links, and can further realize regional propagation loss prediction. In 2000, Levy [8] made a relatively complete summary of the derivation, terrain, and boundaries of electromagnetic wave propagation calculated by the parabolic equation method, and preliminarily deduced the three-dimensional parabolic equation algorithm. In 2003, Janaswamy [9] used 3D vector parabolic equations to predict path loss on flat terrain with buildings. In 2005, Hu et al. [10] established a macrocell radio propagation model for mobile communication based on 3DPE. In 2019, Rasool et al. [11] deduced the solution of 3DPE, and used 3DPE for field strength prediction for flat and irregular forest environments. In this paper, the PE model was improved to simulate the possible CCI of the over-the-horizon transmission of the communication signal on the intercity link.

When electromagnetic waves encounter irregular terrain during transmission, such as wedge-shaped boundaries, more intense reflection will occur, so the modeling of complex terrain is also particularly important. In 1987, Dockery and Konstanzer [12] first used PE for tropospheric wave propagation problems. In 1996, Kuttler and Dockery [13,14] proposed the method of continuous hybrid Fourier transform under the impedance boundary to solve PE, which greatly improved the efficiency of the SSFT algorithm. In 2000, Donohue and Kuttler [15] applied the SSFT PE algorithm to irregular terrain, extended the terrain shift map algorithm proposed by Beilis and Tappert to WAPE, and deduced a new impedance boundary condition for electromagnetic wave incidents on the surface of a finite conductor, such that the WAPE solution could be adapted to the previous MFT algorithm. Numerical examples demonstrate its effectiveness on impedance surfaces with terrain slopes of 10–15°

and discontinuous slope changes of 15–20°. To accommodate terrains with larger slopes, it is recommended to combine the terrain shift map with terrain masking (knife-edge diffraction) approximations. In this paper, the WAPE-based terrain shift map algorithm and the new impedance formula are applied to the real inter-city link distribution to evaluate the atmospheric loss distribution.

Since the influence of troposcatter cannot be ignored, scholars have begun to turn their focus to the influence of atmospheric ducts on troposcatter communications. In 1993, Hitney [16] proposed the scattering parabolic equation method based on the radio physical optics (RPO) model. This method can simply and effectively calculate the tropospheric scattering caused by atmospheric turbulence on the basis of PE. In 2009, Ivanov et al. [17] used this method to further study the boundary layer leakage of tropospheric scattering and the propagation characteristics in evaporation ducts. In 2013, Wang et al. [18] carried out experiments with three unequal-distance cross-sea circuits to study the signal fading characteristics of tropospheric scattering in the sea area where evaporation ducts occur. The evaporation duct phenomenon was found to significantly alter the fading level of the signal.

In 2015, Song et al. [19] found that the atmospheric meteorological state has a great influence on the scattering loss through the study of the scattering parabolic equation. In 2016, Wanger et al. [20] evaluated the refractive index of the lower atmosphere and gave a propagation loss (PL) measurement; that is, using height-independent PL measurements over a range of 10–80 km to infer information about the existence and potential parameters of atmospheric ducts in the lowest 1 km of the atmosphere. Its main improvement includes turbulence-induced distance-dependent fluctuations in the forward-propagation model, where the inhomogeneous turbulence model uses parameters that are height-dependent on the mixing ducts. In this paper, Hitney's RPO basic scattering model was applied to the transmission over inter-city links with evaporation ducts, and the non-uniform turbulence model proposed by Wanger was used for transmission calculations over inter-city links with mixed ducts.

In this paper, combined with open source data from multiple sources, the probability of occurrence of ducts in the selected area was verified by WRF. The land types in inter-city links were extracted based on MCD12Q1 from MODIS. The empirical values of complex dielectric parameters suitable for these land types were found from various references or measured values. In the end, we substituted impedance correction, real terrain relief, and turbulence correction into intercity links to evaluate the probability of possible interference when ducts and turbulence occurred between different cities.

2. A Multi-Source Real Data Model for Intercity Link Communication Propagation Computing

2.1. WRF Models Regional Duct Distribution

The new generation of the mesoscale advanced research WRF model system is a flexible and advanced atmospheric simulation system. It is suitable for a wide range of electromagnetic applications from meters to thousands of kilometers. The WRF modeling system is shown in Figure 1 and mainly consists of the following main programs: 1. WRF pre-processing system (WPS); 2. WRF data assimilation (WRF-DA) assimilation system; 3. advanced research WRF (ARW) solver; 4. post-processing and visualization tools.

We used the WRF pre-processing system (WPS) to specify the static terrain data path of the processing area and perform interpolation on the terrain data (including terrain, land use, and soil type). The background field and boundary field data used in WPS were NCEP FNL (final) business global analysis data [21,22]. The global meteorological data in GRIB2 (Gridded Binary2) format were downloaded. The grid data resolution was $1° \times 1°$ and the time interval was 6 h.

The latitude and longitude coordinate ranged of the selected area were (25.846°N–36.224°N, 117.472°E–123.528°E). We set the number of coarse grids and fine grids, and their relative positions. There were several cities that we simulated in the fine grids, of which, the latitude and longitude coordinates were Wuxi (120.3166°E, 31.5072°N), Hangzhou (120.1602°E, 30.2856°N), Nanjing (118.7988°E, 32.0651°N), Shanghai (121.4802°E, 31.2389°N), and Jiaxing (120.7635°E, 30.7509°N), respectively. Using the Lambert conformal projection, the simulation area was used as a coarse grid, and the key selected urban area was a fine grid. The schematic diagram of the simulation area is shown in Figure 2, and the parameter settings are shown in Table 1. After the initialization parameters were set, we executed *geogrid.exe* to define the model domain to be processed, and imported the static terrain data into the grid; secondly, we commanded *ungrib.exe* to extract the meteorological fields from the imported grid-formatted terrain fields; finally, we executed *metgrid.exe* to horizontally interpolate the meteorological fields extracted by ungrib.exe in the second step into the grid model defined by *geogrid.exe* in the first step. The vertical interpolation of the meteorological field to the WRF eta level was realized in the actual program.

Table 1. Parameter settings of the WRF numerical simulation.

Options		Settings
Nested Levels		**Double Nesting**
Coarse grid settings	Longitude and latitude coordinate of grid center	(120.5°E, 31.035°N)
	Number of grids	192 × 384
	Horizontal resolution	3 km × 3 km
Fine grid settings	Position in the coarse grid	(32, 20)
	Number of grids	123 × 246
	Horizontal resolution	1 km × 1 km

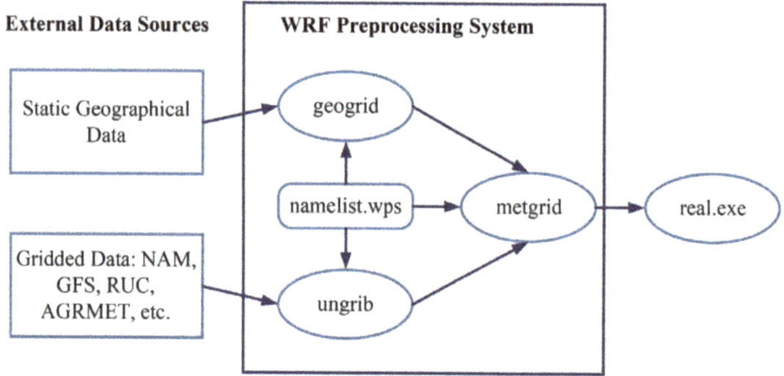

Figure 1. Procedures of the WRF pre-processing system.

The background field data of 1 March 2018, 1 June 2018, 1 September 2018, and 1 December 2018UTC were selected, respectively, to simulate the duct distribution in the selected plain area by seasons. The duct distribution located in East China is shown in Figure 3.

As shown in Figure 3, it can be seen that the ducts were distributed in the sea area throughout the year, while the inland ducts were less distributed and scattered in March. The inland ducts were densely distributed in June and December, and the inland ducts were also widely distributed in September, but not continuous. Due to the large amount of data to be analyzed in the selected area, only the first day of each quarter was selected for numerical simulation. The actual occurrence of the duct was also related to sudden changes in weather, such as rainfall and snowfall, or the diurnal temperature variation. However, based on the numerical simulation results of WRF in the East China Plain, it is fully proven that the ducting phenomenon occurs in this area all year around. Therefore, the surface duct with the base layer is used to evaluated the CCI of the intercity in East China.

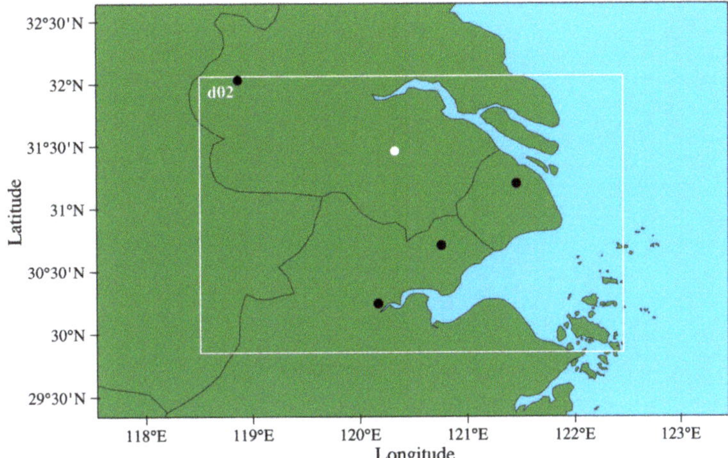

Figure 2. The nested grids of the simulated areas in WRF modes.

Figures 4 and 5—we calculated the duct height and duct intensity distributions on 1 June and 1 December 2018 using WRF mesoscale numerical modeling software.

As shown in Figure 4, the average distribution of duct strengths between cities in June is more than 60 m, the intercity link from Shanghai, Jiaxing, to Wuxi also has a hybrid duct phenomenon of up to 240 m, and its strength can reach up to 35 M-units or so, which is shown in Figure 5a. In December, the marine ducts in this area were densely distributed, and the heights and strengths of the ducts were significantly higher than those in coastal cities. In inland areas, the duct heights on the Nanjing–Wuxi link were basically distributed at about 40–80 m, while duct heights on the intercity link of Shanghai–Wuxi and Jiaxing–Wuxi could reach more than 200 m; the average duct strength was about 20 M-units. Considering the limitation of the installation height of the communication signal antenna, based on the duct distribution results simulated by the real data using WRF, this paper studied the over-the-horizon transmission of the 5G communication signal on the intercity link when the non-uniform surface-based duct with the trapping layer height was 40–80 m.

2.2. Extraction of Digital Elevation Data on Intercity Link

This paper used DEM digital elevation data with a resolution of 90 m for the analysis and calculation. The data set was provided by geospatial data cloud site, Computer Network Information Center, Chinese Academy of Sciences (http://www.gscloud.cn). We accessed the data in 17 July of 2022. The data came from SRTM (Shuttle Radar Topography Mission) and were jointly measured by the National Aeronautics and Space Administration (NASA) and the National Imagery and Mapping Agency (NIMA).The data volume of radar images acquired by the SRTM system was about 9.8 trillion bytes. After various pre-processing steps, such as editing, gross error removal, determining water surface elevation, and defining coastlines, etc., the results were cut into 15,000 pieces according to geographic coordinates. Each map frame was stored in latitude and longitude, and a digital elevation model (DEM) was made. The data accuracy that could be used for scientific research was the STRM-3 version, which was sampled every 3 arc seconds, and the data of 3 arc seconds were collected and generated from 1 arc second of data. They were divided in $1° \times 1°$ of the latitude and longitude grid in the geographic projection, and the horizontal resolution was about 90 m. SRTM terrain data obtained by CIAT (the International Center for Tropical Agriculture) using a new interpolation algorithm, this method better fills the data gap of SRTM 90. The interpolation algorithm is from Reuter et al. [23].

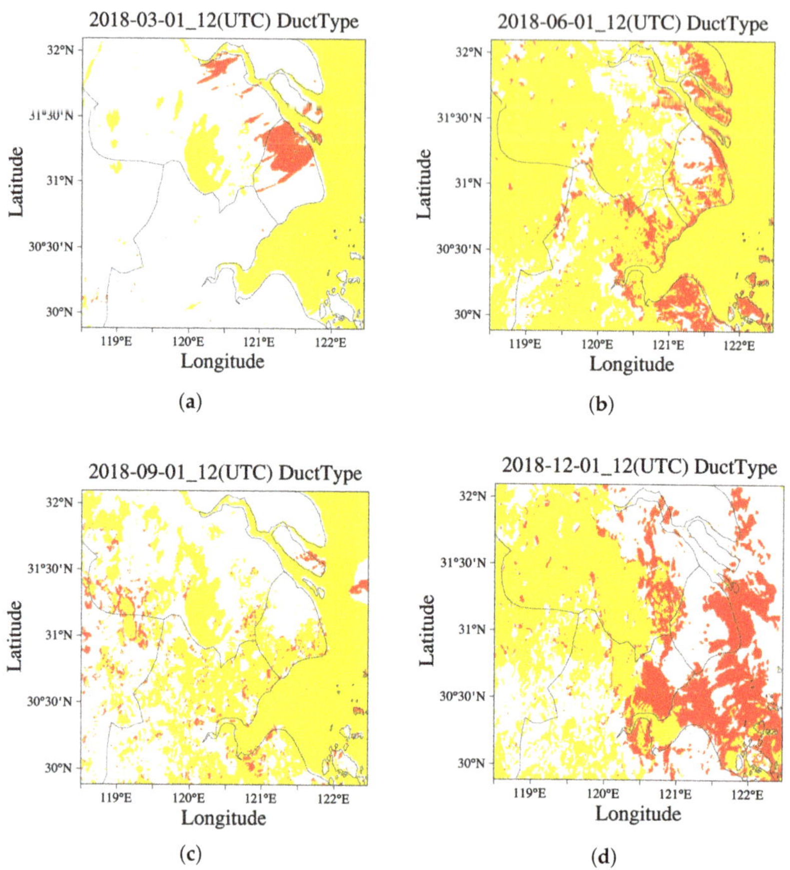

Figure 3. The distribution of ducts in selected plain areas by seasons, where the yellow area represents the surface duct and the red area represents tge elevated duct. (**a**) March. (**b**) June. (**c**) September. (**d**) December.

SRTM-3 data are divided into files every 5 degrees of longitude and latitude, with a total grid data of 24 rows (including the area that latitude ranges from 60°N to 60°S), 72 columns (longitude ranged from −180°W to 180°E). According to DEM data product, the elevation data of several intercity links from the area of interest (25.846°N–36.224°N, 117.472°E–123.528°E), could be determined with a resolution of 90 m. Interpolating to fill the null zone, the global elevation distribution of the selected area is obtained as shown in Figure 6.

We extracted the link elevations from four surrounding cities to city A. Considering that the minimum distance between cities was about 90 km, which was about 0.81° when it was represented by the longitude and latitude coordinates. To not reduce the accuracies of DEM data on intercity links, we extracted the required link DEM data using traversal and interpolation. The processing steps are in Figure 7.

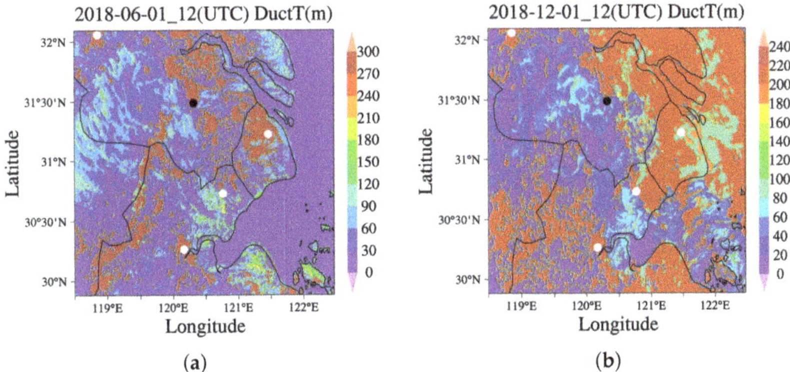

Figure 4. Duct height (m) calculated by WRF on 1 June and 1 December 2018. (**a**) Duct height (m) on 1 June 2018. (**b**) Duct height (m) on 1 December 2018.

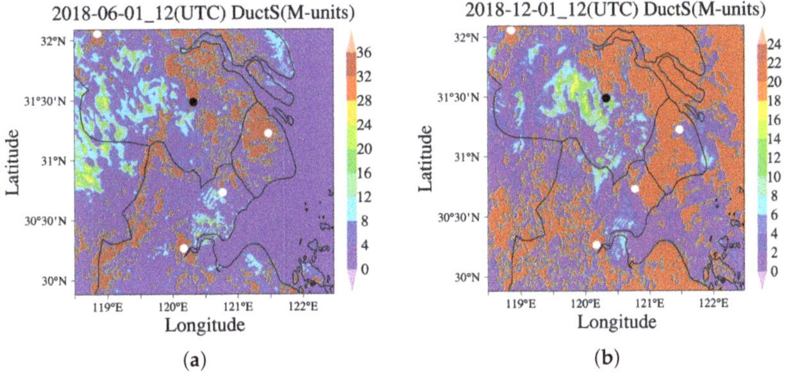

Figure 5. Duct Strength (M-units) calculated by WRF on 1 June and 1 December 2018. (**a**) Duct strength (M-units) on 1 June 2018. (**b**) Duct strength (M-units) on 1 December 2018.

2.3. Data Extraction of Land Cover Distribution Types of Intercity Links from MODIS

2.3.1. Land Cover Type Distribution on Intercity Links of East China

The MODIS Land cover product MCD12Q1 provides a series of 13 sets of scientific data sets with 500 m of resolution global land cover projections, which were published in an annual span, including five different land cover classification schemes of IGBP/UMD/LAI/BGC/PFT, and a new three-layer classification standard based on the land cover classification system (LCCS) from the Food and Agriculture Organization, including a quality assurance (QA) layer, posterior probability of the LCCS three-layer standard, and binary land water cover. MCD12Q1 was secondary validation datum based on cross-validation of the training dataset generated from projections.

MCD12Q1 was derived from observations from the Terra and Aqua data inputs and was generated using MODIS reflectance data supervised classifications. The primary land cover program identified 17 land cover categories as defined by the International Geosphere–Biosphere Program (IGBP), including 11 natural vegetation categories, 3 developed and planted land categories, and 3 non-vegetated land categories, as listed in Table 2.

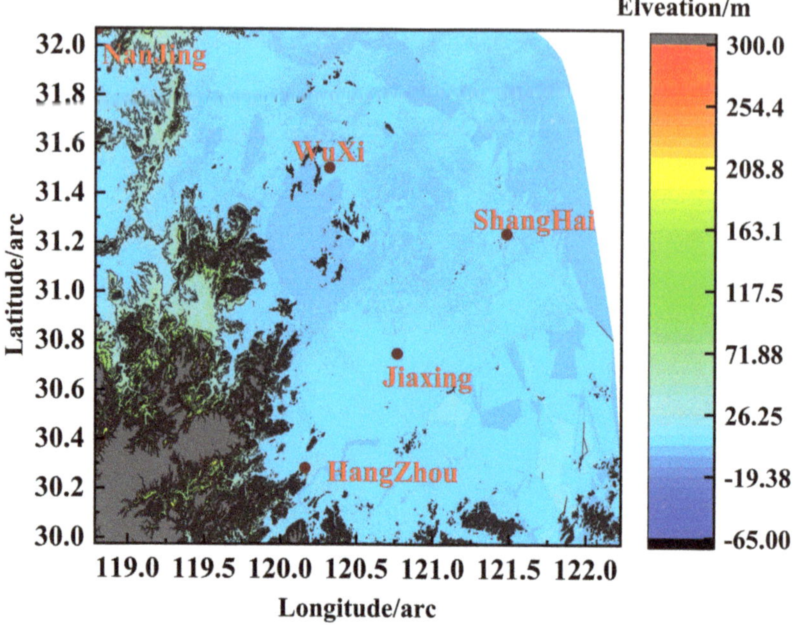

Figure 6. DEM of the selected area.

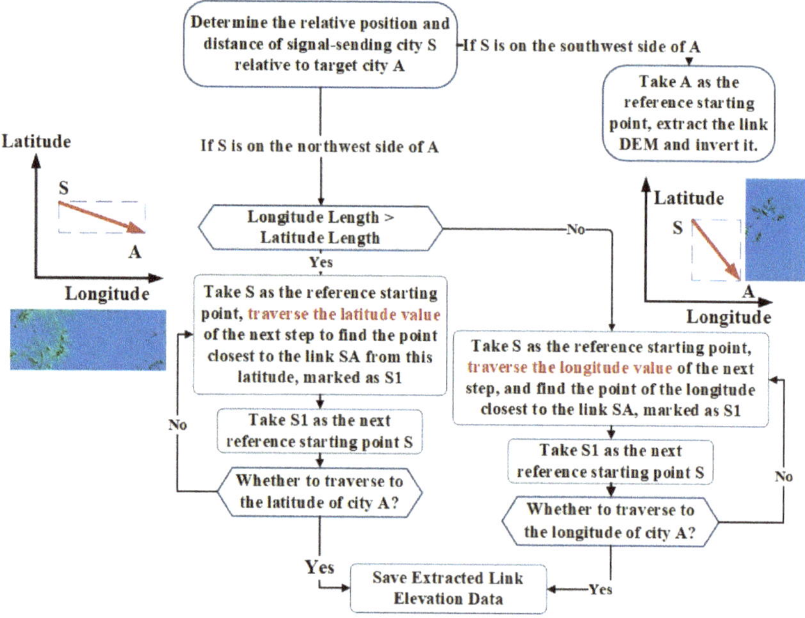

Figure 7. Flow chart of the extracting line-distributed DEM data from polygon-distributed DEM data.

Table 2. IGBP Land Cover Type.

Number	IGBP Land Cover Type
1	Evergreen needleleaf forest (ENF)
2	Evergreen broadleaf forest (EBF)
3	Deciduous needleleaf forest (DNF)
4	Deciduous broadleaf forest (DBF)
5	Mixed forest (MF)
6	Closed shrublands (CSH)
7	Open shrublands (OSH)
8	Woody savannas (WSA)
9	Savannas (SAV)
10	Grasslands (GRA)
11	Permanent wetlands (WET)
12	Croplands (CRO)
13	Urban and built-up (URB)
14	Cropland/natural vegetation mosaic (CVM)
15	Snow and ice (SNO)
16	Barren or sparsely vegetated (BSV)
17	Water bodies (WAT)

In this paper, the vegetation distribution in the latitude and longitude range areas was converted into WGS84 projection by positive selection projection and expressed in Figure 8:

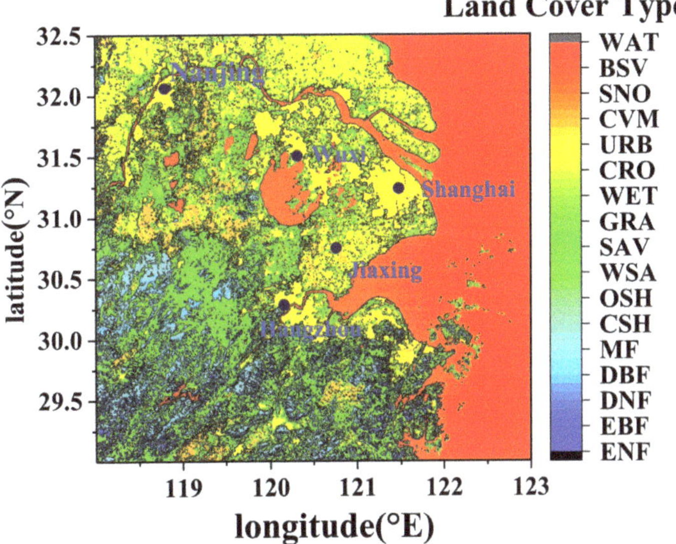

Figure 8. MCD12Q1-LC1 distribution of the selected region.

We converted the original sinusoidal projection of the MCD12Q1 to WGS84 projections. The original data of MCD12Q1 was tile distribution, which was similar to that of DEM. After converting the projection of MCD12Q1, the original square tile became a diamond distribution. Then the required intercity land type data were extracted from the adjacent converted diamond tile data. The value of the land type of the latitude and longitude of the intercity link was uniformly interpolated according to the equidistant latitude and longitude interval. Finally, we traversed the data coordinates in the shorter dimension direction (longitude direction or latitude direction), found the point closest to the longitude and latitude slope of the link, and performed a one-dimensional intercity link land type on the standard of the IGBP arrangement. The land cover types classified in the IGBP of the links from the four surrounding cities to Wuxi, marked in Figure 8, are shown in Figure 9:

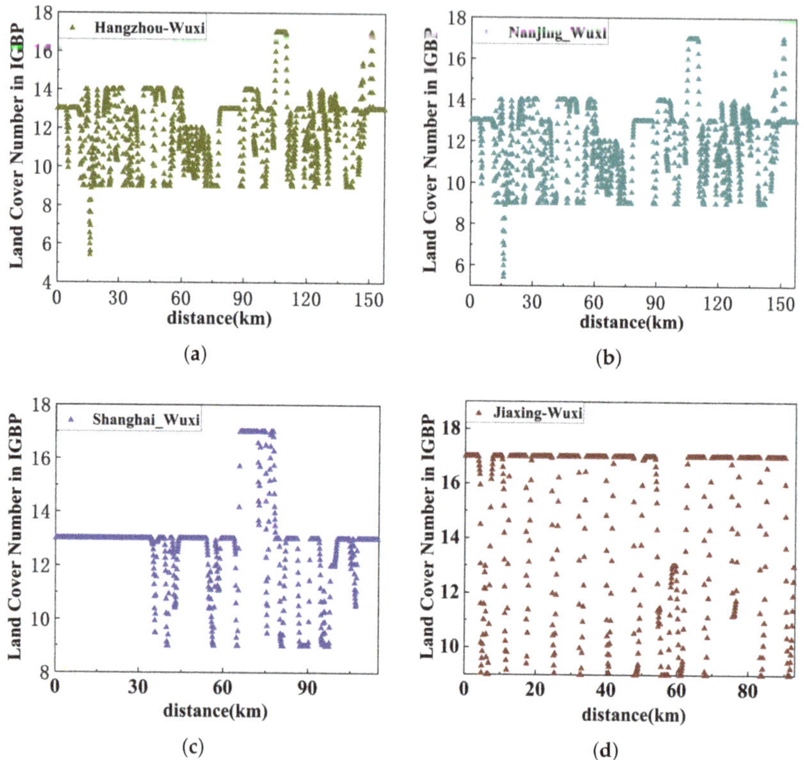

(a)

(b)

(c)

(d)

Figure 9. Land Cover types of IGBP on intercity links from four different cities to Wuxi. (**a**) LC distribution on Hangzhou–Wuxi. (**b**) LC distribution on Nanjing–Wuxi. (**c**) LC distribution on Shanghai–Wuxi. (**d**) LC distribution on Jiaxing–Wuxi.

2.3.2. Determination of Corresponding Dielectric Parameters in the Area Covered by the Intercity Link

The classification of the IGBP features was relatively fine, and there have been many studies on the empirical model and test results of several kinds of vegetation dielectric permittivities. After consulting a large number of literature studies, we chose to use the test results in the literature to solve the simulation in this paper. The land types involved in several intercity links selected in this paper mainly include: 9 Savannas, 10 grasslands, 11 permanent wetlands, 12 croplands, 13 urban and built-ups, 15 snow and ice, 16 barren or sparsely vegetated, 17 water bodies.

According to [24], we chose the following experimental results: the permittivity of $18.2 + 4.5i$ of sandy loam from a frequency of 3 GHz and water content of 0.3 g/cm^3 as the permittivity of barren or sparsely vegetated; the no. 16 classification in the IGBP land cover (LC) rule. Silty clay in Figure 10 was selected as the no. 11 classification in the IGBP LC rule when the frequency was 1.4 GHz and the water content $m_v = 0.45$; that is, the permittivity of the permanent wetland WET was assumed as $26.5 + 1.2i$.

By referring to Reference [25], the crops with large soil fertility contribution rates distributed in the selected area (north China region) were mainly rice. The dielectric permittivities of rice leaves of 50 rice growth days of were selected as the permittivities of crops (CRO) in the IGBP classification (i.e., $4.0 + i0.93$). Moreover, the permittivity of rice straw grown for 10 days was taken as the permittivity of the savanna (SAV) in IGBP classification, which was 2.85 + i0.33.

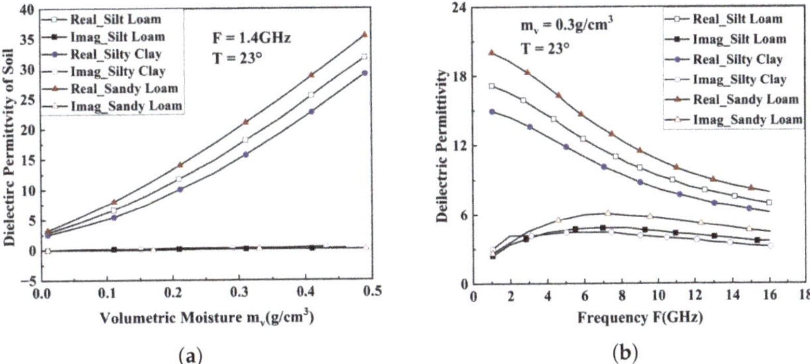

Figure 10. Dielectric permittivity of wet soil at different volumetric moisture or frequencies. (**a**) Dielectric constant vs. Volumetric Moisture. (**b**) Dielectric constant vs. frequency.

According to Reference [26], combined with the test results, the permittivity of asphalt in the oil-generating area was generally $\epsilon' = 5\sim15$, $\epsilon'' = 0.070\sim0.30$. Then we chose the dielectric permittivity of the asphalt of the intermediate state as the permittivity of urban and built-up (URB) areas in the IGBP classification, which was 5 + i0.070.

In Reference [27], from the relationship that the complex dielectric permittivity of wet snow versus water content or frequency at natural conditions (as shown in Figure 11), we choose the snow and ice (SNO, 15th classification in IGBP LC) dielectric permittivity $15.8 - i1.8$, when the frequency was 3 GHz, the water content was 0.3, and the temperature was $-5\,^\circ$C.

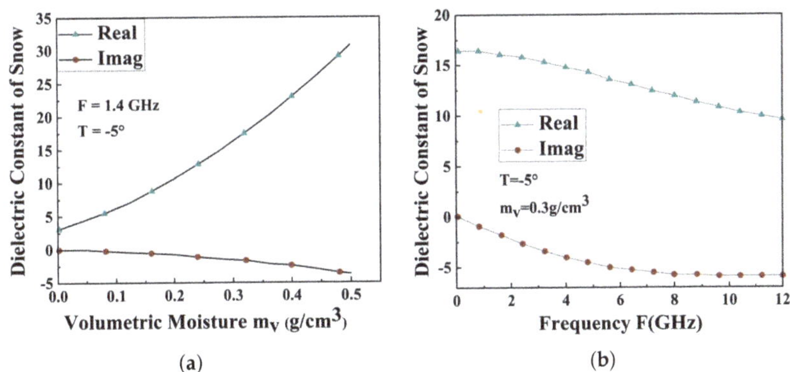

Figure 11. Dielectric permittivity of snow at different volumetric moisture/frequencies. (**a**) Dielectric constant vs. Volumetric moisture. (**b**) Dielectric constant vs. Frequency.

The reference dielectric permittivity values used for the land cover types ultimately involved in this paper are listed in Table 3.

This paper used the WRF database source, MCD12Q1 land cover classification from MODIS, terrain relief from the DEM to model the over-the-horizon propagation of communication signals at a sub-6 frequency band, comprehensively considering atmospheric ducts and tropospheric inhomogeneous turbulence. The schematic diagram of the over-the-horizon propagation of intercity links is shown in Figure 12.

Table 3. Dielectric constant used on simulated intercity links.

IGBP Land Cover Type	Dielectric Constant
Savannas (SAV)	2.85 + i0.33
Grasslands (GRA)	40 + i0.085
Permanent wetlands (WET)	26.5 + i1.2
Croplands (CRO)	4.0 + i0.93
Urban and built-up (URB)	5 + i0.070
Snow and ice (SNO)	15.8 − i1.8
Barren or sparsely vegetated (BSV)	18.2 + i4.5
Water Bodies (WAT)	81 + i0.22

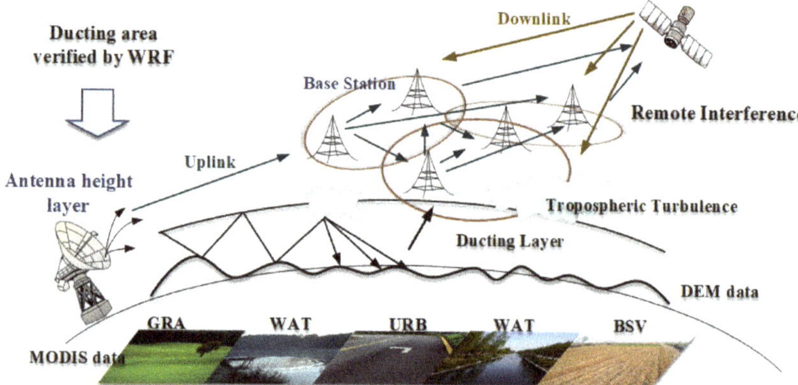

Figure 12. Schematic diagram of over-the-horizon forward-propagation of intercity links based on multi-source data fusion.

3. Scattering Parabolic Equation Algorithm for Irregular Terrain and Inhomogeneous Turbulent Atmosphere

When 5G signals are transmitted between cities, irregular terrain relief and dielectric roughness exist on the lower boundary of the atmospheric duct. At the same time, there may be non-uniform turbulence in the duct layer and even its upper and lower boundaries. This paper considers the actual situation and studies the radio wave propagation based on the complex meteorological environment when ducting and turbulence occurs. Compared with many existing terrain models, after research and comparison, a wide-angle shift-map (WASM) algorithm was determined in the terrain boundary processing of the WAPE model. The over-the-horizon propagation effects caused by duct considering the irregular terrain, different surface dielectric parameters at the lower boundary, and non-uniform tropospheric turbulence at the upper boundary of the duct layer were simulated on intercity links. The simulation results were compared with the results of numerical modeling software, AREPS (advanced refractive effects prediction system).

The tropospheric atmosphere uses the refraction index n to describe the refraction characteristics of electromagnetic waves. The propagation of electromagnetic waves on the earth's surface needs to consider the influence of the earth's curvature. The atmospheric refraction index n is expressed by the modified refraction index m as

$$m(h) = n(h) + h/a, \tag{1}$$

The modified refractive index M is expressed as

$$M = (m - 1) \times 10^6 = N + 10^6 \times h/a = N + 157h, \tag{2}$$

When simulating the tropospheric duct propagation, the refractive index profile used a hybrid four-parameter three-line model to represent a hybrid duct with a base layer:

$$M(z) = \begin{cases} M_0 + k_1 z & 0 \le z \le z_1 \\ M_1 + k_2(z - z_1) & z_1 < z \le z_1 + z_2 \\ M_2 + 0.118(z - z_1 - z_2) & z > z_1 + z_2 \end{cases} \tag{3}$$

where z_1 is the height of the duct base layer and z_2 is the duct thickness.

When tropospheric ducts appear over land due to changes in meteorological conditions, the electromagnetic wave wavelength and emission elevation angle satisfy the following conditions:

Maximum cut-off wavelength that can trap the wave in the duct

$$\lambda_{max} = C \int_{h_L}^{h_U} [M(h) - M_U]^{\frac{1}{2}} dh \tag{4}$$

where M_U is the modified refractive index of the top of the duct trapping layer. When studying the propagation through the duct structure with the base layer, the above formula is simplified as

$$\lambda_{max} = \frac{2}{3} Ct\sqrt{\Delta M} \tag{5}$$

where t is the duct thickness in m and is the modified refractive index difference corresponding to the duct layer height and the surface.

The critical trapping angle is simplified by derivation using Snell's law, as:

$$\theta_c \approx \sqrt{2 \times 10^{-6} \cdot (M_s - M_U)} \tag{6}$$

When the wavelength of the electromagnetic wave was less than the cut-off wavelength, and the emission elevation angle was within the trapping angle, the electromagnetic wave was trapped in the duct and propagated over a long distance beyond the horizon.

In the two-dimensional Cartesian coordinate system, it was assumed that the electromagnetic wave propagated along the x-axis, the electromagnetic field component was independent of the y-axis, the refractive index n of the propagating medium was isotropic, and the field component satisfied the two-dimensional scalar wave equation.

$$\frac{\partial^2 \psi(x,z)}{\partial x^2} + \frac{\partial^2 \psi(x,z)}{\partial z^2} + k_0^2 + n^2 \psi(x,z) = 0 \tag{7}$$

where k_0 is the electromagnetic wavenumber in vacuum, ψ is the electronic or magnetic field. $\psi(x,z) = E_y(x,z)$ for horizontally polarized, and $\psi(x,z) = H_y(x,z)$ for vertically polarized.

We numerically approximate the wave equation to obtain a wide-angle parabolic equation (WAPE) [5] form suitable for computing irregular terrain

$$\frac{\partial u(x,z)}{\partial x} = ik\left[\sqrt{1 + \frac{1}{k^2}\frac{\partial^2}{\partial z^2}} - 1\right] u(x,z) + ik[n(x,z) - 1]u(x,z) \tag{8}$$

The piecewise linear terrain shift map (PLTSM) algorithm proposed by Donohue and Kuttler was used to model the terrain boundary of the PE propagation model. The D–K algorithm is based on the continuous terrain shift map (CTSM) algorithm of Beilis–Tappert, which improves the difficulty of calculating the terrain curvature data.

The core idea of the PLTSM algorithm is to establish a new coordinate system on the irregular terrain section, and translate the PE calculation of each step into a flat ground form for solution. The terrain translation coordinates are expressed as

$$x = u, z = v - T(u) \tag{9}$$

Let the field components expressed in the new coordinate system be equal to the field components expressed in the irregular terrain coordinate system $\phi(x,z) = \Phi(u,v)$. We introduce simplified functions $\phi(x,z) = e^{i\theta}\psi(x,z)$. According to the coordinate map formula and wave equation, the forward-propagation PE considering irregular terrain is obtained

$$\left[\left\{ \frac{\partial}{\partial x} + i\frac{\partial\theta}{\partial x} - T'\left(\frac{\partial}{\partial z} + i\frac{\partial\theta}{\partial z} \right) \right\} - i\sqrt{\left(\frac{\partial}{\partial z} + i\frac{\partial\theta}{\partial z} \right)^2 + k^2 n^2} \right] \psi = 0 \tag{10}$$

In the piecewise linear terrain, the phase function is $\theta(x,z) = kzT'(x) + f(x)$. Let $f'(x) = k(1 + T')/2$, we obtain the WAPE in the coordinate system expressed in terms of the terrain curvature:

$$\frac{\partial}{\partial x} = i\sqrt{k^2 + \frac{\partial^2}{\partial z^2}} \cdot \psi + ik(n - zT'')\psi \tag{11}$$

In this paper, digital elevation model (DEM) data with 90 m resolution were used, which were downloaded from the National Geospatial data cloud. Taking into account the resolution, it was more accurate to use the terrain slope to calculate the long-distance propagation. To avoid dealing with the terrain curvature directly, the PLTSM method was used. The phase factor is expressed in the form of

$$\theta(x,z) = K_0 zT''(x) + f(x) \tag{12}$$

Let

$$f'(x) = K_0(T'^2 - 1) \tag{13}$$

where K_0 is the undetermined constant. Substituting Equations (11)–(13) into the forward-propagation PE (10) after the coordinate transformation, the simplified approximation of PE can be obtained:

$$\frac{\partial\psi}{\partial x} = i\sqrt{\frac{k^2}{1 + T'^2} + \frac{\partial^2}{\partial z^2}} \cdot \psi + ik\sqrt{n^2 - \frac{T'^2}{1 + T'^2}} \cdot \psi \tag{14}$$

The above equation is the WAPE corresponding to the PLTSM method. We abbreviate this method as TWPE (terrain wide-angle parabolic equation).

The atmospheric refraction profile is expressed using the modified refraction index $m(x,z) = n(x,z) - 1 + zT''$ in Equation (14); Equation (14) is solved by the split-step Fourier transform (SSFT) method. The relationship between the transforming field and the physical field could be expressed as $\phi = e^{i\theta}\psi$. For the segmented irregular terrain surface, the terrain inflection point changes due to different terrain slopes on adjacent segments, while the phase factor on adjacent segments need to be processed continuously to satisfy the realistic terrain simulation. The required phase correction item when the signal reflects from segment 1 to segment 2 is as follows:

$$\psi_2(x_{1,2}, z) = \psi_1(x_{1,2}, z) \exp\left\{ ik\left[\left(\frac{T_1'}{1 + T_1^2} \right) - \left(\frac{T_2'}{1 + T_2^2} \right) \right] \right\} \tag{15}$$
$$= \psi_1(x_{1,2}, z) \exp[ikz(\sin\beta_1 - \sin\beta_2)]$$

where $x_{1,2}$ is the reflection point from segment 1 to segment 2; $T' = \tan\beta$, where β is the angle that the local terrain slop makes with the horizontal. In this modification of the phase factor, this phase item makes the controlling factor of height-dependent sequences of transforming filed expressed by $\psi(x_{1,2}, jdz)$. The controlling angle happens to be the physics angle corresponding to the slope difference across the reflection point $x_{1,2}$. Then when controlling the phase of the transform field, in order to suppress the side lobes on the height sequence, the height step size must satisfy the following condition:

$$dz < \frac{2\pi}{k(1 + \sin\theta_M)} \tag{16}$$

where θ_M is the maximum angle of the controlling factor. The limiting condition on the height step size above is much more strict than the Nyquist sampling theorem. Considering the impedance boundary condition suitable for irregular terrain, Donohue and Kuttler deduced the impedance boundary condition on the terrain surface, which PLTSM needs on the basis of the standard Leontovich impedance boundary condition. It is expressed as follows:

$$\frac{\partial\psi}{\partial z} = -ik\cos\beta\left[\sin\gamma\left(\frac{1-\Gamma}{1+\Gamma}\right) + \tan\beta(1-\cos\gamma)\right]\psi \tag{17}$$

where Γ is the Fresnel reflection coefficient, which forms differently according to different polarizations on radio waves. Moreover, $\gamma = \xi + \beta$, where γ represents the grazing angle on the terrain slope, and ξ represents the angle that the propagation direction makes with the horizontal. Compared with the standard impedance boundary condition, the new impedance factor could be expressed as:

$$\alpha' = \cos\beta[\alpha + \tan\beta(1-\cos\gamma)] \tag{18}$$

In addition, we used an inhomogeneous turbulence model on the upper boundary of the duct with the based layer to express the complex tropospheric turbulence. Wanger et al. [20] considered the turbulence-induced range-dependent fluctuation of the lowest 1 km from the surface in the forward-propagation model of the duct. The maximum likelihood (ML) estimation of the atmospheric refractive index had good accuracy, and with prior information about ducting, the maximum a prior (MAP) refractivity estimate could be found.

Based on the empirical model of the three-line four-parameter refractive index profile of the duct with a base layer, the influence of the non-uniform turbulence was superimposed. The refractive index fluctuation is expressed as follows, and is related to the height of the duct base layer m_3 and the duct thickness m_4:

$$n_{ih}(z) = \tilde{n}(z)\left\{1 + K\exp\left[\frac{-(z-(m_3+m_4/2))^2}{(m_4/2)^2}\right]\right\} \tag{19}$$

where $\tilde{n}$ is generated by formula $C_n^2 = C_s$, for the relationship between the structure constant C_n^2 and refractive index, see [20].

4. Propagation Loss Calculation Using the Improved TWPE Model

In this section, we simulate intercity link ducting propagation with turbulence. The terrain relief on links were extracted from real DEM data. The LC types are classified based on the IGBP standards from MODIS data. Some dielectric permittivities of LC types relative to the intercity links are referenced from different literature studies, and measured or modeled on empirical formulae. The frequency was 3 GHz. The antenna height was 60 m, the antenna beam width was $3°$, and the antenna elevation angle was $0°$. The surface-based duct with turbulence in the upper boundary of duct is assumed here. We describe the surface-based duct with the tri-fold four-parameter model, of which the four-parameter combination is [40 40 0.12 −0.2].

In this intercity link from Hangzhou to Wuxi, the propagation distance was as far as 136 km. The relief from Hangzhou to Wuxi reached more than 200 m, which exceeded the radio-wave elevation angle limits of WPE (so was the intercity link from Nanjing to Wuxi). As the results show in Figures 13 and 14, the obstacles greatly increased the propagation loss. The radio-wave could not travel further with the presence of very high obstacles.

The effects of ducting and turbulence were cut off. Therefore, no CCI would occur on the intercity link when there is high topography.

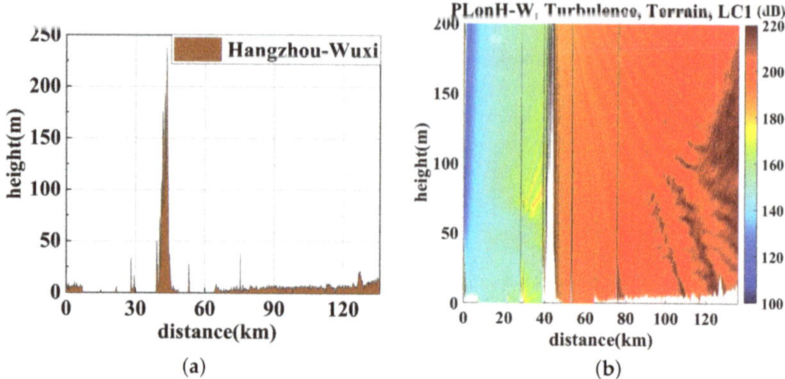

Figure 13. The (**a**) DEM and (**b**) propagation loss with tropospheric scattering, IGBP LC, and DEM considered on Hangzhou–Wuxi. (**a**) DEM form Hangzhou to Wuxi. (**b**) Propagation loss on Hangzhou–Wuxi.

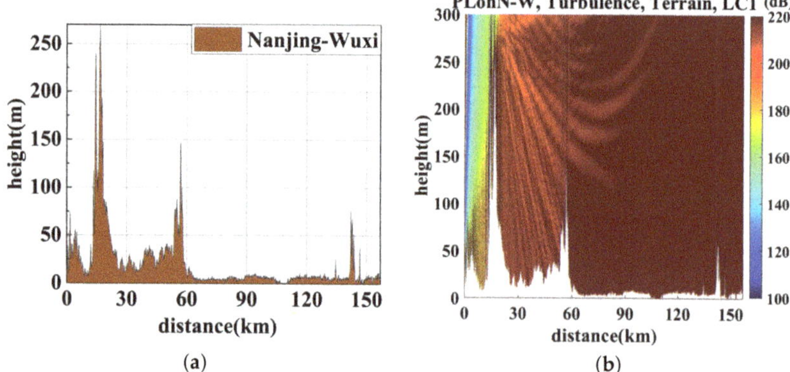

Figure 14. The (**a**) DEM and (**b**) propagation loss with tropospheric scattering, IGBP LC, and DEM were considered on Nanjing–Wuxi. (**a**) DEM from Nanjing to Wuxi. (**b**) Propagation Loss on Nanjing–Wuxi.

Taking into account the above analysis, the Hangzhou–Wuxi and Nanjing–Wuxi intercity links had large obstacles regarding cutting-off the ducting propagation. Therefore, the following selected the relatively gentle Shanghai–Wuxi and Jiaxing–Wuxi links to the intercity links. When the duct occurred on the intercity link, it was simulated whether to consider the non-uniform tropospheric turbulence or the influence of the PL distribution—of the distribution of real land covers—on the possible intercity CCI.

4.1. Joint Effects of Tropospheric Turbulence and Duct

The same parameters as the previous section are used in this section to simulate the situation of Shanghai–Wuxi and Jiaxing–Wuxi without considering tropospheric turbulence.

It can be seen from the comparison in Figures 15b and 16b of the range-dependent PL at the antenna level (60 m) that the non-uniform turbulence in the troposphere mainly affected the middle and rear sections of the intercity link. When considering the joint effect of the tropospheric inhomogeneous turbulence and the duct, in the range of 0–40 km, the tropospheric turbulence made no obvious effects on ducting propagation. At the end of the intercity link of 40–70 km, the tropospheric turbulence effect reduced the PL of the duct. After 70 km of the intercity link, the joint effect of tropospheric inhomogeneous turbulence and ducting on PL was reduced relative to the middle part compared with the ducting-only propagation situation.

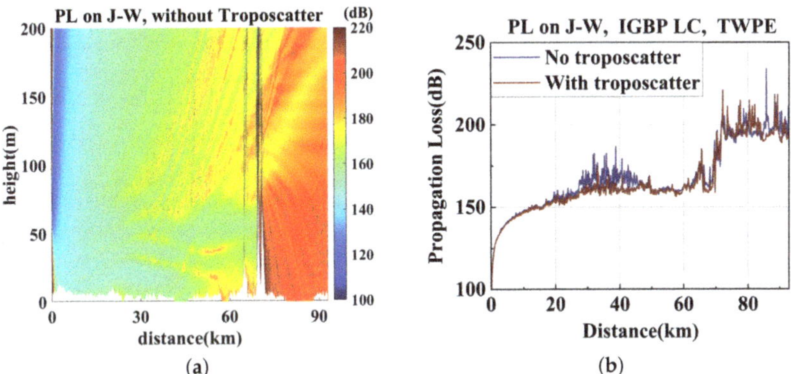

Figure 15. The PL on Jiaxing–Wuxi (**a**) in space (**b**) at a height of 60 m. (**a**) PL on Jiaxing–Wuxi without the troposcatter. (**b**) PL on J–W at a height of 60 m with or without the troposcatter.

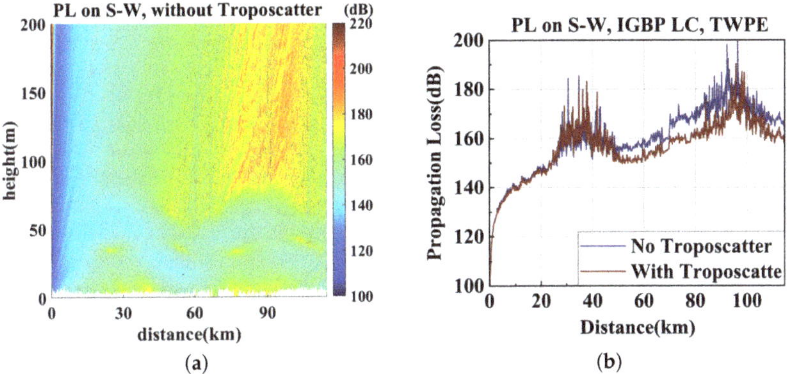

Figure 16. The PL on Shanghai–Wuxi (**a**) in space (**b**) at a height of 60 m. (**a**) PL on Shanghai–Wuxi without the troposcatter. (**b**) PL on S–W at a height of 6 0m with or without the troposcatter.

Based on this, it can be concluded that the non-uniform turbulence model superimposed on the non-uniform tri-fold four-parameter hybrid duct has a boosting effect on the inter-city propagation at 40 km and further. This may lead to the fact that, on the urban link from surrounding cities to Wuxi, such as the Shanghai–Wuxi link, the downlink signal of the remote base station will likely interfere with the uplink signal at the first 20–60 km of the link, affecting the random uplink access user. After 60 km, or due to sudden changes in terrain, or being too far away, the PL of the two links at the antenna height of 60 m exceeds 160 dB, which means the sensitivity that can be distinguished by the receiver. Therefore, the possibility of remote interference is not considered, but simulated by the improved scattering PE model, the PL of this middle section of the path link is reduced, the signal of the transmitting section may propagate to the user area of this section, and remote interference may still occur. When considering the farthest distance that useful signals can propagate between intercity links, the PL calculated by adding the influencing factors of tropospheric turbulence in this section also shows that it is possible to deploy base stations as precise as possible based on ensuring accurate communications and no remote interference. For example, a base station relay can be set up at 20 km where similar terrain and LCs are distributed, such as Shanghai–Wuxi, and a base station relay can be assumed at 60 km where similar terrain and LCs are distributed, such as Jiaxing–Wuxi link. In addition, because the two intercity links are close together, if Wuxi receives a useful signal from the Jiaxing link under the influence of a stronger duct, it may also receive a remote signal from Jiaxing at the same time, causing user communication interference. It could be concluded that the influence of tropospheric turbulence has an important reference value for deploying base stations and avoiding remote interference between base stations.

4.2. Verification of TWPE

Propagation losses from Shanghai–Wuxi using the same environment settings and radar parameters as intercity links of Hangzhou–Wuxi are shown in Figure 17. Different LC types were considered to evaluate the PL distribution on the S–W link using the TWPE model, as shown in Figure 17a,b. The AREPS result with the IGBP LC type is shown in Figure 17c. The range-dependent PLs at a ducting layer of 60 m on the S–W link of different LCs using TWPE and AREPS are compared in Figure 17d.

The propagation distance from Shanghai to Wuxi is about 115 km. Moreover, the terrain relief of the intercity link between Shanghai to Wuxi was kept below 10 m, which fits the application condition of the paraxial approximation of PWE. Based on the plain cover characteristics of East China, the land types of terrain on this intercity link are classified into grassland and water bodies at first. We conducted WAPE considering terrain relief, tropospheric turbulence, and dielectric permittivity on the first assumption and the land types of IGBP from MODIS data.

From Figure 17a,b, we used the IGBP cover type form MODIS with a resolution of 250 m, the PL distribution basically remains at the same level as the conducted form TWPE with LC type using grassland and water bodies. Figure 17d shows the PL distribution at the transmitting height of 60 m in the trapping layer of the surface-based duct. The PL at a height of 60 m using the LC type of IGBP from MODIS between 60 and 105 km is larger than that of using the grassland and water bodies as terrain features. On the remaining links, the PL using the two LC classifications are reversed or on the same level. AREPS uses the RPO model to divide the long-distance ducting propagation into four models to improve the calculation speed and facilitate engineering applications. There was not much to consider on the tropospheric scatter and terrain scatter in AREPS. Therefore, the PL trend was similar but AREPS result was 20 dB lower than TWPE result.

Given that the industry can achieve GPS receivers with tracking sensitivity below -160 dBm, at the height of the surface-based duct within the receiver, the realistic LC classification could be an essential environmental condition to evaluate whether CCI would happen on this intercity link. The possibility of CCI could be an essential decision support of 5G base station deployment.

Propagation loss from Jiaxing to Wuxi using the same environment settings and radar parameters, the same as in the previous section, are shown in Figure 18. Different LC types were considered to evaluate the PL distribution on the J–W link using TWPE model, as shown in Figure 18a,b. The AREPS result with the IGBP LC type is shown in Figure 18c. The range-dependent PLs at a ducting layer of 60 m on the J–W link of different LCs using TWPE and AREPS are compared in Figure 18d.

The distance of the intercity link from Jiaxing to Wuxi was about 93 km. Moreover, the terrain relief was around 20 m, besides a peak, which had a maximum height of 80 m. The peak was not high enough to cut off ducting and diffraction, and the radio wave elevation angle fit the maximum limit of TWPE.

From Figure 18b, we can see that when we considered the LC types more specifically, the PL on the intercity link Jiaxing–Wuxi increased slightly. Due to the limitation of the

RPO model in AREPS using a range-independent RO model in the range-dependent environment, in extreme cases this can cause discontinuities along the RO/PE boundary [28]. This discontinuity occurred in the peak of the intercity link of Jiaxing–Wuxi.

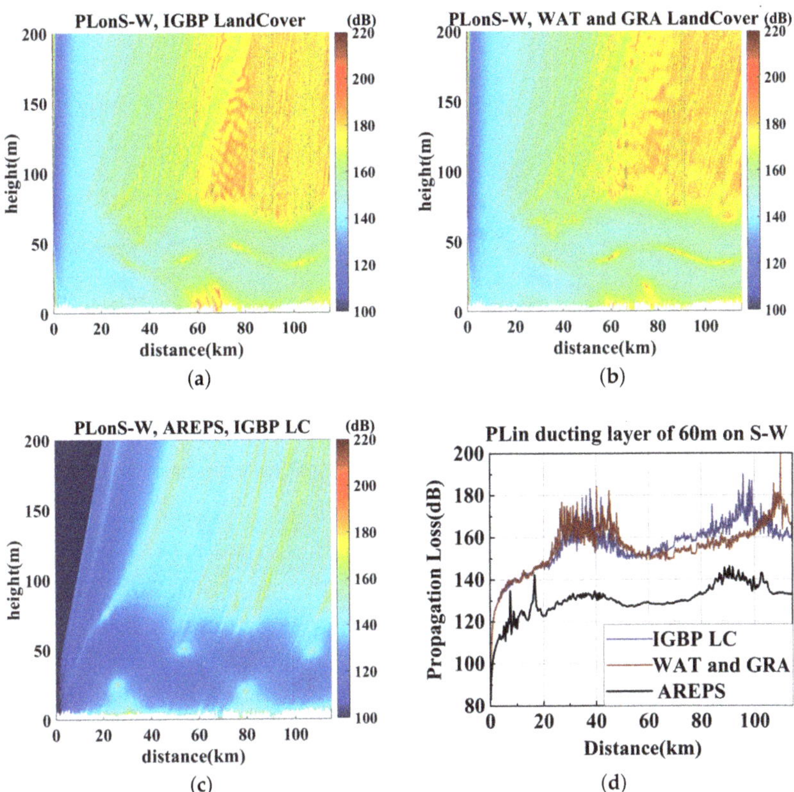

Figure 17. The propagation loss with tropospheric scattering, IGBP LC, and DEM considered on Shanghai–Wuxi. (**a**) PL from Shanghai–Wuxi using TPWE and IGBP LC. (**b**) PL from Shanghai–Wuxi using TPWE, WAT, and GRA LC. (**c**) PL from Shanghai–Wuxi using AREPS and IGBP LC. (**d**) PL at the duct-trapping layer of 60 m of different LCs.

The two classifications of LC types do not give obvious differences probably due to the classifications of grasslands and waterbodies being close to those of the real LC type distributions on the intercity link. It can be seen from Figure 9d that the real LC data on the Jiaxing–Wuxi link extracted by the IGBP LC classification also consisted of most WAT—17 in IGBP LC1—and several other kinds of vegetation categories. Moreover, the classifications from MODIS data just gave more specific resolutions on the LC types. Additionally, when we used the LC type classification in the resolution of 250 m from MODIS, the fast fading was much more obvious.

4.3. Effects of Different Land Covers

The previous section shows (assuming that the LC is grassland and water bodies) the PL trend and numerical range of PL at the antenna height layer when mixed ducts and tropospheric turbulence occur on the plain area. The accuracy of the results of the TWPE model with tropospheric turbulence was verified. The PL situation covered by different LCs will be discussed in this section.

In this section, three types of LCs, barren or sparsely vegetated (BSV), urban and built-up (URB), and savannas (SAV), were selected, respectively. The over-the-horizon propagation was studied on the Shanghai–Wuxi and Jiaxing–Wuxi links considering the joint effect of tropospheric turbulence and the duct. Figure 19 shows the PL distribution on Shanghai–Wuxi link. The parameters are the same as in the previous section, except that the LC dielectric parameters use the values corresponding to Table 3, respectively.

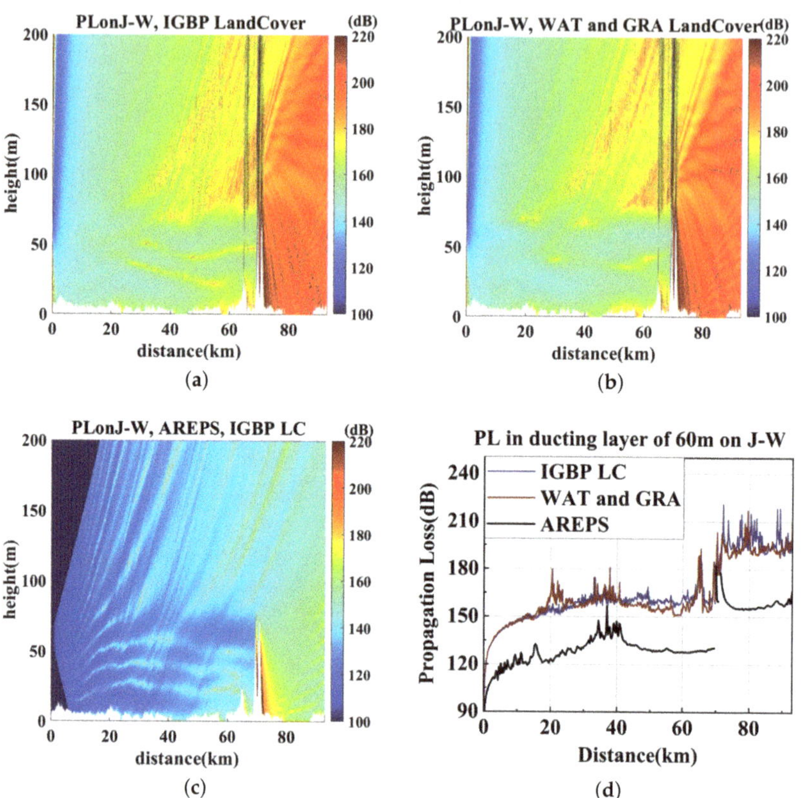

Figure 18. The Propagation loss with tropospheric scattering, IGBP LC, and DEM on Jiaxing–Wuxi. (**a**) PL from Jiaxing–Wuxi using TPWE and IGBP LC. (**b**) PL from Jiaxing–Wuxi using TPWE, WAT, and GRA LC. (**c**) PL from Jiaxing–Wuxi using AREPS and IGBP LC. (**d**) PL at the duct-trapping layer of 60 m of different LCs.

We used MSE to compare the PLs on Shanghai–Wuxi under three hypothetical LC types with the PLs on the S–W link under real IGBP LCs; see Figure 17a. The results are shown in Figure 20.

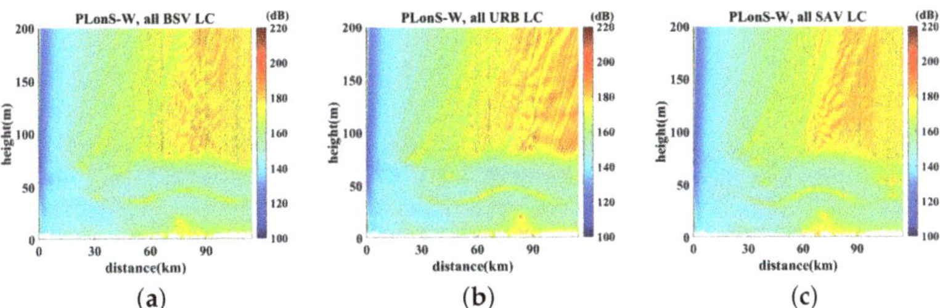

Figure 19. The PL using the improved TWPE with (**a**) BSV, (**b**) URB, (**c**) SAV LC, on Shanghai–Wuxi. (**a**) PL on S–W with BSV LC. (**b**) PL on S–W with URB LC. (**c**) PL on S–W with SAV LC.

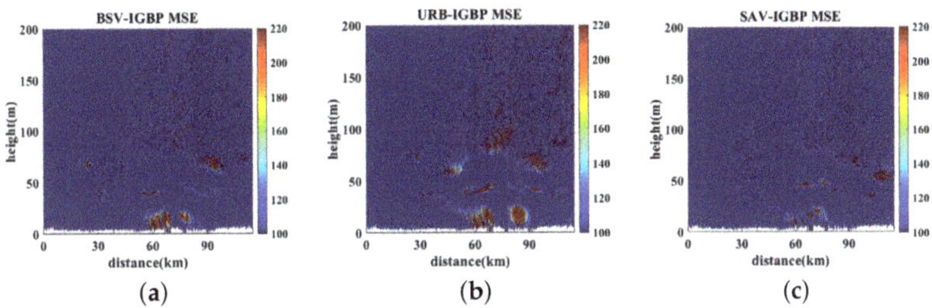

Figure 20. The MSE of PL using the improved TWPE between IGBP LC and different LCs, on Shanghai–Wuxi. (**a**) BSV-IGBP MSE of PL on S–W. (**b**) URB-IGBP MSE of PL on S–W. (**c**) SAV-IGBP MSE of PL on S–W.

To clearly compare the influences of the PLs of the transmission signal in the ducting layer of the Shanghai–Wuxi link under the backgrounds of several different LC types, we compared the PLs of the link at a height of 60 m in the ducting layer, as shown in Figure 21.

According to the impedance boundary conditions applicable to irregular terrain, it is assumed that the LCs are BSV/URB/SAV. Figure 19 shows the spatial distribution of the PL on Shanghai–Wuxi link when the three types of hypothetical single LCs are distributed. Figure 20 shows the spatial distribution of the mean square error (MSE) obtained by assuming different LC types and the real LC classified by IGBP on the Shanghai–Wuxi link. It can be seen that different LCs have a great influence on the PL accuracy, particularly at the antenna propagation height near the surface within the duct trapping layer. From the PL distribution of the antenna height layer shown in Figure 21, it can be seen that on the Shanghai–Wuxi link, the difference between the PL coverage of different LCs is mainly reflected in the front section and the rear section. The 20–40 km and 70–100 km sections both show that the IGBP LC results are significantly larger than the PL covered by a single LC type. The research in [29] shows that terrain fluctuation has a great influence on the PL distribution of the signal's over-the-horizon propagation. The results in this section show that the changes of different LCs may increase the PL, especially in the section with sudden and large terrain fluctuations, the actual situation may be a sudden change of LC, so it is more necessary to consider the coverage of real terrain.

Figure 22 shows the over-the-horizon PL distribution of different LCs on the Shanghai–Wuxi link when the receiving antenna height is 40 m. Different from the results obtained with the receiving antenna height of 60 m in Figure 21, there is a significant difference on the range-dependent PL in various LCs at 40 km; that is, in the middle of the link. However, under the backgrounds of the distribution of real LCs, the calculated PL still exceeds the sensitivity threshold of the receiving antenna at a height of 40 m at range of 45 km, blocking the long-distance propagation of the signal. Therefore, the possibility of users receiving remote interference on the road section beyond the 45 km Shanghai–Wuxi link is small. If the road section undergoes urban renovation (biased towards the dark cyan line in URB,

Figure 22), or vegetation green development (biased towards the dark yellow line in SAV, Figure 22), the propagation distances of the communication signal could vary according to the specific meteorological conditions and changes in the distribution of LC features. Therefore, according to the changes of the annual live updates of the LC types, the more finely the classification of LCs that may develop in the future, and the dielectric parameter values corresponding to the more accurate LC types, the reception range of communication signals could be estimated through a more accurate model.

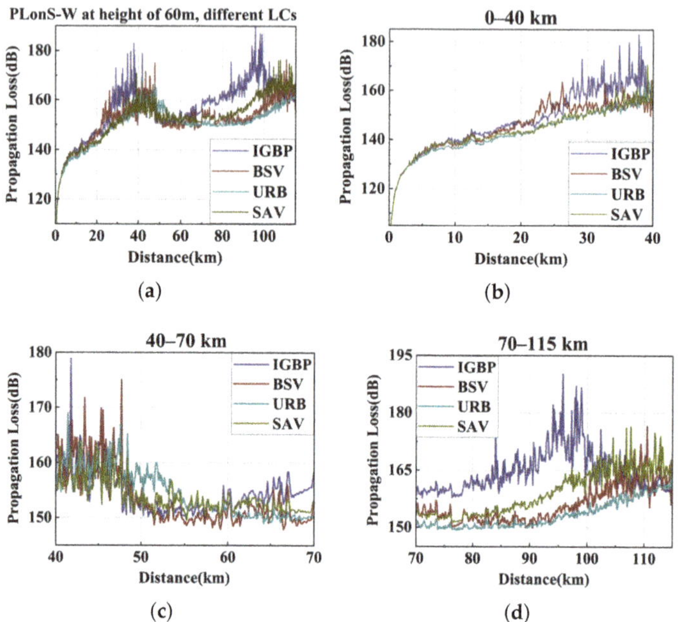

Figure 21. The propagation loss using the improved TWPE at a height of 60 m using different land covers, on Shanghai–Wuxi. (**a**) PL on S–W at different LCs at a antenna height of 60 m. (**b**) PL on S–W at a range of 20–40 km at different LCs. (**c**) PL on S–W at a range of 40–70 km at different LCs. (**d**) PL on S–W at a range of 70–115 km at different LCs.

Figure 23 shows the PLs of over-the-horizon propagation on Jiaxing–Wuxi with three hypothetical LCs considering the joint effect of the tropospheric turbulence and duct. The parameters are the same as in the previous section.

We used MSE to compare the PLs under three hypothetical LC types with the PLs under real IGBP LCs on the Jiaxing–Wuxi link, as shown in Figure 18a. The results are shown in Figure 24.

On the path link of Jiaxing–Wuxi, the propagation distance was 93 km, but there was a high terrain fluctuation at 70 km. Figure 24 shows the MSE between the parameter conditions of the assumed single LC feature distribution and the real IGBP feature distribution. As can be seen from Figure 24, the MSE between the PL calculated on the assumed single LC type parameters and the PL calculated on the real LC parameters classified by IGBP started to grow from the top of the duct, which was also the energy leaking area at a distance of 70 km from Jiaxing. It was obvious that the MSE of PL at the antenna height was the smallest when the assumed LC was SAV, which was closely related to the distribution of LCs on the intercity link. According to IGBP terrain classification in Figure 9d, Jiaxing–Wuxi was mostly covered by water bodies, and also contained many types of landform switching.

In order to clearly compare the influence of the PLs of the transmission signal in the ducting layer on Jiaxing–Wuxi link under the backgrounds of several different LC types, we compared the PLs of the link at a height of 60 m in the ducting layer, as shown in Figure 25.

Figure 25 shows the range-dependent PL at the antenna height of 60 m on different LC types. As shown in Figure 21c, the PL increased sharply, which was around 200 dB. Therefore, the difference of the PL calculated using TWPE between a single LC and on IGBP LC was not obvious at 0–20 km. From the 20–40 km PL on the J–W of different LCs shown in Figure 25a, the over-the-horizon PL on IGBP LC was close to that on BSV LC, but higher than the results of the URB or SAV LC. The PL of IGBP LC was slightly smaller in the 0–20 km section of Jiaxing–Wuxi. The PL distribution of different LCs at 0–40 km on the antenna height layer was similar to that of the Shanghai–Wuxi link, which is helpful for the deployment of short-distance 5G base stations. The middle section at a range of 40–70 km was mainly affected by terrain fluctuations and ducts, and the PL was kept within a certain range. The LC of this section can be roughly assumed using the simplified dielectric permittivity to simulate the PL distribution and could be performed to save time and costs. From the PL distribution of 70–93 km, it can be inferred that when the terrain fluctuation exceeds 80 m, the PL might exceed the minimum resolution value of the GPS receiver, so that the possibility of remote interference on the link can be ignored. Moreover, when long-distance communication is required, the corresponding 5G relay base station deployment can be reasonably arranged with reference to this result.

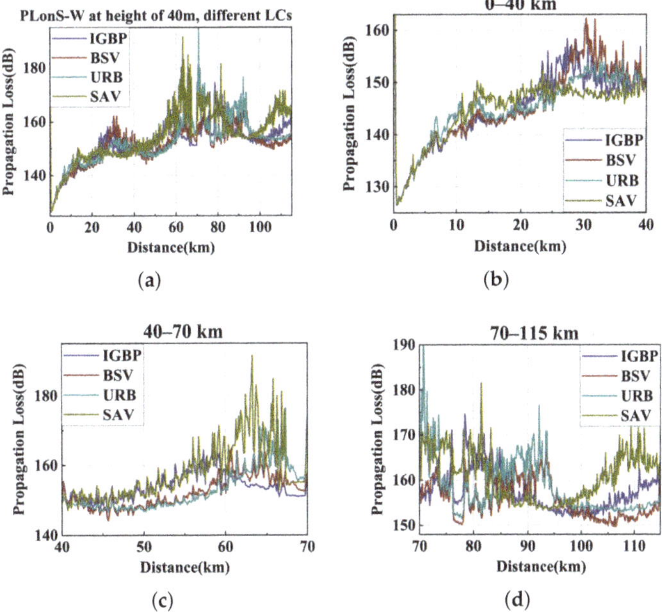

Figure 22. The propagation loss using the improved TWPE at a height of 40 m using different land covers, on Shanghai–Wuxi. (**a**) PL on S–W at different LCs at an antenna height of 40 m. (**b**) PL on S–W at a range of 20–40 km at different LCs. (**c**) PL on S–W at a range of 40–70 km at different LCs. (**d**) PL on S–W at a range of 70–115 km at different LCs.

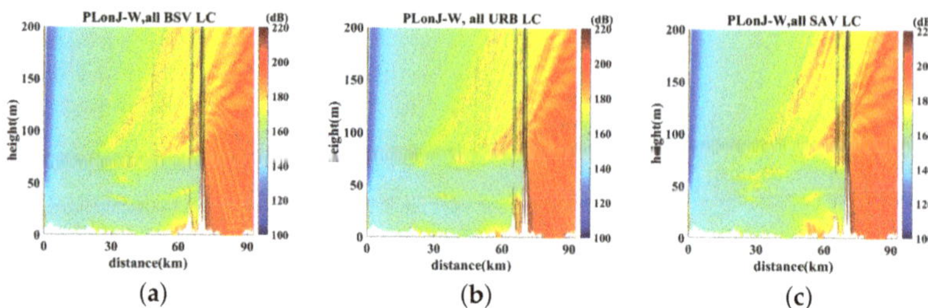

Figure 23. The MSE of PL using the improved TWPE between IGBP LC and different LCs on Shanghai–Wuxi. (**a**) PL on J–W with BSV LC. (**b**) PL on J–W with URB LC. (**c**) PL on J–W with SAV LC.

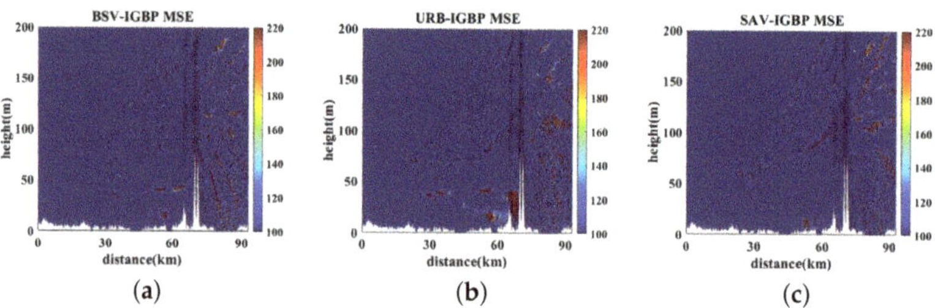

Figure 24. The MSE of PL using the improved TWPE between IGBP LC and different LCs on Shanghai–Wuxi. (**a–c**) BSV-IGBP MSE of PL on J–W.

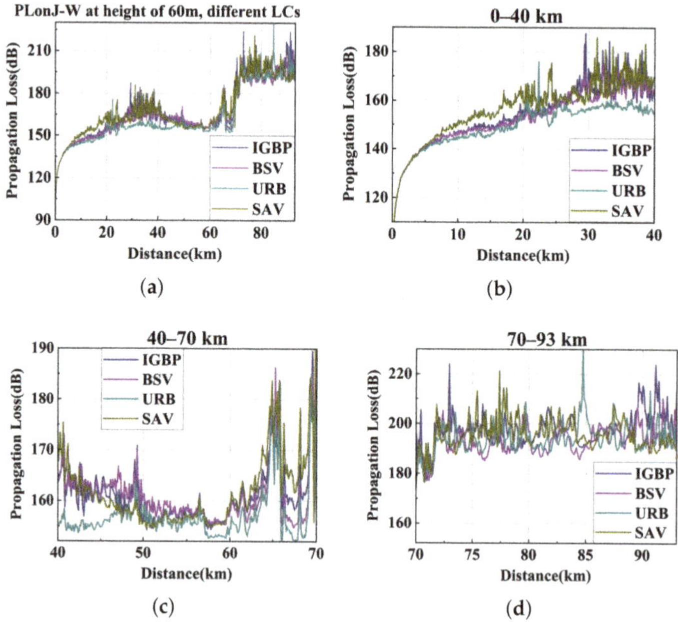

Figure 25. The Propagation loss using the improved TWPE at a height of 60 m using different land covers, on Jiaxing–Wuxi. (**a**) PL on J–W at different LCs at at antenna height of 60 m. (**b**) PL on J–W at range of 0–40 km at different LCs. (**c**) PL on J–W at a range of 40–70 km at different LCs. (**d**) PL on J–W at a range of 70–93 km at different LCs.

Figure 26 shows the range-dependent PL distribution at the height of the receiving antenna when the receiving antenna is 40 m on the Jiaxing–Wuxi link. At 40–70 km in the middle section of the link, the PLs simulated by different types of LC begin to show obvious differences. Compared with Figure 25 when the receiving antenna height is 60 m, the overall PL is reduced; that is, when the receiving antenna height is lower, the propagation distance of the signal increases. Within the range of signal-noise-ratio (SNO) that the antenna can distinguish, when the receiving antenna is 40 m, the signal can be transmitted to about 45 km. In addition, if the LC on the link develops in a unified manner, such as highway development, greening, desertification, etc.; that is, if the LC on the link does not change much, as shown in Figure 23b, the overall PL is reduced, and the propagation distance may increase by 10–20 km.

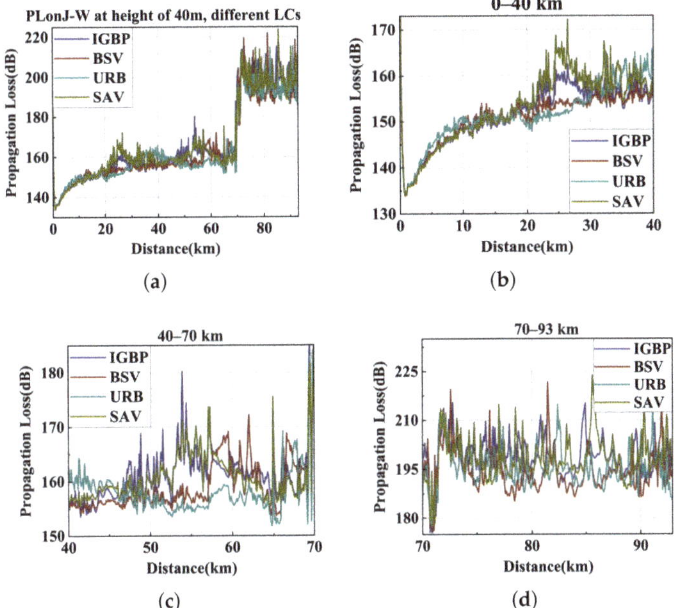

Figure 26. The propagation loss using the improved TWPE at a height of 40 m using different land covers, on Jiaxing–Wuxi. (**a**) PL on J–W at different LCs at an antenna height of 40 m. (**b**) PL on J–W at a range of 0–40 km at different LCs. (**c**) PL on J–W at a range of 40–70 km at different LCs. (**d**) PL on J–W at a range of 70–93 km at different LCs.

4.4. Analysis of Remote Interference between Different Intercity Links

Comparing Figures 22–25, the Shanghai–Wuxi link is relatively close to the Jiaxing–Wuxi link, and base stations on one of the links from 0 to 45 km might receive an uplink signal from another city, resulting in remote signal interference. Therefore, this section compares the PL of the four links with the receiving antenna heights of 60 m and 40 m, respectively (the heights are both within the trapping layer of the surface-based duct assumed in this paper), as shown in Figure 27. The probability of possible remote interference of inter-city signals was analyzed.

As shown in Figure 27, due to large terrain fluctuations, the propagation distance on Hangzhou–Wuxi and Nanjing–Wuxi links were about 15 and 40 km, respectively. From the city orientation shown in Figure 8, it can be seen that Hangzhou, Jiaxing, and Shanghai are all located within a radius of about 100 km southeast of Wuxi, and signal serialization may occur between the links from each city to Wuxi. As shown in Figure 9d, the Jiaxing–Wuxi link had a sudden increase in PL at different receiving antenna heights due to relatively large terrain fluctuations and switching of LCs. In the 0–40 km stage, the signals transmitted from Hangzhou, Jiaxing, and Shanghai were relatively stable under the effect of surface-based duct and tropospheric turbulence. Therefore, within a certain radius close to Wuxi, remote interference may occur among the signals of the three urban links, where a useful signal from one city may become an interfering signal received by another user. Based on the analysis of this phenomenon, it can be considered that when the base stations are deployed among relatively close cities, combined with a variety of possible ducting conditions, the tropospheric turbulence effect, and the annual update of the distribution of LCs, the simulation prediction of the propagation of over-the-horizon signals between cities can be carried out. According to the results, within the relevant radius, methods such as high obstacle blocking, alternating high and low antennas, and flexible site selections of base stations could be adopted to carry out preventive 5G base station deployment between cities with similar terrains to what we simulated in this paper, where remote interference may occur.

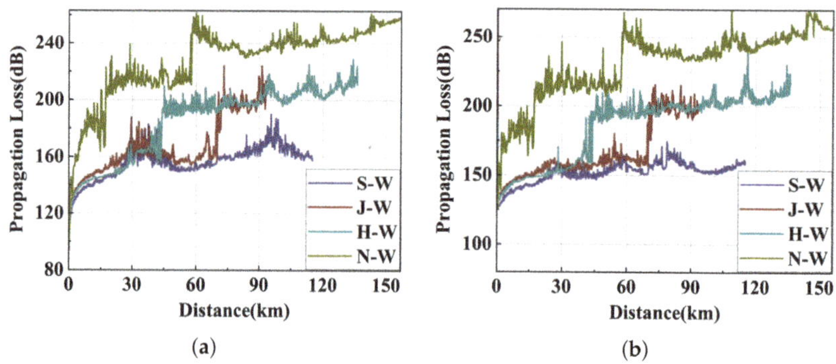

Figure 27. The range-dependent PL on four intercity links at antenna heights of (**a**) 60 m and (**b**) 40 m. (**a**) Range-dependent PL from four cities to Wuxi at ANH = 60 m. (**b**) Range-dependent PL from four cities to Wuxi at ANH = 40 m.

4.5. Deep Learning Model Predicts Land-Based Ducting Propagation Using Geomorphology Data

It was found that the intercity links whose terrains are similar to that of the intercity link analyzed in the previous sections, such as Shanghai–Wuxi or Jiaxing–Wuxi, are prone to the ducting effects caused by meteorological changes, causing remote interference at the antenna height. Since most of the relevant cities and landform data are less available, in this section, we use d1300 randomly generated datasets with different propagation distances, terrains, and land covers to perform physical simulations of over-the-horizon PL distributions under the same radar parameters, duct parameters, and random turbulence as the previous sections. The intercity link maximum distance was randomly generated within 90–100 km. The terrain relief height was randomly generated within 0–20 m. The LC type was randomly generated between 9 and 17 according to the IGBP classification scheme, and the corresponding dielectric parameters are listed in Table 3.

The simulated 1300 groups of terrain, land cover, and dielectric parameters at different distances and the PL at the corresponding antenna height were used for deep learning modeling. We intended to build a deep learning model to predict the PL on the intercity link with input parameters of terrain and land cover characteristics, and the ducting and radar environment settings remained the same as in previous sections.

Before training, because the randomly generated data may have been inconsistent with the actual situation, the obvious errors of the forward-propagation results were eliminated. A total of 56 sets of data were eliminated, and the remaining 1244 sets of data were used for deep learning modeling. After shuffling the order of datasets, the first 900 datasets were used as training sets, and the last 344 sets of data were used to test the training model. The deep multilayer perceptron (DMLP) network was used here for deep learning modeling. We continuously optimized the parameters to obtain the optimal model, and a total of 4000, 2000, 3000, and 999 neural networks were built for training. The schematic diagram of the deep learning forward prediction model is shown in Figure 28.

One of the results calculated by the TWPE scattering model in the test set was used to compare the prediction result by the deep learning model, as shown in Figure 29a, with the antenna height at 60 m, and Figure 30a with the antenna height at 40 m. Using the mean square error as the loss function, the formula of the specific loss function is as follows:

$$loss = \frac{\sum_{i=1}^{n}(pred(i) - y(i))^2}{n} \tag{20}$$

where $pred(i)$ represents the range-dependent PL predicted by the network, $y(i)$ is the true range-dependent PL calculated by the TWPE model, and n is the number of groups of training datasets.

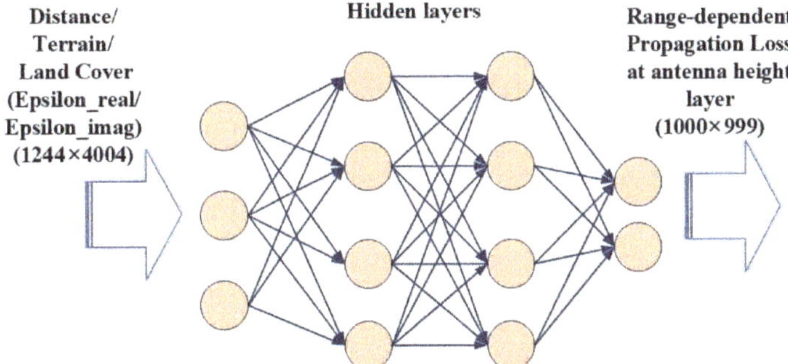

Figure 28. The schematic diagram of the deep learning forward-prediction model.

Using the following definition to give the model prediction accuracy functions to evaluate the quality of the established model

$$acc = 1 - \frac{y_{pred} - y_{true}}{y_{true}} \tag{21}$$

The accuracy of each range step was solved at antenna heights of 60 or 40 m, and the mean value was defined as the predicted accuracy of the sample. Then the predicted accuracy distribution was obtained on each tested sample. The accuracy results of test datasets are shown in Figures 29b and 30b, respectively.

When the antenna height was 60 m, the MSE of the DMLP prediction was 118.202749. When the antenna height was 40 m, the MSE was 108.696593. After the optimized deep learning parameters were well trained, they could quickly and efficiently predict (at a certain fixed antenna height, terrain, and landform) whether there would be communication interference problems between cities around 100 km apart with ducting and turbulence.

It can be seen from Figures 29b and 30b that the prediction accuracy function of the test set of the DMLP model simulation results is mostly kept between 0.8 and 1.0, and a good prediction effect was achieved. We selected two groups of DMLP predictions with relatively good test results and compared them with the TWPE simulation results. Figures 29a and 30a show that the DMLP prediction make large errors after 80 km, and the DMLP prediction results fluctuate greatly.

To improve the prediction effect of the model, we introduced the long- and short-term memory (LSTM) network model to perform deep learning modeling on the original 1244 sets of data, and used two layers of LSTM networks with 80 and 60 neural units, respectively. The LSTM has requirements for the network input dimension. It is a matrix form of [number of samples, number of expansions in time steps, number of sample features at each time step]. To meet the requirement of this parameter dimension, a total of 1244 groups of training and testing data of the original fully-connected layer were rearranged according to the four input features, which was distance, terrain fluctuation, the real part, and imaginary part of the dielectric parameters of the corresponding land covers. Then the input parameter dimensions were converted into the three-dimensional matrix. After the deep training using LSTM network, the prediction range-dependent PLs at antenna heights of 60 and 40 m were obtained, as shown in Figures 31 and 32.

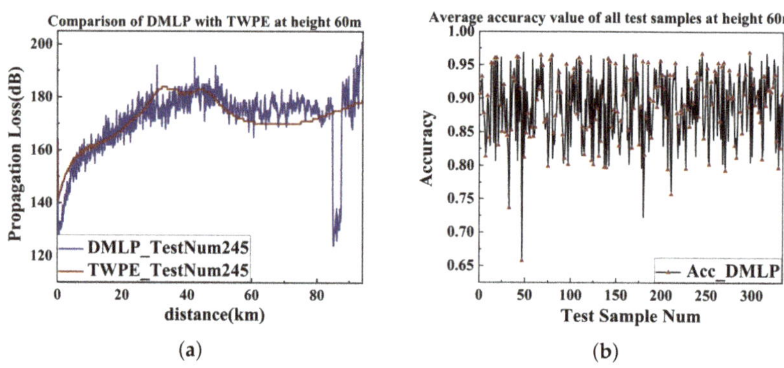

Figure 29. DMLP prediction at a height of 60 m and accuracy distribution. (**a**) PL Prediction using DMLP at an antenna height of 60 m. (**b**) DMLP of PL at a height of 60 m of accuracy on the test samples.

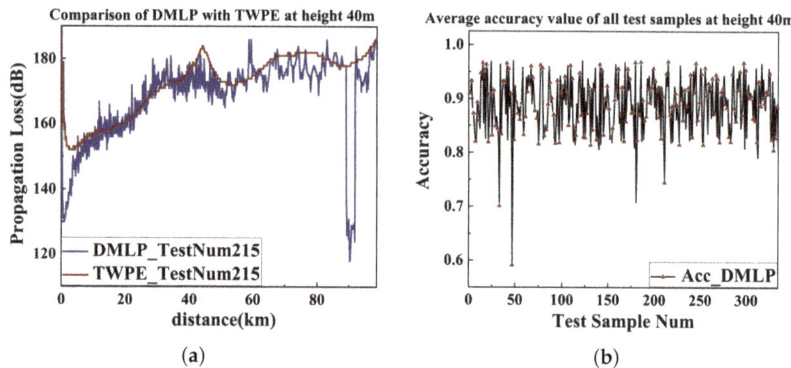

Figure 30. DMLP prediction at a height of 40 m and accuracy distribution. (**a**) PL prediction using DMLP at an antenna height of 40 m. (**b**) DMLP of PL at a height of 40 m accuracy on test samples.

The LSTM model still maintains a good prediction effect. The MSE of LSTM prediction of PLs at an antenna height of 60 m was 93.749120. Moreover, the MSE of LSTM prediction of PLs at an antenna height of 40 m was 90.245694. According to the loss value, the MSE of LSTM was smaller than that of DMLP. The prediction accuracy of LSTM remained mostly between 0.8 and 1.0. However, the shortcoming of the rapid fluctuation and the large error after 80 km of DMLP were greatly improved in LSTM.

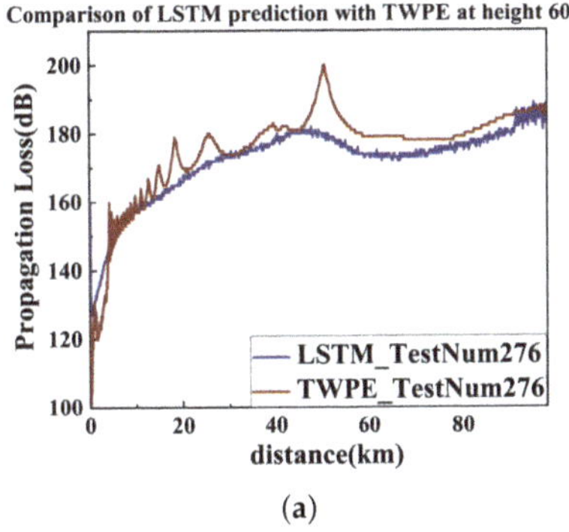
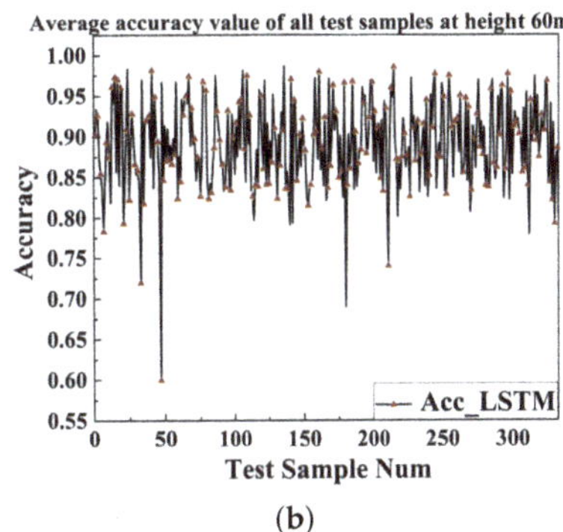

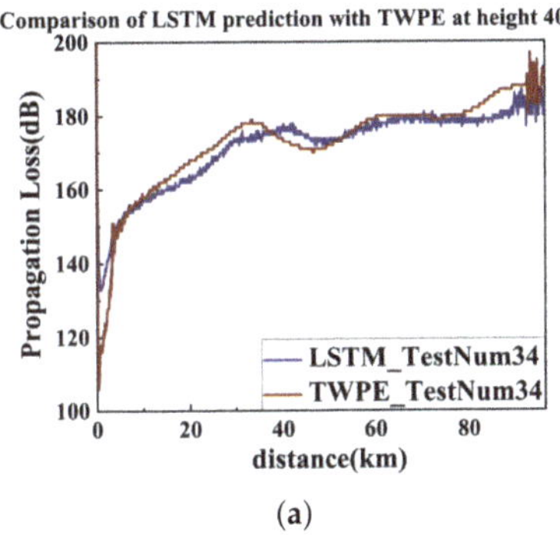
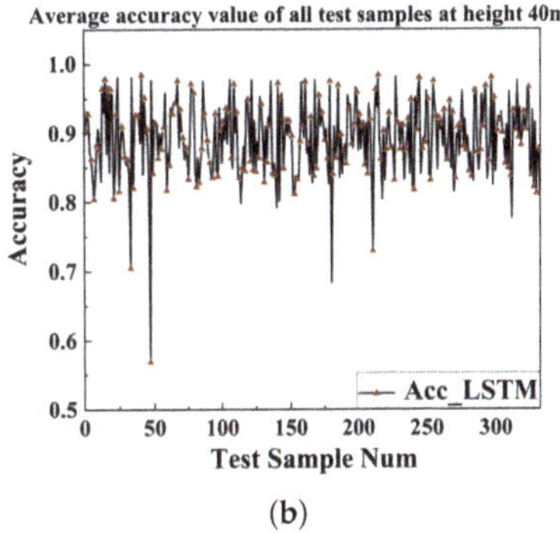

(a)

(b)

Figure 31. LSTM prediction at a height of 60 m and accuracy distribution. (**a**) PL prediction using LSTM at an antenna height of 60 m. (**b**) LSTM of PL at a height of 60 m accuracy on test samples.

(a)

(b)

Figure 32. LSTM prediction at a height of 40 m and accuracy distribution. (**a**) PL prediction using LSTM at an antenna height of 40 m. (**b**) LSTM of PL at a height of 40 m accuracy on test samples.

The terrain and land cover datasets used in the deep learning model were randomly generated. There might have been some random terrain and LCs that did not conform to the actual continuous distribution, so the model training results have certain errors. However, it can be seen that the topography and LC types have great influence on the PL distribution over the horizon. Therefore, it is effective and efficient to use the deep learning model to predict the over-the-horizon PL of the intercity link in the ducting environment with different landforms. The results also show that topography and LC types have great influence on the over-the-horizon propagation when a terrestrial duct happens. Subsequent continuous improvements of parameters to optimize the deep learning model may improve the prediction accuracy.

5. Discussion

On plain areas, tropospheric ducting and turbulence could effect the signals to make an over-the-horizon propagation of 5G communication signals. Due to the dense distribution of 5G base stations, chances of CCI increases between intercity links when ducts form because of unusual changes, such as temperature increases with height occurring in meteorological conditions. Moreover, the tropospheric turbulence related to the distribution of the hybrid duct height had a sub-segment effect on the over-the-horizon propagation. Before 40 km, the tropospheric ducting took the lead in the over-the-horizon transmission; in the middle of the 40–80 km, the tropospheric turbulence boosted the propagation effect by making the propagation signals multi-reflect on the air masses at the top layer of duct. If the PL in this middle section was reduced to the antenna receiving sensitivity range, the transmitted signal might have been received by local users when it reached the middle and rear sections of the urban link, resulting in remote interference.

If there is no major topographic change, the tropospheric turbulence will still play a role in the rear section of the intercity link, which can reduce the PL. However, because the propagation distance is too long, the signal may exceed the antenna receiving sensitivity at a distance of 70 km and will not be received by users in this section. If there are large topographic changes between cities, topographic fluctuations play a dominant role in over-the-horizon propagation. Figure 13b shows that peaks over 200 m make the PL exceed 160 dB immediately. Obstacles beyond the transmit antenna height could cut off the propagation entirely. When no obvious peaks or valleys exist on intercity links, there is the probability of duct occurrences after rain or at night on waterside plain or coastal cities areas. In our simulations on Shanghai–Wuxi and Jiaxing–Wuxi, signals propagating beyond the horizon with propagation loss did not decrease much in the duct-trapping layer when surface-based duct and tropospheric scattering occurred. Therefore, with the dense deployment of 5G communication base stations, tropospheric ducting and scattering effects may jointly cause CCI on intercity links with undulating terrain in plain areas. Therefore, it is essential to accurately model the over-the-horizon propagation of intercity link signals caused by the combined effects of duct and tropospheric turbulence effects, considering real terrain and LCs. The simulation results could be used in preventive deployment of 5G base stations and targeted deployment of relay base station locations, which could greatly save related costs.

A total of 1300 sets of randomly generated data were used to conduct deep learning training based on the TWPE simulated results. The training parameters with better effects after multiple optimizations were selected. The model was built with fixed radar antenna parameters and atmospheric environment parameters. The topography and landform data were entered to predict the PL distribution at the antenna height layer, and achieved good prediction results. When an intercity link whose terrain and land form are similar to the plain area is obtained, the deep learning model can be used to effectively and quickly predict the PL distribution, so as to evaluate the possibility of remote interference and carry out subsequent base station distribution deployment.

6. Conclusions

Due to the air mass flow, the inhomogeneous tropospheric turbulence exists almost all of the time. The tropospheric duct is very likely to be formed after heavy rain in the summer or with day–night temperature change. Researchers have confirmed the truth of over-the-horizon propagation of radio waves over the sea. On the land, when we simulate the over-the-horizon propagation to evaluate whether CCI could form between cities, the terrain relief, land cover type, and dielectric constant of land cover would be essential. The real elevation of intercity links from DEM data, real land cover type from MODIS data, and the corresponding dielectric permittivity of land cover types from different literature were considered in our simulation. Moreover, the method of the terrain shift map on WPE with range-dependent non-uniform turbulence was used to simulate the propagation loss on real intercity links. We also used the deep learning model to predict the distribution of the PL of the over-the-horizon signal cities at different distributions of intercity land forms in the ducting and random tropospheric turbulence environments. The prediction loss value was stable at about 0.6 m. This deep learning model could be applied to inter-city remote interference prediction when ducting occurs in similar plain terrains.

When terrain relief is rather high compared to the duct trapping layer height, the effects of terrain dominate in the propagation. When the terrain relief is relatively flat with no large peaks, ducting and turbulence dominate in the propagation. In our simulation, propagation loss could keep under 160 dB in the trapping layer of duct on intercity link length around 100 km, which could form CCI between cities. Thus, with 5G base stations, frequency band networking, arrangements of large-scale antenna arrays, the effects of tall building block outs, and tropospheric ducts need to be considered.

While the deployment of 5G base stations is still in full swing, the 6G satellite low-orbit internet has also developed simultaneously. Due to the limited number of satellites that can be accommodated in low-Earth orbit, the world is currently undergoing constellation plans in different states, such as networking, testing, and project approval, and is conducting an unprecedented competition that cannot be lost. The development of satellite communication will have higher requirements on the utilization rate of the current communication frequency band, signal propagation efficiency, and anti-interference ability. This requires us to take into account various dynamic conditions, such as meteorology, terrain, and LC types, and accurately model the propagation of useful and interfering signals for communication at current frequency band.

With the continuous development of commercial areas and urban–rural junctions, the topography and land cover types of build-up areas will continue to change. The coastal and plain cities near the water are more prone to form ducts, and the coastal cities are developing rapidly. Urban buildings, greening, natural landscape development, urban–rural integration areas development, crop planting, etc., are all constantly developing and changing. The LC type has a very obvious influence on the propagation loss of the over-the-horizon communication, so the over-the-horizon effect caused by the duct and tropospheric turbulence may trigger remote CCI to coastal cities. The uplink transmission signal may be exceeded by the guard time slot, interfering with the downlink received signal. The work done in this paper could play a guiding role in the deployment of 5G base stations to prevent possible CCI and the selection of base station relay locations.

The analysis of this paper gives a prediction method of communication signals in real mixed duct and non-uniform tropospheric turbulent environments for real and constantly evolving and changing land covers. The dielectric permittivity of constantly changing land coverage could be improved by referencing more literature studies and experimental results.

Author Contributions: Conceptualization, K.Y., Z.W. and J.W.; Data curation, X.G. and K.Z.; Formal analysis, T.W.; Investigation, K.Y.; Project administration, X.G.; Resources, J.W.; Supervision, T.Q.; Validation, K.Z.; Writing—original draft, K.Y.; Writing—review and editing, Z.W., T.W. and L.L. All authors have read and agreed to the published version of the manuscript.

Funding: Project supported by the National Natural Science Foundation of China (grant nos. 62271381, 62005205, 62071359, 61975158, 61901335, and 62001377) and Shaanxi Province Science Foundation for Youth, China (grant no. 2020JQ-329).

Data Availability Statement: The DEM data set is provided by Geospatial Data Cloud site, Computer Network Information Center, Chinese Academy of Sciences. (http://www.gscloud.cn). The MODIS data products are archived and available via FTP from the Land Processes Distributed Active Archive Center(DAAC) at EROS Data Center. (https://ladsweb.modaps.eosdis.nasa.gov/).

Conflicts of Interest: The authors declare no conflict of interest.

Abbreviations

The following abbreviations are used in this manuscript:

5G	The fifth generation
PE	parabolic equation
DEM	digital elevation model
MODIS	moderate-resolution imaging spectroradiometer
IGBP	International Geosphere-Biosphere Program
WAPE	wide-angle parabolic equation

TWPE	terrain wide-angle parabolic equation
AREPS	advanced refractive effects prediction system
CCI	co-channel interference
PL	propagation loss
LC	land cover
MAP	maximum a priori
ML	maximum likelihood
DMLP	deep multilayer perceptron
LSTM	long and short-term memory

References

1. Huang, X.; Dong, J.; Wang, W.; Xu, H. A case study of 5G network planning based on Massive MIMO. *Telecom Eng. Tech. Stand. Abbr.* **2017**, *30*, 50–54.
2. Turton, J.D.; Bennetts, D.A.; Farmer, S.F.G. An introduction to radio ducting. *Meteorol. Mag.* **1988**, *117*, 245–254.
3. Leontovich, M.; Fock, V.A. Solution of the problem of propagation of electromagnetic waves along the earth's surface by the method of parabolic equation. *J. Phys. USSR* **1946**, *10*, 13–23.
4. Hardin, R.H.; Tappert, F.D. Applications of the Split-Step Fourier Method to the Numerical Solution of Nonlinear and Variable Coefficient Wave Equations. *Siam Rev.* **1973**, *15*, 423.
5. Feit, M.D.; Fleck, J.A. Light propagation in graded-index optical fibers. *Appl. Opt.* **1978**, *17*, 3990–3998. [CrossRef]
6. Thomson D.J.; Chapman N. A wide-angle split-step algorithm for the parabolic equation. *J. Acoust. Soc. Am.* **1983**, *74*, 1848–1854. [CrossRef]
7. Kuttler, J.R. Differences between the narrow-angle and wide-angle propagators in the split-step Fourier solution of the parabolic wave equation. *IEEE Trans. Antennas Propag.* **1999**, *47*, 1131–1140. [CrossRef]
8. Levy, M. *Parabolic Equation Methods for Electromagnetic Wave Propagation*, 1st ed.; The Institution of Engineering and Technology: London, UK, 2000; pp. 268–272.
9. Janaswamy, R. Path loss predictions in the presence of buildings on flat terrain: A 3-D vector parabolic equation approach. *IEEE Trans. Antennas Propag.* **2003**, *51*, 1716–1728. [CrossRef]
10. Hu, H.; Chen, Z.; Chai, S.; Mao, J. Research on mobile communication radio propagation characteristic based on 3DPE. In Proceedings of the 2005 Asia-Pacific Microwave Conference, Suzhou, China, 4–7 December 2005.
11. Rasool, H.F.; Pan, X.M.; Sheng, X.Q. A fourier split-step based wide-angle three-dimensional vector parabolic wave equation algorithm predicting the field strength over flat and irregular forest environments. *Appl. Comput. Electromagn. Soc. J.* **2019**, *34*, 874–881.
12. Dockery, G.D.; Konstanzer, G.C. Recent advances in prediction of tropospheric propagation using the parabolic equation. *Johns Hopkins APL Tech. Dig.* **1987**, *8*, 404–412.
13. Dockery, D.; Kuttler, J.R. An improved impedance-boundary algorithm for Fourier split-step solutions of the parabolic wave equation. *IEEE Trans. Antennas Propag.* **1996**, *44*, 1592–1599. [CrossRef]
14. Kuttler, J.R.; Dockery, G.D. Theoretical description of the parabolic approximation/Fourier split-step method of representing electromagnetic propagation in the troposphere. *Radio Sci.* **1991**, *26*, 381–393. [CrossRef]
15. Donohue, D.J.; Kuttler, J.R. Propagation modeling over terrain using the parabolic wave equation. *IEEE Trans. Antennas Propag.* **2000**, *48*, 260–277. [CrossRef]
16. Hitney, H.V. A practical tropospheric scatter model using the parabolic equation. *IEEE Trans. Antennas Propag.* **1993**, *41*, 905–909. [CrossRef]
17. Ivanov, V.; Shalyapin, V.; Levadny, Y.V. Microwave scattering by tropospheric fluctuations in an evaporation duct. *Radiophys. Quantum Electron.* **2009**, *52*, 277–286. [CrossRef]
18. Wang, H.G.; Zhang, L.J.; Guo, X.M.; Wu, Z.S. Study of Fading Characteristics of Evaporation Duct Channel for Microwave Over-the-Horizon Propagtaion. *J. Microw.* **2013**, *29*, 1–5.
19. Song, S.T. Analysis on Influence of Tropospheric Atmosphere State on the Transmission Loss of Troposcatter Communication Link. *Radio Commun. Technol.* **2015**, *41*, 46–48.
20. Wanger, M.; Gerstoft, P.; Rogers, T. Estimating Refractivity from Propagation Loss in Turbulent Media. *Radio Sci.* **2016**, *51*, 1876–1894.
21. Liu, X.; Wu, Z.; Wang, H. Inversion method of regional range-dependent surface ducts with a base layer by Doppler weather radar echoes based on WRF model. *Atmosphere* **2020**, *11*, 754. [CrossRef]
22. Liu, X.; Wu, Z.; Wang, H. Inversion for Inhomogeneous Surface Duct without a Base Layer Based on Ocean-Scattered Low-Elevation BDS Signals. *Remote Sens.* **2021**, *13*, 3914. [CrossRef]
23. Reuter, H.I.; Nelson, A.; Jarvis, A. An evaluation of void-filling interpolation methods for SRTM data. *Int. J. Geogr. Inf. Sci.* **2007**, *21*, 983–1008. [CrossRef]
24. Dobson, M.C.; Ulaby, F.T.; Hallikainen, M.T.; El-rayes, M.A. Microwave Dielectric Behavior of Wet Soil-Part II: Dielectric Mixing Models. *IEEE Trans. Geosci. Remote Sens.* **1985**, *GE-23*, 35–46.

25. Tang, Y.H.; Huang, Y. Spatial Distribution Characteristics of the Percentage of Soil Fertility Contribution and Its Associated Basic Crop Yield in Mainland China. *J. Agro-Environ. Sci.* **2009**, *28*, 1070–1078.
26. Xiao, J.K. Microwave Dielectric Property of Solid Bitumen. *Geochimica* **1983**, *1*, 24–31.
27. Zhang, J.R. The dielectric constant in microwave remote sensing. *Remote Sens. Technol. Appl.* **1994**, *9*, 30–43.
28. Hitney, H.V. Hybrid ray optics and parabolic equation methods for radar propagation modeling. In Proceedings of the 92 International Conference on Radar, International Conference, Brighton, UK, 12–13 October 1992; pp. 58–61.
29. Yang, K.; Wu, Z.; Guo, X.; Wu, J.; Cao, Y.; Qu, T.; Xue, J. Estimation of co-channel interference between cities caused by ducting and turbulence. *Chin. Phys. B* **2021**, *31*, 024102. [CrossRef]

remote sensing

Article

Concrete Bridge Defects Identification and Localization Based on Classification Deep Convolutional Neural Networks and Transfer Learning

Hajar Zoubir [1], Mustapha Rguig [1], Mohamed El Aroussi [1], Abdellah Chehri [2], Rachid Saadane [1] and Gwanggil Jeon [3,*]

[1] Laboratory of Systems Engeneering (LaGes), Hassania School of Public Works, Casablanca 20000, Morocco
[2] Department of Mathematics and Computer Science, Royal Military College of Canada, Kingston, ON K7K 7B4, Canada
[3] Department of Embedded Systems Engineering, Incheon National University, Incheon 22012, Korea
* Correspondence: gjeon@inu.ac.kr

Abstract: Conventional practices of bridge visual inspection present several limitations, including a tedious process of analyzing images manually to identify potential damages. Vision-based techniques, particularly Deep Convolutional Neural Networks, have been widely investigated to automatically identify, localize, and quantify defects in bridge images. However, massive datasets with different annotation levels are required to train these deep models. This paper presents a dataset of more than 6900 images featuring three common defects of concrete bridges (i.e., cracks, efflorescence, and spalling). To overcome the challenge of limited training samples, three Transfer Learning approaches in fine-tuning the state-of-the-art Visual Geometry Group network were studied and compared to classify the three defects. The best-proposed approach achieved a high testing accuracy (97.13%), combined with high F1-scores of 97.38%, 95.01%, and 97.35% for cracks, efflorescence, and spalling, respectively. Furthermore, the effectiveness of interpretable networks was explored in the context of weakly supervised semantic segmentation using image-level annotations. Two gradient-based backpropagation interpretation techniques were used to generate pixel-level heatmaps and localize defects in test images. Qualitative results showcase the potential use of interpretation maps to provide relevant information on defect localization in a weak supervision framework.

Keywords: concrete bridge; visual inspection; defect; deep convolutional neural network; transfer learning; interpretation techniques; weakly supervised semantic segmentation

Citation: Zoubir, H.; Rguig, M.; El Aroussi, M.; Chehri, A.; Saadane, R.; Jeon, G. Concrete Bridge Defects Identification and Localization Based on Classification Deep Convolutional Neural Networks and Transfer Learning. *Remote Sens.* **2022**, *14*, 4882. https://doi.org/10.3390/rs14194882

Academic Editor: Lefei Zhang

Received: 31 May 2022
Accepted: 26 September 2022
Published: 30 September 2022

Publisher's Note: MDPI stays neutral with regard to jurisdictional claims in published maps and institutional affiliations.

1. Introduction

Bridges are key elements of a road network and play a critical role in the functional operation of the transportation system. During their service life, they are subjected to multiple deterioration mechanisms induced by material aging, variable loading, aggressive environmental actions, and extreme weather conditions. As a result, various types of damage (e.g., crack and corrosion [1,2]) occur over time and alter the structural behavior of bridges. Therefore, it is essential to accurately and timely detect and evaluate the damage to prevent failure and maintain structural safety and serviceability.

Structural Health Monitoring (SHM) has attracted much attention and has been the subject of several works in recent decades. Numerous techniques based on sensors (e.g., accelerometers, velocimeters, and displacement sensors) [3], non-destructive testing (e.g., ground-penetrating radar, infrared and ultrasonic techniques) [4], and visual inspection [5,6] have been deployed to identify, localize, and quantify damage in bridges. However, visual inspection has been the predominant practice for bridge condition assessment [7–9]. Trained inspectors conduct an in situ examination of bridge elements based on established guidelines and evaluate the condition of the entire bridge. However, the conventional framework of this practice is time-consuming, labor-intensive, and error-prone due to the subjective judgment of inspectors. Moreover, it requires access equipment and vehicles to reach areas of the bridge with low accessibility, which incurs additional costs to the monitoring operation [8].

In recent years, technological advances in civil engineering and related disciplines have promoted the emergence of innovative tools to manage civil infrastructures. Within this context, bridge owners and managers have shown increasing interest in Unmanned Aerial Vehicles (UAVs) as an assistive, efficient, and cost-effective means offering great potential for inspection automation [8,10]. However, one of the major challenges associated with this inspection scheme lies in deploying an efficient method to process the large amount of image data collected by the UAVs' sensors. To this end, several vision-based techniques have been extensively explored to automate defect detection in different civil engineering structures. These methods include traditional Image Processing Techniques (IPTs) [11], Machine Learning algorithms [12], and Deep Convolutional Neural Networks (DCNNs) [13].

In the particular context of concrete damage detection, cracks are the primary type of damage investigated by researchers. IPTs are used to extract representative properties of cracks from input images by applying various filters and morphological operations (e.g., Edge Detectors [14], Thresholding [15], Percolation [16,17], and Principal Component Analysis [18]). Then, the extracted features are fed to Machine Learning models, such as Support Vector Machines [19,20] and Nearest Neighbor Classifiers [20], to perform the classification task. However, IPTs provide hand-crafted features for training [21] and present limited learning capabilities that do not represent the complexity of the concrete texture and the challenging conditions of image acquisition, such as lighting, shading, and camera movements [21,22].

On the other hand, DCNNs extract features from a set of training images through the convolution operation and classify them within one learning framework. Owing to their robust feature extraction and learning capabilities, DCNNs have been widely examined in concrete damage classification studies.

For example, Dorafshan et al. [23] demonstrated the superiority of the AlexNet network [24] over six standard edge detectors in classifying concrete crack images of the SDNET dataset [25].

Kim et al. [26] trained and optimized the LeNet-5 network [27] to detect cracks in concrete surfaces using a dataset of 40,000 images. The proposed model achieved an accuracy of 99.8% and could be implemented using low-power computational devices.

Yu et al. [28] developed a method based on DCNNs to detect cracks in image patches of damaged concrete. The authors proposed an architecture consisting of six convolutional layers, two pooling layers, and three fully connected layers and employed the enhanced chicken swarm algorithm to optimize the meta-parameters of the DCNN model.

Mundt et al. [29] proposed the CODEBRIM dataset that features five non-exclusive damage classes in bridges (i.e., crack, spallation, exposed reinforcement bar, efflorescence, and corrosion). In addition, they investigated reinforcement learning approaches to build a DCNN model for the multi-target classification task, and their best meta-learned models yielded a testing accuracy of 72%.

Since training DCNNs requires a significant amount of image data and due to the limited size of concrete damage datasets, researchers have explored Transfer Learning techniques to train deep learning networks for concrete damage classification [30–33]. Pretrained DCNNs (e.g., AlexNet [24], VGG [34], ResNet [35], Inception [36]) on large benchmark datasets (e.g., ImageNet [37], MNIST [27], CIFAR100 [38]) are used to transfer knowledge from a source domain (e.g., ImageNet dataset) to a target domain (e.g., a small-scale concrete damage dataset) through different settings and learning approaches [39].

Yang et al. [21] developed a low-cost automated inspection approach based on UAVs and deep learning. They constructed the CSSC database and used a fine-tuned VGG16 model to classify cracks and spalling in concrete bridge elements and achieved a mean accuracy of 93.36% with the CSSC dataset.

Hüthwohl et al. [40] used a pre-trained inception-V3 network to define a hierarchical multi-classifier for reinforced concrete bridge defects (i.e., cracks, efflorescence, spalling, exposed reinforcement, and rust staining). Experimental results showed that the multi-classifier could assign class labels with an average F1-score of 83.5%.

Yang et al. [33] proposed an end-to-end-based Transfer Learning method for crack detection using three knowledge transfer approaches (i.e., sample, model, and parameter transfer knowledge), a fine-tuned VGG16 model, and three crack datasets. Their experiments showed that by training 13 convolutional and two fully connected layers of the pre-trained VGG16 model on the three datasets, crack detection was improved and achieved a testing accuracy of 97.07% on the SDNET dataset.

Bukhsh et al. [31] investigated cross-domain and in-domain Transfer Learning approaches. They compared the performance of the VGG16, InceptionV3, and the ResNet50 models in different Transfer Learning strategies to detect damages in six binary and multi-label concrete damage datasets. Their experiments demonstrated that combined representations of in-domain and cross-domain Transfer provide considerable performance gain, particularly with tiny datasets.

Zhu et al. [41] built a robust classifier to detect four defects, including cracks, pockmarks, spalling, and exposed rebar. They used the pre-trained inceptionV3 model to extract features from input images and a fully connected network to classify defects. The proposed model was trained on 1180 images with arbitrary sizes and resolutions for 374.1 s and recorded a testing accuracy of 97.8%.

On the other hand, Gao and Mosalam [32] proposed the concept of structural ImageNet and manually labeled 2000 images for four recognition tasks: component type identification (binary), spalling condition check (binary), damage level evaluation (three classes), and damage type determination (four classes). They applied two different strategies of Transfer Learning based on the pre-trained VGG16 model. For the damage type multi-classification task, a 68.8% accuracy with 23% overfitting was obtained by retraining the last two convolutional blocks of the network.

In the aforementioned works, the performance of the proposed methods has varied according to the size and complexity of the datasets and the adopted Transfer Learning approach. Most studies have re-trained more than two or all convolutional layers and update a high number of the network parameters to achieve a higher detection accuracy. However, this approach is computationally expensive, requires more training time, and is also subject to overfitting in the context of heavily parameterized networks and small datasets.

In a bridge condition assessment framework, defect localization is crucial to evaluate damage's impact on the bridge's structural integrity. For this purpose, deep learning based-semantic segmentation algorithms have been deployed to provide pixel-level classification results to improve damage detection accuracy.

Zhang et al. [42] designed a fully convolutional model to detect and group image pixels for three types of concrete surface defects (i.e., crack, spalling, and exposed rebar). The authors prepared a dataset with mask labeling of 1443 images to train and test the model. Their proposed method achieved a semantic segmentation accuracy of 75.5%.

Fu et al. [43] introduced a crack detection method based on an improved DeepLabv3+ semantic segmentation algorithm. They established a concrete bridge crack segmentation dataset to train and test the proposed model. The experimental results proved the effectiveness of the trained algorithm that reached an average intersection over union ratio of 82.37%.

Wang et al. [44] constructed a crack dataset of 2446 manually labeled images to train and evaluate the performance of five deep networks for semantic segmentation. The best model achieved an F1-score of 77.32% and an intersection over union ratio of 62.98%. The authors also discussed the influence of dataset choice and image noise on the detection performance.

Dung and Anh [45] developed a fully convolutional network-based method and annotated 600 crack-labeled images for semantic segmentation. The proposed model reached approximately 90% for the average precision score. The authors demonstrated their method's effectiveness by accurately identifying and capturing crack path and density variation in a crack opening video.

The above studies have shown very promising results in detecting damages. However, the fully supervised semantic segmentation deep networks are complex and are faced with a common major challenge associated with data scarcity. These models require training labeled images with pixel-level annotations that are expensive and necessitate the empirical knowledge of field experts. Furthermore, most publicly available concrete damage datasets only provide image-level annotations.

To alleviate the heavy workload associated with data annotation in a fully supervised learning framework, weakly supervised segmentation methods consider different weak annotations (e.g., image-level and bounding box labels) as the supervision condition [46]. Within the context of damage detection, Dong et al. [47] designed a patch-based weakly supervised semantic segmentation network to detect cracks in construction materials. In their proposed method, an input image is cropped, and the resulting patches are annotated at an image level. Class activation maps of cracks are obtained for each patch. They are fed to a fully connected conditional random field to generate the corresponding synthetic labels, which are used to train a segmentation network.

König et al. [48] presented a weakly supervised segmentation approach leveraging classification labels to detect surface cracks. To obtain pixel-level segmentation pseudo labels, the authors utilized a patch-threshold segmentation combined with coarse localization maps generated by a Convolutional Neural Network trained on images with classification annotations. The generated pseudo labels were used to train a standard semantic segmentation network to perform crack segmentation.

Zhu and Song [49] developed a weakly supervised network for crack segmentation in asphalt concrete bridge decks. Based on an autoencoder, the original data generates a weakly supervised start point for convergence, and image feature extraction and segmentation are performed under weak supervision.

This paper investigates a weakly supervised framework based on interpretation techniques and leveraging image-level annotations to generate pixel-level maps. The goal is to provide a coarse localization of three distinct types of damage in concrete bridge images.

The main contributions of this work are the following:

1. A multi-class labeled dataset with more than 6900 images was constructed. The dataset features three common types of defects in concrete bridges (i.e., cracks, spalling, and efflorescence) and covers their diverse representation in the real world of bridge inspection.
2. Three classification schemes using the pretrained Visual Geometry Group (VGG) network with its 16 learning layers (i.e., VGG16 [34]), Transfer Learning, and the proposed dataset were compared. The experiments investigated the effect of the number of layers to be retrained on the model's performance in terms of classification measures (i.e., accuracy, precision, recall, and F1 score), computational time, and generalization ability.
3. Based on the best classification scheme, the effectiveness of interpretable neural networks was explored in the context of weakly supervised semantic segmentation (i.e., image-level supervision). Two gradient-based backpropagation interpretation techniques (i.e., Gradient-weighted Class Activation Mapping (Grad-CAM) [50] and Grad-CAM++ [51]) were used to generate pixel-level heatmaps and localize defects. Qualitative results of test images showcase the potential of interpretation heatmaps to provide localization information in a weak supervision framework.

The rest of the paper is organized as follows: Section 2 presents an overview of the methodology followed in this paper, the proposed dataset, the VGG16 model, and the interpretation techniques studied in this work. Section 3 details the experimental setup, and the experimental results are presented and discussed in Section 4. Conclusions are provided in the final section of the paper.

2. Methodology and Materials

An overview of the adopted methodology in this work is shown in the flowchart in Figure 1. The first module corresponds to dataset construction's image acquisition and preparation process. The second module represents the implementation of the pre-trained VGG16 network using Transfer Learning to classify three types of defects in concrete bridges. Finally, interpretation techniques were deployed in the third module to generate pixel-level heatmaps to localize concrete damage.

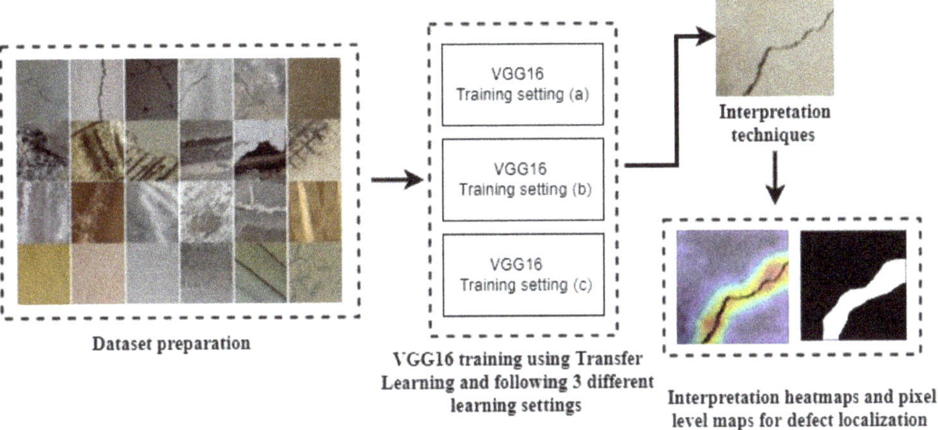

Figure 1. Overview of the proposed method.

2.1. Dataset

The dataset constructed in this paper consisted of 6952 RGB images with a 200×200 px resolution of concrete cracks (1304), concrete spalling (1100), concrete efflorescence (1029), and non-defective background (3519). Cracks and background images were extracted from the dataset established by the authors of [52].

More than 1200 images of Moroccan bridges representing decks and piers with concrete spalling and efflorescence were collected and processed according to the same experimental setup and procedure in [52]. The images were captured using two 20-MP consumer digital cameras with 5 mm of focal length, a sensitivity of 100 ISO, and a maximum resolution of 5152×3864. They were gathered at varying distances from bridges and a maximum $8\times$ optical zoom was applied. Moreover, the images were taken under different weather and lighting conditions, and a flash was used to illuminate the dark bridge areas containing defects. It is noteworthy that the original images have not undergone any processing operations other than the manual cropping using the inbac tool [53].

The dataset in [52] was expanded with the concrete spalling and efflorescence classes, and the resulting dataset is publicly available at [54] for academic purposes.

Various colors, textures, surface conditions of concrete and defect representations were included in the constructed dataset to cover the variation of defect appearance, extent, and severity level in the real world of bridge inspection. Figure 2 presents sample images of the proposed dataset.

Figure 2. Sample images of the constructed dataset (row 1: concrete cracks, row 2: concrete spalling with exposed reinforcement, row 3: concrete efflorescence, row 4: different representations of the non-defective background class).

2.2. VGG16 and Transfer Learning

The Visual Geometry Group introduced VGG16 in 2014 [34]. The algorithm is very efficient and won first place in object localization and second place in image classification in the ImageNet Large Scale Visual Recognition Challenge. This model trained on the ImageNet dataset achieved a top-1 accuracy of 71.5% and a top-5 accuracy of 90.1% in image classification.

The network contains 13 convolutional layers with 3×3 filters (i.e., convolution kernels) to extract features. In addition, the network contains five max-pooling layers to reduce the number of learnable parameters (i.e., weights and biases) and three fully connected layers to map the flattened features to the Softmax layer where target class probabilities are calculated. In addition, the Rectified Linear Unit (ReLU) activation function is used to increase the non-linearity of the model. The network takes 224×224 RGB images as inputs and has more than 138 million learning parameters. Figure 3 presents the architecture of the VGG16 model.

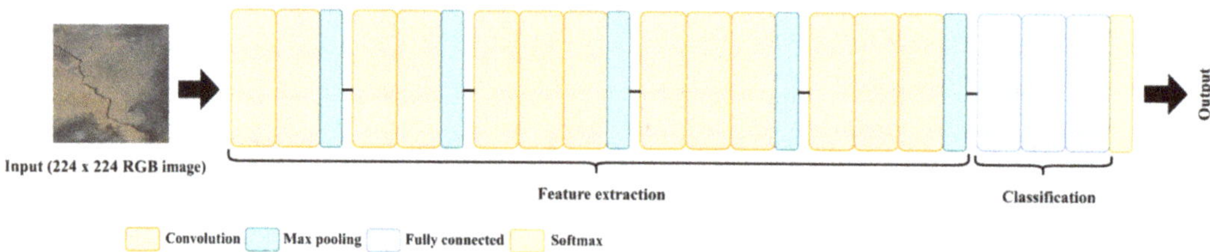

Figure 3. Architecture of the VGG16 model.

The learning layout of the VGG16 network, and DCNNs in general, is based on optimizing a loss function (e.g., Binary and Multi-Class Cross-Entropy loss) that measures the discrepancy between the predicted outputs and ground truth through back-propagation.

The optimization scheme generally uses gradient descent optimizers (e.g., Stochastic Gradient Descent and adaptive optimizers) to update the learning parameters of the network.

For image classification, VGG16 and other state-of-the-art DCNNs are usually trained on the ImageNet dataset that contains millions of images belonging to thousands of classes. However, since the size of domain-specific datasets (e.g., concrete defects datasets in the case of this study) is limited, Transfer Learning techniques are applied to overcome the scarcity of labeled data.

In a Transfer Learning approach, pre-trained models on large datasets (e.g., ImageNet) are fine-tuned and partially retrained on the small target dataset. In this learning framework, the weights of the lower-level layers are generally maintained since they represent generic features. In contrast, the high-level layers are more sensitive to the target dataset and must be retrained to update their learning parameters [23]. The Transfer Learning settings examined in this paper are detailed in the experimental setup section.

2.3. Interpretation Techniques

In the context of image classification, interpretation techniques are intended to explain the predictions of trained models by visualizing the regions of the inputs that contributed to the final classification result. Thus, they can provide a coarse localization of target objects using image-level annotations.

Hereafter, a simplified explanation of the intuition behind the two gradient-based back-propagation techniques used in this paper (i.e., Grad-CAM and Grad-CAM++) is presented.

2.3.1. Gradient-Weighted Class Activation Mapping (Grad-CAM)

The Grad-CAM approach is based on the gradient information for the last convolutional layer of a trained network [50].

The gradients of the score for class c (y^c) with respect to the feature maps A^k of the convolutional layer are computed via back-propagation and then global-average pooled to obtain the weights w_c^k [50]:

$$w_c^k = \frac{1}{Z} \sum_i \sum_j \frac{\partial y^c}{\partial A_{ij}^k} \tag{1}$$

where, Z is the number of pixels in the activation map.

The weight w_c^k expresses the importance of feature map k for the class c. The class discriminative localization map Grad-CAM $L_{Grad\text{-}CAM}^c$ is obtained by computing a weighted sum of the forward feature maps A^k of the last convolutional layer [50]:

$$L_{Grad\text{-}CAM}^c = ReLU\left(\sum_k w_k^c A^k\right) \tag{2}$$

where $ReLU$ is the Rectified Linear Unit activation function. It is used to focus only on the features that have a positive influence on the target class [50].

2.3.2. Grad-CAM++

Grad-CAM++ is a generalization to Grad-CAM and provides better visualizations of the network decisions [48]. In Grad-CAM++, the weights w_c^k are computed as follows [51]:

$$w_c^k = \sum_i \sum_j \alpha_{ij}^{kc} . ReLU\left(\frac{\partial y^c}{\partial A_{ij}^k}\right) \tag{3}$$

where α_{ij}^{kc} are weighting coefficients for the pixel-wise gradients for class c and feature map A^k and are defined as follows:

$$\alpha_{ij}^{kc} = \frac{\dfrac{\partial^2 y^c}{\left(\partial A_{ij}^k\right)^2}}{2\dfrac{\partial^2 y^c}{\left(\partial A_{ij}^k\right)^2} + \sum_a \sum_b A_{ab}^k \left\{\dfrac{\partial^3 y^c}{\left(\partial A_{ij}^k\right)^3}\right\}} \tag{4}$$

3. Experimental Setup

This work aims to define an efficient and automatable method to identify and localize damage in concrete bridge images using DCNNs and Transfer Learning. This section presents the Transfer Learning schemes followed to train the pre-trained VGG16 model on the proposed dataset. In addition, the weakly supervised semantic segmentation framework based on the above-explained interpretation techniques is also discussed.

3.1. VGG16 Fine-Tuning and Training

The VGG16 model can capture high-level features [21] and has the ability to generalize to other datasets [32]. Moreover, it has shown an excellent performance in many studies on damage classification in concrete surfaces [31–33]. Therefore, it was chosen as a base model for the learning approach proposed in this paper.

Training this deep network from scratch requires enormous computational resources, significantly labeled data, and excessive training time. Thus, three Transfer Learning settings were explored in this work and compared based on standard classification metrics, training time, and generalization ability.

First, the pre-trained VGG16 with the ImageNet weights was uploaded, and the last fully connected layers of the model were adjusted to the number of classes (i.e., four classes). Then, based on the assumption that the high-level layers of DCNNs are more sensitive to the target dataset, the last layers of the pre-trained model were retrained on the constructed dataset to update their learning parameters.

Gradient descent and back-propagation were used following three different approaches in this work:

- Retraining the classification layers (a)
- Retraining the classification layers and the last convolutional layer (b)
- Retraining the classification layers and the last two convolutional layers (c)

Figure 4 presents the three Transfer Learning-based training settings investigated in this paper.

The dataset presented in Section 2 was randomly split into three subsets: 70% of the images were used in the training set, 10% in the validation set, and 20% in the testing set. The number of images per subset and per class is shown in Table 1.

Table 1. Number of images per subset per class.

	Background	Cracks	Efflorescence	Spalling
Training set	2463	912	720	770
Validation set	351	130	102	110
Testing set	705	262	207	220

In addition, data augmentation techniques (i.e., random horizontal and vertical flips and random rotations) were applied to the training set to avoid overfitting.

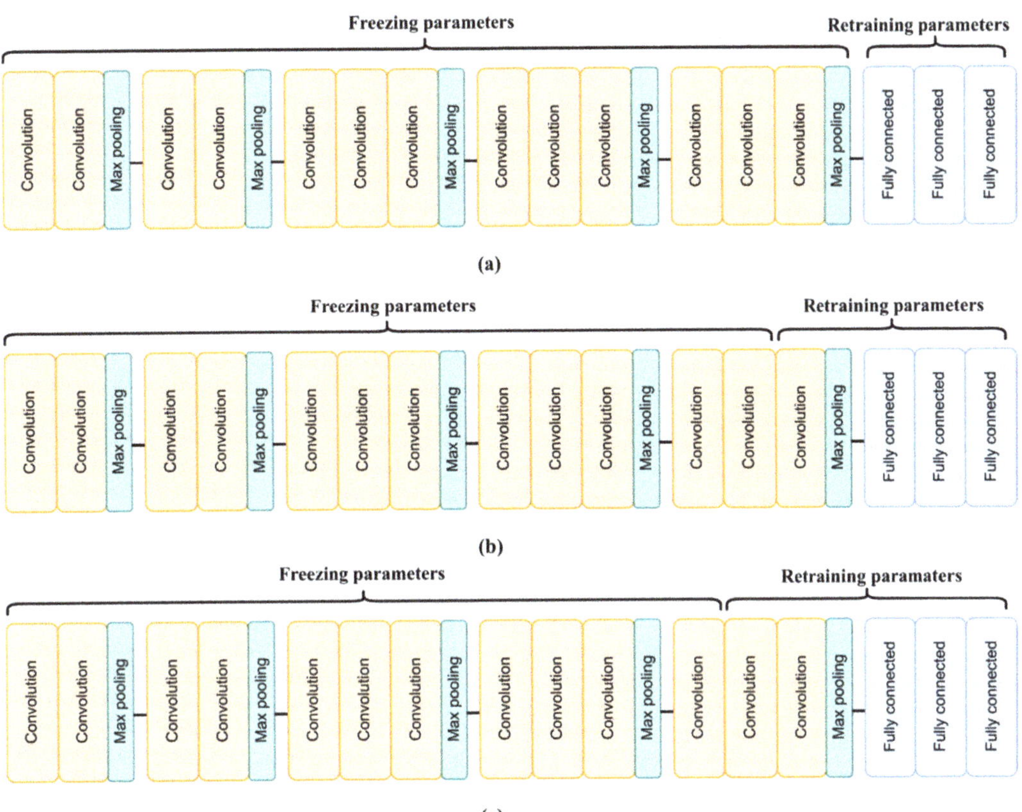

Figure 4. Transfer Learning configurations: (**a**) retraining only the classification layers, (**b**) retraining the classification layers and the last convolutional layers, (**c**) retraining the classification layers and the last two convolutional layers.

The optimization method recommended in [32] based on Stochastic Gradient Descent (SGD) with momentum and a small learning rate was used in this work's experiments. The SGD with momentum optimizer reduces the computational load and accelerates the training convergence. The training was conducted for 25 epochs as the results converged. In addition, low training and validation errors were achieved while mitigating overfitting. The cross-entropy loss function was optimized using the SGD with a learning rate of 0.001, a momentum of 0.9, and a mini-batch size of 32. All the experiments were carried out using Pytorch in Google Colaboratory (Colab) with the 12GB NVIDIA Tesla K80 GPU provided by the platform.

3.2. Evaluation Metrics

In each learning configuration, the performance of the model was evaluated using the following metrics:

$$\text{Accuracy} = \frac{\text{TP} + \text{TN}}{\text{TP} + \text{TN} + \text{FP} + \text{FN}} \tag{5}$$

$$\text{Precision} = \frac{\text{TP}}{\text{TP} + \text{FP}} \tag{6}$$

$$\text{Recall} = \frac{\text{TP}}{\text{TP} + \text{FN}} \tag{7}$$

$$F1_{Score} = 2\left(\frac{1}{Recall} + \frac{1}{Precision}\right)^{-1} \tag{8}$$

TP (True Positives) refer to the number of correctly classified images as defects.

TN (True Negatives) refer to the number of background images that are correctly classified as background.

FP (False Positives) refer to the number of background images that are incorrectly identified with defects.

FN (False Negatives) refer to the number of images incorrectly identified as background images.

The Root Mean Squared Error (RMSE) was also used to assess the model's performance in the three different training schemes. It is defined by Equation (9):

$$RMSE = \sqrt{\sum_i (1 - y_i)^2 / n} \tag{9}$$

where y_i is the calculated probability of the image i (from the testing subset) belonging to the ground truth class.

3.3. Weakly Supervised Semantic Segmentation

Based on the best learning scheme, feature maps of the last convolutional layer of the trained model were used to provide visual explanations of classification results using Grad-CAM and Grad-CAM++. The implementation of these interpretation techniques was based on the publicly available repository in [55]. Pixel-level heatmaps were generated for test images, and a threshold of 0.5 was applied to each image to localize the regions with a target class probability above 50%.

4. Results and Discussions

This section presents and discusses the results obtained after training and testing the model following the three Transfer Learning schemes on the constructed dataset. In addition, some representative localization maps of cracks, spalling, and efflorescence generated by the two interpretation techniques are also presented.

4.1. Training and Testing Results

Figure 5 plots the model's learning and loss curves following the three Transfer Learning settings. It can be seen that in all three training approaches, the model converged quickly from the early epochs; this is mainly attributed to the reduced number of layers to retrain and the fixed parameters of the non-trainable layers.

The training and validation accuracies in the three learning settings generally increase over time, reaching a plateau around the last three epochs. The loss curves in (b) and (c) show a slight tendency to over-fitting due to the increasing number of parameters to update in their corresponding learning schemes. Therefore, it is believed that training more than two convolutional layers will lead to a higher possibility of overfitting and, consequently, will decrease the model's generalization ability.

Table 2 lists the three learning schemes' best training, validation, testing accuracies, and RMSE. The results show that the model in scheme (c) presents a better performance in training compared to settings (a) and (b). The achieved training accuracy was 98.34% in (c) over 94.62% and 91.10% in (b) and (a), respectively. The model in setting (c) also yielded a higher testing accuracy of 97.13% over 94.61% and 91.10% in (b) and (a), respectively. Furthermore, the results were obtained with only 1.21% overfitting (calculated as the difference between training and testing accuracies). These exciting results show that approach (c) enables the model to have better generalizability to extend the learning from the training subset to unseen test data. These results also demonstrate that more important features representing the target dataset were learned in the last two convolutional layers of the pre-trained model.

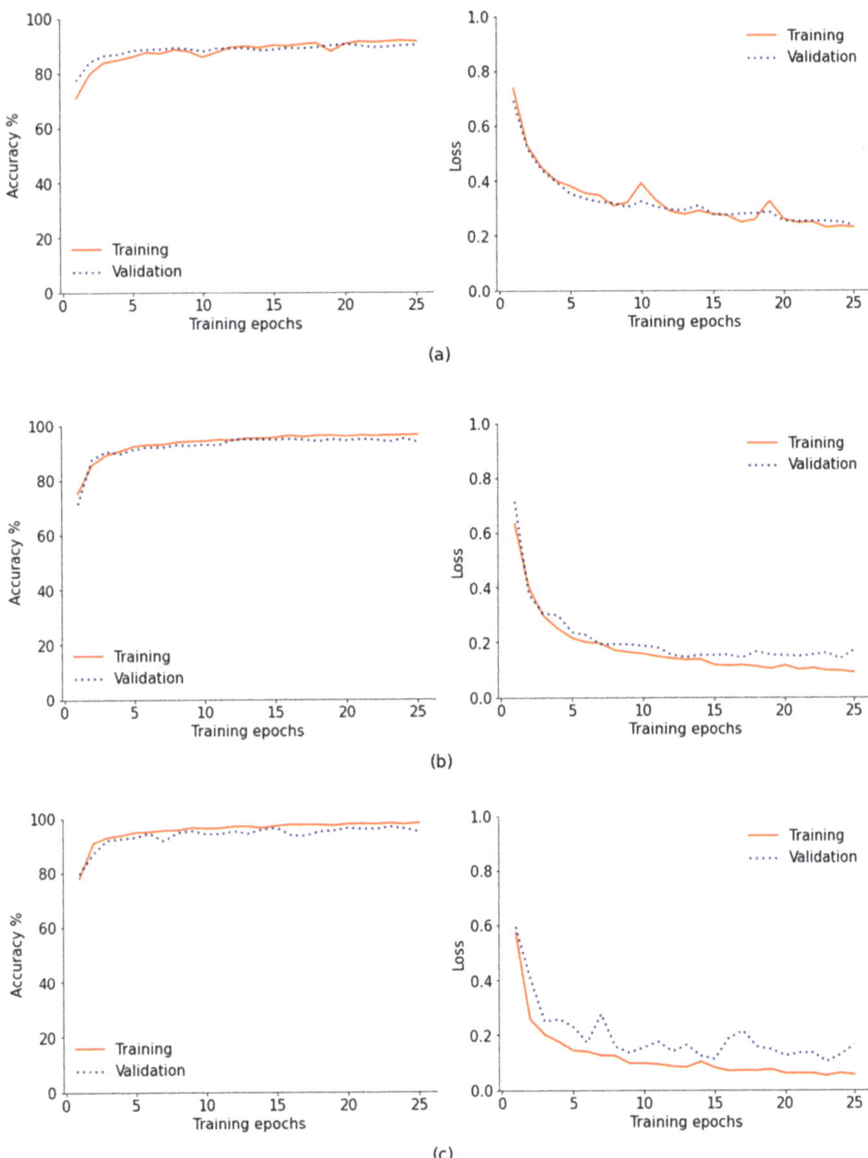

Figure 5. Learning and loss curves: (**a**) retraining only the classification layers, (**b**) retraining the classification layers and the last convolutional layers, (**c**) retraining the classification layers and the last two convolutional layers.

In addition, the Root Mean Squared Error decreased with more retrained layers (0.15 in (c) over 0.20 and 0.27 in (b) and (a), respectively), denoting that the predictive capacity of the model improves with updating more learning parameters.

It is also essential to mention that the training time corresponding to the three classification methods is reasonable. Moreover, the difference between the three approaches in terms of training time is significantly low (e.g., 8 s between (b) and (c)). At the same time, a considerable gain in performance was observed (e.g., a 1.61% gain in training accuracy between (b) and (c)). Therefore, fine-tuning the classification layers and the last two convolution layers of the pre-trained VGG16 is efficient as it balances prediction performance, training time, and overfitting.

Table 2. Best training, validation, and testing accuracies and training times of the three training settings.

Transfer Learning Scheme	Best Training Accuracy	Best Validation Accuracy	Testing Accuracy	RMSE	Training Time
Retraining only the classification layers (a)	91.61%	90.62%	91.10%	0.27	6 min 35 s
Retraining the classification layers and the last convolutional layers (b)	96.73%	94.80%	94.62%	0.20	6 min 47 s
Retraining the classification layers and the last two convolutional layers (c)	98.34%	96.68%	97.13%	0.15	6 min 55 s

To further visualize the performance of the trained DCNN on the test subset, normalized confusion matrices for the three learning schemes are presented in Figure 6.

The results reflect confusion between background and efflorescence images and background and crack pictures. This confusion was particularly observed in schemes (a) and (b). For example, 7% and 6% of background images were classified as cracks or efflorescence in methods (a) and (b), respectively. However, this confusion was notably reduced in the scheme (c) as less than 3% of background images were predicted as cracks or efflorescence.

The observed confusion is mainly related to the complexity of the concrete surface in terms of colors and textures. In addition, some surface alterations in the training dataset (e.g., stains, markings, minor defects such as scaling and segregation) represent noisy defect-like features in concrete images and, as a result, make feature learning more challenging. For example, some background images contain concrete joints representing straight lines in the concrete surface and hence are likely to be misclassified as cracks. Generally, this confusion between classes can be further handled by adding more labeled samples to the training dataset and integrating an additional denoising process into image data.

Figure 7 illustrates some misclassification examples corresponding to the learning setting used in (c).

The precision, recall, and F1-scores of each defect class were computed using the confusion matrices. The results are summarized in Table 3.

The model achieved higher precision, recall, and F1-score results in learning scheme (c) compared to the other learning settings. For example, 86.64%, 94.03%, and 97.38% are the cracks F1-scores achieved in the learning settings (a), (b), and (c), respectively.

By comparing the F1-scores of the three classes in the learning scheme (c), the efflorescence class yielded a lower score than the other defects, which showed nearly similar performance (95.01% for efflorescence over 97.38% and 97.35% for cracks and spalling, respectively). This can be attributed to the wide representations of the efflorescence in concrete and the complexity of features required to describe this defect class. This issue can be solved by adding more training data that extensively covers the diverse representations of this type of damage.

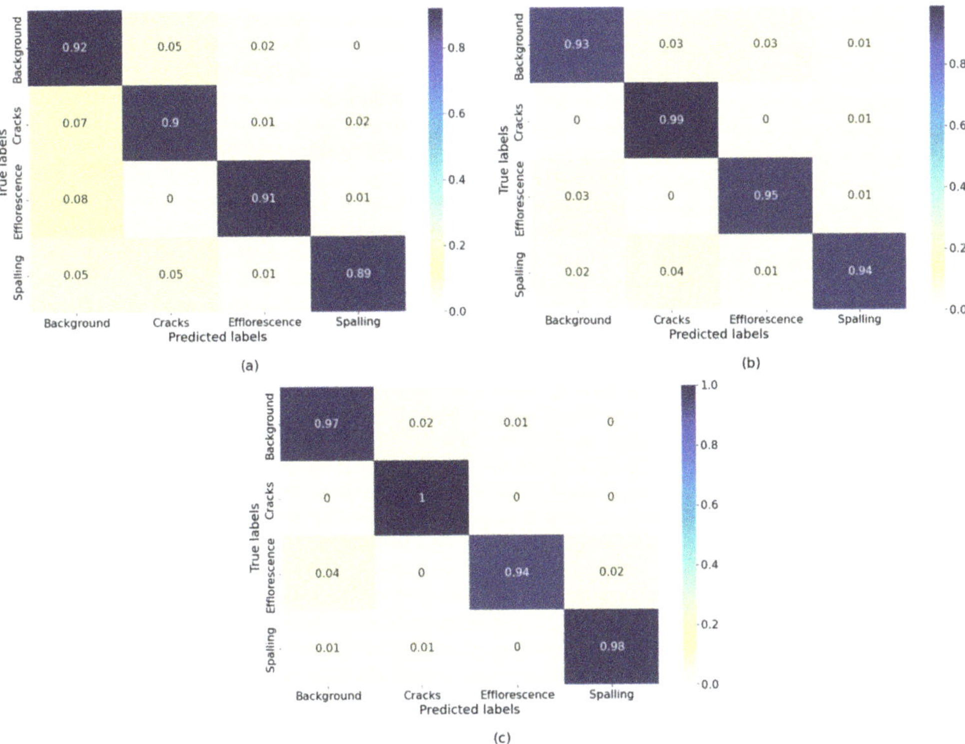

Figure 6. Confusion matrices of the three learning settings: (**a**) retraining only the classification layers, (**b**) retraining the classification layers and the last convolutional layers, (**c**) retraining the classification layers and the last two convolutional layers.

Figure 7. Misclassification examples (row 1: background images containing concrete joints misclassified as cracks, row 2: background images with surface alteration and different concrete colors misclassified as efflorescence).

Table 3. Precision, recall, and F1-scores of the three models.

	Damage Type	Learning Scheme (a)	Learning Scheme (b)	Learning Scheme (c)
Precision	Cracks	83.04%	89.66%	94.89%
	Efflorescence	89.62%	89.29%	95.69%
	Spalling	96.92%	95.89%	96.92%
Recall	Cracks	90.57%	98.86%	100%
	Efflorescence	90.91%	95.24%	94.34%
	Spalling	88.89%	93.75%	97.78%
F1-Score	Cracks	86.64%	94.03%	97.38%
	Efflorescence	90.26%	92.17%	95.01%
	Spalling	92.59%	94.81%	97.35%

4.2. Defect Localization Results

The trained model using learning scheme (c) was employed to implement the interpretation techniques presented in Section 2.

Images of the testing subset were used to visualize the implementation results. Figure 8 shows sample examples of the obtained results.

As intended, the resulting heatmaps highlight the discriminative image regions that contributed to image classification. These heatmaps show the probability of the target class at each pixel. By analyzing the qualitative results in Figure 8, the active regions are primarily consistent with the defect area. Grad-CAM++ provided better visualization results for cracks and efflorescence examples compared to Grad-CAM.

The pixel-level maps generated after applying a threshold of 0.5 provide a coarse localization of the concrete defects and offer semantically meaningful discrimination at the pixel level between defects and background. Therefore, it is believed that in the context of weakly supervised semantic segmentation, interpretation methods can provide relevant pixel-level maps using only image annotations as the supervision condition. The proposed method has reasonably captured a coarse localization of defects while avoiding the annotation workload of the fully supervised semantic segmentation-based frameworks.

However, since the visualization results using these interpretation techniques depend on the feature space learned by the classifier, some highlighted areas do not represent the target classes in the test images, and other regions representing damage were not captured.

As a result, it would be challenging to localize and quantify the damage precisely (e.g., crack path and density). This can be attributed to the underlying complexity of the training dataset, its limited size, and the limited learning capabilities of the pre-trained network due to the difference between the source domain (ImageNet dataset) and the target domain (the proposed concrete damage dataset). Thus, to further examine the potential of interpretation techniques in weakly supervised semantic segmentation, more customized networks tailored to the damage classification task and trained on more comprehensive datasets should be explored.

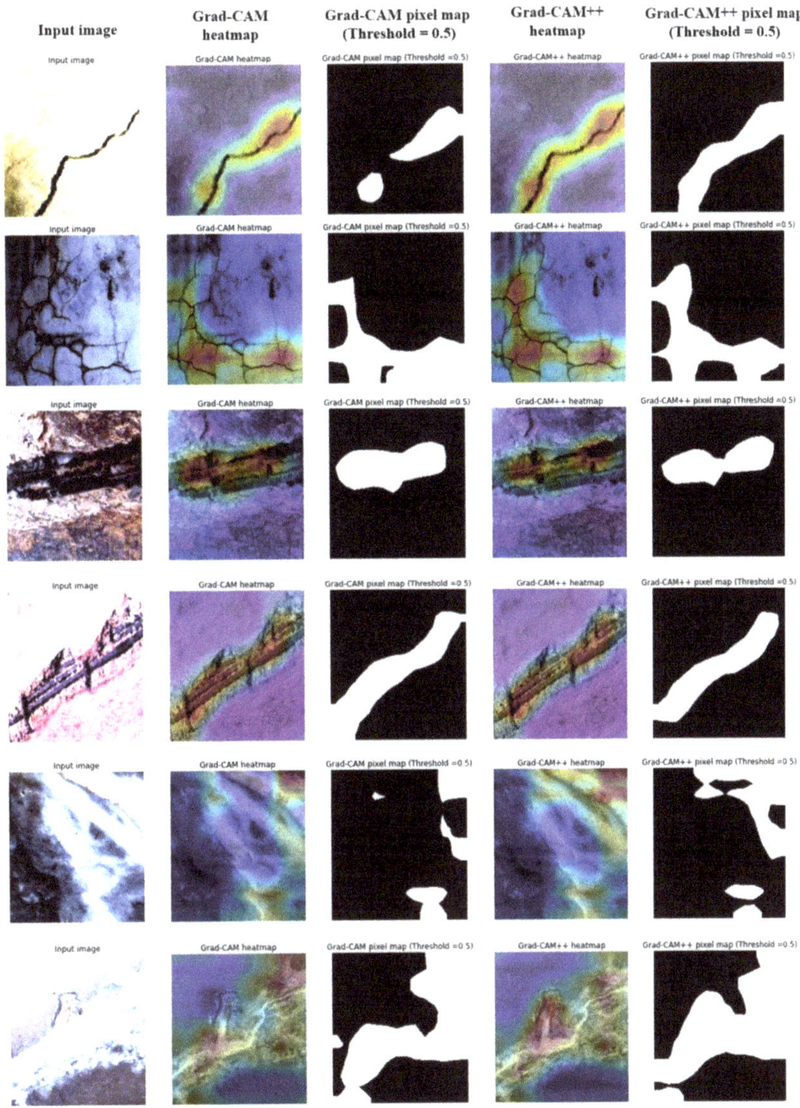

Figure 8. Sample results of the interpretation techniques implementation (rows 1–2: cracks, rows 3–4: concrete spalling, rows 5–6: efflorescence).

5. Conclusions and Perspectives

Structural Health Monitoring (SHM) is gaining increasing importance in assessing bridge conditions as it allows for identifying, localizing, and evaluating damage severity. Therefore, SHM is part of an economic strategy since it intervenes in the definition of maintenance actions and participates in the optimization of its allocated resources.

This paper presented a benchmark image dataset featuring three common defects (i.e., cracks, efflorescence, and spalling) of concrete bridges. The dataset covers different appearances of the three defects and the concrete surface in the real world of bridge inspection. A VGG16 network was trained on the proposed dataset following three Transfer Learning schemes with varying layers. In each learning configuration, the performance of the model was evaluated based on classification metrics, computational time, and generalization ability. Experiments showed a significant gain in classification measures when retraining the classification layers and the last two convolutional layers of the VGG16 network. The trained model yielded a high testing accuracy of 97.13%, combined with high F1-scores of 97.38%, 95.01%, and 97.35% for cracks, efflorescence, and spalling, respectively. In addition, a slight tendency to overfitting was observed in the corresponding learning scheme, which means that increasing the number of layers to be retrained will lead to the degradation of the model's generalization performance. These experimental results show the robustness of the proposed learning setting as it ensures a balance between classification metrics, computational time, and generalization ability.

This work also explored the potential of interpretation techniques to localize the three defects in the context of weakly supervised semantic segmentation. To this end, two gradient-based back-propagation methods were used to generate pixel-level heatmaps of test images leveraging the above-discussed learning setting. The resulting maps highlight the regions contributing to the classification result and then provide relevant pixel-level maps to localize defects using a model trained on image-level annotations. However, since these techniques rely on the feature space learned by the model, their results are limited by the representativity of target classes in the training dataset, the challenging complexity of the concrete surface texture and condition in bridge inspection images, and the learning capability of the model.

Therefore, in another attempt to solve the damage localization task, more advanced object detection models and datasets with different annotation levels will be investigated in future works.

Author Contributions: Conceptualization, H.Z. and M.R.; methodology, H.Z., M.R. and M.E.A.; software, H.Z.; validation, H.Z., M.R., M.E.A., A.C. and R.S.; formal analysis, H.Z., M.R. and M.E.A.; investigation, H.Z. and M.R.; resources, H.Z.; data curation, H.Z.; writing—original draft preparation, H.Z., M.R., M.E.A., A.C. and R.S.; writing—review and editing, A.C. and G.J.; visualization, H.Z., M.R., M.E.A., A.C. and R.S.; supervision, M.R.; project administration, M.R.; funding acquisition, A.C. and G.J. All authors have read and agreed to the published version of the manuscript.

Funding: The financial support from the NSERC Discovery Grant program.

Data Availability Statement: Not applicable.

Conflicts of Interest: The authors declare no conflict of interest.

References

1. Ma, Y.; Guo, Z.; Wang, L.; Zhang, J. Probabilistic Life Prediction for Reinforced Concrete Structures Subjected to Seasonal Corrosion-Fatigue Damage. *J. Struct. Eng.* **2020**, *146*, 04020117. [CrossRef]
2. Wang, L.; Dai, L.; Bian, H.; Ma, Y.; Zhang, J. Concrete cracking prediction under combined prestress and strand corrosion. *Struct. Infrastruct. Eng.* **2019**, *15*, 285–295. [CrossRef]
3. Pourzeynali, S.; Zhu, X.; Ghari Zadeh, A.; Rashidi, M.; Samali, B. Comprehensive Study of Moving Load Identification on Bridge Structures Using the Explicit Form of Newmark-β Method: Numerical and Experimental Studies. *Remote Sens.* **2021**, *13*, 2291. [CrossRef]
4. Kot, P.; Muradov, M.; Gkantou, M.; Kamaris, G.S.; Hashim, K.; Yeboah, D. Recent Advancements in Non-Destructive Testing Techniques for Structural Health Monitoring. *Appl. Sci.* **2021**, *11*, 2750. [CrossRef]
5. Zollini, S.; Alicandro, M.; Dominici, D.; Quaresima, R.; Giallonardo, M. UAV Photogrammetry for Concrete Bridge Inspection Using Object-Based Image Analysis (OBIA). *Remote Sens.* **2020**, *12*, 3180. [CrossRef]
6. Galdelli, A.; D'Imperio, M.; Marchello, G.; Mancini, A.; Scaccia, M.; Sasso, M.; Frontoni, E.; Cannella, F. A Novel Remote Visual Inspection System for Bridge Predictive Maintenance. *Remote Sens.* **2022**, *14*, 2248. [CrossRef]
7. Omar, T.; Nehdi, M. Condition Assessment of Reinforced Concrete Bridges: Current Practice and Research Challenges. *Infrastructures* **2018**, *3*, 36. [CrossRef]
8. Dorafshan, S.; Maguire, M. Bridge inspection: Human performance, unmanned aerial systems and automation. *J. Civ. Struct. Health Monit.* **2018**, *8*, 443–476. [CrossRef]
9. Alsharqawi, M.; Zayed, T.; Abu Dabous, S. Integrated condition rating and forecasting method for bridge decks using Visual Inspection and Ground Penetrating Radar. *Autom. Constr.* **2018**, *89*, 135–145. [CrossRef]
10. Seo, J. Drone-enabled bridge inspection methodology and application. *Autom. Constr.* **2018**, *94*, 112–126. [CrossRef]

11. Mohan, A.; Poobal, S. Crack detection using image processing: A critical review and analysis. *Alex. Eng. J.* **2018**, *57*, 787–798. [CrossRef]

12. Hsieh, Y.-A.; Tsai, Y.J. Machine Learning for Crack Detection: Review and Model Performance Comparison. *J. Comput. Civ. Eng.* **2020**, *34*, 04020038. [CrossRef]

13. Sony, S.; Dunphy, K.; Sadhu, A.; Capretz, M. A systematic review of convolutional neural network-based structural condition assessment techniques. *Eng. Struct.* **2021**, *226*, 111347. [CrossRef]

14. Abdel-Qader, I.; Abudayyeh, O.; Kelly, M.E. Analysis of Edge-Detection Techniques for Crack Identification in Bridges. *J. Comput. Civ. Eng.* **2003**, *17*, 255–263. [CrossRef]

15. Talab, A.M.A.; Huang, Z.; Xi, F.; HaiMing, L. Detection crack in image using Otsu method and multiple filtering in image processing techniques. *Optik* **2016**, *127*, 1030–1033. [CrossRef]

16. Fast Crack Detection Method for Large-Size Concrete Surface Images Using Percolation-Based Image Processing | SpringerLink. Available online: https://link.springer.com/article/10.1007/s00138-009-0189-8 (accessed on 4 March 2022).

17. Yamaguchi, T.; Nakamura, S.; Saegusa, R.; Hashimoto, S. Image-Based Crack Detection for Real Concrete Surfaces. *IEEJ Trans. Electr. Electron. Eng.* **2008**, *3*, 128–135. [CrossRef]

18. Abdel-Qader, I.; Pashaie-Rad, S.; Abudayyeh, O.; Yehia, S. PCA-Based algorithm for unsupervised bridge crack detection. *Adv. Eng. Softw.* **2006**, *37*, 771–778. [CrossRef]

19. Prasanna, P.; Dana, K.J.; Gucunski, N.; Basily, B.B.; La, H.M.; Lim, R.S.; Parvardeh, H. Automated Crack Detection on Concrete Bridges. *IEEE Trans. Automat. Sci. Eng.* **2016**, *13*, 591–599. [CrossRef]

20. Jahanshahi, M.R.; Masri, S.F. A Novel Crack Detection Approach for Condition Assessment of Structures. In Proceedings of the International Workshop on Computing in Civil Engineering 2011, Miami, FL, USA, 19–22 June 2011; pp. 388–395. [CrossRef]

21. Yang, L.; Li, B.; Li, W.; Liu, Z.; Yang, G.; Xiao, J. Deep Concrete Inspection Using Unmanned Aerial Vehicle Towards CSSC Database. In Proceedings of the 2017 IEEE/RSJ International Conference on Intelligent Robots and Systems, Vancouver, BC, Canada, 24–28 September 2017; p. 9.

22. da Silva, W.R.L.; de Lucena, D.S. Concrete Cracks Detection Based on Deep Learning Image Classification. *Proceedings* **2018**, *2*, 489. [CrossRef]

23. Dorafshan, S.; Thomas, R.J.; Maguire, M. Comparison of deep convolutional neural networks and edge detectors for image-based crack detection in concrete. *Constr. Build. Mater.* **2018**, *186*, 1031–1045. [CrossRef]

24. Krizhevsky, A.; Sutskever, I.; Hinton, G.E. ImageNet Classification with Deep Convolutional Neural Networks. *Commun. ACM* **2017**, *60*, 84–90. [CrossRef]

25. Dorafshan, S.; Thomas, R.J.; Maguire, M. SDNET2018: An annotated image dataset for non-contact concrete crack detection using deep convolutional neural networks. *Data Brief* **2018**, *21*, 1664–1668. [CrossRef] [PubMed]

26. Kim, B.; Yuvaraj, N.; Sri Preethaa, K.R.; Arun Pandian, R. Surface crack detection using deep learning with shallow CNN architecture for enhanced computation. *Neural Comput. Appl.* **2021**, *33*, 9289–9305. [CrossRef]

27. Lecun, Y. Gradient-Based Learning Applied to Document Recognition. *Proc. IEEE* **1998**, *86*, 47. [CrossRef]

28. Yu, Y.; Rashidi, M.; Samali, B.; Mohammadi, M.; Nguyen, T.N.; Zhou, X. Crack detection of concrete structures using deep convolutional neural networks optimized by enhanced chicken swarm algorithm. *Struct. Health Monit.* **2022**, *21*, 14759217211053546. [CrossRef]

29. Mundt, M.; Majumder, S.; Murali, S.; Panetsos, P.; Ramesh, V. Meta-Learning Convolutional Neural Architectures for Multi-Target Concrete Defect Classification with the Concrete Defect Bridge Image Dataset. In Proceedings of the 2019 IEEE/CVF Conference on Computer Vision and Pattern Recognition (CVPR), Long Beach, CA, USA, 16–20 June 2019; pp. 11188–11197. [CrossRef]

30. Su, C.; Wang, W. Concrete Cracks Detection Using Convolutional Neural Network Based on Transfer Learning. *Math. Probl. Eng.* **2020**, *2020*, 7240129. [CrossRef]

31. Bukhsh, Z.A.; Jansen, N.; Saeed, A. Damage detection using in-domain and cross-domain transfer learning. *Neural Comput. Appl.* **2021**, *33*, 16921–16936. [CrossRef]

32. Gao, Y.; Mosalam, K.M. Deep Transfer Learning for Image-Based Structural Damage Recognition: Deep transfer learning for image-based structural damage recognition. *Comput.-Aided Civ. Infrastruct. Eng.* **2018**, *33*, 748–768. [CrossRef]

33. Yang, Q.; Shi, W.; Chen, J.; Lin, W. Deep convolution neural network-based transfer learning method for civil infrastructure crack detection. *Autom. Constr.* **2020**, *116*, 103199. [CrossRef]

34. Simonyan, K.; Zisserman, A. Very Deep Convolutional Networks for Large-Scale Image Recognition. *arXiv* **2015**, arXiv:1409.1556. Available online: http://arxiv.org/abs/1409.1556 (accessed on 16 April 2021).

35. He, K.; Zhang, X.; Ren, S.; Sun, J. Deep Residual Learning for Image Recognition. *arXiv* **2015**, arXiv:1512.03385.

36. Szegedy, C.; Vanhoucke, V.; Ioffe, S.; Shlens, J.; Wojna, Z. Rethinking the Inception Architecture for Computer Vision. *arXiv* **2015**, arXiv:1512.00567.

37. Deng, J.; Dong, W.; Socher, R.; Li, L.-J.; Li, K.; Li, F.-F. ImageNet: A large-scale hierarchical image database. In Proceedings of the 2009 IEEE Conference on Computer Vision and Pattern Recognition, Miami, FL, USA, 20–25 June 2009; pp. 248–255. [CrossRef]

38. Krizhevsky, A. Learning Multiple Layers of Features from Tiny Images. 2009. Available online: https://www.cs.toronto.edu/~||kriz/learning-features-2009-TR.pdf (accessed on 30 May 2022).

39. Qi, Z.; Duan, K.; Xi, D.; Zhu, Y.; Zhu, H.; Xiong, H.; He, Q. A Comprehensive Survey on Transfer Learning. *Proc. IEEE* **2021**, *109*, 43–76. [CrossRef]

40. Hüthwohl, P.; Lu, R.; Brilakis, I. Multi-classifier for reinforced concrete bridge defects. *Autom. Constr.* **2019**, *105*, 102824. [CrossRef]
41. Zhu, J.; Zhang, C.; Qi, H.; Lu, Z. Vision-based defects detection for bridges using transfer learning and convolutional neural networks. *Struct. Infrastruct. Eng.* **2020**, *16*, 1037–1049. [CrossRef]
42. Zhang, C.; Chang, C.; Jamshidi, M. Simultaneous pixel-level concrete defect detection and grouping using a fully convolutional model. *Struct. Health Monit.* **2021**, *20*, 2199–2215. [CrossRef]
43. Fu, H.; Meng, D.; Li, W.; Wang, Y. Bridge Crack Semantic Segmentation Based on Improved Deeplabv3+. *J. Mar. Sci. Eng.* **2021**, *9*, 671. [CrossRef]
44. Wang, J.-J.; Liu, Y.-F.; Nie, X.; Mo, Y.L. Deep convolutional neural networks for semantic segmentation of cracks. *Struct. Control Health Monit.* **2022**, *29*, e2850. [CrossRef]
45. Dung, C.V.; Anh, L.D. Autonomous concrete crack detection using deep fully convolutional neural network. *Autom. Constr.* **2019**, *99*, 52–58. [CrossRef]
46. Zhang, M.; Zhou, Y.; Zhao, J.; Man, Y.; Liu, B.; Yao, R. A survey of semi- and weakly supervised semantic segmentation of images. *Artif. Intell. Rev.* **2020**, *53*, 4259–4288. [CrossRef]
47. Dong, Z.; Wang, J.; Cui, B.; Wang, D.; Wang, X. Patch-based weakly supervised semantic segmentation network for crack detection. *Constr. Build. Mater.* **2020**, *258*, 120291. [CrossRef]
48. König, J.; Jenkins, M.; Mannion, M.; Barrie, P.; Morison, G. Weakly-Supervised Surface Crack Segmentation by Generating Pseudo-Labels using Localization with a Classifier and Thresholding. *arXiv* **2021**, arXiv:2109.00456. [CrossRef]
49. Zhu, J.; Song, J. Weakly supervised network based intelligent identification of cracks in asphalt concrete bridge deck. *Alex. Eng. J.* **2020**, *59*, 1307–1317. [CrossRef]
50. Selvaraju, R.R.; Cogswell, M.; Das, A.; Vedantam, R.; Parikh, D.; Batra, D. Grad-CAM: Visual Explanations from Deep Networks via Gradient-Based Localization. In Proceedings of the 2017 IEEE International Conference on Computer Vision (ICCV), Venice, Italy, 22–29 October 2017; p. 9.
51. Chattopadhyay, A.; Sarkar, A.; Howlader, P.; Balasubramanian, V.N. Grad-CAM++: Improved Visual Explanations for Deep Convolutional Networks. In Proceedings of the 2018 IEEE Winter Conference on Applications of Computer Vision (WACV), Lake Tahoe, NV, USA, 12–15 March 2018; pp. 839–847. [CrossRef]
52. Zoubir, H.; Rguig, M.; Elaroussi, M. Crack recognition automation in concrete bridges using Deep Convolutional Neural Networks. *MATEC Web Conf.* **2021**, *349*, 03014. [CrossRef]
53. Inbac. Available online: https://github.com/weclaw1/inbac (accessed on 10 March 2022).
54. MCBDD-ZRE/Concrete-Bridge-Defects-Dataset. GitHub. Available online: https://github.com/MCBDD-ZRE/Concrete-Bridge-Defects-Dataset (accessed on 30 May 2022).
55. Lin, V. Vickyliin/Gradcam_Plus_Plus-Pytorch. 2022. Available online: https://github.com/vickyliin/gradcam_plus_plus-pytorch (accessed on 21 July 2022).

MDPI
St. Alban-Anlage 66
4052 Basel
Switzerland
Tel. +41 61 683 77 34
Fax +41 61 302 89 18
www.mdpi.com

Remote Sensing Editorial Office
E-mail: remotesensing@mdpi.com
www.mdpi.com/journal/remotesensing